ROUTLEDGE HANDBOOK OF THE ECONOMICS OF CLIMATE CHANGE ADAPTATION

Climate change is one of the greatest challenges facing humankind owing to the great uncertainty regarding future impacts, which affect all regions and many ecosystems. Many publications deal with economic issues relating to mitigation policies, but the economics of adaptation to climate change has received comparatively little attention. However, this area is critical, and a central pillar of any adaptation strategy or plan, which therefore merits the increase in attention it is receiving.

This book deals with the difficulties that face the economics of adaptation. Critical issues include: uncertainty; baselines; reversibility, flexibility and adaptive management; distributional impacts; discount rates and time horizons; mixing monetary and non-monetary evaluations and limits to the use of cost-benefit analysis; and economy-wide impacts and cross-sectoral linkages. All of these are addressed in the book from the perspective of economics of adaptation. Other dimensions of adaptation are also included, such as the role of low- and middle-income countries, technology and the impacts of extreme events.

This timely book will prove essential reading for international researchers and policy makers in the fields of natural resources, environmental economics and climate change.

Anil Markandya is Scientific Director of the Basque Centre for Climate Change (BC3), an Ikerbasque Research Professor and an affiliate of the University of the Basque Country (UPV/EHU).

Ibon Galarraga is Deputy Director and Research Professor at the Basque Centre for Climate Change (BC3) and an affiliate of the University of the Basque Country (UPV/EHU).

Elisa Sainz de Murieta is a junior researcher at the Basque Centre for Climate Change (BC3) and the University of the Basque Country (UPV/EHU).

'Anil Markandya, one of the world's experts on the economics of climate change, has assembled a group of his peers to analyze the complex, under-researched issues involved in deciding how best to adapt to the now unavoidable changes in climate. Adaptation to a warmer world is a very critical item on our policy agenda, one that does not have nearly the prominence that it merits. This volume is a step in redressing that imbalance.'

Geoffrey Heal, Donald C. Waite III Professor of Social Enterprise,
Columbia Business School, USA; Coordinating Lead Author,
IPCC WGII Fifth Assessment

'In future years, this edited volume will be seen as a landmark for its timely and important contribution to our understanding of the economics of climate change adaptation.'

Edward B. Barbier, John S. Bugas Professor of Economics,
Department of Economics and Finance, University of Wyoming, USA

'This book represents a state of the art on the economics of adaptation and will be of great use to all those working in this field.'

Martin Parry, Visiting Professor at Grantham Institute and Centre
for Environmental Policy, Imperial College London;
Co-Chair of IPCC WGII Fourth Assessment

ROUTLEDGE HANDBOOK OF THE ECONOMICS OF CLIMATE CHANGE ADAPTATION

Edited by
Anil Markandya, Ibon Galarraga
and Elisa Sainz de Murieta

First published 2014
by Routledge
2 Park Square, Milton Park, Abingdon, Oxon, OX14 4RN

and by Routledge
711 3rd Avenue, New York, NY 10017

Routledge is an imprint of the Taylor & Francis Group, an informa business

British Library Cataloguing-in-Publication Data
A catalogue record for this book is available from the British Library

Library of Congress Cataloging-in-Publication Data
Routledge handbook of the economics of climate change adaptation /
[edited by] Anil Markandya, Ibon Galarraga, Elisa Sainz de Murieta.
pages cm — (Routledge international handbooks)
Includes bibliographical references and index.
1. Climate change mitigation—Economic aspects—Handbooks, manuals, etc.
2. Climatic changes—Economic aspects—Handbooks, manuals, etc. 3. Environmental economics—Handbooks, manuals, etc. I. Markandya, Anil, 1945– II. Galarraga, Ibon.
QC903.R695 2013
363.738'74—dc23
2013026363

ISBN: 978-0-415-63311-6 (hbk)
ISBN: 978-0-203-09520-1 (ebk)

Typeset in Bembo
by Apex CoVantage, LLC

CONTENTS

FIGURES, TABLES AND BOXES

Figures

Tables

Boxes

PART I

Economics and adaptation

1

AN INTRODUCTION TO THE ECONOMICS OF ADAPTATION TO CLIMATE CHANGE

Elisa Sainz de Murieta,[1,2] *Ibon Galarraga*[1,2] *and Anil Markandya*[1,2,3]

[1] BASQUE CENTRE FOR CLIMATE CHANGE (BC3), BILBAO, SPAIN

[2] UNIVERSITY OF THE BASQUE COUNTRY (UPV/EHU)

[3] IKERBASQUE, BILBAO

1.1 Introduction

Climate change is one of the greatest challenges facing humankind for several reasons but especially the potential of major life-changing and life-threatening impacts that it creates and the fact that there is a great deal of uncertainty about the magnitude of these future impacts in different regions and across different ecosystems. The Intergovernmental Panel on Climate Change (IPCC) has clearly stated that there is sufficient scientific evidence regarding the unequivocal warming of the climate on all continents (IPCC, 2007a). This change has been more intense during the last century and continues to speed up. The rises in sea level, the disappearance of ice, changes in precipitation and increased tropical cyclone activity seem to endorse this fact. The warming is affecting nearly all marine and terrestrial ecosystems, beginning with the Arctic and Antarctic areas and including tropical marine environments. Regional climate changes already affect (or will affect) nearly all human and natural systems.

While mitigation efforts have been dominating the climate change policy debate for more than 20 years, it is more recently, since the beginning of the 21st century, that the United Nations Framework Convention on Climate Change (UNFCCC) has been trying to mainstream adaptation policies. Adaptation policies refer to those groups of instruments designed and implemented to prepare for the changes that are already occurring and will occur; as IPCC (2001:881) states:

> adaptation is adjustment in ecological, social, or economic systems in response to actual or expected climatic stimuli and their effects or impacts. This term refers to changes in processes, practices, or structures to moderate or offset potential damages or to take advantage of opportunities associated with changes in climate.

The growing importance of adaptation has been mainly due to: (1) the lack of satisfactory mitigation efforts and (2) the fact that even if mitigation policies were successful, severe impacts will occur in the following century (Parry et al., 1998; Pielke et al., 2007).

It is with the Nairobi Work Programme (NWP) formally approved at the COP 12 in Nairobi (Kenya) in 2006 that the international community acknowledged adaptation policy as a cornerstone of the effort to fight climate change. The NWP was agreed on to support all countries, especially developing countries, to: (1) improve their understanding and assessment of impacts, vulnerability and adaptation to climate change and (2) decide on specific adaptation actions and measures taking as a basis the sound scientific, technical and socioeconomic information (UNFCCC, 2007a). The constitution of the Adaptation Fund (AF) was another decisive issue on the adaptation agenda in Nairobi (Sterk et al., 2007), which aims at financing specific adaptation projects and programmes in developing countries that are Parties to the Kyoto Protocol. Following these efforts, it was not until 2010 at the Cancun Submit (COP 16) that the Cancun Accords stated that adaptation "must be addressed with the same priority as mitigation" (UNFCCC, 2011).

The fact that adaptation policies can follow a wide range of typologies, from hard adaptation engineering interventions to softer behavioural alternatives, makes the range of options available very wide. In the literature, various classifications of adaptation policies can be found (e.g., Agrawala and Fankhauser, 2008; Burton, 1993; Carter et al., 1994; Markandya and Galarraga, 2011; Smit et al., 2000, 1999; Stakhiv, 1993; Tompkins and Eakin, 2012; UKCIP, 2007) based on different criteria, as shown in Table 1.1.

The growing policy interest in this area has triggered most of the efforts to understand the economic dimension of climate change adaptation, starting with systematic reviews of the existing evidence (Markandya and Watkiss, 2009; Parry et al., 2009; World Bank, 2010a), followed by methodological developments to support decision-making efforts in adaptation planning. Estimating the economic impacts of climate change proves to be a starting point to assess the cost-effectiveness of the policies. We first need to understand the benefits of the measures relative to the baseline and the impacts line, estimating later the cost of the policies aimed at identifying those with the highest net benefit.

However, several problems arise in this rather simple conceptual setting, which have attracted the interest of many scholars and policy makers towards the economics of adaptation. Apart from the methodological issues explored later in this chapter, there are two important concepts to be addressed. One is maladaptation, which can be defined as "action taken ostensibly to avoid or reduce vulnerability to climate change that impacts adversely on, or increases the vulnerability of other systems, sectors or social groups" (Barnett and O'Neill, 2010:211). According to the authors, there could be five types of maladaptation options: the first refers to a measure that increases emissions; the second groups those actions that affect most those at greater risk; the third includes actions whose economic, environmental or social costs are higher than other alternatives (economists often use this last definition, indicating a misuse of resources [Markandya and Galarraga, 2011]); the fourth category comprises those measures that reduce the incentive to adapt (e.g., by encouraging unnecessary dependence on others) and the last group gathers actions that are neither flexible nor reversible but that set dependency paths, such as great infrastructures.

The second problem that arises when assessing the economics of adaptation is related to the adaptation deficit, which represents the lack of adaptive capacity to current climate variability, without considering climate change (Burton, 2004). It is, therefore, closely linked with

Table 1.1 Different classifications for adaptation policies according to various criteria

Classification of adaptation policy	*Criteria*	*Examples*
Private vs. public	Nature of agents involved	• Public adaptation takes place when there is an intervention from the government to support adaptation. E.g., infrastructure building or land-planning. • Private adaptation occurs when the measures are implemented by individuals or private entities. E.g., insurances for flooding (Tompkins and Eakin, 2012).
Localised vs. widespread	Spatial scope	• Local adaptation refers to those options or measures implemented at a local (or regional) scale. E.g., dike building. • Health National Plans dealing with climate-related affections or early warning systems are examples of widespread adaptation.
Short- vs. long-term	Temporal scope	• Short-term adaptation measures are those urgent measures needed to avoid or moderate the effects of current climate change and variability. E.g., agricultural insurance. • Long-term adaptation deals with reducing vulnerability and enhancing resilience to cope with impacts expected in the long-term, for instance, sea-level rise.
Infrastructural, behavioural, institutional, financial and informational	Type of measure	• Infrastructural measures involve climate-resilient constructions and engineering options, such as sea dikes • Behavioural options are those focused on changing conducts or common practices, such as farming practices or methods. • Institutional responses to climate change include policy, planning or regulatory measures. • Financial options include measures like economic or fiscal incentives for adaptation. • Informational systems include, for example, early warning systems.
No regret, low regret and win-win	Ability to face uncertainty and/or to address other associated benefits	• No regret options are those measures that represent net benefits even in the absence of climate change; low regret measures are those having comparatively low costs and large benefits. E.g., improving efficiency on water irrigation. • Win-win measures, apart from contributing to adaptation, also provide other social, economic or environmental benefits. For example, improving air quality can contribute to both adaptation and mitigation strategies.

development deficit (Parry et al., 2009). The diverse funds established to contribute to adaptation[1] are not directed to cover the costs associated to the adaptation deficit and only take into account the extra costs as a result of climate change (Parry et al., 2009). However, authors like Burton (2004) consider that without a successful adaptation to current climate, it would not be possible to implement an effective adaptation to climate change. Hence mainstreaming climate change into development activities is a key imperative.

The rest of the chapter is organised as follows. Section 1.2 summarises the main methodological issues of adaptation economics, covering the sectoral aspects in Section 1.3 and other dimensions of economics in Section 1.4. These sections follow the general structure of this book of 21 chapters provided by well-known experts in the field with in-depth discussion of the each of the main topics highlighted here.

1.2 Methodological issues on economics of adaptation

The economics of adaptation to climate change entails a number of methodological difficulties that have yet to be properly resolved. The problems involved are not new to economics, but the fact that a great many of them are present in the same discipline presents environmental economists and policy makers with a major challenge.

All these issues, although they are closely related, can be grouped according to Markandya and Watkiss (2009) under the following generic topic headings: uncertainty, fairness and economic valuation (see Figure 1.1). This is the general structure that we will follow in this section. Note that some issues fall under more than one topic heading: for instance, baseline scenarios affect both uncertainty and financial valuation. Others, like ancillary benefits, limits of adaptation or public- versus private-sector adaptation affect both the equity and financial valuation, and they will be addressed in this chapter as "cross cutting issues".

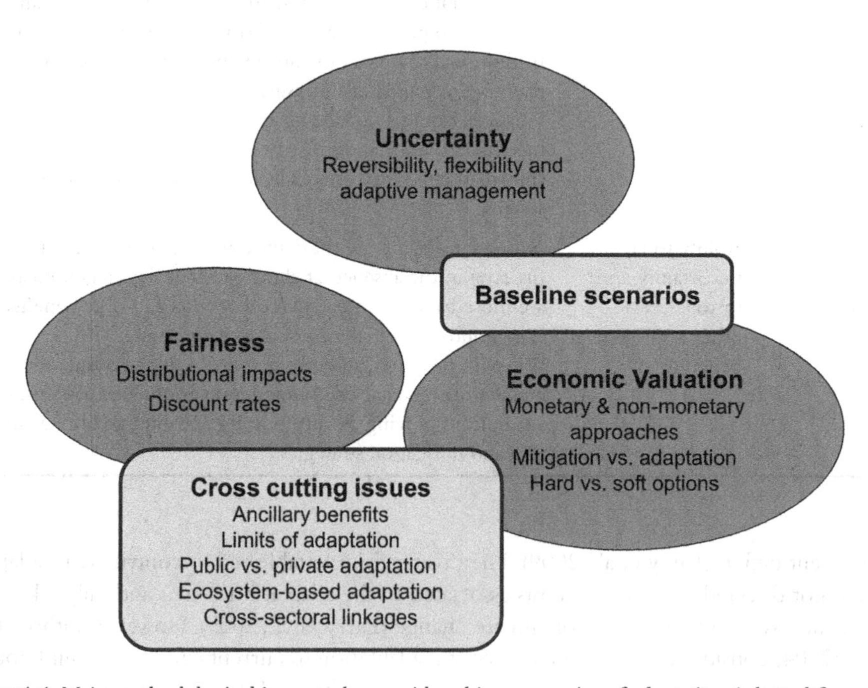

Figure 1.1 Main methodological issues to be considered in economics of adaptation (adapted from Markandya and Watkiss, 2009)

1.2.1 Uncertainty

Uncertainty is one of the main problems when addressing climate change related issues. To begin with, uncertainty is associated with *emission scenarios* as these not only depend on the selected emission scheme, but the socioeconomic development associated to that emission scenario as well as on global and regional models. And this is not only a matter of lack of knowledge; there is evidence in the literature that a better knowledge does not necessarily mean narrower projection ranges. In fact, uncertainty would not disappear even if global circulation models to which emissions are inputs were entirely accurate (Hallegatte, 2009).

Certainly, most available climate models and observations often cannot offer what current decision-making frameworks need to devise adaptation policies, so *statistical downscaling* is needed to translate those projections into regional climate data, and this operation adds additional uncertainty to the results. Finally, uncertainty is also related to the way both *natural and socioeconomic systems* will respond to climate change and the resulting *impacts*. According to Pindyck (2007), environmental problems, climate change impacts in our case, may involve three levels of uncertainty. First, that related to the evolution and response of ecological or biophysical processes; second, the uncertainty about the economic effects due to the direct and indirect impacts; and third, the uncertainty about the technological changes that could lessen both ecological and economic impacts.

Finally, we should also be aware that each step (input data, climate modelling, impacts and adaptation . . .) adds uncertainty to the previous one, so cumulative uncertainties should also be addressed when measuring the impacts of adaptation (Refsgaard et al., 2013).

Uncertainty represents a serious problem on both the local and global scales, and is a significant stumbling block in decision-making. In this context, advocating a preventive approach seems to suggest that especially unfavourable scenarios must be considered when deciding on what actions to take (Weitzman, 2007).

Chapter 5 in this book, by Markandya, examines the different kinds of uncertainty and its sources in the context of climate change and offers a review of the tools and proposals to deal with uncertainty when it is a key issue in the adaptation processes.

Reversibility, flexibility and adaptive management

There are several strategies that decision-makers should consider to address uncertainty. With the purpose of keeping as low as possible the cost of being wrong about future impacts of climate change, it is wise to consider *reversible* and *flexible* policies (Hallegatte, 2009). Reversible choices refer to the possibility of going back to the situation that existed before a policy measure was implemented. The value of the reversibility of a strategy or measure is often known as *option value*, and this could be used as a tool when assessing the economics of adaptation. The option value refers to the value arising from delaying a decision. An example of this strategy can be the decision not to build a preventative infrastructure. However, it will often be impossible to delay a decision indefinitely, so it might be useful to compare adaptation options with different degrees of reversibility (Hallegatte, 2009).

Flexibility allows for adjusting or adapting in the light of new available information, to cope with impacts more or less severe than predicted. If the situation changes progressively, it also allows for incremental adaptation (HMT, 2009). *Increasing resilience* is another way to deal with uncertainty by designing *robust strategies* or measures, able to cope with a wide range of future climate conditions. This is closely linked with *adaptive management*, which "involves an iterative process in which managers learn from experimental management actions. Management actions are applied as experiments, the system is monitored, and actions are then potentially changed to address changes in the state of the system" (Lawler, 2009:85).

Low-regret measures, defined as those that involve net social and/or economic benefits irrespective of whether or not anthropogenic climate change occurs (IPCC, 2007b), represent another group of actions to incorporate uncertainty into decision-making. Examples of such measures include developing a crop insurance scheme or improving water infrastructure. Another option to take uncertainty into account is defining *win-win measures*, which address climate change but deliver other benefits as well, such as improving air quality or building flood-proofing buildings (HMT, 2009).

Even if there are a variety of methods to take uncertainty into account, we should bear in mind that they will also lead to serious problems for analysis, planning and decision-making (HMT, 2009; Lempert and Schlesinger, 2000; Markandya, Chapter 5 this volume).

Baseline scenarios

One of the most challenging aspects when analysing the economics of adaptation to climate change is the definition of the baseline scenario. Ideally, the baseline represents what would happen to the fundamental variables in the absence of climate change. Since one is looking forward several decades, the future baseline is not a business-as-usual scenario based on historic trends, but a complex reconstruction where it is necessary to estimate changes in population, economy, development, behaviour patterns and a number of other factors that are highly uncertain and even more so in the long term (Markandya and Galarraga, 2011:32). Baseline scenarios are, therefore, inexorably associated with high levels of uncertainty (Markandya and Watkiss, 2009). The HMT Green Book (2007) proposes several methodologies to deal with uncertainty related to baselines: probability distribution, and when these are not available, sensitivity analysis, the consideration of scenarios for baselines and, sometimes, the use of Monte Carlo Analysis as well (Chapter 5, p. 10).

1.2.2 Fairness and equity

Distributional impacts

Climate change raises difficult issues of justice and equity, particularly with respect to the distribution of responsibility, burdens and benefits among poor and wealthy nations (Posner and Sunstein, 2009). Since developed countries have been responsible for the largest share of historical and actual global emissions, it seems unfair to require developing countries to have to undertake strong commitments to reduce global greenhouse gas (GHG) emissions (Paavola and Adger, 2006). However, developing countries will suffer the most severe climate change impacts, and at the same time, stabilizing GHG emissions will not be feasible without their participation. Finding the way for reducing global GHG in an acceptable way for both developed and developing countries represents one of the biggest challenges of international climate negotiations (Markandya, 2010).

In the past two decades, debates on climate justice have focused on mitigation, and especially the allocation of greenhouse gas emissions among countries, due to the urgency to take action to reduce anthropogenic climate change (Paavola and Adger, 2006). Recently, however, as the adaptation is becoming a more important issue in climate change policy, fairness and equity-related issues have also emerged in decision-making on adaptation. For example, there is evidence that a fair global adaptation funding would have positive consequences for both developed and developing countries. This is due to the fact that improving fairness would contribute to building trust among those who feel unfairly treated (some developing countries). Then, this new confidence of developing countries in industrialized ones is expected to have a positive effect on international agreements on mitigation (Rübbelke, 2011). Paavola and Adger (2006) argue that there are four major justice dilemmas of adaptation, which include: 1) determining the responsibility of developed countries for climate change impacts, 2) defining the assistance that these countries should grant the developing countries for adaptation, 3) distributing this assistance among the different vulnerable countries and adaptation measures and 4) implementing procedures to ensure a fair adaptation policy.

Distributional impacts, however, do not exist just in an international context as the adverse effects and benefits of climate change will also affect differently within societies, at national and subnational levels. In the final analysis, it will be the poorest and most vulnerable groups and sectors of society who will suffer most from climate change (Thomas and Twyman, 2005). Some authors argue that, from the point of view of equity and justice, the actual adaptation deficit should also be addressed (Adger et al., 2005). This is, adaptation should reduce existing inequalities and poverty and improve income distribution as a lower vulnerability today will also result in less vulnerability to climate change in the future. Obviously, not all impacts can be avoided, for technical, economic or social reasons. These impacts that will not be avoided are known as "residual damages", and addressing them will be a key issue as they could represent up to two-thirds of all potential impacts in the long term (Parry et al., 2009).

Distributional impacts, therefore, call for the utmost rigour and precision in analysis if they are to be taken into account properly. Chapter 6 of this volume by Hunt and Ferguson provides a comprehensive review of the role of fairness and equity regarding adaptation to climate change from an economic perspective.

Discount rates

In the previous section, we discussed distributional impacts at the international and societal (intra-generational and inter-sectoral) levels. In the long term, considerations involving intergenerational equity and safeguarding the rights of future generations are key factors. In particular, discount rates play a key role with respect to both equity and efficiency (Munasinghe, 2000) as they represent the extent to which today's society values costs and benefits to future generations. Zeckhauser and Viscusi (2008) edited a comprehensive special issue entirely devoted to this topic that can be consulted for more information.

Given the length of the periods involved in the climate change context (between 100 and 200 years), the use of discount rates equivalent to those on the market is a critical issue. Some authors suggest rates below 2%, or even close to zero (Cline, 1992; Stern, 2007), while others favour rates of around 6% (Nordhaus, 1994). Others, like Weitzman (2001) and Evans et al. (2004), instead, suggest using decreasing rates as impacts become more long-term, and in fact it seems that during the last decade a consensus was reached on social discount rate, which should decline with the longer time horizon. Another approach is presented by Chiabai et al. (2012), who propose a simple rule to estimate variable discount rates for valuing natural assets in the long term, as many ecosystems will most probably be scarcer in the future. Both Gollier (2012) and Chapter 7 (this volume) by Groom provide further details about discounting in a climate change context.

The valuation of adaptation options, by their nature, will be very much affected by discounting. For instance, if we look at flood prevention, adaptation measures will require strong short-term investments, while the benefits will most probably be received in the mid or long term. Also, it is well known that climate change will affect developing countries (and therefore, the global poor) the most, so already existing inequalities will be exacerbated, unless adaptation policies counteract this effect. It is clear that discounting opens a whole new field of discussion on ethical issues involving intra-generational and international equity, depending on the considered values or methodologies: the lower the rate used, the higher the valuation of future generations and the greater the present value of expected future benefits from reduced impacts as a result of adaptation measures.

Many other technical aspects are considered in the literature of discounting, as summarised in Chapter 7.

1.2.3 Economic valuation and efficiency

From an economic perspective, adaptation is normally evaluated by determining whether and by how much the benefits of such measures exceed the costs incurred (Agrawala and Fankhauser, 2008). Such an approach is referred to as a cost-benefit analysis (CBA) and essentially involves calculating in monetary terms all of the costs and benefits. An adaptation option would represent a good investment if the aggregate benefits exceed the aggregate costs (Markandya and Watkiss, 2009).

Although CBA is important when designing economic policy, other criteria are also considered when making a decision. This is because CBA, in its simple form, does not cover all aspects: it ignores the distribution of the costs and benefits of adaptation options, and it fails to account for those costs and benefits that often cannot be reflected in monetary terms, such as ecological impacts and impacts on health, as well as concerns that influence welfare, such as peace and security. For these reasons CBA should be only one input into the decision-making process, and other approaches are frequently used as a complement or a substitute. These include cost-effectiveness analysis, multi-criteria analysis (MCA) and other approaches (Galarraga et al., 2011b).

Monetary and non-monetary valuations

Valuation of the economic benefits of adaptation has focused on estimating the reduction of damages in monetary terms. However, not all impacts can be valued in such terms or could only be partially so valued. This is particularly the case for non-market damages. In view of this, it has been argued that adaptation decisions should use other metrics or combine a measure of net benefits with an estimate of impacts in physical terms.

This being so, the mixture of monetary and non-monetary values is necessary, but it greatly complicates the analysis. Such analysis is limited by this and other issues, such as the availability of (partial) information and the distribution of impacts. Other methods such as cost effectiveness, the risk approach and a multi-criteria approach should also be considered in order to incorporate qualitative information (Barbier et al., 1990).

Interrelationships between mitigation and adaptation

The relationship between adaptation and mitigation is complex and highly important, and it needs to be taken into consideration. In fact, the IPCC addressed this issue in its AR4 report where a chapter on the interrelationship between the two forms of climate action was included (Klein et al., 2007). Assessing the links between adaptation and mitigation was an important step forward of AR4.

Due to the long history of past emissions and the inertia of the climate system, it will not be possible to completely avoid the effects of climate change even with the effective implementation of the most ambitious mitigation policies (Parry et al., 1998). Adaptation is, thus, not only necessary but unavoidable. However, if there is no progress in mitigation, the amount of climate change that is likely to happen will make adaptation impossible or extremely costly for some natural and human systems (Klein et al., 2007; Stern, 2007). Therefore, an effective climate policy should include a portfolio of both adaptation and mitigation options.

Adaptation and mitigation policies work at different spatial (local or regional versus international), time and institutional scales, hence it will not be always possible to develop synergies among them. In some cases, adaptation and mitigation might represent a trade-off when designing the climate policy and distributing climate funds (Abadie et al., 2013) whereas in other situations synergies might exist. Opportunities for such synergies are present in some sectors (agriculture and forests, urban infrastructure), while in others the options are more limited (coastal, energy) (Klein et al., 2007). Chapter 4 of this volume studies the potential for achieving synergies between adaptation and mitigation through REDD+ (Reduction of Emissions by Deforestation and forest Degradation) schemes.

The author discusses the implementation of REDD+ mechanisms by protecting forests in Tanzania. This will first contribute to mitigation, as emissions from deforestation will be reduced, while through afforestation there will be an increase of carbon stored in the biosphere. With regard to adaptation, an increase of forested areas will play an important role in reducing vulnerability, especially by addressing irrigation problems, limiting the impacts of floods and landslides and improving biodiversity. However, as the forest in question is being actually used by farmers for charcoal production, forest protection would reduce their income and make them more vulnerable. Additionally, charcoal users would have to find an alternative, most likely kerosene. The authors analyse these interrelationships to see the extent at which REDD+ programs can actually represent a successful option for both mitigation and adaptation (Chapter 4, by Aaheim and García).

Hard versus soft adaptation measures

Hard measures involve the construction of infrastructures and protective barriers, while soft measures are associated with changes in the behaviour and habits of socioeconomic agents. According to Markandya and Watkiss (2009), there has been a tendency to focus on hard engineering options as they are easier to cost than behavioural or policy measures, but soft measures need to be given greater consideration. In fact, this preference towards structural measures may create a distortion adversely affecting potentially critical soft measures needed to enable adaptation (such as better land-use planning, early warning systems or insurance schemes) and lead to inappropriate and expensive adaptation measures. It could also result in overestimation of adaptation costs (Agrawala and Fankhauser, 2008). Nevertheless, the classification of hard vs. soft does not necessarily imply that hard options are technology-based while soft options are not. In fact, there are many examples of green infrastructure (wetland restoration, landslide prevention by tree planting . . .) that can be considered soft and at the same time have an important innovative and technological component (Markandya and Galarraga, 2011).

An important aspect of this discussion is that soft adaptation options are easier to reverse, which makes them more suitable to deal with a highly uncertain future (Hallegatte, 2009).

1.2.4 Cross cutting issues

Ancillary benefits

Adaptation to climate change often has benefits apart from lessening the impacts of climate change. For instance, many adaptation measures contribute to reducing actual vulnerability with respect to current climate variability or extreme events. Other examples include improvements in air quality, the positive impact on health of changes in transportation and life habits, etc. These benefits, which accrue as a positive side effect of adaptation measures and were not among the main objects of those measures, are known as ancillary or secondary benefits (EEA, 2007).

According to some authors (for example, Van Ierland et al., 2007) a slight distinction should be made between *no regret options* and *ancillary benefits* that is worth mentioning. No regret options are those adaptation measures for which non-climate-related benefits arise in such a way that the implementation of those measures would be beneficial irrespective of future climate change taking place. Ancillary benefits, on the other hand, may or may not be enough to justify the measures on their own.

Although many ancillary benefits are difficult to assess in monetary terms (improved health or increased ecosystem resilience, for example), these should definitely be considered when assessing costs and benefits of adaptation. As future climate change will particularly affect the more disadvantaged and vulnerable groups of society, the distribution of ancillary benefits should also be taken into account in the analysis (Markandya and Watkiss, 2009).

The limits of adaptation

The limits of adaptation and the need to reduce the exposure of ecosystems to impacts and increase their resistance or resilience must also be taken into account. There is an incipient body of literature concerned with defining this concept, which is open to important nuances depending on what systems are being analysed. In fact, we can analyse the limits of adaptation from two different perspectives: the *exogenous* or analytical approach, which primarily examines the objective limits of adaptation, including ecological or biophysical limits, economic limits and technological limits. Most studies conducted so far focus on this approach. In contrast, Adger et al. (2009) propose an *endogenous* approach that "emerges from inside society". Considering how societies are organised, social limits of adaptation "depend on the goals, values, risk and social choice", and therefore are "mutable, subjective and socially constructed" (Adger et al., 2009:338). Likewise, Dow et al. (2013) propose an actor-centred and risk-based (*endogenous*) approach for defining limits to social adaptation (Markandya and Galarraga, 2011).

Other references that need to be considered on this issue include the papers of Holling (1973), which first defined the concept of resilience in environmental terms, Perrings (1998), which redefined the concept, and Walker et al. (2010), which provides an example of operationalizing resilience in the Australian context.

Public versus private-sector adaptation

Adapting to climate change may involve different levels of society, from individuals, firms and civil society, to public institutions and governments at local, regional, national and international scales (Adger et al., 2005). Even if many studies focus on public adaptation, both public and private adaptation need to be considered (Markandya and Watkiss, 2009).

Sectors such as engineering or construction most likely will have to adapt, at least to some extent, their building methods, and they will most certainly contribute to climate-proofing existing infrastructure as well. Insurance policies are another important form of private-sector adaptation. Telecommunication companies, media and new technologies can play an important role in relation to the monitoring and communication of hazards, while companies of the primary sector should also consider how to adapt to new climate conditions to ensure food security, and the financial sector could play a main role, for instance, in relation to the financing of adaptation measures. These are just a few examples that show the key role that the private sector could have regarding adaptation (Agrawala and Fankhauser, 2008).

In some cases, public-sector adaptation can be considered as a disincentive to private adaptation, particularly to autonomous (or unplanned) adaptation, so there is clearly a need for coordination between the public and private sectors (Markandya and Watkiss, 2009). Clearly, public policies have an important responsibility in providing adaptation as a public good where private activities might not occur due to externalities or other failures. At the same time, there are many cases where the effectiveness of adaptation could be improved if public and private sectors would act jointly (Agrawala and Fankhauser, 2008).

The role of economic modelling

The study of climate change has some features that make it particularly challenging as we have already noted. First, there is still great uncertainty regarding the future impacts of global, regional and local levels; second, the impacts, although some of them are already visible today, will have very long term consequences, to 2050, 2100 and beyond; finally, the impacts will not be evenly distributed, neither among countries nor between different social groups (Galarraga and Markandya, 2009). All these make the analysis and modelling effort particularly difficult.

Despite this, it is possible to estimate important features of the social dimension of climate change through integrated assessment models (IAMs) that combine economic, environmental and climate information at different scales. The role of economic models in relation to adaptation and mitigation policies is comprehensively reviewed in Chapter 8 by Bosello.

Among the top-down integrated assessment approach, there are three major models used in the economic analysis of climate change: computable general equilibrium models (CGEMs), which offer an explicit representation of domestic and international trade; dynamic growth models, which have been used primarily in the context of energy and emissions reduction policies; and macro-econometric models that assess the relationship between different variables from past observations (time series) rather than economic theory. Top-down IAMs have been traditionally used in mitigation policies.

Bottom-up models, also known as engineering or partial equilibrium models, have been usually applied to estimate the cost of mitigation policy, but research on the costs and benefits of adaptation has been also conducted through bottom-up studies due to its predominantly local/regional nature. However, the growing interest in assessing the economic impacts of both mitigation and adaptation policies has resulted in a very recent but growing body of literature that addressed these issues through macroeconomic approaches (Agrawala et al., 2011).

Ecosystem-based adaptation

Ecosystem-based adaptation (EbA) considers the use of biodiversity and ecosystem services as an instrument to help human communities adapt to the adverse impacts of climate change. Ecosystem-based approaches include activities of sustainable management, conservation and restoration of ecosystems (Colls et al., 2009).

So far, most adaptation initiatives have focused (as noted) on the use of "hard" adaptation options, but the role of EbA as a complement – and sometimes substitute – of infrastructure investments is being increasingly recognized (Colls et al., 2009; McKinnon and Hickey, 2010). EbA contributes to reduce vulnerability to climate impacts across several human sectors, including disaster risk reduction (e.g., salt marsh ecosystems provide coastal protection against storm surge-events, coastal erosion and sea-level rise), livelihood diversification and food security (ensuring access to natural resources) and sustainable water management (flood protection, improving water quality) (Colls et al., 2009; Jones et al., 2012). EbA often represents a more cost-effective alternative than hard adaptation initiatives. According to Moberg and Rönnbäck (2003), the value of coastal protection from erosion provided by reefs can reach US$1million/km coastline over a 25-year period in areas with major infrastructure, while artificial ways of coastal protection can cost from US$246,000 up to $5 million/km.

Ecosystem-based approaches have additional advantages over hard initiatives: first, there are multiple ancillary or co-benefits they can deliver, such as carbon storage, biodiversity conservation or sustainable economic development (Munang et al., 2013). On the contrary, few hard interventions provide extra benefits beyond the adaptation option for which they were designed; furthermore, sometimes they can even generate negative impacts on surrounding natural and human systems. Second, while hard adaptation measures are usually permanent, EbA approaches are potentially more flexible and have no-regret character (Jones et al., 2012). Consider, for instance, coastal wetlands that can migrate upwards (Cearreta et al., 2013) and inland as sea level rises (assuming that there is available land for migration). Third, EbA adaptation initiatives can create synergies with other adaptation initiatives, development goals and mitigation strategies (Munang et al., 2013).

EbA, however, also faces a range of barriers – such as insufficient finance, land use conflict or community opposition – and limitations related to the degree of future climate change and the ecological limits of nature (Colls et al., 2009). Mainstreaming EbA as a basic approach to climate change adaptation and decision making, therefore, still remains a challenge (Munang et al., 2013).

Chapter 9 of this volume by Ojea is devoted to reviewing the state of the art on the growing field of ecosystem-based adaptation.

1.3 Sectoral approach to economics of adaptation

Existing studies regarding the economics of adaptation can be classified into two main groups: first, global aggregated assessments that are relatively abstract and include some simplifying assumptions that are difficult to apply when devising adaptation policies at a smaller scales; second, more disaggregated sectoral- and project-level studies that provide much more detailed and accurate spatial (regional, national, local) and sectoral scales (Galarraga et al., 2011b). According to Agrawala and Fankhauser (2008), sectoral level studies should be the base upon which higher-order assessments are carried out.

From an economic perspective, there is a large amount of information available at the sectoral level, especially in relation to the costs and benefits of adaptation. However, it is also true that this information is unevenly distributed across sectors (Agrawala and Fankhauser, 2008), as discussed below. Next, we provide a brief introduction to the economics of adaptation for each of the main sectors involved in climate change adaptation, which will be subsequently discussed in detail in the chapters included in the section of this Handbook addressing *Adaptation in Activity Sectors*.

1.3.1 Energy

The energy sector is one of the main contributors to GHG emissions with 64% of global emissions related to human activities (Emberson et al., 2012) and, therefore, to climate change. Hence most of the studies so far have focused on mitigation (see Ansuategi, Chapter 10 of this volume, for further details on the energy sector). However, this sector is not only a contributor to climate change but also vulnerable to its impacts (Ebinger and Vergara, 2011). First, climate change can potentially affect energy availability. This will be especially relevant in cases such as hydropower or biofuels. The results from a recent study on the hydropower sector in Costa Rica showed a significant reduction in the hydropower production, estimated between 41% and 43%, by 2100 in all future scenarios considered (A2, A1B, B1) (Sainz de Murieta and Chiabai, 2013). The potential impact on other renewable sources – wind, solar, waves – will be site dependent. Obviously, fossil fuels will not be directly affected, although exploration and access to them could also be affected, for instance, by the strike of extreme events. Second, climate change and especially extreme events, may also affect energy supply systems, both at energy production sites or disruption of transmission infrastructures (for further details on extreme events see IPCC (2012) and Chapter 21 of this volume. Last, consumption patterns may also be altered by climate change. Isaac and van Vuuren (2009) estimated a global reduction of 34% in the demand for heating under climate change by the end of the century, while the demand for cooling would increase by 72%.

All these effects could lead to major economic impacts worldwide. In the USA, energy costs could increase due to climate change up to US$140.7 billion by 2100 in a business-as-usual scenario (Ackerman and Stanton, 2008). Thus, it is clearly a priority to include the energy sector in the context of adaptation policies to climate change more than it has been so far. Further research should also be carried out to fill in the gaps of knowledge regarding this important issue.

1.3.2 Agriculture

There is a lot of literature analysing the economic benefits of adaptation in the agricultural sector (Agrawala and Fankhauser, 2008). The impacts of climate change (CC) include a mixture of positive and negative impacts depending on where the agricultural zone is, the crop analysed and planting techniques. Generally speaking, climate change may lead to increases in yields at mid and high latitudes and to decreases in tropics and subtropics, although many exceptions exist, particularly where increases in monsoon intensity increase precipitation. Particularly in

South Asia and Africa, the risk of hunger appears to increase as a result of CC, depending, of course, on the number of vulnerable people in these regions among many other factors. Studies such as those of Parry et al. (2004, 2007) are some good research papers on the area.

There is still quite a lot of uncertainty related, for instance, to the beneficial effects of CO_2 on crop growth. Most of the knowledge comes from field experiments that have near optimal application of fertilizer, pesticide and water, while in practice this is not the case.

Adaptation strategies will be very site-specific but can include changing planting seasons, and/or crops, new infrastructure such as irrigation systems, improved insurance system or other type of support for farmers.

Markandya and Watkiss (2009) state that "in practice, support to farmers will be strongly driven by a combination of cost benefit analysis and the needs of providing sustainable livelihoods to poor farmers – i.e., distributional considerations will be very important".

Nainggolan et al. in Chapter 12 of this volume offer a very interesting review of the topic, concluding that "the adoption of a wide range of adaptation options will necessarily form the strategy to capitalize on the opportunities on the one hand and to deal with the negative consequences of changing climate on the other. At farm level, some adaptation options are agronomic and technical in nature while others are focused more on managerial and financial aspects. While to a certain degree farm-level adaptation can be implemented autonomously, in many other situations, initiatives and measures at higher levels are crucial to support and incentivise farm level action".

1.3.3 Coastal areas

There are many impacts related to climate change in coastal areas where most of the population worldwide live. Chapter 13 of this volume offers a wide perspective on the existing studies. The main driver of these impacts is sea-level rise (SLR) and an increase in the intensity and frequency of extreme events, including storms. There are a number of potential effects that complicate comparison, as Nicholls et al. (2006) show.

The impacts are largely in terms of loss of the services of land and several estimates of loss of services under climate change (see Hinkel and Klein [2009] for the methodology; Hinkel et al. [2010] for Europe, Markandya and Mishra [2009] for India, and Galarraga et al. [2011b] for a local case study of the Basque Country). Sea-level rise and coastal damage is perhaps one of the better quantified areas of impacts. Agrawala and Fankhauser (2008) show that many studies on the costs and benefits of adaptation in coastal zones exist, although as Losada and Diaz-Simal (Chapter 13) argue, most of the analysis do not consider regional relative sea-level rise. Of course, adaptation measures can reduce this loss of services and the net benefits are then measured in terms of the cost per hectare of the adaptation measures relative to the increase in services per hectare as a result of the measures.

There are also distributional impacts in poor countries where SLR can affect livelihoods of poor people. Here, decisions may need to compare alternative livelihoods.

But SLR is not the only climate associated impact on coastal zones, as IPCC (2007b) shows. Other important issues exist, such as extreme sea events, seawater intrusion, storm surges, coastal erosion and increasing sea temperature among others. How to incorporate all these potential impacts into decision-making in order to design effective adaptation policies is a challenge.

"There is still a lot of work to be done in order to reduce the currently existing high uncertainties", state Losada and Diaz-Simal (Chapter 13, this volume).

1.3.4 Health

From infectious diseases to malnutrition and disaster-related injuries, climate change will affect human health in several ways; most of them adverse, even if a few would be beneficial (e.g., milder winters would reduce seasonal mortality in high latitude developed countries). The climate-health connection associated with thermal stress, extreme weather events, physical hazards and some infectious diseases are the easiest to analyse. This is the reason why most of the research produced so far has given preferential attention to these issues, together with future regional food yields and hunger prevalence. However, there is an emerging approach that considers health risks in a broader way, including those related to social, demographic and economic impacts of climate change (McMichael et al., 2006:859–860).

According to the IPCC (Confalonieri et al., 2007), health impacts will be greater in low-income countries, but the more vulnerable social groups in developed countries will also be affected (e.g., the elderly and children, the urban poor, etc.). Hence, adaptive capacity needs to be addressed globally; impacts of recent hurricanes (Katrina, Sandy) and heat waves (e.g., the 2003 heat wave in Europe) show that even high-income countries are already in need of improving their response to extreme events (Confalonieri et al., 2007:393).

Different approaches can be used to prioritize goals in health adaptation and identify the most appropriate set of measures, the most popular being the cost-effectiveness analysis (CEA), cost-benefit analysis (CBA), and the multi-criteria analysis (MCA) (see Chapter 14 for a critical analysis of these methods in the context of environmental and health economics). Current estimates of the costs of adaptation in the health sector vary greatly depending on the methodology, the available information, the health issues assessed as well as the world regions under analysis. For instance, a study by the UNFCCC (2007b) estimated that health adaptation costs would range from US$ 4–12 billion per year in 2030. However, even if the study was based in the best available information for developing countries, it clearly underestimates the total health costs, as not all countries, activities nor diseases were included in its scope (Parry et al., 2009).

1.3.5 Flood risk

The frequency and intensity of river floods is inevitably linked to rainfall. In the context of climate change, the amount and distribution of precipitation is expected to influence the frequency and severity of floods. In fact, more intense precipitation events have already been observed, and this trend is expected to continue in the future (Milly et al., 2002). Note that even if projections point to a reduction in total rainfall, increases in extreme precipitation may occur, which will raise the risk of river flooding (Kundzewicz and Schellnhuber, 2004).

Yet, climate is not the only factor that can exacerbate flood risk: land-use planning, the existence – or not – of early warning systems and the overvaluation of structural defences (such as dikes and channellings) are examples of other type of factors that intensify flood hazards (Kundzewicz and Schellnhuber, 2004). However, adaptation can contribute to reducing most of the climate change-induced increases in river flooding risks at relatively low costs (EEA, 2007). A case study from the Netherlands shows that optimal flood defence investments could reduce climate-induced flood damage from 39.9 billion euros to 1.1 billion over the 21st century at a relatively modest cost of around 1.5 billion euros. This case study is also especially interesting as it represents an example of a new policy approach to flood risk management: together with more traditional technical measures, spatial solutions are also identified with the objective to create "room for the river". Another interesting example of adaptation to flood hazards is the TE2100 Plan[2] that "sets out the strategic direction for managing flood risk in the Thames estuary to the end of the century and beyond". It includes real options analysis, which allows for the consideration of uncertainty and flexibility (Galarraga et al., 2011b).

Chapter 16 by Foudi and Oses-Eraso analyses the methodology to carry out flood risk assessments and proposes cost-benefit analysis as a tool to assist in decision-making linked to flood prevention.

1.3.6 Economic impacts and inter-sectoral relationships

As already mentioned, sectoral assessments have the advantage of providing detailed information about the economic impacts of climate change, but they have a more limited scope when assessing national or global adaptation policies. Indirect and induced impacts must not be forgotten, nor must the usefulness of general equilibrium models that take into account relations between different sectors of activity. Multi-sectoral estimates have been carried out in three fronts: (1) at the national level, especially among least developed countries (LDC) as part of the NAPAs; (2) at the regional level through projects like PESETA,[3] which assesses the impacts of climate change across several sectors in Europe; and (3) at the global level, where international agencies, such as UNFCCC, World Bank, UNDP or Oxfam, have valued the global costs of adaptation (Agrawala and Fankhauser, 2008).

Chapter 15 of this volume by Perry and Ciscar reviews several studies based on multi-sectoral, bottom-up approaches to understanding the influence of climate change in the economy and how adaptation could minimise its impacts.

1.4 Other dimensions of adaptation

1.4.1 International cooperation

As we have noted, the early work on climate change focused on mitigation; adaptation started to receive some attention mostly after COP16 in Kenya in 2006 and the publication of IPCC 4AR where adaptation was recognized as meriting the same level of attention as mitigation (Mertz et al., 2009). Unlike mitigation, that requires global cooperation, adaptation has a strong local component, and this is one of the reasons explaining why it has also been virtually absent in international negotiations.

However, as Pickering and Rübbelke discuss in Chapter 3, an efficient and fair adaptation policy may need a certain level of international cooperation. This is especially true considering that low- and middle-income countries that contributed less to GHG emissions will suffer the most severe impacts from climate change; at the same time, most of these countries also have great difficulties in obtaining the necessary resources for adapting to climate change. Additionally, international cooperation on adaptation may also have positive side effects, contributing to a climate of trust among developing countries with respect to developed countries, which in turn could contribute to global mitigation efforts. And the other way around: developing countries have asked for adaptation reinforcement as a previous requirement for getting involved in a global compromise for emissions reduction. Therefore, a lack of clear international support for adaptation could also obstruct an international agreement on mitigation (see Rübbelke [2011] and Chapter 3 for further information on this issue).

1.4.2 Fast Growing Countries and adaptation

The term Fast Growing Countries (FGC), although there is no universally accepted definition, refers to those nation states with particularly good performance in terms of economic growth. According to Virmani (2012:5), "most studies of fast growing economies use 5 per cent average growth in per capita GDP to identify fast growing economies", but the author proposes an alternative definition by which FGC would include those "countries that had an average growth rate of per capita GDP of 7 per cent or more, for a continuous period of 10 years or more". BASIC (Brazil, South Africa, India, China) or BRIC (Brazil, Russia, India, China) are common FGC groups (see Chapter 17 for a full assessment on this issue).

Most FGCs have not the same level of human development as the so called developed countries, thus institutions and governance need to be strengthened, along with industrial policy and urban planning; these policies and measures will involve a certain energy and resource consumption (Pan and Zheng, 2011). However, overpopulation and rapid growth may involve an increase in the exposure and vulnerability to climate impacts. For instance, the most densely populated coastal zones developed in recent years are those with highest risk for extreme events (Shi, 2011). In recent periods, more than 70% of the economic loss due to natural disasters in China was climate-related (Luo et al., 2011).

The general increase in climate vulnerability has generated a strong need for adaptation in FGCs. In fact, adaptation can contribute to achieving development goals in the short term, together with reductions in vulnerability over the long term (World Bank, 2010a). Therefore, even if international funding might be a constraint for implementing adaptation policy, all the FGCs have reached a consensus on the idea that adapting to climate change may lower the cost of development (see Chapter 17 for further details on adaptation in FGCs).

1.4.3 Adaptation in low-income countries

As mentioned before, climate change impacts are expected to be higher in developing countries. There are several reasons for this: first, the already existing high level of poverty in these countries; second, many of these economies largely depend on weather and climate (for example, the agricultural sector); and third, they have lower adaptive capacity and, often, lack of political will and difficulty in accessing resources and funding as well (Atta-Krah, 2012; Tol, 2005).

Despite this unfavourable starting point, some adaptation initiatives at the national and community levels are already underway. However, the implementation of these measures requires a major economic effort, which will increase as the impacts of climate change get more severe. According to the World Bank (2010b), the cost to developing countries of adapting to a 2° C warmer climate by 2050 could range between US$75 billion to US$100 billion per year. This cost represents the economic impact of additional adaptive capacity needed to face future climate change and is, therefore, additional to the costs of development.

In a situation of financial constraints and big development/adaptation challenges, economic tools such as cost-benefit analysis (CBA) can be very useful to help decision- and policy makers identify priorities (Chambwera and Stage, 2010). In Chapter 18, Lunduka et al. propose a stakeholder-based CBA for low- and middle-income countries. This approach allows consideration of the qualitative aspects of CBA; for instance, how cost and benefits of adaptation are distributed among the vulnerable groups.

In any case, adaptation should be considered as an opportunity to enhance development, rather than a separate issue (Chambwera and Stage, 2010). In this way, low- and middle-income countries should prioritise those adaptation measures that contribute to achieving the development goals over others addressing solely the impacts of climate change.

1.4.4 The role of regional and local governments in adaptation

Climate policy requires global compromises and coordination. However, regional (subnational) and local governments can play an important role in setting up the measures and policies to address climate change, especially if we look at adaptation. According to the UN, many regions are already involved in climate policy design, both at national and international scales, more than ever before (UNDP-UNEP-EMG-ISDR, 2008).

Regional governments in many parts of the world are responsible for many of the policy areas involved in climate policies (i.e., energy, special planning, transport, industry, housing, environment, etc.) so they are key actors in implementing the policy actions for both adaptation and mitigation (Galarraga et al., 2011a).

If we focus on adaptation, there are several reasons to underline the role of regional and local governments with regard to this issue: first, climate change impacts can vary significantly from one region to another, thus adaptation should be especially designed to respond to regional vulnerability. Second, regions are closer to their citizens, therefore are able to identify the specific context, the strengths and difficulties that may arise during the implementation process. Third, this proximity is also a good tool for facilitating social participation and public information and awareness. Actually, this could even guarantee a better implementation process (Galarraga et al., 2011a).

Nevertheless, some difficulties may also arise when working at the regional and local scales, difficulties in most cases related to state-region and interregional coordination.

1.4.5 The role of technology in adaptation

In the literature on climate change, technological advances have been traditionally addressed from the perspective of reducing emissions. However, technology can also be very useful for adapting to climate change and, in fact, most adaptation options include some form of technology or technique (Christiansen et al., 2011). Efficient cooling systems, desalination technologies or increasingly advanced weather forecasts that anticipate extreme events are examples of how technology can enhance adaptive capacity.

Although technological capacity and innovation can be considered a fundamental element of adaptive capacity, many technological solutions are specific to a sector or are associated with a particular impact (Adger et al., 2007). For example, Clements et al. (2011) provide a range of technological systems to improve adaptive capacity in developing countries from an agro-ecological approach. The proposed series of technologies contribute to building long-term resilience while enhancing productivity at the same time. Klein et al. (2001:534) identify four main groups of technological options for reducing the vulnerability of coastal zones: technologies for information and awareness, for planning and designing adaptation strategies, for implementing those and, finally, for monitoring and evaluating their effectiveness. Elliott et al. (2011) address adaptation technologies and practices for developing countries in the water sector. The authors propose an Integrated Water Resource Management (IWRM) approach as an overall decision-making framework, since adaptation in the water sector should be addressed as part of an inter-sectoral strategy for guaranteeing the sustainable and safe use of water resources. The role of technology from an economic perspective is analysed in depth in Chapter 20 of this volume.

Nevertheless, technology by itself is not enough to respond to the challenges of climate change adaptation for several reasons: first, the framework for decision-making in situations of great uncertainty may constrain the development or implementation of technological options. Second, some technological solutions might not be suitable options from a cultural, local or economic perspective. Third, the local context is very important when addressing adaptation; consequently, technologies proven to be successful in one place may not be so in another (Adger et al., 2007).

Finally, it is important to stress that the adaptive capacity is not only influenced by the economy and technology, but also by social factors such as human capital and institutional capacity (Adger et al., 2007).

1.4.6 Adaptation and extreme events

According to the IPCC, climate disasters are defined as:

> severe alterations in the normal functioning of a community or a society due to hazardous physical events interacting with vulnerable social conditions, leading to widespread adverse human, material, economic, or environmental effects that require immediate emergency response to satisfy critical human needs and that may require external support for recovery.
>
> *(IPCC, 2012:3)*

As follows from the definition, the severity of the disaster not only depends on the characteristics of the extreme event itself, but also the exposure and vulnerability of the affected areas, which in turn depend upon various other factors, such as anthropogenic climate change, natural climate variability and socioeconomic development. Adaptation focuses on reducing exposure and vulnerability by increasing resilience to the potential impacts of extreme events due to climate change (IPCC, 2012).

Disasters that occurred around the globe during the last decades have driven a growing interest of researchers and policy makers in assessing the economic impacts of extreme events. Significant progress has been made since, but research has been mostly carried out in the developed country context (Okuyama, 2008). However, there is a general consensus that macroeconomic effects are expected to be more severe in low- and middle-income countries (Lal et al., 2012).

The specific characteristics of each hazard and its associated damages represent a major challenge for economic modelling of the impacts of extreme events. Okuyama (2008) offers an exhaustive review of methodologies for disaster impact analysis, such as input-output (IO), social accounting matrix (SAM), computable general equilibrium (CGE) and econometric models, each method showing strong and weak features (e.g., IO and SAM tend to overestimate economic impacts while CGE may lead to underestimations). Mitchell, Mechler and Peters in Chapter 21 of the volume assess the way disaster losses threaten future economic development in a climate change context and how mainstreaming risk management into economic and fiscal policy can be a key issue in increasing resilience and reducing the exposure and vulnerability to extreme events.

1.5 Concluding remarks

Decision makers in both the public and private sectors and in developed and developing countries are becoming increasingly aware of the need to adapt to climate change. As this chapter shows, an important contribution to this area is an economic analysis of the different adaptation options. The literature on how economic tools can be used, the difficulties they present and how these difficulties can be overcome is relatively young but growing fast. Particularly important are the boundaries of the economic analysis and how it can complement other methods of assessment. We cover these developments in this book, drawing on the work of leading practitioners in the field. To be sure, there are many gaps that are not addressed. The field is still developing (and new results are coming in all the time), so it is impossible to be completely up to date. Nevertheless the reader will find a good guide to the thinking in the field.

Given space limitations, we have also not been able to cover all areas or sectors where adaptation actions are relevant. A particular gap for example is tourism, where major changes in demand can be expected as a result of climate change. Nevertheless, even for such sectors the discussions on methods in the book are valuable and useful.

The material assembled here should be of great interest to policy makers, researchers and students working in this area. We hope you find this to be the case and welcome any comments you may have on the contents.

Notes

1 Least Developed Countries Fund (LCD Fund), the Special Climate Change Fund (SCCF) and the Adaptation Fund.
2 Available at: http://www.environment agency.gov.uk/homeandleisure/floods/125045.aspx
3 PESETA is a project coordinated by the European Commission Joint Research Centre (JRC). More information available at: http://peseta.jrc.ec.europa.eu/

References

Abadie, L., Galarraga, I. and Rübbelke, D. 2013. An Analysis of the Causes of the Mitigation Bias in International Climate Finance. *Mitigation and Adaptation Strategies for Global Change*, 18 (7): 943–955.

Ackerman, F., Stanton, E.A., 2008. *The cost of climate change: What we'll pay if global warming continues unchecked*. Natural Resources Defense Council, New York.

Adger, W.N., Agrawala, S., Mirza, M.M.Q., Conde, C., O'Brien, K.L., Pulhin, J., Pulwarty, R., Smit, B., Takahashi, K., 2007. Assessment of adaptation practices, options, constraints and capacity. In *Climate change 2007: Impacts, adaptation and vulnerability*, edited by Parry, M., Canziani, O., Palutikof, J., van der Linden, P., and Hanson, C., 719–743. Contribution of Working Group II to the Fourth Assessment Report of the IPCC. Cambridge, UK: Cambridge University Press.

Adger, W.N., Arnell, N.W., Tompkins, E.L., 2005. Successful adaptation to climate change across scales. *Global Environmental Change* 15, 77–86.

Adger, W.N., Dessai, S., Goulden, M., Hulme, M., Lorenzoni, I., Nelson, D.R., Naess, L.O., Wolf, J., Wreford, A., 2009. Are there social limits to adaptation to climate change? *Climatic Change* 93, 335–354.

Agrawala, S., Bosello, F., Carraro, C., de Cian, E., Lanzi, E., 2011. Adapting to climate change: Costs, benefits, and modelling approaches. *International Review of Environmental and Resource Economics* 5, 245–284.

Agrawala, S., Fankhauser, S., 2008. Economic aspects of adaptation to climate change: Costs, benefits and policy instruments. OECD Publishing.

Atta-Krah, A., 2012. Challenges in building resilience to respond to climate change: A focus on Africa and its least developed countries. UN Economic Commission for Africa.

Barbier, E.B., Markandya, A., Pearce, D.W., 1990. Sustainable agricultural development and project appraisal. *European Review of Agricultural Economics* 17, 181–196.

Barnett, J., O'Neill, S., 2010. Editorial: Maladaptation. *Global Environmental Change* 20, 211–213.
Burton, I., 1993. The environment as hazard. New York: Guilford Press.
Burton, I., 2004. Climate change and the adaptation deficit. Occasional Paper 1, *Environment Canada*.
Carter, T.R., Parry, M.L., Harasawa, H., Nishioka, N., 1994. IPCC technical guidelines for assessing climate change impacts and adaptations. London: University College London.
Cearreta, A., García-Artola, A., Leorri, E., Irabien, M.J., Masque, P., 2013. Recent environmental evolution of regenerated salt marshes in the southern Bay of Biscay: Anthropogenic evidences in their sedimentary record. *Journal of Marine Systems*, Supplement, S203–S212, 109–110.
Chambwera, M., Stage, J., 2010. *Climate change adaptation in developing countries: Issues and perspectives for economic analysis.* London: IIED.
Chiabai, A., Galarraga, I., Markandya, A., Pascual, U., 2012. The equivalency principle for discounting the value of natural assets: An application to an investment project in the Basque coast. *Environ Resource Econ* 1–16.
Christiansen, L., Olhoff, A., Trærup, S. (Eds.), 2011. Technologies for adaptation perspectives and practical experiences. Roskilde: UNEP Risø Centre.
Clements, R., Haggar, J., Quezada, A., Torres, J., 2011. Technologies for climate change adaptation. Agriculture sector, TNA Guidebook Series. Roskilde: UNEP Risø Centre.
Cline, W.R., 1992. *The economics of global warming.* Peterson Institute.
Colls, A., Ash, N., Ikkala, N., 2009. Ecosystem-based adaptation: A natural response to climate change. Gland, Switzerland: IUCN.
Confalonieri, U., Menne, B., Akhtar, R., Ebi, K.L., Hauengue, M., Kovats, R.S., Revich, B., Woodward, A., 2007. Human health. In Climate change 2007: Impacts, adaptation and vulnerability, edited by Parry, M., Canziani, O., Palutikof, J., van der Linden, P., and Hanson, C., 391–431. Contribution of Working Group II to the Fourth Assessment Report of the IPCC. Cambridge: Cambridge University Press.
Dow, K., Berkhout, F., Preston, B.L., Klein, R.J.T., Midgley, G., Shaw, M.R., 2013. Limits to adaptation. *Nature Clim. Change* 3, 305–307.
EEA, 2007. Climate change: The cost of inaction and the cost of adaptation (EEA Technical report No. No 13/2007). EEA.
Elliott, M., Armstrong, A., Lobuglio, J., Bartram, J., 2011. Technologies for climate change adaptation. The water sector. TNA Guidebook Series. Roskilde: UNEP Risø Centre.
Emberson, L., He, K., Rockström, J., 2012. Energy and environment. In *Global energy assessment: Toward a sustainable future*, edited by Johansson, T.B., Patwardhan, A., Nakicenovic, N., and Gomez-Echeverri, L., pp. 191–253. Cambridge (UK), New York (NY, USA) and Laxenburg (Austria): Cambridge University Press and International Institute for Applied Systems Analysis (IIASA).
Evans, E., Ashley, R., Hall, J., Penning-Rowsell, E., Saul, A., Sayers, P., Thorne, C., Watkinson, A., 2004. Foresight. Future flooding. Scientific summary 1 (2004). Office of Science and Technology, London.
Galarraga, I., Gonzalez-Eguino, M., Markandya, A., 2011a. The role of regional governments in climate change policy. *Environmental Policy and Governance* 21, 164–182.
Galarraga, I., Markandya, A., 2009. El cambio climático y su importancia socioeconómica. *Ekonomiaz* 71, 14–39.
Galarraga, I., Osés, N., Markandya, A., Chiabai, A., Khatun, K., 2011b. Aportaciones desde la economía de la adaptación a la toma de decisiones sobre Cambio Climático: Un ejemplo para la Comunidad Autónoma del País Vasco. *Economía Agraria y Recursos Naturales* 11, 113–142.
Gollier, C., 2012. *Pricing the planet's future: The economics of discounting in an uncertain world.* Princeton University Press.
Hallegatte, S., 2009. Strategies to adapt to an uncertain climate change. *Global Environmental Change* 19, 240–247.
Hinkel, J., Klein, R.J.T., 2009. The DINAS-COAST project: Developing a tool for the dynamic and interactive assessment of coastal vulnerability. *Global Environmental Change* 19 (3): 384–395.
Hinkel, J., R. Nicholls, A. Vafeidis, R. Tol, Avagianou, T., 2010. Assessing risk of and adaptation to sea-level rise in the European Union: An application of DIVA. *Mitigation and Adaptation Strategies for Global Change* 5 (7): 1–17.
HMT, 2007. *The Green Book: Appraisal and evaluation in central government, treasury guidance.* London.
HMT, 2009. *Accounting for the effects of climate change, supplementary Green Book guidance.*

Holling, C.S., 1973. Resilience and stability of ecological systems. *Annual Review of Ecology and Systematics* 1–23.

IPCC, 2001. *Climate change 2001: Impacts, adaptation, and vulnerability*. Contribution of Working Group II to the Third Assessment Report of the Intergovernmental Panel on Climate Change. Cambridge, UK: Cambridge University Press.

IPCC, 2007a. *Climate Change 2007: Synthesis Report*. Contribution of Working Groups I, II and III to the Fourth Assessment Report of the IPCC. R.K. Pachauri and A. Reisinger (Eds.). Geneva, Switzerland.

IPCC, 2007b. *Climate Change 2007: Impacts, adaptation and vulnerability*. Contribution of Working Group II to the Fourth Assessment Report of the IPCC. Cambridge, UK: Cambridge University Press.

IPCC, 2012. Summary for policymakers. In *Managing the risks of extreme events and disasters to advance climate change adaptation: A special report of Working Groups I and II of the Intergovernmental Panel on Climate Change*, edited by Field, C.B., Barros, V., Stocker, T.F., Qin, D., Dokken, D.J., Ebi, K.L., Mastrandrea, M.D., Mach, K.J., Plattner, G.K., Allen, S.K., Tignor, M., and Midgley, P.M., 1–19. Cambridge, UK, and New York, NY, USA: Cambridge University Press.

Isaac, M., van Vuuren, D.P., 2009. Modeling global residential sector energy demand for heating and air conditioning in the context of climate change. *Energy Policy* 37, 507–521.

Jones, H.P., Hole, D.G., Zavaleta, E.S., 2012. Harnessing nature to help people adapt to climate change. *Nature Clim. Change* 2, 504–509.

Klein, R.J., Huq, S., Denton, F., Downing, T.E., Richels, R.G., Robinson, J.B., Toth, F.L., 2007. Interrelationships between adaptation and mitigation. In *Climate change 2007: impacts, adaptation and vulnerability*, edited by Parry, M., Canziani, O., Palutikof, J., van der Linden, P., and Hanson, C., 745–777. Contribution of Working Group II to the Fourth Assessment Report of the IPCC. Cambridge: Cambridge University Press.

Klein, R.J.T., Nicholls, R.J., Ragoonaden, S., Capobianco, M., Aston, J., Buckley, E.N., 2001. Technological options for adaptation to climate change in coastal zones. *Journal of Coastal Research* 531–543.

Kundzewicz, Z.W., Schellnhuber, H.J., 2004. Floods in the IPCC TAR perspective. *Natural Hazards* 31, 111–128.

Lal, P.N., Mitchell, T., Aldunce, P., Auld, H., Mechler, R., Miyan, A., Romano, L.E., Zakaria, S., 2012. National systems for managing the risks from climate extremes and disasters. In *Managing the risks of extreme events and disasters to advance climate change adaptation. Special report of the intergovernmental panel on climate change*, edited by Field, C.B., Barros, V., Stocker, T.F., Qin, D., Dokken, D.J., Ebi, K.L., Mastrandrea, M.D., Mach, K.J., Plattner, G.K., Allen, S.K., Tignor, M., and Midgley, P.M., 339–392. Cambridge, UK, and New York, NY, USA: Cambridge University Press.

Lawler, J.J., 2009. Climate change adaptation strategies for resource management and conservation planning. *Annals of the New York Academy of Sciences* 1162, 79–98.

Lempert, R.J., Schlesinger, M.E., 2000. Robust strategies for abating climate change. *Climatic Change* 45, 387–401.

Luo, Y., 2011. Annex VIII: China's climate disasters. In *Green Book of climate change. Annual report on actions to address climate change: Durban dilemma and China's strategic options*, edited by Wang, W.G., Zheng, G.G., Luo, Y., Pan, J.H., and Chao, Q.H. Beijing: Social Sciences Academic Press.

Markandya, A., 2010. Involving developing countries in global climate policies. In *Climate Change policies: Global challenges and future prospects*, edited by Cerdá, E. and Lavandeira, X, 187–199. Cheltenham, UK: Edward Elgar.

Markandya, A., Galarraga, I., 2011. Technologies for adaptation: An economic perspective. In *Technologies for adaptation perspectives and practical experiences*, edited by Christiansen, L., Olhoff, A., and Trærup, S. Roskilde: UNEP Risø Centre.

Markandya, A., Mishra, A., 2011. *Costing adaptation: Preparing for climate change in India*. New Delhi: Teri Press.

Markandya, A., Watkiss, P., 2009. *Potential costs and benefits of adaptation options: A review of existing literature*. UNFCCC Technical Paper. FCCC/TP/2009/2 80.

McKinnon, K., Hickey, V., 2010. *Convenient solutions to an inconvenient truth: Ecosystem-based approaches to climate change*. International Bank for Reconstruction and Development/The World Bank, Washington D.C., USA.

McMichael, A.J., Woodruff, R.E., Hales, S., 2006. Climate change and human health: Present and future risks. *The Lancet* 367, 859–869.
Mertz, O., Halsnæs, K., Olesen, J.E., Rasmussen, K., 2009. Adaptation to climate change in developing countries. *Environmental Management* 43, 743–752.
Milly, P.C.D., Wetherald, R.T., Dunne, K.A., Delworth, T.L., 2002. Increasing risk of great floods in a changing climate. *Nature* 415, 514–517.
Moberg, F., Rönnbäck, P., 2003. Ecosystem services of the tropical seascape: Interactions, substitutions and restoration. *Ocean & Coastal Management* 46, 27–46.
Munang, R., Thiaw, I., Alverson, K., Mumba, M., Liu, J., Rivington, M., 2013. Climate change and ecosystem-based Adaptation: A new pragmatic approach to buffering climate change impacts. *Current Opinion in Environmental Sustainability* 5, 67–71.
Munasinghe, M., 2000. Development, equity and sustainability (DES). In guidance papers on the cross cutting issues of the third assessment report of the IPCC, IPCC Supporting Material, edited by Pachauri, R., Taniguchi, T., and Tanaka, K. Geneva, Switzerland.
Nicholls et al. (2006). *Metrics for assessing the economic benefits of climate change policies: Sea level rise.* ENV/EPOC/GSP(2006)3/FINAL. OECD, 2006.
Nordhaus, W.D., 1994. *Managing the global commons: The economics of climate change* (Vol. 31). Cambridge, MA: MIT press.
Okuyama, Y., 2008. *Critical review of methodologies on disaster impacts estimation.* (Background paper for EDRR report).
Paavola, J., Adger, W.N., 2006. Fair adaptation to climate change. *Ecological Economics* 56, 594–609.
Pan, J.H., Zheng, Y., 2011. The concept and theoretical implications of carbon emissions rights based on individual equity. *Chinese Journal of Population, Resources and Environment* 9.
Parry, M., Arnell, N., Berry, P., Dodman, D., Fankhauser, S., Hope, C., Kovats, S., Nicholls, R., Satterthwhite, D., Tiffin, R., Wheeler, T., 2009. Assessing the costs of adaptation to climate change: A review of the unfccc and other recent estimates. *International Institute for Environment and Development and Grantham Institute for Climate.* London: Change.
Parry, M., 2007. The implications of climate change for crop yields, global food supply and risk of hunger. *SAT eJournal* | ejournal.icrisat.org 4, 1.
Parry, M.L., Rosenzweig, C., Iglesias, A., Livermore, M., Fischer, G., 2004. *Effects of climate change on global food production under SRES emissions and socio-economic scenarios, Global Environmental Change*, 14, 53–67.
Parry, M., Arnell, N., Hulme, M., Nicholls, R., Livermore, M., 1998. Adapting to the inevitable. *Nature* 395, 741–741.
Perrings, C., 1998. Resilience in the dynamics of economy-environment systems. *Environmental and Resource Economics* 11, 503–520.
Pielke, R., Prins, G., Rayner, S., Sarewitz, D., 2007. Climate change 2007: Lifting the taboo on adaptation. *Nature* 445, 597–598.
Pindyck, R.S., 2007. Uncertainty in environmental economics. *Review of Environmental Economics and Policy* 1, 45–65.
Posner, E.A., Sunstein, C.R., 2009. Justice and climate change. In *Post-Kyoto international climate policy: Summary for policymakers*, edited by Aldy, J.E. and Stavins, R.N., 93–95. Cambridge University Press.
Refsgaard, J.C., Arnbjerg-Nielsen, K., Drews, M., Halsnæs, K., Jeppesen, E., Madsen, H., Markandya, A., Olesen, J.E., Porter, J.R., Christensen, J.H., 2013. The role of uncertainty in climate change adaptation strategies – A Danish water management example. *Mitig Adapt Strateg Glob Change* 18, 337–359.
Rübbelke, D.T.G., 2011. International support of climate change policies in developing countries: Strategic, moral and fairness aspects. *Ecological Economics* 70, 1470–1480.
Sainz de Murieta, E., Chiabai, A., 2013. Climate change impacts on the water services in Costa Rica: A production function for the hydroenergy sector. *BC3 Working Papers. BC3.* Bilbao.
Shi, P.J., 2011. *Atlas of natural disaster risk in China.* Beijing: China Science Press.
Smit, B., Burton, I., Klein, R.J.T., Street, R., 1999. The science of adaptation: A framework for assessment. *Mitigation and Adaptation Strategies for Global Change* 4, 199–213.

Smit, B., Burton, I., Klein, R.J.T., Wandel, J., 2000. An anatomy of adaptation to climate change and variability. In *Societal adaptation to climate variability and change*, edited by Kane, S.M. and Yohe, S.W., 223–251. Springer Netherlands.
Stakhiv, E., 1993. Evaluation of IPCC evaluation strategies (draft report). Institute for Water Resources, U.S. Army Corps of Engineers, Fort Belvoir, VA.
Sterk, W., Ott, H.E., Watanabe, R., Wittneben, B., 2007. The Nairobi climate change summit (COP 12 – MOP 2): Taking a deep breath before negotiating post-2012 targets? *Journal for European Environmental & Planning Law* 4, 139–148.
Stern, N., 2007. *The economics of climate change: the Stern review.* Cambridge University Press.
Thomas, D.S.G., Twyman, C., 2005. Equity and justice in climate change adaptation amongst natural-resource-dependent societies. *Global Environmental Change* 15, 115–124.
Tol, R.S.J., 2005. Emission abatement versus development as strategies to reduce vulnerability to climate change: An application of FUND. *Environment and Development Economics* 10, 615–629.
Tompkins, E.L., Eakin, H., 2012. Managing private and public adaptation to climate change. *Global Environmental Change* 22, 3–11.
UKCIP, 2007. Identifying adaptation options (UKCIP technical report). Oxford: United Kingdom Climate Impacts Programme.
UNDP-UNEP-EMG-ISDR, 2008. *UN and regions' partnership for sustainable development and to address climate change.* Brussels: UNDP-UNEP-EMG-ISDR.
UNFCCC, 2007a. *Report of the subsidiary body for scientific and technological advice on its twenty-fifth session.* No. FCCC/SBSTA/2006/11.
UNFCCC, 2007b. *Background paper on analysis of existing and planned investment and financial flows relevant to the development of effective and appropriate international response to climate change.* Bonn: UNFCCC.
UNFCCC, 2011. *Report of the conference of the parties on its sixteenth session; addendum: Part two: Action taken by the conference of the parties on its sixteenth session.* No. FCCC/CP/2010/7/Add.1.
Van Ierland, E.C., de Bruin, K., Dellink, R.B., Ruijs, A., 2007. *A qualitative assessment of climate adaptation options and some estimates of adaptation costs.* Netherlands Policy Programme ARK-Routeplanner projects.
Virmani, A., 2012. *Accelerating and sustaining growth: Economic and political lessons.* No. ID 2169730, IMF Working Papers. Rochester, NY.
Walker, B., Pearson, L., Harris, M., Maler, K.-G., Li, C.-Z., Biggs, R., Baynes, T., 2010. Incorporating resilience in the assessment of inclusive wealth: An example from South East Australia. *Environmental and Resource Economics* 45, 183–202.
Weitzman, M.L., 2001. Gamma discounting. *The American Economic Review* 91, 260–271.
Weitzman, M.L., 2007. A review of the Stern Review on the economics of climate change. *Journal of Economic Literature* 45, 703–724.
World Bank, 2010a. *Economics of adaptation to climate change. Synthesis report.* The International Bank for Reconstruction and Development/The World Bank, Washington D.C., USA.
World Bank, 2010b. *The cost to developing countries of adapting to climate change. New methods and estimates.* Washington D.C.: Consultation draft.
Zeckhauser, R.J., Viscusi, W.K., 2008. Discounting dilemmas: Editors' introduction. *J Risk Uncertain* 37, 95–106.

2

STATE OF THE ART ON THE ECONOMICS OF ADAPTATION

Clemens Heuson, Erik Gawel and Paul Lehmann[1]

[1] HELMHOLTZ CENTRE FOR ENVIRONMENTAL RESEARCH – UFZ, LEIPZIG, GERMANY, DEPARTMENT OF ECONOMICS

2.1 The economics of adaptation to climate change – guidance for a new field of policy

Besides mitigation, adaptation to climate change is a natural component within any climate policy strategy that aims at minimizing total costs associated with climate change (Tol, 2005). In the view of lacking success of international negotiations on mitigation, the adaptation option becomes even more important (see e.g., Fankhauser, 2009; Pielke et al., 2007). Numerous countries have already initiated a process of adaptation by drafting strategies or action plans for public adaptation measures.[1] Hence, there is a particular need for scientific support. The discipline of economics has a key role to play in this context since it provides tools and methods for answering questions that are crucial to the adaptation policy process, such as forming and operationalizing adaptation goals, identifying adequate actors and instruments or scrutinizing interdependencies between adaptation and mitigation. The still relatively young field of research is growing at a considerable pace and already exhibits a wide range of methodological approaches and research questions. Against this background, this chapter aims to provide a systematic overview of the state of the art on the economics of adaptation – with a focus on conceptual and analytical work – and, building upon this, to outline key pointers for future research.

The structure of this chapter is intended to capture the logical sequence of research questions. Section 2.2 lays the foundations by defining the boundaries of the research field and framing the concept of adaptation in economic terms. Section 2.3 focusses on the basic economic insights on the relation between mitigation and adaptation. This is of particular importance since first, both strategies are inevitably linked to each other in being substitutes and complements, respectively, to confront climate change and thus require integrated analysis (Agrawala et al., 2011); second, adaptation plays an important strategic role regarding international cooperation on mitigation. Probably, the major task of the economic discipline is to provide guidance on how to allocate scarce resources between different adaptation options as well as between mitigation and adaptation (Section 2.4). This entails first identifying and analysing the optimal extent and pathway of adaptation from a theoretical perspective. Second, in order to operationalize the respective findings and feed them into the policy process, an empirical assessment of adaptation costs and benefits is required. Care has to be taken when it comes to the identification of the adequate actors for actually conducting adaptation measures (Section 2.5). From the economic perspective, publicly provided adaptation is only legitimate in case of market failure that cause inefficiencies in autonomous adaptation efforts. Another ground for government intervention is given by goals that are beyond the optimality concept, such as distributive justice or security of supply. Having identified cases for public adaptation, Section 2.6 addresses possible means and instruments of adaptation policy that allow for overcoming barriers to optimal private adaptation and realizing non-optimality-related goals, respectively. In this regard, it is important to see that adaptation policy itself can also be subject to optimality barriers. Thus, Section 2.7 introduces approaches that screen the adaptation policy process from a positive perspective, with a special focus on self-interest driven barriers of political actors. Finally, Section 2.8 concludes and proposes directions for future research.

2.2 Framing adaptation as an object of economic research

2.2.1 A system-based conceptualization of adaptation

The starting point for framing adaptation as an object of (economic) research is given by the definition of the Intergovernmental Panel on Climate Change (IPCC, 2001), which has been broadly accepted in the economic literature (see e.g., Tol, 2005):

> Adaptation is adjustment in ecological, social, or economic systems in response to actual or expected climatic stimuli and their effects or impacts (. . .) to moderate or offset potential damages or to take advantage of opportunities associated with changes in climate. (. . .)

In line with this definition, Smit et al. (1999) develop from an interdisciplinary point of view the three dimensions of adaptation, depicted by the questions 'Adaptation to what?' (climate-related stimuli), 'Who/what performs the adaptation?' (adaptation system) and 'How does adaptation occur?' (measures). The grounds for adaptation are provided by climate-related stimuli, i.e., altered climate conditions (e.g., precipitation or temperature) and the resulting ecological or economic impacts (e.g., droughts, crop failures, income losses) that are linked to the sensitivity of the (adaptation) system under observation. First, this system is to be defined according to the level at which the adaptation takes place. For example, adaptation at the level of an agricultural holding comprises crop diversification, whereas at a global level it can manifest as a shift in the international food trade structure. Furthermore, the definition refers to the nature of the adaptation system, which may be ecological, political, social or economic or may encompass a combination of these components. Finally, the system has to be differentiated according to who performs the adaptation (e.g., coastal protection managers) and what modifies itself or is modified (e.g., coastal settlements). The last adaptation dimension focuses on the question of how, i.e., with what measures, the adaptation system confronts climate-related stimuli (structure of adaptation). Table 2.1 shows the characterization of adaptation measures according to Smit et al. (1999).

Table 2.1 Categorisation of adaptation measures (Smit et al., 1999, p. 208)

General differentiating concept or attribute	*Examples of terms used*	
Purposefulness	autonomous spontaneous automatic natural passive	planned purposeful intentional policy active strategic
Timing	anticipatory proactive ex ante	responsive reactive ex post
Temporal scope	short term tactical instantaneous	long term strategic cumulative
Spatial scope	localised	widespread
Function/Effects	retreat – accommodate – protectprevent – tolerate – spread risk – change – restore	
Form	structural – legal – institutional – regulatory – financial – technological	
Performance	optimality – efficiency – implementability – equity	

2.2.2 An action-based conceptualization of adaptation

However, for economic sciences, this comprehensive system-oriented conceptualization of adaptation might be inappropriate. Rather, a stronger focus on economic systems that are subject to climate change impacts or even more on related actors might by necessary. For this reason, Eisenack and Stecker (2012) offer an alternative framework that conceptualizes climate change adaptations as actions, restricting themselves to adaptations made by human actors (see Figure 2.1).

Similarly to Smit et al. (1999), the starting point is given by a stimulus, i.e., a statistical change in meteorological variables due to climate change. The stimulus is only relevant for adaptation when it affects an exposure unit, which may comprise actors or all kinds of systems (i.e., social, ecological and so forth) that depend on climatic conditions. When combined, the stimulus and exposure unit form a climate change impact. The response to the latter is carried out by individual or collective actors, called operators. In this sense, the operator's activities only qualify for being an action when serving a certain purpose. Finally, the actor or system being the target of an action is called the receptor. Note that exposure unit, operator and receptor may either coincide or (partly) fall apart. Finally, the implementation of adaptation measures requires means in a broad sense, ranging from financial or material resources through to legal power and information.

The purpose of this framework is on the one hand to structure the analysis of adaptation with a focus on related actors and institutions. In this way, barriers of adaptation can be revealed and

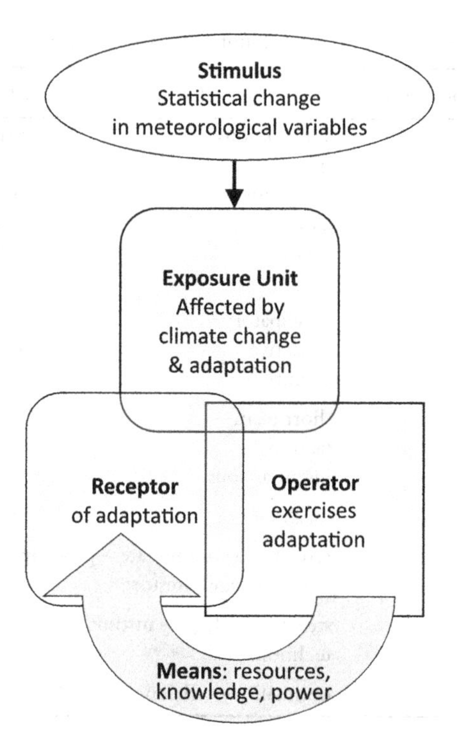

Figure 2.1 Framework for analysing adaptation as actions* (Eisenack and Stecker, 2012, p. 246)

*Boxes with rounded corners can be both actors or biophysical units, while operators are always actors. Operator, receptor and exposure unit are not necessarily identical (indicated by overlapping boxes). The straight arrow indicates a causal relation, and the large arrow a teleological relation.

systemized. Barriers typically emerge when there is a mismatch of the actors' functions described above. On the other hand, the framework allows for clarifying the concept of adaptation and for getting over shortcomings of the so far established frameworks. For instance, some of the questions raised by Smit et al. (1999) in order to point out the adaptation dimensions can be clarified by using the introduced concepts ('Adaptation to what?' – impact, i.e., stimulus combined with exposure unit; 'Who or what adapts?' – operator, receptor and their relation to the exposure unit; and 'How does adaptation occur?' – linkage of means and purpose, consideration of processes vs. actions).

2.2.3 The basic approach of economic adaptation research

Aaheim and Aasen (2008) demonstrate how, building upon such frameworks, adaptation can be tackled as object of economic research. The basic idea is that "(. . .) the impacts of climate change can be analysed with the same economic tools used for analysing the impacts of changing economic conditions" (Aaheim and Aasen, 2008,

p. 1). Simply speaking, the basic task of the economic discipline is to provide tools for investigating adaptation to changing economic constraints, whereas specific behavioural assumptions concerning the economic agents are taken for granted. Thus, provided that it is known how climate change affects these constraints, adaptation can be tackled and understood by the economic discipline. The advantage of this approach rests upon its capability of making understandable to which extent and in what way economic agents autonomously adapt. In this respect, Aaheim and Aasen (2008, p. 2) distinguish between direct autonomous adaptation, referring to the changes that economic actors make when confronted with climate change (e.g., increased demand for air conditioners), and indirect autonomous adaptation, referring to the market response resulting from the direct adaptation (e.g., increased price for air conditioners leading to a new market equilibrium). Additionally, the approach allows for revealing cases in which autonomous adaptation may be insufficient and thus requires public adaptation efforts (see Section 2.5).

2.3 Basic economic insights on the relation between mitigation and adaptation

2.3.1 Needs for and limits of an integrated assessment of mitigation and adaptation

Basically, there are two reasons requiring an integrated consideration and analysis of adaptation and mitigation. First, both are strategies to confront the adverse impacts of climate change and thus need to be taken into account at the same time when it comes to the question of how to allocate scarce resources such that total climate change costs are minimized (Agrawala et al., 2011; cf. Section 2.4). Second, the option of adaptation plays a crucial role in influencing the countries' strategic behaviour in international cooperation on mitigation (Heuson et al., 2012b; Marrouch and Chaudhuri, 2011; cf. Chapter 3). However, Tol (2005) argues that mitigation and adaptation should largely be considered separately, because they generally do not fit into the same analysis framework for three reasons. First, there is a discrepancy in the basic scope of action. While mitigation efforts are part of the area of competence of national governments against the backdrop of international climate protection negotiations, adaptation measures are primarily implemented by local managers of natural resources, households and firms. Furthermore, the tools for decision support relating to the planning and implementation of mitigation and adaptation measures are directed at different addressees (mitigation – e.g., ministries for energy or environment; adaptation – e.g., local water management or farmers). Finally, there is also a discrepancy in terms of the temporal scope of decision support (mitigation – short-term measures aimed at long-term effects; adaptation – short-term measures targeted at short- to long-term developments). Nevertheless, Tol (2005) sees so-called facilitative adaptation – measures for building up adaptive capacity – as an exception, i.e., as suitable for a joint analysis with mitigation, since it has similar scales. Referring to this case, now the two reasons for joint analysis stated above are examined in detail.

2.3.2 *Mitigation and adaptation as inevitably interlinked components of a comprehensive climate policy*

Adaptation is now widely acknowledged as an indispensable component besides mitigation in any policy that ameliorates climate change-related damages at the least possible cost (Agrawala et al., 2011). Buob and Stephan (2011a) point out that, a priori, the relation between the two strategies in economic terms is not unambiguous. From a static perspective, mitigation and adaptation are substitutes in protecting a country or region against climate change damages. However, from a dynamic perspective, the strategies may be complements in the sense that mitigation slows down climate change. Consequently, societies gain time and are likely to face lower future costs of adaptation (Ingham, 2005). In any case, a comparison of the fundamental characteristics of mitigation and adaptation is recommended in order to achieve a clear distinction of both strategies, which is essential for any further in-depth economic analysis, especially in terms of optimization (Section 2.4). Such a comparison is, amongst others, carried out by Füssel and Klein (2006) – see Table 2.2. Traditionally, mitigation receives greater attention than adaption, both from the scientific and political angle for the following reasons: mitigation can avert negative effects of climate change on all climate-sensitive systems, whereas in many systems the scope for adaptation is limited – think of small island states that are virtually defenceless against rising sea levels. Moreover, the benefits of mitigation are certain because mitigation combats climate change-related problems directly at the source. In contrast, the effectiveness of (proactive) adaptation frequently depends on predictions about the regional vulnerability situation and the related, typically highly uncertain, climate change impacts. Also, mitigation complies with the polluter-pays principle, contrary to adaptation: developing countries generally demonstrate the greatest need for adaptation even though they have contributed far less to climate change than the industrial nations. Finally, obtaining quantitative data on greenhouse gas emissions is relatively unproblematic, which allows for monitoring the success of mitigation efforts. It is much more difficult to measure the effectiveness of adaptation since, due to its heterogeneity, no universal measure of success exists (Cimato and Mullan, 2010).

However, some characteristics favour stronger consideration of adaption. According to the scope of their effects, adaptation measures can be implemented at local or regional level. The situation is different for mitigation, the effectiveness of which depends on collective global efforts. Thus, adaptation is typically a private good,[2] mitigation a public good, which is subject to the free-rider problem. Moreover, compared with mitigation, adaptation measures are often

Table 2.2 Fundamental characteristics of mitigation and adaptation (following Füssel and Klein, 2006)

	Mitigation	*Adaptation*	
Benefitted system	All systems	Selected systems	Advantage mitigation
Effectiveness/benefits	Certain	Generally uncertain	
Polluter pays principle	Typically yes	Not necessarily	
Monitoring success	Relatively easy	More difficult	
Scale of the effect	Global	Local to regional	Advantage adaptation
Payer benefits	Only little	Almost fully	
Added benefit	Sometimes	Frequently	
Lead time	Decades	None up to decades	

associated with an added benefit, in particular in terms of reducing the risks of current climate variability. With regard to the lead time, the benefits of (reactive) adaptation measures are often immediately effective, whereas the effect of mitigation will only kick in after a delay of several decades due to the inertia of the climate system.

2.3.3 The strategic role of adaptation in international negotiations on mitigation

The second rationale for jointly analysing mitigation and adaptation is given by the fact that adaptation has crucial impacts on the outcome of international cooperation on mitigation. Most of the related studies are based on game theory models. Auerswald et al. (2011) point out the basic strategic role of adaptation: by adapting to climate change, a country reduces its associated residual damage, which leads to an enhanced pay-off in the non-cooperative equilibrium. Consequently, the country improves its threat point within international negotiations on mitigation and thus can influence the related burden sharing in its own interest. Starting from these basic insights, several contributions study the countries' strategic efforts in mitigation and adaptation with varying framework conditions. Buob and Stephan (2011a) apply a model with several world regions having available a limited budget to be invested in mitigation and adaptation. They discover that the budget allocation crucially depends on the regions' initial endowment concerning environmental quality and financial means. A special setting is studied by Barrett (2008) in restricting to the case of fixed adaptation and mitigation costs caused by related investments. This constellation provokes corner solutions, i.e., countries solely invest in adaptation (mitigation) in case of non-cooperative or cooperative behaviour. Zehaie (2009) introduces a static, non-cooperative framework comprising two countries, focusing on the role the sequence of mitigation and adaptation decisions plays for the countries' behaviour. When adaptation is timed before mitigation, countries strategically raise their adaptation effort in order to commit to a lower mitigation level in the second stage. In the opposite case, there is no scope for strategic behaviour since adaptation is a private good. Ebert and Welsch (2012) apply a similar framework to study the impact of productivity, pollution sensitivity and adaptive capacity on the countries' emission and adaptation decisions.

A recently evolving branch of literature investigates institutional aspects of climate finance. Buob and Stephan (2011b) analyse the basic incentives for industrialized countries to contribute to adaptation funds.[3] Pittel and Rübbelke (2011) demonstrate that adaptation funding increases the developing countries' fairness perception and which promotes their willingness to contribute to joint mitigation efforts. Heuson et al. (2012b) study the strategic effects, such as incentives for strategic mitigation, arising from the various funding instruments that are about to be implemented in the post-Kyoto process. Moreover, it is shown that some of the instruments fail in fulfilling a minimum requirement for sustainable funding since they do not induce the recipient country to increase its contribution to the public good of mitigation. However, this is the only way for the donor country to profit from funding.

Besides these non-cooperative settings, a few contributions deal with the role of adaptation within given mitigation agreements. Benchekroun et al. (2011) argue that the basic strategic role of adaptation depicted above naturally reduces the stability of such agreements. Principally, this problem can be healed by explicitly including adaptation in the negotiation process. In this way, adaptation loses its commitment function, which boosts the free-riding incentives in terms of mitigation, and the agreement's stability increases. However, the effect of such an extended mitigation-adaptation-agreement on the global emission level is ambiguous a priori, since the countries might agree on substituting mitigation through adaptation, depending on the cost-benefit-ratio.

2.4 Optimal extent and pathway of adaptation

The core question to be answered within the adaptation process is to which extent and at what point of time a society should engage in adaptation – especially against the background that scarce resources which are directed to adaptation compete with the other climate policy strategy, mitigation, but also with completely different societal uses, such as health or education. Economics as discipline of the rational handling of scarcity is predestined to offer orientation in this respect, which basically entails two steps. First, the optimal extent and pathway of adaptation has to be analysed from a theoretical perspective. Second, in order to operationalize the respective findings and feed them into the policy process, an empirical assessment of adaptation costs and benefits is required. Naturally, as adaptation is highly heterogeneous and context-dependent, the precise manifestation of 'optimal adaptation' and the capturing of the respective costs and benefits depend on the analysis' degree of aggregation, see Figure 2.2. In this respect, a joint analysis of adaptation and mitigation implies the maximum level of aggregation,

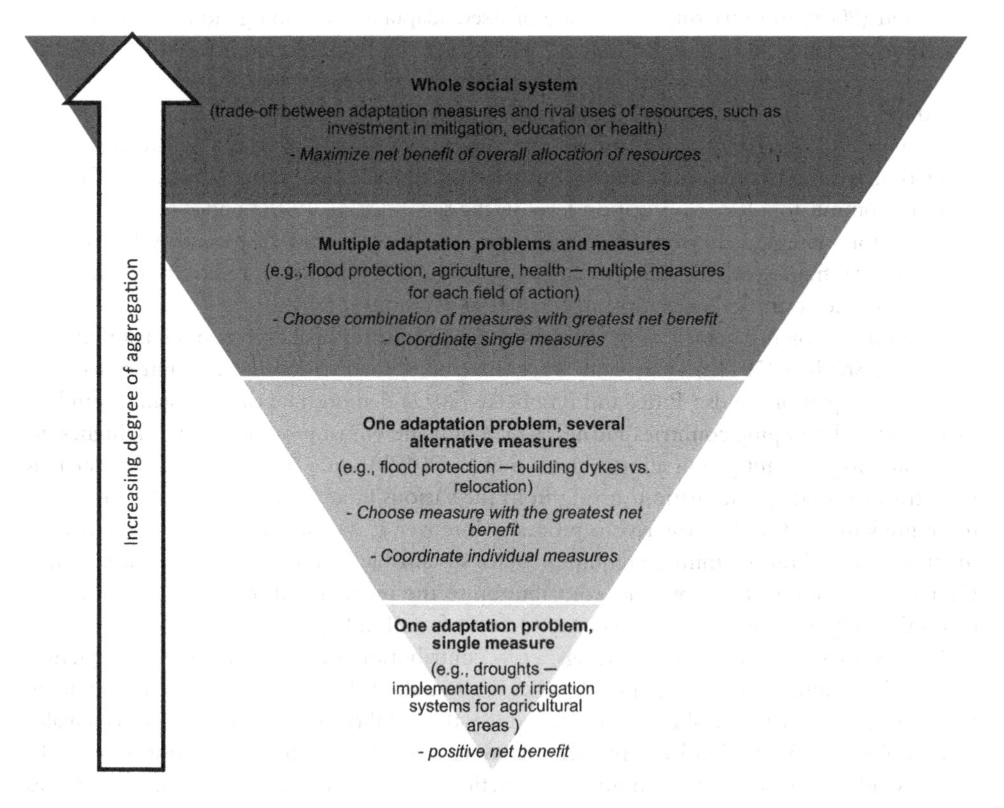

Figure 2.2 Dimensions of 'optimal adaptation' (Heuson et al., 2012a, p. 17)

because this is the only case in which the strategies exhibit the same scales and thus fit into the same analysis framework (cf. Tol, 2005).

2.4.1 Theoretical considerations

The literature on the theoretical foundations of optimal adaptation is subdivided into two strands. The first strand examines the problem on the basis of an isolated assessment of adaptation. The second strand undertakes an integrated assessment of adaptation and mitigation and thus (implicitly) refers to the exceptional case of building adaptive capacity set out by Tol (2005).

Isolated assessment of adaptation

The criterion of optimality can either apply to the intra- (static analysis) or intertemporal allocation of adaptation (dynamic analysis). In this context, Adger et al. (2005) emphasize that a static analysis is sufficient for sectors that adapt rapidly due to a low capital intensity, whereas for capital-intensive, rigid sectors a dynamic analysis is essential.

Clearly, static, microeconomic assessments of optimal adaptation are rather straightforward. This can be seen from the approach by Mendelsohn (2000), using a simple deterministic model to derive the conditions for the efficient provision of private and public adaptation goods. In case of private goods, efficiency is achieved when marginal costs are balanced with (private) marginal benefits; in case of public goods, the efficient provision follows the well-known Samuelson rule. One of the rare macroeconomic contributions to the research on isolated static adaptation optimality is given by Aaheim et al. (2009), who use a computable general equilibrium model with multiple sectors and regions. Adaptation occurs directly via shifts in technologies and preferences and indirectly via the corresponding market reactions. The main finding is that adaptation indeed reduces the macroeconomic climate change impacts, but that the share of costs that is reduced by adaptation is inversely proportional to the global mean temperature.

The dynamic analysis is dominated by microeconomic approaches. Fankhauser et al. (1999) use a simple model to formalize the conditions for the optimal timing of adaptation investments. The basic related trade-off is explained as follows: A delay in investments initially leads to saved adaptation costs which, however, are offset by additional damage costs in the future. Dobes (2010) and Osberghaus et al. (2010a) point out that this trade-off is actually much more complex in view of massive uncertainties about the climate change impacts and potentially irreversible investments. The prospect of learning now favours the stalling of investment. The benefit expected to result from this – the option value – is, of course, to be included in the benefit-cost calculation on the investment decision. In this respect, Wright and Erickson (2004) demonstrate that decreased predictability of climate change impacts results in postponed adaptation and increased pre-adaptation damage costs.

Another largely microeconomics-oriented branch of the literature deals less with the optimal timing of investments than with the problem of how the intertemporal allocation of adaptation efforts should best be designed. The question here is basically how much resources should be channelled into anticipatory adaptation measures and into measures that react to impacts that have already occurred. In this respect, Fankhauser et al. (1999) argue that due to uncertain and changing climate conditions, (adaptation) capital should either be more rapidly offset and exchanged or it should be assigned greater robustness and flexibility. Finally, Dumas and Ha-Duong (2008) undertake one of the rare attempts to address the optimal intertemporal allocation with a macroeconomic approach. Using a growth model with perfect information and foresight of the actors, they show that the optimal adaptation pathway favours proactive compared to reactive adaptation, because measures implemented today are not only designed for current climatic conditions, but also anticipate future conditions. Thus, delaying adaptation after damages is inefficient.

Integrated assessment of adaptation and mitigation

Similarly to the isolated assessment of optimal adaptation, the integrated assessment is broken down into static and dynamic components.

Tol (2005) outlines the basic static marginal condition that characterizes the optimal adaptation-mitigation-mix: marginal costs of mitigation have to be balanced with the marginal utility of mitigation. The latter arises from equalizing residual marginal damage costs and marginal costs of adaptation. McKitrick and Collinge (2002) question the existence of an unambiguously optimal level of mitigation against the background of adaptation measures. They demonstrate that, due to the non-convexities caused by adaptive behaviour, this is not necessarily the case. Kane and Shogren (2000) frame the problem with the theory of endogenous risk. They consider a country that can either engage in mitigation to reduce damage risk or in adaptation to reduce (residual) damage costs. Auerswald et al. (2011) and Tulkens and Steenberghe (2009) investigate the adaptation-mitigation-mix within the framework of similar models, but under the condition of an exogenous risk of damage, putting a special focus on risk aversion.

In the integrated dynamic analysis, only few microeconomic approaches are to be found. The Scientific Advisory Board of the German Federal Ministry of Finance (BMF) argues that, given the uncertain consequences of climate change and potential irreversibilities, adaptation displays an optional character (BMF, 2010): in principle, delaying adaptation would be desirable in order to reduce the uncertainty surrounding the impacts of climate change through learning; however, the situation is complicated by the well-known problem of irreversibility. The adaptation option alleviates this problem, since it can attenuate potential negative impacts, thus providing scope for manoeuvre for the above-described waiting strategy. Ingham et al. (2007) analyse this problem in a stochastic two-period partial market model and come to the conclusion that the prospect of learning or information gain in conjunction with the adaptation option causes a lower optimal level of mitigation in the present. Several works are based on classical macroeconomic modelling approaches: Aalbers (2009) uses a dynamic stochastic general equilibrium model to calculate the optimal discount rates for adaptation, mitigation and other investments. He demonstrates that these generally differ since the investment types are exposed to different degrees of climate-related and non-climate-related risks. Brechet et al. (2010) examine the optimal accumulation of physical capital, adaptation capital and greenhouse gases in a deterministic growth model. In doing so, they establish a correlation between the optimal adaptation-mitigation-ratio and productivity.

Integrated Assessment Models (IAMs) clearly dominate the dynamic analysis of the optimal adaptation-mitigation-mix. These numeric models combine knowledge from a range of disciplines in order to gain insights relevant to policy (Patt et al., 2010). IAMs are also used – in relation to specific countries or regions – to estimate the benefit and costs of adaptation. Typically, physical climate models illustrating the cause and effect chain of climate change are coupled with economic growth models. Having long been limited to mitigation, individual attempts have been made to depict the adaptation option implicitly using IAMs. The starting point is the Ricardian approach, where adaptation is modelled as a shift in production. As IAMs of this type are only of secondary importance, they will not be addressed further (see Patt et al., 2010 for more information).

The latest generation of IAMs treats adaptation as an explicit control variable. The focus here is both on the optimal intra- and intertemporal adaptation-mitigation mix.[4] Agrawala et al. (2011) summarize the key findings that have been found to be robust across the different types of IAMs. Various model calculations underpin the strategic complementarity of adaptation and mitigation in dynamic terms. Since both strategies reduce the damages of climate change, the calculations replicate the theoretical result saying that two instruments pursuing the same goal cannot do worse than a single instrument. In fact, it has been demonstrated that welfare can be increased if the adaptation option is applied additionally to mitigation (Bahn et al., 2010; Bruin et al., 2009). In this context adaptation turns out to be the favoured option in which more resources are invested than in mitigation, and which also makes the greater contribution to damage reduction (Bosello et al., 2010). However, this result is largely due to the fact that the possibility of catastrophic events which have a very low probability of occurrence and which can only be avoided through mitigation cannot be captured in the IAMs (Settle et al., 2007). Despite the strategic complementarity, there is a trade-off between adaptation and mitigation, since both compete for scarce resources (Hof et al., 2009). Furthermore, expansion of adaptation diminishes the marginal utility of mitigation, and vice versa (Fankhauser, 2010). The reciprocal crowding-out of both strategies is, however, characterized by asymmetry: adaptation displaces mitigation to a greater degree than in the reverse situation because in the medium to long term, the extent of the damage avoided by mitigation is too small to reduce the need for adaption. With regard to the optimal timing, the following rule has emerged: whereas investment in mitigation should be made as early as possible, adaptation expenditures should follow the dynamics of the expected damages, i.e., they should occur at first with a certain time lag, but thereafter the rate of spending should increase (Bosello et al., 2010).

Referring to these approaches, several critical opinions have been put forward that question the validity and political relevance of IAMs and point out their limitations (e.g., Jotzo, 2010). The basic tenor is that IAMs are insufficient to take the heterogeneous character of adaptation into account and are therefore unsuitable for delivering concrete policy recommendations. At best, qualitative insights into the relationship between adaptation and mitigation can be expected. Patt et al. (2010) argue that many of the characteristic features of adaptation can only be reproduced insufficiently in IAMs. As a result, the models tend to overestimate the net benefit of adaptation and propagate insufficient mitigation efforts. Specifically, the authors put forward the following points of criticism: IAMs are not capable of reflecting the bottom-up character of adaptation. This is demonstrated by the fact that the costs and the benefits of adaptation are linked to local measures and born by, or are of benefit to, local stakeholders (Tol, 2005). Statements on the optimal level and distribution of adaptation expenditures therefore require knowledge of the respective local circumstances. IAMs with a global focus cannot, however, meet these requirements. Uncertainty is another problem that is difficult to solve. This applies in particular to adaptation, since it is primarily tied to the spatial and temporal distribution of climate change impacts (Patt et al., 2010).

2.4.2 Empirical assessment of adaptation costs and benefits

An empirical assessment of adaptation benefits and costs is vital for the actual implementation of optimal adaptation (mitigation) strategies. These data not only form the basis for project-related decision-making at regional or local levels, they also serve as a price signal for politicians at the international level, especially in terms of financing adaptation measures (Agrawala and Fankhauser, 2008). Concerning the collection of regional and/or local benefit and cost data, there are two fundamental approaches (Gebhardt et al., 2011; cf. Figure 2.3).[5] The starting point for the bottom-up approach is formed by climate projections and estimates of impacts for the region of interest. Based on this, and with the involvement of decision makers and stakeholders, appropriate adaptation measures tailored to the regional (or local) context are identified and subjected to economic assessment. In contrast, top-down approaches aim at breaking down global, IAM-based benefit and cost estimates to the regional or local level by so-called dynamic downscaling (cf. Heuson et al., 2012a).

Due to massive uncertainties about climate change impacts as well as methodological problems related to the evaluation of nonmonetary goods (e.g., the recovery function of a landscape), the benefit of adaptation is not easy to record or quantify. Thus, a significantly smaller proportion of the relevant contributions deal with the benefit side. Dobes (2009) draws up

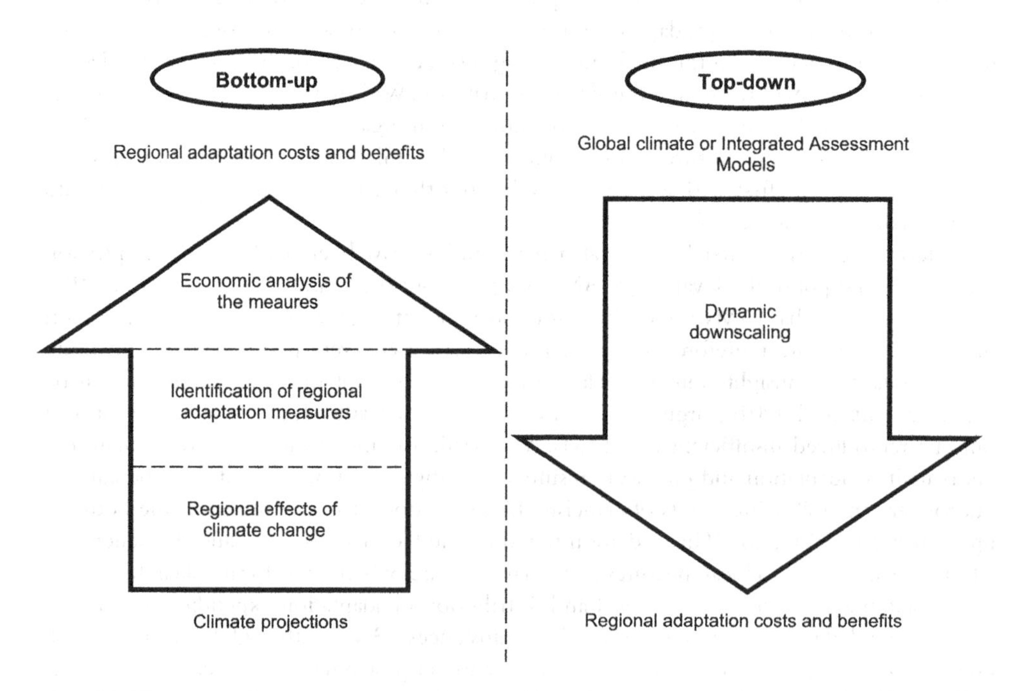

Figure 2.3 Two ways to determine the regional costs and benefits of adaptation (Heuson et al., 2012a, p. 25)

a multi-stage plan for identifying the priorities and preferences of individuals with regard to adaptation measures as a first step towards a monetary evaluation of benefit. Glenk and Fischer (2010) examine the general public's evaluation of government-implemented adaptation measures. At the centre of their study is the question of how preferences for the said measures are constituted by hierarchical networks of values and convictions. They come to the conclusion that constitutive values such as efficiency, sustainability or solidarity are more robust evaluation criteria than the marginal willingness-to-pay.

Most studies focus exclusively on the cost side using a top-down approach: for instance, Deschênes and Greenstone (2007) demonstrate how adaptation costs in the health sector can be estimated using physical climate and health economics models, while Osberghaus and Reif (2010) make the first attempt to examine the fiscal impact of adaptation measures. However, there is quite some criticism on the top-down-based cost estimates. Fankhauser (2010) identifies the weaknesses of previous adaptation cost estimates. He finds fault with, among other things, the breadth and depth of the studies as well as the omission of relevant costs – most studies capture the investment costs only and not the total lifetime costs of the respective measure. The study by Gebhardt et al. (2011) takes up on this criticism: taking the state of Saxony-Anhalt as an example, the study describes a first attempt to develop and elaborate a bottom-up approach in a regional setting. The resulting consideration of specific regional conditions allows a more precise and detailed illustration of the adaptation costs relevant for regional or local decision makers.

2.5 Legitimacy of public adaptation

Adaptation to changing (economic, societal or biophysical) framework conditions is a standard challenge for economic actors. From a mainstream economic point of view, markets in general are considered to cope best with both uncertainty and change processes due to dynamic efficiency of market-oriented use of decentralized information. Hence, it is necessary to justify when and why public adaptation should be executed rather than autonomous adaptation conducted by private actors themselves: as long as market allocation performs efficiently with respect to adaptation, government interventions or government-implemented adaptation measures are neither necessary nor legitimate (Mendelsohn, 2000, Aaheim and Aasen, 2008). The opposite applies when the individual deviates from the social benefit-cost calculation or when forms of market failure – public goods, externalities, information asymmetry or market power – are present (OECD, 2009; Dannenberg et al., 2009; Osberghaus et al., 2010a; BMF, 2010 or Hallegatte et al., 2011).

2.5.1 Conventional market failure

Many adaptation measures display the characteristics of *public goods* leading to a suboptimally low level of private or autonomous adaptation (Dannenberg et al., 2009, Fankhauser et al., 1999, BMF, 2010, Osberghaus et al., 2010a). Cimato and Mullan (2010) differentiate between global (e.g., the provision of information or basic research into drought-resistant crops), national (e.g., adaptation measures in the areas of infrastructure or healthcare) and local public goods (e.g., dams for the protection of specific areas). Dobes et al. (2010) highlight the key importance of provision of information by the government in order to guarantee greater planning certainty in the case of private investments into adaptation. Furthermore, adaptation measures can be accompanied by positive or negative *external* effects which need to be internalised through government interventions. For example, it is possible that interdependencies exist between several local adaptation measures such as, e.g., competition for a single water source used for different irrigation systems, which must be coordinated by the government (Dannenberg et al., 2009). In addition, adaptation measures can cause negative environmental externalities such as, e.g., increased CO_2 emissions due to the use of air-conditioning systems, which in this case would imply a suboptimally high level of adaptation (Tol, 2005). Moreover, the *asymmetric distribution* of climate- or adaptation-relevant information leads – for instance in the real estate or insurance markets – to the well-known problem of adverse selection or moral hazard thus requiring government relief (Dannenberg et al., 2009; Schwarze, 2008 or Schwarze and Wagner, 2007). Finally, *market power* can occur in the case of adaptation measures, particularly in the adaptation of infrastructural goods (or in their creation for adaptation purposes) such as rail or power networks where efficient production implies a natural monopoly (Fankhauser, 2009).[6]

2.5.2 Forms of extended market failure

Optimal market outcomes are linked to certain institutional conditions and frameworks (Dannenberg et al., 2009; BMF, 2010). A functioning system of *property rights* is of key importance. Without such a system, long-term investments, which play a decisive role in the adaptation process, would not be forthcoming and the success of both government and private adaptation actions would be endangered (Osberghaus et al., 2010a). For instance, the implementation of an agricultural irrigation system can only take place when the property rights to the water sources in question are clearly defined. Moreover, the judicial system is indispensable as a public service to ensure the enforceability of those rights. Institutions that guarantee the financing of individuals or companies with insufficient budgets, i.e., an appropriate banking and credit system (e.g., micro-credits), also play an important role in enabling the implementation of necessary adaptation measures (Cimato and Mullan, 2010). In relation to institutional conditions, inertia and/or *path dependencies* that stand in the way of the timely modification or creation of adaptation-relevant institutions also prove problematic. Taking the US water sector as an example, Libecap (2011) demonstrates the reasons for these path dependencies. The relevant institutions, e.g., ordinances, were created at a time when water was used primarily for agricultural purposes and are therefore no longer adequate in the current context of mainly industrial water use. However, a corresponding modification of the institutions is sometimes associated with significant costs and is therefore not always possible. Consequently, the existing institutional framework conditions make it difficult to implement the required adaptation measures in the water sector.

Inhibitory framework conditions can also manifest as *technological path dependencies*. These stem from the fact that many vulnerable system components, such as real estate or energy networks, are based on long-term investments. These investments become sunk costs ex post, resulting in corresponding rigidities and thus, ultimately, in suboptimal adaptation because the necessary shifting or substitution of these components cannot be carried out when required (Aaheim and Aasen, 2008).

The reasons for suboptimal adaptation behaviour may also lie with the adapting actors who, contrary to neoclassical economic theory, do not act with perfect rationality and perfect foresight. Limited rational behaviour may, alternatively, be evoked by cultural aspects, such as certain moral concepts or traditions (IPCC, 2007b). Furthermore, human behaviour is often characterised by temporal inconsistencies and the phenomenon of hyperbolic discounting, which can lead to behaviour such as inertia, procrastination or strategic ignorance (Cimato and Mullan, 2010 as well as Carrillo and Mariotti, 2000). This inertia is largely due to the fact that the benefits of the adaptation measures, in contrast to the costs, often accrue well beyond the planning horizon of the respective actor because of the time-lag in the impacts of climate change.

Grothmann et al. (2009) identify important psychological or behavioural economic determinants of individuals' adaptation behaviour that may be responsible for insufficient incentives for autonomous adaptation: problem-related (e.g., perception of the problem or environmental awareness) and action-related factors (e.g., convictions about effectiveness or subjective norms), while with regard to the execution of the action, customs/habits and other factors (e.g., clarity of the target or emotions) are decisive to adaptation behaviours. In view of these diverse influences, it becomes clear that the mere provision of information by the government does not necessarily guarantee efficient autonomous adaptation. Similarly, Osberghaus et al. (2010b) use econometric methods to demonstrate that certain psychological aspects and risk perceptions can have an inhibitory effect on information processing. Gifford (2011) identifies various psychological factors (e.g., insufficient problem recognition, ideological world views which exclude environmentally friendly behaviour, adherence to customs/habits or lack of trust in experts) that could explain the inertia of actors in the area of adaptation.

2.5.3 Regulatory barriers

Barriers to autonomous adaptation are not only rooted in the framework conditions and characteristics of the market. Existing regulatory and political interventions, even when aimed at other, non-adaptation-related goals (e.g., intervention in the areas of biodiversity, water or agriculture), can significantly influence decisions on autonomous adaptation (Cimato and Mullan, 2010). For example, agricultural policy measures are reflected in the resilience of natural systems used for agricultural production. Keskitalo (2009) argues that in view of increasing globalisation, individual regional or national regulatory interventions are not the only ones that need to be considered in the context of autonomous adaptation decisions – sometimes whole sets of measures, right through to governance networks, influence the decision situation at different levels. Fankhauser et al. (1999) stress that virtually all policy areas (health, education, environment, etc.) make a significant contribution to shaping the framework conditions for autonomous adaptation. Hence it is essential to consider adaptation-relevant aspects in all pertinent decisions and interventions so that regulatory barriers can be prevented from the outset. In the literature, this aspect is often summarised under the concept of 'mainstreaming'. Using case studies, OECD (2009) demonstrates how adaptation can be integrated into development policy.

2.5.4 Pursuing non-efficiency-related goals

In addition, government intervention is also necessary to implement the non-efficiency-related policy goals, especially since these are not conveyed through the market mechanism (Dannenberg et al., 2009; Osberghaus et al., 2010a). Several studies highlight from their point of view the most important goals and aspects, such as social insurance against catastrophic events (Fankhauser, 2009), overcoming poverty (Hallegatte et al., 2011) or equity and justice (Paavola and Adger, 2002).

2.6 Means and instruments of adaptation policy

2.6.1 Descriptive analysis: Governance structures, institutions and instruments

The literature aimed at drawing up an inventory of possible and existing adaptation instruments, institutions and governance structures is divided into two sections. The first provides various general systematisations. The second comprises numerous analyses aimed at individual sectors or regions (see for an overview Heuson et al. 2012a, p. 42 ff.). In the literature, various systematisations have been developed that account for the specific characteristics of government interventions or instruments. For example, OECD (2009) systematises the possibilities for government intervention in the area of adaptation according to the purpose or intention of the intervention, such as 'sharing of losses' (e.g., reconstruction through public funds), 'modification of the threat' (e.g., flood protection), 'avoidance of climate change impacts' by means of structural/technological measures, judicial or regulatory instruments, institutional or administrative measures or market-based instruments etc.

In contrast, the systematisations of Cimato and Mullan (2010) and Hallegatte et al. (2011) are based on the type of government intervention:

- direct regulation (e.g., technology or process-related restrictions or prohibitions),
- market-based (economic) instruments (e.g., taxes or tradable usage rights),
- research and monitoring programmes,
- provision of information,
- investment in infrastructure (e.g., dyke construction),
- reallocation measures (e.g., compensation or credit programmes) and
- institutional reforms (e.g., in the water or agricultural sector).

Finally, Goklany (2007) concentrates on institutions and measures that are targeted specifically at building adaptive capacity and that at the same time comply with sustainable development goals. Lastly, the IPCC (2007a) and Aaheim and Aasen (2008) choose the impacts of climate change and the sectors affected by climate change, respectively, as their categorisation criteria.

2.6.2 Normative analysis: The designU⊠U⊠ of adaptation policy and instrument choice

Given the various instruments and design options for adaptation policy, economic literature provides policy recommendations on the normative level, which are expected to help overcome barriers to public adaptation and achieve the goal set. In this context, Balbi and Guipponi (2009), as well as Patt and Siebenhüner (2005), discuss the potential of Agent Based Models to realistically reflect the process or system of adaptation policy, taking barriers into account, so that appropriate recommendations for political action can be derived. On the one hand, they target concrete instruments and measures. However, more holistic recommendations that address the governance of adaptation as a whole, i.e., all the relevant institutions, instruments and regulations in a specific field of action, are clearly in the majority.

Normative analysis of government adaptation instruments and measures are mainly context-dependent or problem-focussed and are also limited primarily to the choice, but not to the design of said instruments and measures. An exception here is Grothmann et al. (2009), who make general recommendations on the application of instruments, which they suggest can be used to overcome psychological barriers to adaptation. Furthermore, target groups should be approached in phases when they are particularly open to change in order to guarantee the greatest possible effectiveness of the instruments. Barr et al. (2010) and Dellink et al. (2009), on the other hand, derive recommendations for the design of international adaptation funding for developing countries by means of an index-based operationalization of the equity goal. Agrawala and Carraro (2010) also deal with the topic of financial aid. They make the case for the instrument of micro-credits, in particular, to fund short-term adaptation measures with a low volume of investment. Macro-financing instruments, on the other hand, should be used to fund long-term, resource-intensive measures. In the context of developing countries, various government adaptation measures for the agricultural sector are proposed and prioritised, whereby investment in education and research and the provision of information take priority (Popp et al., 2009; Hassan and Nhemachena, 2008 as well as Paavola, 2004).

Other studies place their focus on industrial countries. Dannenberg et al. (2009) and Osberghaus et al. (2010a) identify actual or potential inefficiencies of private adaptation in Germany's most important economic sectors and, on this basis, formulate sector-specific policy measures for implementation. Covich (2009) drafts proposals for US drinking water management, while Cimato and Mullan (2010) do the same for the areas of insurance, real estate, public infrastructure and ecosystems, thereby explicitly taking potential barriers to public adaptation into account.

2.6.3 Choice of instruments and measures

Normative analysis of instrument choice has to refer to decision criteria. Which criteria should be applied to evaluate the different measures depends on the respective form of the adaptation (Smit et al., 2000; cf. Figure 2.4). On the one hand, spontaneous (also: autonomous and reactive) adaptation measures are primarily evaluated in relation to its cost-benefit ratio, i.e., the damage savings potential (or potential gains) are compared with the costs of the adaptation measures. Estimates of this type are usually carried out within the framework of Integrated Assessment Models or Impact Models (Tol et al., 1998). The results of the evaluation are not only relevant in order to be able to better predict the impacts of climate change; they are also a necessary precondition for modelling reference scenarios that are free of political interventions and serve as a basis for developing adaptation policies (Smit et al., 2000).

In contrast, for planned (also: government-induced, anticipatory) adaptation measures, numerous evaluation criteria are used (e.g., Smit et al., 1999). The IPCC (1994) has put forward fundamental preliminary considerations on the selection of these criteria. In particular, it stresses that the criteria must be as specific as possible and clearly verifiable. The most elementary criterion is effectiveness, i.e., the degree to which the goals of the measure are achieved (Cimato and Mullan, 2010). If only one goal is pursued, then the application of the effectiveness criterion is trivial. In the practice of adaptation policy, however, the aim usually is to simultaneously achieve several goals, which frequently exhibit conflicting relationships (Klein et al., 2005), so that the clarity of the effectiveness criterion is lost. This necessitates a weighting or ranking of the individual goals, based on which trade-offs between the goals can be compared and the adaptation options can be selected according to specific methods (see IPCC, 1994).

The criterion of cost-efficiency is aimed at achieving a politically defined goal or an adaptation goal emerging from technical or scientific premises at minimal social cost (cf., e.g., Smit

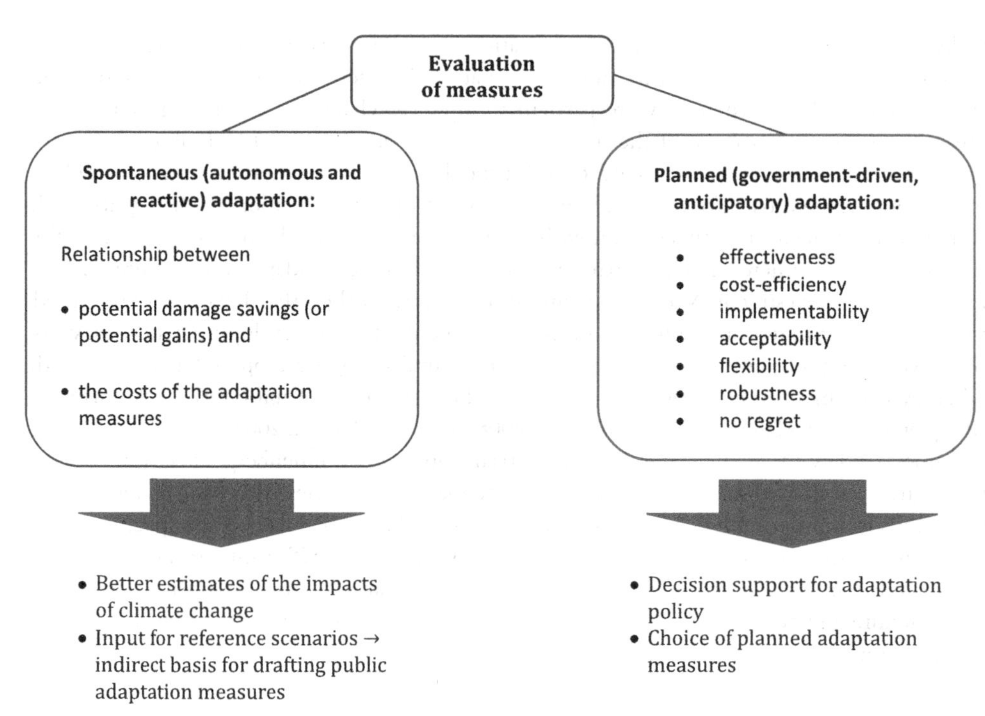

Figure 2.4 Criteria for evaluating adaptation measures

et al., 1999 and 2000 as well as Sharma and Sharma, 2010). In contrast to optimality, the benefit of adaptation is explicitly not taken into account in order to avoid the associated information problems. Several studies deal with the operationalization of the cost-efficiency criterion in the context of various fields of activity. Wheeler (2011) develops a quantitative method, which supports decision makers and/or donor countries in determining the cost-efficient allocation of adaptation resources in developing countries. This study focuses on the problem areas of weather-induced catastrophes, sea-level rise and agricultural production. Cai et al. (2011) design a simulation model for defining cost-efficient adaptation strategies in the energy sector. Within this framework, multiple energy sources, technologies and subsectors can be observed. Furthermore, the model allows the consideration of uncertainties on different levels, e.g., in relation to climate change impacts or adaptation planning.

In addition, a range of complementary and/or alternative criteria have been introduced. Smit et al. (1999) and Tol (1996) stress that the implementability or the political enforceability of measures as well as their acceptability among the general public determine the success of the adaptation. Smith (1997) and Smith and Lenhart (1996) argue that, in view of the uncertainty surrounding the impacts of climate change, measures should be designed to be as flexible as possible, i.e., they should be adjustable to altered climatic conditions at short notice or be able to react to unexpected (extreme) events. Another criterion that takes uncertainty into account is the robustness of the measures. This demands that the effect and functionality of the measures be guaranteed for the entire range of possible and/or probable climatic conditions (Lempert and Schlesinger, 2000). The no-regret criterion links the criteria of optimality and robustness.

Accordingly, measures should generate a non-negative net benefit under all possible climate scenarios (Hallegatte, 2009).

On the basis of these evaluation criteria for planned adaptation measures, the respective decision maker now has to select the "best" measure or the "best" set of measures. If only one goal is being pursued and no other criteria are to be applied, e.g., flexibility, then the selection can simply be made using the criterion of effectiveness. However as soon as multiple goals are to be realised and/or additional criteria are to be used, a special selection procedure has to be implemented. The IPCC (2001) provides a comprehensive summary of these procedures.[7] Here, a distinction is made between procedures that contribute directly to the decision-making process, in other words, those that result in the generation of an explicit ranking of the possible courses of action (decision theories, game theory). On the other hand, there are procedures only supporting the decision maker indirectly, by clarifying the implications and trade-offs between the options or by narrowing down the number of options that come into question (e.g., safe landing approach, Bayesian networks, etc.). However, they do not produce any clear recommendations for action.

In decision theory, benefit-cost, cost-efficiency and multi-criteria analysis methods play an important role in decision support in the context of adaptation (Klein and Tol, 1997). Benefit-cost analysis makes it possible to consider multiple goals simultaneously by breaking them down to the same monetary scale so that possible synergies and conflicts between the goals can be balanced out. Ultimately, the action alternative that demonstrates the greatest net benefit, i.e., the greatest difference between the benefit and the cost of the option, is the one to be selected. If the benefit and the costs (e.g., in the case of investment decisions) span over several periods, then they are to be related to the present value by discounting. In practice, however, massive uncertainties about the impacts of climate change and informational problems in the assessment of damages in relation to non-monetary goods (e.g., biodiversity loss) make it difficult to quantify the benefit side (Klein et al., 2005). Cost-efficiency analysis, which explicitly excludes the benefit side, offers a way out of this dilemma. It concentrates on selecting the measure that achieves the set targets at the lowest possible cost, thereby getting around the information problems. Benefit-cost and cost-efficiency analyses do make it possible to consider multiple goals, but they evaluate the response options according to just one criterion – the optimality or cost-efficiency criterion. If additional criteria are to be considered in the selection (e.g., flexibility or acceptability), it is necessary to resort to multi-criteria analysis (Hallegatte, 2009). Here, the qualitative and quantitative criteria are weighted according to specific considerations and the available measures are ranked accordingly (see Klein and Tol, 2007 for more details).

2.6.4 Governance architecture of adaptation policies

Against this background, a number of studies give general recommendations for the overall design of adaptation governance. Adger et al. (2009) call for adaptation policy to be oriented along an ethical model that respects and considers primarily those affected or threatened by the adverse effects of climate change and their the cultural background. Brunner et al. (2005) also favour an ethical model-based approach, whereby they attach special importance to the integration of science, politics and the people affected. Furthermore, the leitmotif of integration is likewise demanded in several other areas. For instance, Adger (2001) call for the integration of adaptation and mitigation policy, Dang et al. (2003) additionally call for the integration of development policy. The active involvement of stakeholders and those affected is seen to play an important role (Dulal et al., 2009; Hulme et al., 2007). Dobes (2010) and Dulal et al. (2009) advocate an integrated view of climate change-related and other aspects of equity in the political process. Increasing importance is being attached to the concept of mainstreaming, which envisages the integration of adaptation into all political and regulatory areas (beyond climate policy in the proper sense) (Cimato and Mullan, 2010; Klein et al., 2009). The same applies to the concept of multilevel governance, which stands for close cooperation, integration and coordination between decision makers and authorities on the different government hierarchy levels (Keskitalo, 2009). In this context, a decentralised implementation of adaptation measures is recommended, which ideally begins where the concern and the know-how are greatest. If there are interdependencies between several measures, a higher-level authority must take over the coordination (BMF, 2010; Hulme et al., 2007). Finally, a further key requirement lies in building institutional capacity in the area of adaptation (Adger et al., 2009; Cimato and Mullan, 2010; Klein et al., 2005).

In addition to these general policy recommendations, several studies exist that focus alternatively on specific problems, sectors or regions. Many of these concentrate on adaptation governance in developing countries (Hardee and Mutunga, 2009; Koch et al., 2007; World Bank, 2010; Vignola et al., 2009), whereby there is broad agreement on the key suggestions for decision makers: multi-level governance, less hierarchy, integration of stakeholders and local communities and integration of family and health policy as well as the sciences. With regard to cross-border adaptation problems, the studies recommend a clear definition of responsibilities and international cooperation (PRC, 2009 – EU Coastal Protection), as well as the coordination and integration of national legal systems, knowledge transfer and cooperation in the area of funding (EEA, 2009 – *The Alps Facing the Challenge of Changing Water Resources*; Chrischilles, 2011 – *Research Cooperation with Developing Countries*). In relation to public infrastructure, Neumann and Price (2009) encourage the integration of area-specific planning. Amundsen et al. (2010), on the other hand, identify multilevel governance as well as the building of local institutional capacity as playing a key role in adaptation policy of Norwegian municipalities. In relation to adaptation funding, Przylski and Hallegatte (2010) recommend that the focus be placed first on building institutional capacity and only in the second stage on concrete measures. Ayers and Huq (2009) call for the integration of financial aid instruments in the areas of development and adaptation. Finally, Mercer (2010) sees synergies in the integration of adaptation policy and measures to reduce disaster risk.

2.7 Positive analysis of adaptation policy

A positive analysis of adaptation policy helps to shed light on possible barriers to public adaptation. Several studies provide classifications of barriers (e.g., Eisenack and Stecker, 2012; Füssel, 2008) and empirical evidence for specific regional and sectoral case studies (e.g., Amundsen et al., 2010; Arnell and Delaney, 2006). To synthesize experiences from barriers research, Lehmann et al. (2012) propose a framework to cluster variables influencing government decision-making on adaptation (see Figure 2.5). Characteristics of these variables that hamper (or promote) proper public adaptation may constitute a barrier (or driver). A first tier of relevant variables includes information, incentives and resources. These variables are functions of an underlying, second-tier set of variables encompassing the natural and socioeconomic environment, actor-specific characteristics and the institutional environment.

To take decisions on adaptation, governments first need proper information about regional and local climate change and associated impacts as well as available adaptation options and their

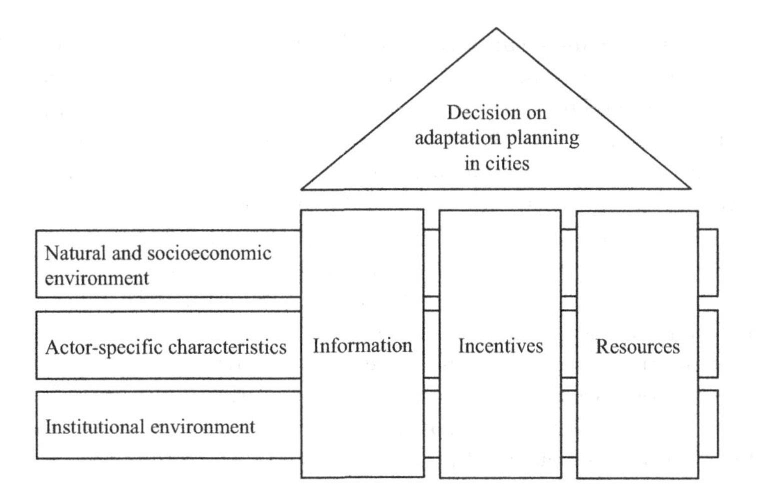

Figure 2.5 Variables influencing government decision-making on adaptation (Lehmann et al., 2012)

costs and benefits. Empirical studies show that particularly local governments often lack this information (Measham et al., 2011).

Second, the willingness to promote public adaptation is also contingent on the incentives related to adaptation, i.e., the cost-benefit ratio of a measure and its consistency with other policy objectives, such as economic development or public health (Hallegatte et al., 2011; Yohe, 2001). Moreover, Gawel et al. (2012) point out that public decisions on adaptation are also guided by the particular interests of politicians (e.g., reelection), bureaucrats (e.g., budget maximization), voters (e.g., reduction of political burdens) and interest groups from industry and society (e.g., rent-seeking). These competing interests may have an impact on the extent of adaptation (over- as well as under-adaptation is possible) as well as the types of adaptation measures (structural may be preferred over nonstructural measures).

Third, adaptation decisions depend on the needed and available resources for planning and implementing adaptation measures (Moser and Ekstrom, 2010). Again, particularly local decision makers are often found to be particularly resource-constrained (Crabbé and Robin, 2006; Measham et al., 2011).

The level of information, incentives and resources can be explained by the characteristics of three underlying sets of variables:

The natural and socioeconomic environment encompasses a set of noninstitutional factors guiding public adaptation decisions, such as the characteristics of the climate system (e.g., intensity and scale of climate change), the natural environment in general (e.g., geographical patterns of site) and the socioeconomic system (e.g., demography and level of development). The complexity of the natural and socioeconomic environment typically impairs the certainty of information about impacts from climate change, especially at the local scale (Hunt and Watkiss, 2011). The natural and socioeconomic environment also influences incentives to promote adaptation as it determines a society's exposure and sensitivity to climate change (Moser and Ekstrom, 2010). Moreover, the availability of resources for adaptation is highly dependent on the level of economic development.

In addition, actor-specific characteristics of decision makers are decisive, such as perceptions, preferences, experiences and knowledge, which form actors' mental models (Reser and Swim, 2011). They are sensitive to interactions with the natural environment, for example, personal experience with extreme weather events. Mental models are filters determining which information is actually perceived and how it is interpreted and valued (Eisenack and Stecker, 2012). Personal preferences and risk attitudes also drive the incentives to address adaptation (Adger et al., 2009). The combination of mental models and more general personal skills (e.g., creativity, openness, communication) has an impact on whether certain actors are accepted as leaders in adaptation and can obtain necessary resources and support (Measham et al., 2011).

Finally, the institutional environment is another important determinant of public adaptation action, i.e., the formal (laws and regulations) and informal rules (routines and cultures) that guide the interactions of actors and organisations (North, 1990). Several institutional patterns have been found to constrain public adaptation:

- In a multi-level governance context, national regulation may impair the incentives for proper public adaptation at the local level by (1) failing to provide a clear mandate for local action (Measham et al., 2011), (2) failing to internalize external effects of local adaptation (e.g., basin-wide benefits from flood management) (Osberghaus et al., 2010) and (3) setting adverse incentives (e.g., national insurance schemes reducing the need for local adaptation) (Corfee-Morlot et al., 2011).
- A lack of mainstreaming, i.e., the failure to integrate adaptation concerns into other policy fields such as public health, may produce another barrier as it conceals co-benefits of and increases resources needed for adaptation (Measham et al., 2011). This lack may be associated with organisational routines within administrative units (Berkhout et al., 2006) but also the self-interest of bureaucrats (Gawel et al., 2012).
- A lack of participation may restrict access to local knowledge and financial and nonfinancial support from different stakeholders (Hunt and Watkiss, 2011).
- More generally, culture, value systems and traditions of a society influence the priorities of and capacities of political decision makers and the acceptance of public adaptation measures by general population (Adger et al., 2009).

2.8 Future research

Clearly, providing guidance and decision support for policy makers is the primary task of economic adaptation research. Consequently, many attempts have been made to translate the resulting findings into guidelines on the planning, ranking, implementation and evaluation of adaptation policy measures (cf. Füssel, 2008; Hallegatte et al., 2011). Nevertheless, there is the need for scrutinizing whether the answers given so far cover all the essential questions arising from the adaptation policy process.

Basically, the economic discipline has given the standard response to the newly emerging research topic of climate change adaptation (cf. Aaheim and Aasen, 2008): established economic tools and methods have been applied to investigate how climate change affects economic constraints and in which way this translates into behavioural changes of economic actors. In doing so, the major topics of the adaptation (policy) process, such as the interlinkage between mitigation and adaptation, the optimal extent and pathway of adaptation, legitimization of public adaptation, means, instruments and finally positive aspects of adaptation policy, indeed have been addressed, as can be seen from this chapter. However, there is scope for extending and revising this research in two respects. First, there are some specific gaps concerning the topics mentioned. Second, according to a recent contribution of Downing (2012), it might be useful to revise or readjust, respectively, the basic approach of the economic adaptation.

2.8.1 Specific scope for further research

Not surprisingly, in terms of adaptation goals, the economic discipline so far has concentrated on optimality. However, as policy makers similarly need to take into account non-efficiency related goals (e.g., distributive justice) when implementing adaptation measures or instruments, a stronger weight should be put on operationalizing these goals and revealing possible synergies and conflicts between them. First steps into this direction are given by Dellink et al. (2009), who develop principles for a fair sharing of adaptation costs and operationalize these with an index-based system, or studies aimed at identifying synergies between adaptation, mitigation and development policy goals (Dang et al., 2003; Goklany, 2007). Apart from this, there is considerable need for further research with respect to the goal of optimality itself. The top-down-based estimates of adaptation costs and benefits are subject to immense uncertainties and thus can hardly fulfil their task of serving as a guardrail for policy decisions (Fankhauser, 2009). For this reason, methodological improvements allowing for adequately capturing the heterogeneous, context-dependent character of adaptation within the IAMs are due. Additionally, insurmountable uncertainties and the respective implications for the interpretation and meaning of the provided data have to be communicated clearly by the economic discipline. On the contrary, bottom-up-based estimates promise to be more reliable and precise given the adaptation's regional rootedness. However, being rather cumbersome and requiring a lot of effort, they are hardly suited for quantifying costs and benefits for a whole region. Against this background, the question arises whether an ingenious integration of top-down and bottom-up approaches could help to overcome their weaknesses and bring together their strengths, respectively. Basically, the economic discipline does well in raising awareness of the need for government intervention in the adaptation process when it comes to classical forms of market failure, such as the provision of public goods. However, further research is urgently due concerning extended forms of market failure in the adaptation context. On the one hand, behavioural barriers to efficient autonomous adaptation should be addressed, e.g., with the behavioural economic methods: if these are neglected, adaptation policy measures, like the provision of information, run the risk of being ineffective (cf. Osberghaus et al., 2010b). On the other hand, barriers rooted in existing (non-adaptation) regulatory, frameworks deserve closer attention, e.g., through approaches of institutional economics. In light of the uncertainties that surround the climate change impacts and the costs and benefits of adaptation policy measures, the advancement of decision support tools that explicitly address and reveal these uncertainties takes a prior position on the research agenda. Moreover, as being fraught with risk, the cost-benefit criterion should be secured by supporting criteria, like flexibility or robustness of measures. Only narrowly considered so far has been the positive analysis of adaptation policy, despite that this is of special importance in order to reveal and overcome barriers to efficient public adaptation – for instance due to self-interest-driven of political actors (Gawel et al., 2012). Finally, given the lacking success of the recent climate summits, further research on the strategic role for global cooperation on mitigation played by the adaptation option and upcoming instruments of international climate finance with game theoretic and/or experimental economic methods is essential (Heuson et al., 2012b).

2.8.2 Readjusting the basic approach of economic adaptation research

In a recent contribution, Downing (2012) argues that the so far given standard approach of the economic discipline to adaptation, which mainly relies on comparative statics of reference and climate impact scenarios, falls short of coping with the adequate representation of decision processes and uncertainty. For this reason, the author promotes a shift from framing adaptation "... as vulnerability-impacts to adaptation pathways". Similarly, this involves a shift from approaches of predicting and provide to understanding dynamic processes. Despite that generally this transition towards dynamic pathway approaches seems to be widely accepted, appropriate methods and tools for supporting its execution are still rare. According to Downing (2012), in this respect case studies are especially valuable for exploring how various kinds of uncertainties associated with climate change impacts can be valued and coped with in decision processes. A full appraisal and understanding of the dynamic adaptation process and related policy responses naturally cannot be accomplished by economic sciences alone. Rather, a close collaboration with other social and natural sciences is required – but, the economic discipline can and has to deliver an essential contribution this multidisciplinary canon.

Notes

1 For an overview of national adaptation strategies within the European Union, see PEER (2009).
2 For exceptions and barriers in the private provision of adaptation measures, see Section 2.5.
3 For a detailed view on the international cooperation on adaptation see Chapter 3.
4 The following IAMs are mainly applied here: AD-WITCH, AD-DICE, AD-RICE and FUND. A description and a comparison of the various model types would exceed the scope of this chapter. For further information, see Agrawala et al. (2011); Bosello et al. (2010); Patt et al. (2010).
5 Due to the limited scope, this chapter rather focusses on comparing the basic approaches for the empirical assessment of adaptation costs and benefits than on providing concrete numbers. For the latter purpose, see Agrawala et al. (2011).
6 Eisenack (2010) takes on another market power-related problem that is unique in the literature: in a microeconomic model, he provides proof that Cournot oligopolists achieve suboptimal adaptation performance in an endogenous market structure.
7 Although this refers to the mitigation context, the procedures described therein can be similarly applied within the framework of adaptation decisions.

References

Aaheim, A., Aasen, M., 2008. *What do we know about the economics of adaptation?* CEPS Policy Brief No. 150, Centre for European Policy Studies.

Aaheim, A., Amundsen, H., Dokken, T., Ericson, T., Wei, T., 2009. *A macroeconomic assessment of impacts and adaptation to climate change in Europe.* CICERO Report 2009:06, Center for International Climate and Environmental Research, Oslo.

Aalbers, R., Rob F.T., 2009. *Discounting investments in mitigation and adaptation: A dynamic stochastic general equilibrium approach of climate change.* CPB Discussion Paper No. 126, CPB Netherlands Bureau for Economic Policy Analysis, The Hague, Netherlands.

Adger, W.N., 2001. Scales of governance and environmental justice for adaptation and mitigation of climate change, *Journal of International Development* 13, 921–931.

Adger, W. N, Arnell, N.W., Tompkins, E.L., 2005. Successful adaptation to climate change across scales. *Global Environmental Change* 15, 77–86.
Adger, W.N., Dessai, S., Goulden, M., Hulme, M., Lorenzoni, I., Nelson, D.R., Naess, L.O., Wolf, J., Wreford, A., 2009. Are there social limits to adaptation to climate change? *Climatic Change* 93, 335–354.
Agrawala, S., Bosello, F., Carraro, C., Cian, E. de, Lanzi, E., 2011. Adapting to climate change: Costs, benefits, and modelling approaches. *International Review of Environmental and Resource Economics* 5, 245–284.
Agrawala, S., Carraro, M. 2010. *Assessing the role of microfinance in fostering adaptation to climate change.* Nota di Lavoro 82.2010, Fondazione Eni Enrico Mattei, Milan.
Agrawala, S., Fankhauser, S., 2008. *Economic aspects of adaptation to climate change: Costs, benefits and policy instruments.* OECD Publishing.
Amundsen, H., Berglund, F., Westskog, H., 2010. Overcoming barriers to climate change adaptation – a question of multilevel governance? *Environment and Planning C: Government and Policy* 28, 276–289.
Arnell, N.W., Delaney, E.K., 2006. Adapting to climate change: Public water supply in England and Wales. *Climatic Change* 78, 227–255.
Auerswald, H., Konrad, K.A., Thum, M.P., 2011. *Adaptation, mitigation and risk-taking in climate policy.* SSRN eLibrary.
Ayers, J.M., Huq, S., 2009. Supporting adaptation to climate change: What role for official development assistance? *Development Policy Review* 27, 675–692.
Bahn, O., Chesney, M., Gheyssens, J., 2010. *The effect of adaptation measures on the adoption of clean technologies.* Les Cahiers du GERAD, G-2010-19, HEC Montreal, Canada.
Balbi, S., Giupponi, C. 2009. *Reviewing agent-based modelling of socio-ecosystems: A methodology for the analysis of climate change adaptation and sustainability.* Working Paper No. 15, Department of Economics, University of Venice, Italy.
Barr, R., Fankhauser, S., Carraro, C., Enrica, C. de, 2010. Adaptation investments: A resource allocation framework. *Mitigation and Adaptation Strategies for Global Change* 15, 843–858.
Barrett, S., 2008. Dikes v. windmills: Climate treaties and adaptation. Unpublished manuscript. School of Advanced International Studies, John Hopkins University.
Benchekroun, H., Marrouch, W., Ray Chaudhuri, A., 2011. *Adaptation effectiveness and free-riding incentives in international environmental agreements.* Tilburg University, Center for Economic Research.
Berkhout, F., Hertin, J., Gann, D.M., 2006. Learning to adapt: Organisational adaptation to climate change impacts. *Climatic Change* 78, 135–156.
BMF, 2010. Klimapolitik zwischen Emissionsvermeidung und Anpassung – Gutachten des Wissenschaftlichen Beirats beim Bundesministerium der Finanzen (BMF).
Bosello, F., Carraro, C., de Cian, E., 2010. Climate policy and the optimal balance between mitigation, adaptation and unavoided damage. *Climate Change Economics* 1, 71–92.
Brechet, T., Hritonenko, N., Yatsenko, Y., 2010. *Adaptation and mitigation in long-term climate policies.* CORE Discussion Paper No. 2010065, Université catholique de Louvain, Center for Operations Research and Econometrics (CORE).
Bruin, K.C. de, Dellink, R.B., Tol, R.S.J., 2009. AD-DICE: an implementation of adaptation in the DICE model. *Climatic Change* 95, 63–81.
Brunner, R.D., Steelman, T.A., Coe-Juell, L., Cromley, C.M., Edwards, C.M., Tucker, D.W., 2005. *Adaptive governance. Integrating science, policy and decision making.* New York: Columbia University Press.
Buob, S., Stephan, G., 2011a. To mitigate or to adapt: How to confront global climate change. *European Journal of Political Economy* 27, 1–16.
Buob, S., Stephan, G., 2011b. *On the incentive compatibility of funding adaptation.* NCCR Climate Research Paper 2011/02, The National Centre of Competence in Research on Climate (NCCR Climate), Bern, Switzerland.
Cai, Y.P., Huang, G.H., Tan, Q., Liu, L., 2011. An integrated approach for climate-change impact analysis and adaptation planning under multi-level uncertainties: Part I. *Methodology, Sustainable Energy Reviews* 15, 2779–2790.
Carrillo J.D., Mariotti, T., 2000. Strategic ignorance as a self-disciplining device. *Review of Economic Studies* 67, 529–544.
Cimato, F., Mullan, M., 2010. *Adapting to climate change: Analysing the role of government.* Department for Environment, Food and Rural Affairs (DEFRA), London, UK.

Corfee-Morlot, J., Cochran, I., Hallegatte, S., Teasdale, P.-J., 2011. Multilevel risk governance and urban adaptation policy. *Climatic Change* 104, 169–197.
Covich, A.P., 2009. *Emerging climate change impacts on freshwater resources. A perspective on transformed watersheds.* RFF Report, Resources for the Future, Washington, D.C.
Crabbé, P., Robin, M., 2006. Institutional adaptation of water resource infrastructures to climate change in Eastern Ontario. *Climatic Change* 78, 103–133.
Chrischilles, E. (2011): Anpassung an den Klimawandel. Reduzierung der Verletzlichkeiten durch internationale Kooperation, *Ökologisches Wirtschaften* 3/2011, 43-46.
Dang, H.H., Michaelowa, A., Tuan, D.D., 2003. Synergy of adaptation and mitigation strategies in the context of sustainable development: The case of Vietnam. *Climate Policy* 3, Supplement 1, S81–S96.
Dannenberg, A., Mennel, T., Osberghaus, D., Sturm, B., 2009. The economics of adaptation to climate change – the case of Germany. Discussion Paper No. 09-057, Centre for European Economic Research (ZEW), Mannheim.
Dellink, R., Elzen, M. den, Aiking, H., Bergsma, E., Berkhout, F., Dekker, T., Gupta, J., 2009. Sharing the burden of financing adaptation to climate change. *Global Environmental Change* 19, 411–421.
Deschênes, O., Greenstone, M., 2007. Climate change, mortality, and adaptation: Evidence from annual fluctuations in weather in the US. Working Paper No. 13178, National Bureau of Economic Research, Cambridge.
Dobes, L., 2009. People versus planners: Social preferences for adaptation to climate change. Environmental Economics Research Hub Research Reports No. 0941, Environmental Economics Research Hub, Crawford School of Public Policy, The Australian National University, Canberra.
Dobes, L., 2010. Notes on applying "real options" to climate change adaptation measures, with examples from Vietnam. CCEP Working Paper 7.10, Centre for Climate Economics & Policy, Crawford School of Economics and Government, The Australian National University, Canberra.
Downing, T.E., 2012. Views of the frontiers in climate change adaptation economics. *Wiley Interdisciplinary Reviews: Climate Change* 3, 161–170.
Dulal, H.B., Shah, K.U., Ahmad, N., 2009. Social equity considerations in the implementations of Caribbean climate change adaptation policies. *Sustainability* 1, 363–383.
Dumas, P., Ha-Duong, M., 2008. *Optimal growth with adaptation to climate change.* Collection HAL, halshs-00207621, Centre International de Recherche sur l'Environment et le Développement, Nogent-sur-Marne, France.
Ebert, U., Welsch, H., 2012. Adaptation and mitigation in global pollution problems: Economic impacts of productivity, sensitivity, and adaptive capacity. *Environmental & Resource Economics* 52, 49–64.
EEA, 2009. *Regional climate change and adaptation. The Alps facing the challenge of changing water resources.* European Environment Agency (EEA) Report No. 8/2009, Copenhagen, Denmark.
Eisenack, K., 2010. *The inefficiency of private adaptation to pollution in the presence of endogenous market structure.* Oldenburg Discussion Papers in Economics, V-328-10, Carl-von-Ossietzky-Universität Oldenburg, Department of Economics.
Eisenack, K., Stecker, R., 2012. A framework for analyzing climate change adaptations as actions. *Mitigation and Adaptation Strategies for Global Change* 17, 243–260.
Fankhauser, S., 2009. *A perspective paper on adaptation as a response to climate change.* Copenhagen Consensus Center, Copenhagen Business School, Denmark.
Fankhauser, S., 2010. The costs of adaptation. *Wiley Interdisciplinary Reviews: Climate Change* 1, 23–30.
Fankhauser, S., Smith, J.B., Tol, R.S.J., 1999. Weathering climate change: Some simple rules to guide adaptation decisions. *Ecological Economics* 30, 67–78.
Füssel, H.-M., 2008. Assessing adaptation to the health risks of climate change: What guidance can existing frameworks provide? *Int J Environ Health Res* 18, 37–63.
Füssel, H.-M., Klein, R.J.T., 2006. Climate change vulnerability assessments: An evolution of conceptual thinking. *Climatic Change* 75, 301–329.
Gawel, E., Heuson, C., Lehmann, P., 2012. *Efficient public adaptation to climate change: An investigation of drivers and barriers from a Public Choice perspective.* UFZ Discussion Paper No. 14/2012. Helmholtz Centre for Environmental Research – UFZ, Leipzig.
Gebhardt, O., Kumke, S., Hansjürgens, B., 2011. Kosten der anpassung an den klimawandel – eine ökonomische analyse ausgewählter sektoren in sachsen-anhalt. In *UFZ-Bericht* 05/2011. Helmholtz-Zentrum für Umweltforschung – UFZ, Leipzig.

Gifford, R., 2011. The Dragons of inaction – psychological barriers that limit climate change mitigation and adaptation. *American Psychologist* 66, 290–302.
Glenk, K., Fischer, A., 2010. Insurance, prevention or just wait and see? Public preferences for water management strategies in the context of climate change. *Ecological Economics* 69, 2279–2291.
Goklany, I., 2007. Integrated strategies to reduce vulnerability and advance adaptation, mitigation, and sustainable development. *Mitigation and Adaptation Strategies for Global Change* 12, 755–786.
Grothmann, T., Werner, J., Krömker, D., Werg, J. Stolberg, A., Homburg, A., Reckien, D., Egli, T., Buchert, M., Zimmer, W., Hoffmann, C., Siebenhüner, B., 2009. Förderung von klimaschutz und klimaanpassung in privathaushalten – erfolgsfaktoren, instrumente, strategie. In Klimaschutz und Anpassung an die Klimafolgen. Strategien, Maßnahmen und Anwendungsbeispiele, edited by Mahammadzadeh, M. and Biebeler, H. Köln: Institut der deutschen Wirtschaft.
Hallegatte, S., 2009. Strategies to adapt to an uncertain climate change. *Global Environmental Change* 19, 240–247.
Hallegatte, S., Lecocq, F., Perthuis, C.D., 2011. Designing climate change adaptation policies: An economic framework. Policy Research Working Paper No. 5568, The World Bank Sustainable Development Network.
Hardee, K., Mutunga, C., 2009. Strengthening the link between climate change adaptation and national development plans: Lessons from the case of population in National Adaptation Programmes of Action (NAPAs). *Mitigation and Adaptation Strategies for Global Change* 15, 113–126.
Hassan, R., Nhemachena, C., 2008. Determinants of African farmers' strategies for adapting to climate change: Multinomial choice analysis. *African Journal of Agricultural and Resource Economics* 2, 83–104.
Heuson, C., Gawel, E., Gebhardt, O., Hansjürgens, B., Lehmann, P., Meyer, V., Schwarze, R., 2012a. *Fundamental questions on the economics of climate adaptation: Outlines of a new research programme.* UFZ Reports No. 05/2012, Helmholtz Centre for Environmental Research – UFZ, Leipzig.
Heuson, C., Peters, W., Schwarze, R., Topp, A.-K., 2012b. *Which mode of funding developing countries' climate policies under the post-Kyoto framework?* UFZ Discussion Paper No. 10/2012, Helmholtz Centre for Environmental Research – UFZ, Leipzig.
Hof, A.F., de Bruin, K.C., Dellink, R.B., den Elzen, M.G.J., van Vuuren, D.P., 2009. The effect of different mitigation strategies on international financing of adaptation. *Environmental Science & Policy* 12, 832–843.
Hulme, M., Adger, W.N., Dessai, S., Goulden, M., Lorenzoni, I., Nelson, D., Naess, L.-O., Wolf, J., Wredford, A., 2007. Limits and barriers to adaptation: Four propositions, Tyndall Briefing Note No. 20, Tyndall Centre for Climate Change Research, UK.
Hunt, A., Watkiss, P., 2011. Climate change impacts and adaptation in cities: A review of the literature. *Climatic Change* 104, 13–49.
Ingham, A., 2005. *How do the costs of adaptation affect optimal mitigation when there is uncertainty, irreversibility and learning?* Tyndall Centre Working Paper No. 74, Tyndall Centre for Climate Change Research, UK.
Ingham, A., Ma, J., Ulph, A., 2007. Climate change, mitigation and adaptation with uncertainty and learning. *Energy Policy* 35, 5354–5369.
IPCC, 1994. *IPCC Technical Guidelines for assessing climate change impacts and adaptations.* Part of the IPCC Special Report to the First Session of the Conference of the Parties to the UN Framework Convention on Climate Change, Department of Geography, University College London, London, UK.
IPCC, 2001. *Climate change 2001: Impacts, adaptation and vulnerability, Chapter 10.* Third Assessment Report of the Intergovernmental Panel on Climate Change (IPCC).
IPCC, 2007a. *Climate change 2007: Synthesis report.* Contribution of Working Groups I, II and III to the Fourth Assessment Report of the Intergovernmental Panel on Climate Change (IPCC).
IPCC, 2007b. *Climate change 2007: Impacts, adaptation and vulnerability, Chapter 17.* Fourth Assessment Report of the Intergovernmental Panel on Climate Change (IPCC).
Jotzo, F., 2010. *Prerequisites and limits for economic modelling of climate change impacts and adaptation.* Research Report No. 55, Environmental Economics Research Hub, Crawford School of Economics and Government, Australian National University, Canberra Australia.
Kane, S., Shogren, J.F., 2000. Linking adaptation and mitigation in climate change policy. *Climatic Change* 45, 75–102.
Keskitalo, E.C.H., 2009. Governance in vulnerability assessment: The role of globalising decision-making networks in determining local vulnerability and adaptive capacity. *Mitigation and Adaptation Strategies for Global Change* 14, 185–201.

Klein, R.J.T., Nicholls, R.J., Mimura, N., 1999. Coastal adaptation to climate change: Can the IPCC Technical Guidelines be applied? *Mitigation and Adaptation Strategies for Global Change* 4, 239–252.

Klein, R.J.T, Tol, R.S.J., 1997. *Adaptation to climate change: options and technologies. An overview paper.* Institute for Environmental Studies, Vrije Universiteit Amsterdam.

Klein, R.J.T., Schipper, E.L.F., Dessai, S., 2005. Integrating mitigation and adaptation into climate and development policy. *Environmental Science and Policy* 8, 579–588.

Koch, I.C., Vogel, C., Patel, Z., 2007. Institutional dynamics and climate change adaptation in South Africa. *Mitigation and Adaptation Strategies for Global Change* 12, 1323–1339.

Lehmann, P., Brenck, M., Gebhardt, O., Schaller, S., Süßbauer, E., 2012. *What influences progress in planned adaptation? Conceptual framework and evidence from cities in Peru, Chile and Germany.* UFZ Discussion Paper. Helmholtz Centre for Environmental Research – UFZ, Leipzig.

Lempert, R.J., Schlesinger, M.E., 2000. Robust strategies for abating climate change. *Climatic Change* 45, 387–401.

Libecap, G.D., 2011. Institutional path dependence in climate adaptation: Coman's some unsettled problems of irrigation. *American Economic Review* 101, 64–80.

Marrouch, W., Chaudhuri, A.R., 2011. International environmental agreements in the presence of adaptation. CentER Discussion Paper No. 2011-023, Tilburg University, The Netherlands.

McKitrick, R., Collinge, R.A., 2002. The existence and uniqueness of optimal pollution policy in the presence of victim defense measures. *Journal of Environmental Economics and Management* 44, 106–122.

Measham, T., Preston, B.L., Smith, T.F., Brooke, C., Gorddard, R., Withycombe, G., Morrison, C., 2011. Adapting to climate change through local municipal planning: Barriers and challenges. *Mitigation and Adaptation Strategies for Global Change* 16, 889–909.

Mendelsohn, R., 2000. Efficient adaptation to climate change. *Climatic Change* 45, 583–600.

Mercer, J. 2010. Disaster risk reduction or climate change adaptation: Are we reinventing the wheel? *Journal of International Development* 22, 247–264.

Moser, S.C., Ekstrom, J.A., 2010. *A framework to diagnose barriers to climate change adaptation.* Proceedings of the National Academy of Sciences 107, 22026–22031.

Neumann, J.E., Price, J.C., 2009. *Adapting to climate change – the public policy response – public infrastructure.* RFF Report, Resources for the Future (RFF), Washington, D.C.

OECD, 2009. *Integrating climate change adaptation into development co-operation.* Policy guidance, Organisation for Economic Co-operation and Development (OECD), Paris.

Osberghaus, D., Dannenberg, A., Mennel, T., Sturm, B., 2010a. The role of the government in adaptation to climate change. *Environment and Planning C: Government and Policy* 28, 834–850.

Osberghaus, D., Finkerl, E., Pohl, M., 2010b. *Individual adaptation to climate change: The role of information and perceived risk.* ZEW Discussion Paper No. 10-061, Zentrum für Europäische Wirtschaftsforschung (ZEW), Mannheim.

Osberghaus, D., Reif, C., 2010. *Total costs and budgetary effects of adaptation to climate change: An assessment for the European Union.* ZEW Discussion Paper No. 10-046, Zentrum für Europäische Wirtschaftsforschung (ZEW), Mannheim.

Paavola, J., 2004. *Livelihoods, vulnerability and adaptation to climate change in the Morogoro Region, Tanzania.* CSERGE Working Paper EDM 04-12, Centre for Social and Economic Research on the Global Environment (CSERGE), Norwich, UK.

Paavola, J., Adger, W.N., 2002. *Justice and adaptation to climate change.* Working Paper No. 23, Tyndall Centre for Climate Change Research, University of East Anglia, Norwich, UK.

Patt, A., Siebenhüner, B., 2005. Agent based modeling and adaptation to climate change. *Vierteljahrshefte zur Wirtschaftsforschung* 74, 310–320.

Patt, A., van Vuuren, D., Berkhout, F., Aaheim, A., Hof, A., Isaac, M., Mechler, R., 2010. Adaptation in integrated assessment modeling: where do we stand? *Climatic Change* 99, 383–402.

PEER, 2009. *PEER: Europe adapts to climate change comparing national adaptation strategies.* PEER Report No 1. Partnership for European Environmental Research (PEER), Helsinki.

Pielke, R., Prins, G., Rayner, S., Sarewitz, D., 2007. Climate change 2007: Lifting the taboo on adaptation. *Nature* 445, 597–598.

Pittel, K., Rübbelke, D.T.G., 2011. *International climate finance and its influence on fairness and policy.* BC3 Working Paper Series 2011-04, Basque Centre for Climate Change (BC3), Bilbao.

Popp, A. Domptail, S., Blaum, N., Jeltsch, F., 2009: Landuse experience does qualify for adaptation to climate change. *Ecological Modelling* 220, 694–702.

PRC, 2009. *The economics of climate change adaptation in EU coastal areas.* Final Report for the European Commission, Policy Research Corporation (PRC), Brussels.
Przylski, V., Hallegatte, S., 2010. *Climate change adaptation, development and international financial support: Lessons from EU pre-accession and solidarity funds.* Nota di Lavoro 137.2010, Fondazione Eni Enrico Mattei, Milan, Italy.
Reser, J.P., Swim, J.K., 2011. Adapting to and coping with the threat and impacts of climate change. *American Psychologist* 66, 277–289.
Schwarze, R., 2008. Financial risks of natural hazards: Markets and the role of the state. *Journal of Applied Social Science Studies* 128, 545–548.
Schwarze, R., Wagner, G., 2007. The political economy of natural disaster insurance: Lessons from the failure of a proposed compulsory insurance scheme in Germany. *European Environment, The Journal of European Environmental Policy* 17, 403–415.
Settle, C., Shogren, J.F., Kane, S., 2007. Assessing mitigation-adaptation scenarios for reducing catastrophic climate risk. *Climatic Change* 83, 443–456.
Sharma, V., Sharma, P., 2010. *A framework for monitoring and evaluation of climate change adaptation interventions.* Discussion Paper 5, Sambodhi Research & Communications, New Delhi, India.
Shepherd, J.G., 2009. *Geoengineering the climate: Science, governance and uncertainty.* The Royal Society (RS), RS Policy Document 10/29, London.
Smit, B., Burton, I. Klein, R.J.T., Street, R., 1999. The science of adaptation: A framework for assessment. *Mitigation and Adaptation Strategies for Global Change* 4, 199–213.
Smit, B., Burton, I. Klein, R.J.T., Wandel, J., 2000. An anatomy of adaptation to climate change and variability. *Climatic Change* 45, 223–251.
Smith, J.B., 1997. Setting priorities for adapting to climate change. *Global Environmental Change* 7, 251–264.
Smith, J.B., Lenhart. S.S., 1996. Climate change adaptation policy options. *Climate Research* 6, 193–201.
Tol, R.S.J., 1996. A systems view of weather disasters. In *Climate change and extreme events-altered risk, Socio-economic impacts and policy responses*, edited by Downing, T.E. Olsthoorn, A.A. and Tol, R.S.J. Institute for Environmental Studies, Vrije Universiteit and Environmental Change Unit, University of Oxford, Amsterdam and Oxford.
Tol, R.S.J., 2005. Adaptation and mitigation: Trade-offs in substance and methods. *Environmental Science & Policy* 8, 572–578.
Tol, R.S.J., Fankhauser, S., Smith, J.B., 1998. The scope for adaptation to climate change: What can we learn from the impact literature? *Global Environmental Change* 8, 109–123.
Tulkens, H., Steenberghe, V. van, 2009. Mitigation, adaptation, suffering. In *Search of the right mix in the face of climate change.* CESifo Working Paper Series No. 2781, CESifo Group Munich.
Vignola, R., Locatelli, B., Martinez, C., Imbach, P., 2009. Ecosystem-based adaptation to climate change: What role for policy-makers, society and scientists? *Mitigation and Adaptation Strategies for Global Change* 14, 691–696.
Wheeler, D., 2011. *Quantifying vulnerability to climate change: Implications for adaptation assistance.* Working Paper 240, Center for Global Development, Washington, D.C, USA.
World Bank, 2010. *Economics of adaptation to climate change.* Social Synthesis Report, The World Bank, Washington, D.C.
Wright, E.L., Erickson, J.D., 2004. Climate variability, economic adaptation, and investment timing. *International Journal of Global Environmental Issues* 3, 357–368.
Yohe, G.W., 2001. Mitigative capacity – the mirrow image of adaptive capacity on the emissions side. *Climatic Change* 49, 247–262.
Zehaie, F., 2009. The timing and strategic role of self-protection. *Environmental & Resource Economics* 44, 337–350.

3

INTERNATIONAL COOPERATION ON ADAPTATION TO CLIMATE CHANGE

Jonathan Pickering[1] *and Dirk Rübbelke*[2]

[1] PHD CANDIDATE, AUSTRALIAN NATIONAL UNIVERSITY

[2] IKERBASQUE RESEARCH PROFESSOR, BASQUE CENTRE FOR CLIMATE CHANGE (BC3)

3.1 Introduction

Since the benefits of adapting to the adverse impacts of climate change will accrue primarily (albeit not exclusively) to individual countries and communities within them, the task of adaptation is often considered to require primarily domestic policy responses. As such, adaptation seems to stand in contrast to mitigation, where the necessity of international cooperation to protect global atmospheric stability is clearer.

Nevertheless, efficient and equitable adaptation may require a significant degree of international cooperation, not least due to the fact that many of the countries most vulnerable to climate change do not have access to sufficient financial resources or information to adapt to wide-ranging but uncertain climatic impacts and thus require international support. Similarly, international support for adaptation may play an important part in the political economy of multilateral climate change negotiations by building trust among parties and encouraging greater participation by developing countries in global mitigation efforts.

In this chapter, we focus primarily on intergovernmental cooperation on adaptation, which encompasses support through financing, technology transfer and capacity building, typically from developed to developing countries. Whereas in the earlier years of international climate change negotiations adaptation was relatively marginal compared to mitigation, adaptation has become an increasingly prominent issue in recent sessions (Mertz et al. 2009). Notably, industrialized countries have pledged as a group to mobilize climate finance to developing countries covering both adaptation and mitigation reaching US$100 billion per year by 2020. Nevertheless, as we elaborate in the chapter, significant research gaps remain in the field of international cooperation on adaptation (Heuson, Gawel, et al. 2012: 62).

In this chapter, we proceed as follows: in Section 3.2, we discuss potential rationales for countries' contributions to international adaptation funding. In Section 3.3, we describe the role of the international climate change regime in governing cooperation on adaptation. In Section 3.4, we review policy considerations and selected proposals for raising funds for international adaptation support. In Section 3.5 we highlight policy and economic considerations relevant to the delivery of adaptation finance. Section 3.6 concludes.

3.2 Rationales for international cooperation on adaptation

3.2.1 A conceptual framework

International adaptation finance may be provided by private or public (i.e., government) entities. While our focus in this chapter will be on public international adaptation finance, we mention first several brief points about rationales for private international adaptation finance. Private entities may be motivated to become involved in this field by either commercial or philanthropic interests. Philanthropic adaptation finance provided by entities such as nongovernment organizations tends to strive primarily to help the poor or vulnerable, while remittances – which represent considerably larger flows to developing countries than aid – may assist household-level adaptation at a range of income levels (Stern 2007: 561). Non-philanthropic private capital flows predominantly serve investors' profit maximization objectives. While helping poor countries to adapt is not the main motivation of commercial investors in providing funds, it represents a co-benefit of these investments (see Atteridge 2011: 27). Public entities could exploit such co-benefits through initiating public investments or regulatory measures that in turn leverage private investments yielding adaptation benefits. While domestic capital and foreign direct investment may provide an important source of adaptation finance for upper-middle-income countries, international public flows are likely to remain important for lower-income countries (Stern 2007: 561).

We now turn to analyzing the rationales for public entities' involvement in international adaptation finance. Here we draw substantially on R.A. Musgrave's theory of public finance. As Musgrave (1956: 333) stresses, the public sector serves three major budget functions, each of which is addressed by a different fiscal branch or department in his stylized model of government:

- providing for the satisfaction of public wants through the efficient allocation of resources (*allocation branch*),
- providing for adjustments in the distribution of income in order to attain the desired or proper distribution of income and wealth (*distribution branch*), and
- contributing to economic stabilization, i.e., securing price-level stability and full employment (*stabilization branch*).

Government intervention would be justified when market forces alone do not suffice to achieve these objectives. Although R.A. Musgrave's original concept paid little attention to the international sector, it can be comfortably included into this concept, as P.B. Musgrave (2008: 343) argues. That is, individual national governments may also intervene internationally (e.g., by offering transfers) in order to pursue their goals under the three branches.

Stern (2007: 411) states that there are three reasons why market forces are unlikely to achieve efficient adaptation to climate change: (i) uncertainty and imperfect information; (ii) missing and misaligned markets, including underprovision of public goods; and (iii) financial constraints. Although the first two reasons are common reasons for public intervention to improve allocative efficiency, the presence of financial constraints more properly concerns the distribution branch. The stabilization branch does not play a major role in the adaptation context (also see Aakre and Rübbelke 2010a), since stabilization policy should not be pursued by modifications of the levels of public expenditures on goods and services, but should take place in a distribution-neutral way (Musgrave 1959). Nevertheless, adaptation measures may exert positive effects on economic stability (Aakre and Rübbelke 2010b: 769), for example by raising employment levels. In the remainder of this section we will discuss the distributive and allocative rationales for international cooperation on adaptation.

3.2.2 Distributive rationales: Assisting the poor and addressing harm

Financial constraints to adaptation are most intensely perceived by the poor, rendering them more vulnerable to climate change than the rich (World Bank 2009: 42). The greater vulnerability of the poor will in turn worsen their financial situation as climatic impacts materialize. Such negative distributional consequences are widely considered undesirable by the global community. Thus there is a role for the distribution branch in mobilizing adaptation finance.

Distributive justifications for adaptation support for developing countries can largely be split into two components: a) assisting the poor and b) preventing and remedying the imposition of harm on others (compare Moore 2012b: 38). We consider each aspect in turn. First, much research highlights the close links between adaptation and development. The World Bank, for example, has argued that "economic development is perhaps the best hope for adaptation to climate change: development enables an economy to diversify and become less reliant on sectors such as agriculture that are more vulnerable to the effects of climate change" (World Bank 2010: 6; see also Stern 2009: 68). On this basis, some regard international support for adaptation to climate change in poor developing countries as a form of foreign assistance supporting sustainable development (Donner et al. 2011). As discussed below, the idea that adaptation finance should be counted as aid remains controversial and has met strong resistance from developing countries. Nevertheless, it is plausible that altruistic rationales that influence the provision of aid may also play a part in motivating the provision of adaptation finance (Harris and Symons 2010).

Second, harm-based justifications provide a further important distributive rationale for providing adaptation finance. Although precise attribution of causal responsibility for climatic impacts is a complex task, it is nevertheless the case in general that the industrialized world's emissions-intensive economic development has been the main historical driver for global warming (Rive et al. 2006: 192; Höhne et al. 2011). At the same time, as noted above, developing countries tend to be most vulnerable to climate change while having less responsibility for cumulative emissions. Even though developing countries' share of current emissions has now surpassed that of developed countries (Olivier et al. 2012) – and in time developing countries' share of cumulative emissions will also do so (Botzen et al. 2008) – it remains the case that present adaptation needs are the product of cumulative emissions up to the present. Moreover, since per capita emissions in many developing countries are (and will remain) lower than those of many developed countries, there may be grounds for treating emissions required to meet "subsistence" needs as incurring less moral responsibility than those produced for nonessential purposes (Shue 1993; Vanderheiden 2008). Consequently, it could be argued that in addition to their responsibility to prevent future harm through timely mitigation, industrialized countries also have a responsibility to protect developing countries from the negative consequences of their previous and ongoing greenhouse gas emissions (Roberts 2009; Grasso 2010; Pickering and Barry 2012). In this sense, the motivation for providing adaptation assistance would differ from that for providing development assistance.

Whether industrialized countries could be held liable under international law for providing adaptation finance – on the basis, for example, of a breach of obligations of due diligence to avoid transboundary environmental harm – remains a contentious and unsettled question (see Verheyen 2005; Birnie et al. 2009; Faure and Peeters 2011). However, the threat of international liability could strengthen industrialized countries' incentives to seek a negotiated resolution (Gupta 2007: 85), which in turn could provide the basis for more substantial adaptation finance.

3.2.3 Allocative rationales: Public goods and influencing cooperation on mitigation

From an allocative point of view, it is frequently argued that international support for climate change mitigation could help to raise efficiency, since mitigation is a global public good (in that its benefits are non-excludable and non-rival worldwide; see e.g., Buchholz and Peters 2005, Arrow 2007, and Kotchen 2013. Due to non-excludability, the benefits of mitigation can be enjoyed globally regardless of where the mitigation activity takes place. Thus, to achieve an efficient outcome, mitigation should be undertaken in those places where it can be achieved at least cost. International transfers allow industrialized countries to exploit low-cost mitigation options in developing countries so that the former have to invest less in high-cost domestic emissions reductions in order to attain a given national mitigation target.

In contrast to mitigation, adaptation is generally not considered to provide a global public good.[1] Instead, adaptation primarily provides private and public goods at a national or sub-national level. Private goods may include higher agricultural yields due to water efficiency measures or lower damage to buildings from extreme weather events due to better quality construction, while local public goods could include transport infrastructure that is better able to withstand temperature extremes. To that extent, industrialized countries do not gain directly by helping developing countries to adapt to climate change (Barrett 2008). Consequently, from an allocative point of view, the question arises as to why industrialized countries support adaptation internationally.

As Rübbelke (2011) and Pittel and Rübbelke (2013) argue, there may be indirect allocative benefits of adaptation support. First, by improving developing countries' perceptions of the fairness of a global agreement, adaptation support may increase their willingness to contribute to international mitigation efforts. This in turn tends to enhance the total level of the global public good of mitigation that is generated through international negotiations. This effect is enhanced by the current consensus-based decision-making practice of the United Nations Framework Convention on Climate Change (UNFCCC; or "the Convention"), which would allow a relatively small group of vulnerable but economically less powerful states to obstruct major decisions (Eckersley 2012).

Second, some direct or indirect benefits of adaptation could also be enjoyed globally. Stern (2007: 568) notes several global public goods that international adaptation funding could provide directly, including: better monitoring and prediction of climate change; improved modeling of climatic impacts; research to improve drought- and flood-resistant crops; and new methods of addressing land degradation. More indirect effects of measures helping regions to adapt to climate change may include, for example, prevention of the displacement of populations from regions seriously affected by climate-induced drought or extreme weather events, as well as reduced transmission of infectious diseases whose prevalence is exacerbated by higher temperatures (IPCC 2007). Finally, some measures may also simultaneously provide adaptation and mitigation benefits, for example reducing emissions from deforestation and forest degradation (REDD+), which may also help to protect watersheds and biodiversity from climatic impacts.

Abadie et al. (2013) discuss further reasons beyond global public good argumentation that may explain the bias of climate finance towards mitigation. By distinguishing the motivations of public and private sector actors, they note that public sector actors may more readily catalyze private investment in mitigation by creating regulated markets for mitigation, whereas (for reasons discussed in Section 3.5 below), market-based mechanisms to stimulate adaptation are likely to be difficult to establish. Developed countries' governments may have a special interest in mobilizing private finance for mitigation purposes so that their countries can in turn capture benefits from global mitigation efforts.

3.2.4 Comparison with other evaluative criteria

The allocative and distributive rationales outlined above bear important resemblances to other criteria used to evaluate adaptation, as discussed for example by Adger et al. (2005). Their approach includes criteria of effectiveness, efficiency, equity and legitimacy of adaptation. As outlined above, efficiency is applicable to the allocation branch, while equity is related to the distribution branch. If effectiveness is understood as the ability of a policy measure to solve a given problem, the criterion could apply to each branch individually or (if values associated with individual branches could be ranked or weighted) to all branches together (Heuson, Gawel, et al. 2012: 29).[2] Legitimacy – understood as the extent to which the exercise of authority is accepted as justified (see Biermann and Gupta 2011; Pickering et al. 2013) – is relevant to policymaking under all branches. The potential for tradeoffs between rationales is discussed further in Section 3.5.

3.3 Governing adaptation in the international climate regime

3.3.1 Overview

For most of the period since the adoption of the Convention in 1992, adaptation has received much less attention than mitigation. At the time of drafting the Convention, parties generally saw mitigation as the more urgent priority, since many were confident that effective mitigation would diminish the need for adaptation, and significant uncertainty prevailed about the timing and nature of the impacts of climate change (Schipper 2006: 86; Gupta 2010: 642). Some countries were also concerned that focusing on adaptation would provide a perverse incentive for countries to channel less effort into the main task of mitigation (Pielke et al. 2007).

The status of adaptation in the text of the Convention is contested. As outlined below, the UNFCCC sets out clear substantive obligations on adaptation for all parties (Mace 2005: 225). However, the Convention does not define adaptation, nor does it devote any single article to adaptation (Schipper 2006: 89). As a result, in the early years of the Convention's operation it proved difficult to locate adaptation on negotiating agendas (Yamin and Depledge 2004: 213).

Nevertheless, adaptation has grown in prominence in recent negotiations. This has occurred to a large degree as a result of improved scientific understanding and increasing international awareness of current impacts and adaptation needs (Liverman 2011: 404), which have grown while progress on mitigation has remained inadequate (Ciplet et al. 2013). The increased prominence of adaptation finance is also consistent with the concurrent rising urgency of ambitious mitigation in developing as well as developed countries, since (as discussed above) adaptation finance may play an indirect role in encouraging developing countries' willingness to mitigate through enhancing the perceived fairness of the negotiating process. The degree to which the discourse on adaptation has changed since the Convention's adoption is reflected in a recent UNFCCC decision, which provides that adaptation "must be addressed with the same priority as mitigation" (UNFCCC 2011, para. 2(b)). The remainder of this section provides an overview of funding commitments under the UNFCCC, followed by a brief discussion of other aspects of cooperation on adaptation in the international climate regime.

3.3.2 Funding commitments and institutions

Among references to adaptation in the text of the Convention, the funding provisions are among the most explicit. Under Article 4.3, developed countries are required to provide financial resources and transfer of technology needed by developing countries to meet the "agreed full incremental costs" of measures undertaken by developing countries to implement their commitments under the Convention. These commitments include planning and preparation for adaptation and mainstreaming climate change concerns into national decision-making (see Article 4.1). Under Article 4.4, developed countries must also "assist the developing country Parties that are particularly vulnerable to the adverse effects of climate change in meeting costs of adaptation to those adverse effects". Despite both provisions being couched in mandatory rather than voluntary language ("shall" rather than "should"), the absence of any specific amounts or timeframes for the provision of assistance limited their practical effect in the early years of the Convention's operation (Yamin and Depledge 2004: 217–218). In addition, the provision on adaptation finance (Article 4.4) is weaker than Article 4.3 in that it lacks a reference to "agreed full" or "incremental" costs (Yamin and Depledge 2004: 234). One analysis of historical adaptation finance suggests that it comprised less than one percent of total development assistance for most of the period between the early 1970s and mid-1990s (Michaelowa and Michaelowa 2012: 45), although in the absence of consistent coding practices (as discussed below), historical estimates are subject to considerable uncertainty.

Article 11.1 defines a mechanism for providing financial resources under the Convention. The Global Environment Facility (GEF) has for most of the Convention's life functioned as the sole operating entity of the financial mechanism. The Facility channels funds from industrialized countries to meet the "agreed incremental costs" of climate-change related measures in developing countries to achieve "global environmental benefits" (GEF 2011, para. 2). Both the incremental cost and global benefit requirements have impeded the Facility's ability to deliver adaptation funding. Not only is it difficult to establish a baseline against which the incremental costs of adaptation measures can be assessed, but (for reasons outlined above) many adaptation measures may not provide direct global benefits (Mace 2006: 64).

Developing countries' dissatisfaction with existing adaptation finance arrangements and low levels of development assistance channeled towards adaptation led to an expansion of sources under the 1997 Kyoto Protocol and the 2001 Marrakesh Accords (Yamin and Depledge 2004: 232). The Marrakesh Accords agreed to the establishment of three funds dedicated primarily towards adaptation: the Least Developed Countries Fund (LDCF); the Special Climate Change Fund (SCCF); and the Adaptation Fund, financed primarily through a share of proceeds on credits from the Clean Development Mechanism (CDM) established under the Kyoto Protocol. The GEF administers both the LDCF and the SCCF with the World Bank acting as trustee. A major role of the LDCF has been funding the preparation of "National Adaptation Programmes of Action" (NAPAs) to identify urgent and immediate adaptation needs. Following a period during which the GEF's broader climate change focal area supported pilot and demonstration projects on adaptation, all of the GEF's adaptation-related work will be financed solely through the LDCF and SCCF until at least 2014 (GEF 2010). The Adaptation Fund (AF) is governed by a dedicated Board directly accountable to parties to the Kyoto Protocol. The Board was established in 2007 and the Fund itself became fully operational in 2010.

The establishment of new funding mechanisms during the 2000s coincided with rising attention to adaptation. Michaelowa and Michaelowa (2012: 44) observe that the share of global aid used for adaptation rose particularly since the early 2000s. However, it was not until COP 13 in Bali in 2007 that significant momentum arose for a large increase in adaptation funding in tandem with a greater emphasis on mitigation in developing countries. Subsequently, at COP 15 in 2009, the COP took note of the Copenhagen Accord, which included a commitment by developed countries to provide funding approaching US$30 billion in the period 2010–2012 ("fast-start finance") and to mobilize US$100 billion annually by 2020 to support developing countries in their climate change efforts, conditional on the latter's progress on mitigation and transparency, with balanced allocation between adaptation and mitigation.[3]

At a first glance, the 2020 commitment looks remarkably high, especially when it is compared with the only slightly higher net Official Development Assistance (ODA; hereafter "aid") amounting to US$134 billion in 2011 (OECD 2013; see also Pickering and Wood 2011). However, as discussed below, there is no agreement as yet as to how much aid could be counted towards meeting the overall commitment (see Section 3.4), nor the proportion of the overall commitment that will be dedicated to adaptation (see Section 3.5). Moreover, financing needs may be considerably higher than likely adaptation flows under the commitment. For example, recent analysis has estimated the cost for developing countries of adapting to 2 °C warming by 2050 alone to be in the range of US$70-100 billion a year between 2010 and 2050 (World Bank 2010: 19). Estimates of financing needs may vary considerably due to methodological differences (Narain et al. 2011). In particular, estimates of adaptation needs typically cover only the (i) incremental costs of adaptation (compared with a baseline without climate change) without taking into account funding deficits in adapting to existing climatic variability and (ii) costs of planned adaptation by the public sector (that is, excluding autonomous adaptation by private actors) (Narain et al. 2011; Moore 2012a). These reasons contribute to the view of many developing countries that the US$100 billion pledge will be inadequate for meeting overall financing needs.

Current flows of adaptation funding are hard to estimate (let alone to compare with estimates of need on an equivalent footing), in part due to the absence of uniform institutional arrangements and monitoring methodologies. A wide-ranging survey of climate finance estimated that around $4.4 billion a year flowed to adaptation during 2009 and 2010 (mostly in the form of public finance), compared with $92.5 billion a year for mitigation (mostly in the form of private capital; Buchner et al. 2011: 8 – see Table 3.1). However, the apparent absence of commercial finance for adaptation should not be taken for granted, since (due to a lack of fine-grained data) the study simply stipulated that all commercial climate finance was directed to mitigation (Buchner et al. 2011: 44).

The OECD subsequently produced a significantly higher estimate of US$9.3 billion for 2010 using an updated marker for tracking aid flows dedicated to adaptation (OECD 2011).

Table 3.1 Composition of annual flows of international climate finance for developing countries, 2009–2010

USD billion	*Adaptation*	*Mitigation*	*Total*
Public – bi-/multilateral	4.2	35.4	39.6
Offsets (CDM)	0.0	2.3	2.3
Private – philanthropic	0.2	0.2	0.5
Private – commercial	0.0	54.6	54.6
Total	4.4	92.5	96.9

Source: Buchner et al. 2011: 8

This estimate may also need to be treated with some caution, as donors' processes for classifying aid activities as climate-related may be inconsistent and may considerably overestimate funding primarily directed towards adaptation (Michaelowa and Michaelowa 2011; Junghans and Harmeling 2012).

Recent momentum on climate finance has led to the creation of a range of new funding institutions.[4] Of greatest significance for the longer term is likely to be the Green Climate Fund (GCF), whose establishment was agreed upon at COP 16 in Cancún in 2010. While the GCF will fund both mitigation and adaptation needs, it is expected to channel a significant share of new multilateral funding for adaptation (UNFCCC 2011, para. 100). At COP 17 in 2011, parties agreed on a governing instrument for the Fund (see UNFCCC 2012). According to the instrument, the Fund will finance the "agreed full and agreed incremental costs" of activities including adaptation (UNFCCC 2012, para. 35). As of mid-2013, the board of the GCF had held four meetings but was yet to agree upon modalities for funding. Beyond the UNFCCC framework, countries have also established the Pilot Program for Climate Resilience (PPCR), which forms part of the World Bank-managed Climate Investment Funds. To date, the PPCR has attracted a larger share of adaptation funding than any other multilateral funds (see Figure 3.1). This reflects several reasons, including: contributing countries' preference for mainstreaming adaptation into development assistance rather than establishing discrete adaptation projects (as the AF does); their greater confidence in the effectiveness of World Bank institutions compared with UN funds; as well as the innovative but controversial governance structure of the Adaptation Fund Board, as discussed in Section 3.5 below (Harmeling and Kaloga 2011; Seballos and Kreft 2011).

3.3.3 *Other aspects of the UNFCCC's work on adaptation*

Since the early 2000s, other aspects of international cooperation on adaptation have also become more prominent in the UNFCCC. For the most part, these aspects represent "soft" forms of governance aiming to promote functions such as information exchange and networking (Persson 2011: 5), but some have the potential to evolve into more substantive mechanisms or commitments. In 2005, the UNFCCC launched a five-year initiative known as the Nairobi Work Programme on Impacts, Vulnerability and Adaptation to Climate Change (NWP), which aims to assist primarily developing countries in improving their understanding and assessment of impacts, vulnerability and adaptation, and in making informed decisions on adaptation. At COP 16 in 2010, parties established a Cancún Adaptation Framework aimed at enhancing action on

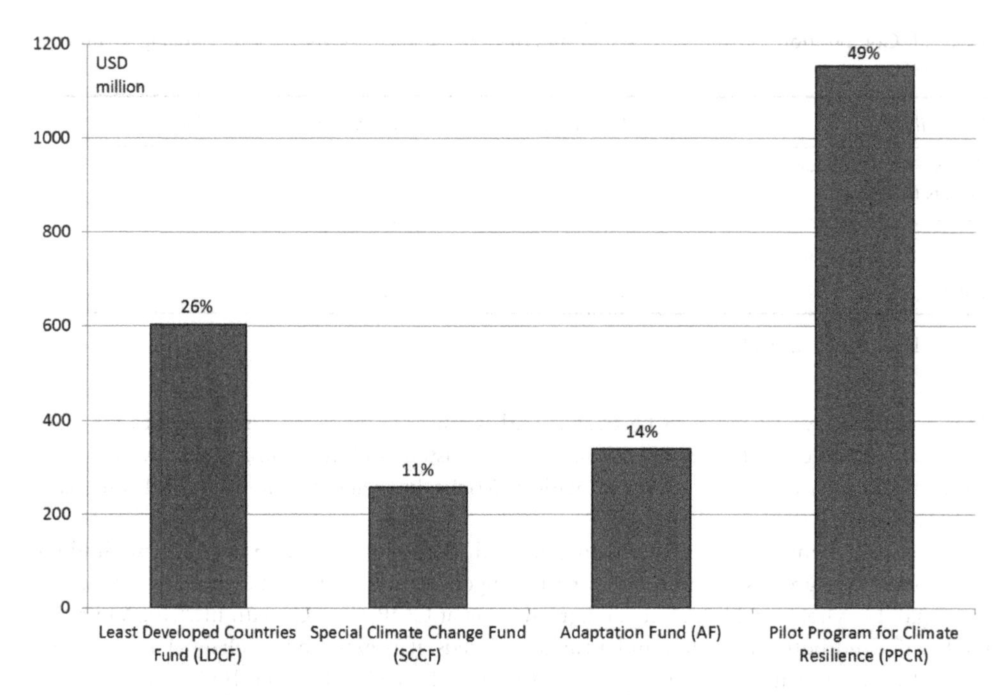

Figure 3.1 Funding pledged to dedicated multilateral adaptation funds
Sources: Adaptation Fund: Adaptation Fund Board 2013, (includes pledged donations and proceeds from CDM levy; other funds: Climate Funds Update 2013 [current as of 6 September 2013])

adaptation under the Convention. This includes the formation of an Adaptation Committee tasked with promoting coherent implementation of enhanced action through functions such as technical support and guidance and information exchange (UNFCCC 2011, para. 20). In addition, at COP 18 in 2012, decided to establish at COP 13 "institutional arrangements, such as an international mechanism" to address "loss and damage" associated with climate change impacts (UNFCCC 2013, para. 9). Such arrangements may inform the development of future mechanisms for insuring against extreme weather events (see also Section 3.5 below).

3.4 Generating international adaptation funding

The analytical framework outlined in Section 3.2 could also be applied to the analysis of potential sources of adaptation finance. Thus we could enquire how particular sources compare from the point of view of their allocative efficiency (that is, whether they distort economic incentives or help to reduce existing distortions) and their distributional impacts (that is, whether the burden or "incidence" of the sources falls heaviest on the wealthy and high-emitting or the poor and low-emitting groups in society). Related criteria specifically relevant to the analysis of adaptation financing sources include: adequacy, reliability or predictability, additionality, technical feasibility, political acceptability and transparency and accountability (see Hof et al. 2011; Bowen 2011; Pickering et al. 2013).

3.4.1 Existing sources of adaptation funding

Most adaptation funding to date has been sourced from industrialized countries' aid budgets, one prominent exception being the CDM levy for the Adaptation Fund discussed below. Both sources of funding have proved controversial, and we briefly discuss each in turn.

Developing countries argue that adaptation funding should be additional to – and separate from – aid budgets, and have expressed concerns about the diversion or double-counting of aid to meet climate finance commitments (Stadelmann et al. 2011; Ciplet et al. 2013). Arguments for distinct treatment of adaptation finance are based on several premises, largely grounded in distributional concerns. First, for the reasons discussed in Section 3.2, developing countries argue that adaptation finance is in principle different from aid. Second, developing countries argue that it is inappropriate to divert funding from one commitment (the UN target of providing 0.7 percent of Gross National Income in aid) to meet another (the UNFCCC climate finance commitments). By contrast, many developed countries – most of whom fall short of the 0.7 percent target – have observed that the UN target is not binding on countries that have not individually committed to meet it (compare Clemens and Moss 2007), and have pointed out the synergies between adaptation and development objectives (World Bank 2009: 17–18). As Stadelmann et al. (2011) argue, alternative baselines could address concerns of diversion without necessarily requiring that climate finance be entirely separate from aid.

The primary source of Adaptation Fund resources is a 2 percent levy on the issuance of CDM credits. Some have criticized the levy on the basis that it constitutes a tax on a good – namely mitigation facilitated through the CDM – when it would be preferable to rely on a Pigouvian tax on a bad, such as a levy on emitting activities (Fankhauser and Martin 2010; Bowen 2011: 1028). In addition, the costs of the levy may not fall exclusively on buyers of credits in developed countries. Fankhauser and Martin (2010: 360) have estimated that unless constraints are placed on the proportion of developed country emissions that can be offset through the CDM, sellers of credits in developing countries may shoulder up to two-thirds of the levy's tax burden. Several studies have demonstrated that even if the levy rate were increased, it would cover at most a limited portion of total adaptation financing needs (Hof et al. 2009; Fankhauser and Martin 2010; Eisenack 2012). At COP 18 in 2012, parties decided to retain the current levy rate, but to extend the levy's coverage to the issuance of Assigned Amount Units (AAU) and Joint Implementation under the Kyoto Protocol (UNFCCC 2013a, para. 21). However, given the smaller group of countries with mitigation commitments in the Protocol's second commitment period, this may not be sufficient to radically affect the adequacy of multilateral funding for adaptation. For these reasons, there is considerable value in considering alternative sources for adaptation funding.

3.4.2 Proposals for generating adaptation funding

The Copenhagen Accord envisages that long-term climate finance will come from "a wide variety of sources, public and private, bilateral and multilateral, including alternative sources of finance" (para. 8). Developing countries, researchers and civil society organizations have put forward a wide range of proposals for raising adaptation funding. Many sources could be applicable to mitigation finance as well, or indeed purposes other than climate finance. However, adaptation is often a prominent focus for proposals to raise new public funding, since it is widely recognized that market-based approaches are more readily suited to raising funds for mitigation rather than adaptation (Bowen 2011: 1022).

In 2010, a UN High-Level Advisory Group on Climate Change Financing (AGF) evaluated a range of financing proposals against many of the criteria outlined above. The AGF supported recourse to a mix of sources rather than relying on any one instrument. It emphasized the dual benefits of carbon pricing and markets in generating finance while also reducing emissions (AGF 2010: 5). Public revenue from carbon pricing could be earmarked for adaptation purposes through measures such as: domestic carbon taxes or emissions trading schemes, international schemes such as the auctioning of Assigned Amount Units (AAUs) under a future multilateral climate agreement or levies on specific emitting sectors such as international shipping and aviation. Germany, for example, has earmarked a portion of its EU Emissions Trading Scheme auction revenues towards its climate finance commitment (Vieweg et al. 2012). Savings from reduced domestic subsidies or tax exemptions for fossil fuel production or use could also be channeled towards adaptation finance. The AGF also supported use of other sources such as private capital and leveraging the resources of multilateral development banks, and indicated that an unspecified amount would likely need to be sourced through developed countries' national budgets.

Each source may have a range of implications for allocative efficiency and distribution. Here we briefly highlight some major issues (see also Pickering et al. 2013). First, the earmarking (or hypothecation) of specific revenue streams for particular purposes is controversial. Theories of public finance generally do not encourage hypothecation, since mismatches may arise between expenditure needs and the revenue-raising capacity of particular sources (Stern 2007: 560). However, others have argued that hypothecation can improve the adequacy and predictability of funding for particular priorities (see Müller 2008: 14). Earmarking may be more politically viable in the case of sources generated at the international level, since such funding is less vulnerable to the "domestic revenue problem", whereby funding raised at the domestic level is susceptible to claims from competing local constituencies (Hof 2011: 616). Hypothecation of international sources for climate-related purposes is likely to remain challenging where the activity is not closely connected to greenhouse gas emissions (as with a global financial transaction tax, which could be claimed for purposes such as reducing poverty or promoting global financial stability). Even where revenue is raised from a source of emissions such as international transport, other constituencies – notably airline and shipping industries – could raise competing claims for a share of funds to offset their adjustment costs. For this reason, most estimates of revenue for such sources assume that only a portion of total revenue will be dedicated to climate finance in developing countries (see e.g., World Bank 2011).

Second, some international fundraising schemes may bring about inefficient outcomes if their coverage is not comprehensive. This is a key concern for funding levied on transactions that could readily be relocated, such as currency transactions (in the case of a financial transaction tax) or refueling (in the case of levies on international shipping) (AGF 2010: 29, 31).

Third, sources may vary in their distributional effects, particularly if the incidence of the source falls on lower-income countries or poorer residents of developed countries. This concern has been prominent in debates about proposals to raise funds by regulating international transport emissions. Distributional concerns could be addressed in this context either by introducing exemptions for smaller emitters or measures to rebate funds to developing countries (Miola et al. 2011; Scott and Rajamani 2012). Earmarking a particular source for adaptation may also help to overcome residual distributional concerns on the part of developing countries since, for reasons outlined in Section 3.2, funding for adaptation is more likely to build developing countries' trust than funding for mitigation. One of the few proposals for generating funding specifically for adaptation – the International Air Travel Adaptation Levy (IATAL) – notes this beneficial aspect of earmarking (Hepburn and Müller 2010: 837).

3.5 Delivering international adaptation funding

A major political and technical challenge in designing institutions for delivering adaptation finance is striking an appropriate balance between the interests and needs of contributors and recipients of funding. Here we discuss four areas of policy choice in delivery, highlighting questions of balance in each area: funding entities responsible for delivering climate finance, relative balance between adaptation and mitigation finance, approaches to prioritizing the allocation of adaptation finance and financial instruments for delivering adaptation finance. In each area, different policy choices may emphasize effectiveness/efficiency or distributive aims, with contributors often focusing greater attention on the former aim and recipients focusing more on the latter (although as noted above, contributors may also have distributive reasons for providing funding).[5]

3.5.1 Governance of funding institutions

The current institutional architecture for delivering adaptation finance is highly fragmented, encompassing not only the dedicated multilateral funds surveyed in Section 3.3 but also the bilateral development agencies of contributing countries as well as multilateral development institutions (Persson et al. 2009; Stadelmann, Brown, et al. 2012). This degree of fragmentation reflects in part the necessity of delivering fast-start finance through existing institutions given the time required to establish new ones. However, it also reflects to a significant extent the interests of contributors of climate finance in retaining control over the use of funds. Accordingly, contributors tend to prefer channeling funds through national development agencies or multilateral agencies that they consider to be more closely aligned with their interests – such as the World Bank – rather than existing UN funds (Harmeling and Kaloga 2011).

The design of new funds under the multilateral framework of the UNFCCC has offered the opportunity to reconfigure the balance between contributors and recipients. The Adaptation Fund Board is unique among multilateral climate funds in that developing countries hold the overall majority of seats. Some had hoped that this would provide a precedent for the Green Climate Fund. However, it appears that the Adaptation Fund is considered exceptional since its funding is primarily derived outside aid channels that are fully controlled by donors (Horstmann 2011: 1089; Harmeling and Kaloga 2011: 25). Instead, the board of the Green Climate Fund (like that of the Climate Investment Funds) has an equal number of developed and developing countries, which still gives developing countries considerably greater influence than they have under many existing channels for development finance.

3.5.2 Ensuring "balanced" allocation between adaptation and mitigation

The Copenhagen Accord requires "balanced" allocation between adaptation and mitigation finance but provides no guidance as to the standards against which the degree of balance should be measured. Developing countries and civil society organizations have argued that balanced allocation requires *equal* allocation between adaptation and mitigation; that is, half of all financing commitments should be allocated to adaptation (Stadelmann, Brown, et al.

2012: 124). However, the vagueness of the Copenhagen Accord wording combined with the decentralized nature of climate funding institutions has meant that contributing countries may largely determine what they see as the requisite degree of balance. In practice, only around 20 percent of fast-start finance has been pledged for adaptation (Stadelmann, Brown, et al. 2012: 121). This trend is consistent with the theoretical claims outlined above that contributing countries' governments will generally favor funding that yields a global rather than local good, and the relative ease of mobilizing private sector for mitigation in contrast to adaptation (see also Abadie et al. 2013 and Table 3.1).

Several options could help to address the issue of balance. First, a plausible view would be that balance between two objectives does not always require an even split between them, but will depend on the level of need for financing the two objectives. Consequently, improved assessments of global financing needs for mitigation and adaptation could help to determine developing countries' overall needs and the proportion of those needs that should be covered by financing commitments. Second, the newly established Standing Committee on Finance under the UNFCCC – which has a mandate to assist the COP in strengthening the coherence and coordination of overall flows (UNFCCC 2011, para. 112) – could make recommendations on how balance could be improved. Third, the adoption of suitable allocation policies within major multilateral funds (such as the Green Climate Fund) could help to improve the balance of overall flows, possibly by counterbalancing the mitigation bias in other sources of global climate finance.

3.5.3 Criteria for allocating adaptation finance

Parties to the UNFCCC have agreed that adaptation finance should be prioritized towards the most vulnerable developing countries (Copenhagen Accord, para. 8). However, there are significant challenges in determining how vulnerability should be measured and whether other factors should also enter into consideration when allocating adaptation finance.

Despite several references to vulnerability, the Convention provides little guidance on which countries are particularly vulnerable (Yamin and Depledge 2004: 227). More recent COP decisions have singled out a smaller list of country groups, such as LDCs, Small Island Developing States (SIDS) and Africa (Copenhagen Accord, para. 3). While the LDCF restricts eligibility to LDCs, other funds have struggled to specify more precise eligibility criteria. Thus under the Adaptation Fund, all developing country parties to the Kyoto Protocol are in principle eligible for funding (Horstmann 2011: 1091).

Evidence-based assessments of vulnerability may help to inform decisions about eligibility for and prioritization of adaptation funding. Researchers generally define vulnerability to climate change as a product of three factors: (i) exposure and (ii) sensitivity to climate impacts, and (iii) the capacity to adapt to those impacts (Adger 2006: 269). Researchers have developed a range of indices to quantify each of these factors for the purposes of resource allocation (see for example Adger et al. 2004; Barr et al. 2010; Wheeler 2011). However, several challenges must be addressed in order to justify using such indices for allocating international finance. First, estimates of exposure and sensitivity are subject to considerable empirical uncertainties. Second, while a country's low adaptive capacity will increase its vulnerability, it is also likely to limit its ability to use funds cost-effectively. In this sense, allocating adaptation finance raises a dilemma similar to that in development assistance, namely whether to allocate funding according to need (whether measured by vulnerability or poverty) or according to capacity for cost-effective use of funds (frequently measured according to indicators of good governance) (Barr et al. 2010: 844). In terms of the conceptual framework outlined above, this could be understood as a tension between distributive and allocative objectives. Some analysis has shown that international allocations will vary considerably depending on how these factors are taken into account (Barr et al. 2010; Wheeler 2011). Little analysis is available on existing patterns of allocation of adaptation finance, but some initial findings from the Adaptation Fund suggest that cost-effectiveness (measured in terms of economic benefits) has been a more prominent factor than the level of vulnerability or poverty (Stadelmann, Persson, et al. 2013).

Since contributors and recipients may have widely varying views on how to construct such an index, multilateral agreement on this issue may be impossible. However, a more ad hoc approach to allocation may be problematic where those determining the allocation (typically contributors and managers of funding) do not have the best information about appropriate means to adapt. Principal-agent problems may therefore arise, causing allocative inefficiencies and/or undesired redistribution of wealth (see e.g., Gibson et al. 2005; Abadie et al. 2013). The managers or contributors (the principals) of funds do not know all individual characteristics that are relevant for delivery decisions, since these characteristics are only known privately to recipients (the agents) of adaptation funding, i.e., developing countries hosting supported adaptation projects.

In response to these concerns, some funds – notably the Adaptation Fund – have experimented with more demand-responsive modes of accessing funds, including enabling domestic institutions in recipient countries "direct access" to some funding rather than requiring them to receive funds through international intermediaries. National adaptation planning – which recent UNFCCC decisions have sought to strengthen (see UNFCCC 2011, para. 15) – can provide a credible basis for identifying adaptation investment priorities and attracting international finance.

Despite the value of demand-driven approaches, they may also have some disadvantages. For example, limited information about adaptation needs may in fact be a by-product of low adaptive capacity; in addition, agents could exploit their discretion to attract higher levels of international funds towards less effective adaptation projects. In either case, scarce international resources would be wasted. For these reasons, international assistance should at a minimum prioritize support vulnerable countries in building their capacity to identify adaptation needs and to access and manage adaptation finance (compare Fankhauser and Burton 2011). Others have emphasized that, in the presence of uncertainty about climatic risks and the close connection between adaptation and development, funding should prioritize reducing societal vulnerability to a broader range of economic and social risks (Khan and Roberts 2013). Moreover, these reasons may suggest the value of a mixed approach combining demand-driven allocation of funding while ensuring coherence in overall funding trends by reference to global indices of vulnerability.

3.5.4 Financial instruments for delivering adaptation finance

Considerable debate exists over the appropriate financial instruments for delivering adaptation finance. Developing countries have argued that adaptation funds should be delivered in the form of grants rather than repayable loans. This argument is based both on the view that payments made to remedy harm – by analogy with compensation payments – should not be repayable as a matter of principle (Schalatek 2011: 82), as well as the view that many adaptation investments will not generate commercial returns. To date, most international adaptation finance has been delivered in the form of grants (Stadelmann, Brown, et al. 2012: 125). However, researchers have proposed some alternative modes of delivery, one of which we briefly outline here: the idea of results-based funding for adaptation, where funding amounts are attached to specific measurable targets and payment is provided once those targets are reached.[6]

The general concept of results-based financing for development has a number of appeals, including promoting greater transparency and incentives for performance (Birdsall and Savedoff 2009), and could be used as a means of delivering public adaptation finance. An alternative results-based approach could involve the creation of market mechanisms for adaptation finance. This would involve commoditizing certain aspects of adaptation, either by restricting the supply of certain adaptive "bads" (such as building permits in high-risk coastal zones), or facilitating the supply of adaptive goods, such as wealth or lives saved (Michaelowa et al. 2012: 190; see also Schultz 2011). Both market- and non-market-oriented results-based funding would present a range of practical hurdles, including establishing reliable measurement of adaptation outcomes, and developing baselines (or "business as usual" scenarios) against which progress would be measured. Both outcome measurement and baseline setting would be complicated by uncertainty about long-term impacts as well as natural climatic variability (Michaelowa et al. 2012: 201).

In addition, market-based mechanisms face a number of political hurdles. A key challenge – and arguably one that limits the feasibility of these mechanisms in the foreseeable future – is how to define adaptation units in such a way as to facilitate trading in a uniform commodity while not undermining efforts to mainstream adaptation into broader development practice (Persson 2011). Michaelowa et al. (2012: 191), while more positive about prospects for adaptation markets, have queried whether a market-based scheme that involved substantial transfers to developing countries – assuming that lower-cost adaptation options are located there – would be politically acceptable to developed countries.

3.6 Conclusions

Our chapter has sought to demonstrate that both allocative and distributive goals should be taken into account in analyzing international support for adaptation. Developing countries' perceptions that climate change burdens are being distributed unfairly may impede their willingness to contribute to international climate protection. Perhaps most importantly, without broad participation of developing as well as developed countries a global agreement cannot be reached, and most developing countries demand adaptation support as a prerequisite for their consent.

Many different agencies and institutions are currently involved in the international adaptation finance arena. This may be due partly to the fact that individual donors tend to prefer those agencies that are more closely aligned with their own interests. The scaling up of climate finance towards 2020 and accompanying efforts to design new institutions such as the Green Climate Fund now offer opportunities to reconfigure the balance between donors and recipients and to make consensus in climate change negotiations – both on mitigation and adaptation – more likely. In designing mechanisms to generate and deliver adequate adaptation funding, policymakers will need to ensure that both allocative and distributive considerations are suitably addressed.

There is still much scope for further research in international adaptation finance. Among the high priority areas for research emerging from our analysis are the evaluation of the geographical distribution of the disbursement of adaptation funds and its possible motivations as well as the identification of feasible innovative sources of adaptation finance.

Acknowledgement

We thank Clemens Heuson, Åsa Persson, María Victoria Román de Lara and Martin Stadelmann for helpful comments.

Notes

1 An alternative perspective is taken by Seo (2010), who considers adaptation as a global public good. However, his perspective seems to be overly generalized, at least for the purposes of this chapter.
2 The related consideration of "cost-effectiveness" is arguably best seen as a form of efficiency, in the sense of attaining a policy goal at least cost, which is in turn a condition for attaining broader economic efficiency with limited resources (compare with Sterner 2003: 136; Black et al. 2009).
3 The Cancún Agreements at COP 16 the following year formalized these commitments as a COP decision.
4 For an analysis of strategic implications of different modes of funding climate policies in developing countries, see Heuson, Peters, et al. 2012.
5 Compare Andresen and Hey (2005: 213), who argue (in relation to global environmental governance) that: "ensuring effectiveness has tended to be the more important concept and goal for the strong and powerful, while securing legitimacy has been more important for the weaker actors".
6 Another modality is insurance against climate impacts such as extreme weather events (see Chapter 21 on disaster risk management and adaptation to extreme events).

References

Aakre, S., and D.T.G. Rübbelke. 2010a. Adaptation to Climate Change in the European Union: Efficiency Versus Equity Considerations. *Environmental Policy and Governance* 20 (3):159–79.

Aakre, S., and D.T.G. Rübbelke. 2010b. Objectives of Public Economic Policy and the Adaptation to Climate Change. *Journal of Environmental Planning and Management* 53 (6):767–91.

Abadie, L., I. Galarraga, and D. Rübbelke. 2013. An Analysis of the Causes of the Mitigation Bias in International Climate Finance. *Mitigation and Adaptation Strategies for Global Change,* 18 (7): 943–955.

Adaptation Fund Board. 2013. Adaptation Fund Trust Fund: Financial Report Prepared by the Trustee as of March 31, 2013. https://www.adaptation-fund.org/sites/default/files/AFB.EFC_.12.8%20AF%20Trust%20Fund%20Financial%20Report%20(as%20of%20Mar%2031,%202013).pdf.

Adger, W.N. 2006. Vulnerability. *Global Environmental Change* 16 (3):268–81.

Adger, W.N., N.W. Arnell, and E.L. Tompkins. 2005. Successful Adaptation to Climate Change across Scales. *Global Environmental Change* 15 (2):77–86.

Adger, W.N., N. Brooks, M. Kelly, S. Bentham, and S. Eriksen. 2004. *New Indicators of Vulnerability and Adaptive Capacity.* Technical Report 7. Norwich: Tyndall Centre for Climate Research. http://www.tyndall.ac.uk/content/new-indicators-vulnerability-and-adaptive-capacity.

AGF. 2010. *Report of the Secretary-General's High-Level Advisory Group on Climate Change Financing [AGF].* New York: United Nations. http://www.un.org/wcm/webdav/site/climatechange/shared/Documents/AGF_reports/AGF%20Report.pdf.

Andresen, S., and E. Hey. 2005. The Effectiveness and Legitimacy of International Environmental Institutions. *International Environmental Agreements: Politics, Law and Economics* 5 (3):211–26.

Arrow, K.J. 2007. Global Climate Change: A Challenge to Policy. *Economists' Voice* June.

Atteridge, A. 2011. *Will Private Finance Support Climate Change Adaptation in Developing Countries? Historical Investment Patterns as a Window on Future Private Climate Finance.* SEI Working Paper No. 2011–05: Stockholm: Stockholm Environment Institute.

Barr, R., S. Fankhauser, and K. Hamilton. 2010. Adaptation Investments: A Resource Allocation Framework. *Mitigation and Adaptation Strategies for Global Change* 15:843–58.

Barrett, S. 2008. *Dikes v. Windmills: Climate Treaties and Adaptation*. Paper presented at the workshop "The Environment, Technology and Uncertainty" of the Ragnar Frisch Centre for Economic Research, Oslo.
Biermann, F., and A. Gupta. 2011. Accountability and Legitimacy in Earth System Governance: A Research Framework. *Ecological Economics* 70 (11):1856–64.
Birdsall, N., and W.D. Savedoff. 2009. *Cash on Delivery: A New Approach to Foreign Aid with an Application to Primary Schooling*. Washington, DC: Center for Global Development.
Birnie, P., A. Boyle, and C. Redgwell. 2009. *International Law and the Environment*. 3rd ed. Oxford: Oxford University Press.
Black, J., N. Hashimzade, and G. Myles. 2009. *A Dictionary of Economics*. 3rd ed. Oxford: Oxford University Press.
Botzen, W.J.W., J.M. Gowdy, and J.C.J.M. van den Bergh. 2008. Cumulative CO_2 Emissions: Shifting International Responsibilities for Climate Debt. *Climate Policy* 8:569–76.
Bowen, A. 2011. Raising Climate Finance to Support Developing Country Action: Some Economic Considerations. *Climate Policy* 11 (3):1020–36.
Buchholz, W., and W. Peters. 2005. A Rawlsian Approach to International Cooperation. *Kyklos* 58 (1):25–44.
Buchner, B., A. Falconer, M. Hervé-Mignucci, C. Trabacchi, and M. Brinkman. 2011. *The Landscape of Climate Finance*. Venice: Climate Policy Initiative. http://climatepolicyinitiative.org/wp-content/uploads/2011/10/The-Landscape-of-Climate-Finance-120120.pdf.
Ciplet, D., J. T. Roberts, and M. Khan. 2013. The Politics of International Climate Adaptation Funding: Justice and Divisions in the Greenhouse. *Global Environmental Politics* 13 (1):49-68.
Clemens, M.A., and T.J. Moss. 2007. The Ghost of 0.7 Per Cent: Origins and Relevance of the International Aid Target. *International Journal of Development Issues* 6 (1):3–25.
Climate Funds Update. 2013. [cited 6 September 2013]. http://www.climatefundsupdate.org/themes/adaptation.
Donner, S.D., M. Kandlikar, and H. Zerriffi. 2011. Preparing to Manage Climate Change Financing. *Science* 334 (6058):908–09.
Eckersley, R. 2012. Moving Forward in Climate Negotiations: Multilateralism or Minilateralism? *Global Environmental Politics* 12 (2):24–42.
Eisenack, K. 2012. Adaptation Financing in a Global Agreement: Is the Adaptation Levy Appropriate? *Climate Policy* 12 (4):491–504.
Fankhauser, S., and I. Burton. 2011. Spending Adaptation Money Wisely. *Climate Policy* 11 (3):1037–49.
Fankhauser, S., and N. Martin. 2010. The Economics of the CDM Levy: Revenue Potential, Tax Incidence and Distortionary Effects. *Energy Policy* 38 (1):357–63.
Faure, M., and M. Peeters, eds. 2011. *Climate Change Liability*. Cheltenham, UK: Edward Elgar.
GEF. 2010. *Strategy on Adaptation to Climate Change for the Least Developed Countries Fund (LDCF) and the Special Climate Change Fund (SCCF)*: Global Environment Facility. http://www.thegef.org/gef/sites/thegef.org/files/publication/GEF-ADAPTION%20STRATEGIES.pdf.
GEF. 2011. Instrument for the Establishment of the Restructured Global Environment Facility. October 2011. http://www.thegef.org/gef/sites/thegef.org/files/publication/GEF_Instrument_Oct2011_final_0.pdf.
Gibson, C.C., K. Andersson, E. Ostrom, and S. Shivakumar. 2005. *The Samaritan's Dilemma: The Political Economy of Development Aid*. Oxford: Oxford University Press.
Grasso, M. 2010. *Justice in Funding Adaptation under the International Climate Change Regime*. Dordrecht: Springer.
Gupta, J. 2007. Legal Steps Outside the Climate Convention: Litigation as a Tool to Address Climate Change. *Review of European Community and International Environmental Law* 16 (1):76–86.
Gupta, J. 2010. A History of International Climate Change Policy. *Wiley Interdisciplinary Reviews: Climate Change* 1 (5):636–53.
Harmeling, S., and A.O. Kaloga. 2011. Understanding the Political Economy of the Adaptation Fund. *IDS Bulletin* 42 (3):23–32.
Harris, P.G., and J. Symons. 2010. Justice in Adaptation to Climate Change: Cosmopolitan Implications for International Institutions. *Environmental Politics* 19 (4):617–36.
Hepburn, C., and B. Müller. 2010. International Air Travel and Greenhouse Gas Emissions: A Proposal for an Adaptation Levy. *World Economy* 33 (6):830–49.

Heuson, C., E. Gawel, O. Gebhardt, B. Hansjürgens, P. Lehmann, V. Mayer, and R. Schwarze. 2012. *Fundamental Questions on the Economics of Climate Adaptation: Outlines of a New Research Programme.* UFZ-Report 05/2012. Leipzig: Helmholtz-Zentrum für Umweltforschung (UFZ).

Heuson, C., W. Peters, R. Schwarze, and A.-K. Topp. 2012. *Which Mode of Funding Developing Countries' Climate Policies under the Post-Kyoto Framework?* UFZ Discussion Paper October, 2012. Leipzig: Helmholtz-Zentrum für Umweltforschung (UFZ).

Hof, A.F., K.C. de Bruin, R.B. Dellink, M.G.J. den Elzen, and D.P. van Vuuren. 2009. The Effect of Different Mitigation Strategies on International Financing of Adaptation. *Environmental Science & Policy* 12 (7):832–43.

Hof, A.F., M.G.J. den Elzen, and A. Mendoza Beltran. 2011. Predictability, Equitability and Adequacy of Post-2012 International Climate Financing Proposals. *Environmental Science & Policy* 14 (6):615–27.

Höhne, N., H. Blum, J. Fuglestvedt, R. Skeie, A. Kurosawa, G. Hu, J. Lowe, L. Gohar, B. Matthews, A. Nioac de Salles, and C. Ellermann. 2011. Contributions of Individual Countries' Emissions to Climate Change and Their Uncertainty. *Climatic Change* 106 (3):359–91.

Horstmann, B. 2011. Operationalizing the Adaptation Fund: Challenges in Allocating Funds to the Vulnerable. *Climate Policy* 11 (4):1086–96.

IPCC. 2007. Summary for Policymakers. In *Climate Change 2007: Impacts, Adaptation and Vulnerability. Contribution of Working Group II to the Fourth Assessment Report of the Intergovernmental Panel on Climate Change,* edited by M. L. Parry, O. F. Canziani, J. P. Palutikof, P. J. van der Linden and C. E. Hanson. Cambridge, UK: Cambridge University Press. 7–22.

Junghans, L., and S. Harmeling. 2012. *Different Tales from Different Countries: A First Assessment of the OECD "Adaptation Marker".* Briefing paper. Bonn: Germanwatch.

Khan, M. R., and J. T. Roberts. 2013. Adaptation and International Climate Policy. *Wiley Interdisciplinary Reviews: Climate Change* 4 (3):171–89.

Kotchen, M.J. 2013. Voluntary- and Information-Based Approaches to Environmental Management: An Impure Public Good and Club Theory Perspective. *Review of Environmental Economics and Policy* 7 (2):276–95.

Liverman, D. 2011. Challenges in Creating an International Regime for Adaptation. In Richardson, K., W. Steffen, D. Liverman, T. Barker, F. Jotzo, D.M. Kammen, R. Leemans, T.M. Lenton, M. Munasinghe, B. Osman-Elasha, H.J. Schellnhuber, N. Stern, C. Vogel, and O. Wæver, *Climate Change: Global Risks, Challenges and Decisions.* Cambridge, UK: Cambridge University Press. 404–05 (Box 14.3).

Mace, M.J. 2005. Funding for Adaptation to Climate Change: UNFCCC and GEF Developments since Cop-7. *Review of European Community & International Environmental Law* 14 (3):225–46.

Mace, M.J. 2006. Adaptation under the UN Framework Convention on Climate Change: The International Legal Framework. In *Fairness in Adaptation to Climate Change,* edited by W. Adger, J. Paavola, S. Huq and M. Mace. Cambridge, MA: MIT Press. 53–76.

Mertz, O., K. Halsnæs, J. Olesen, and K. Rasmussen. 2009. Adaptation to Climate Change in Developing Countries. *Environmental Management* 43 (5):743–52.

Michaelowa, A., M. Köhler, and S. Butzengeiger. 2012. Market Mechanisms for Adaptation: An Aberration or a Key Source of Finance? In *Carbon Markets or Climate Finance? Low Carbon and Adaptation Investment Choices for the Developing World,* edited by A. Michaelowa. London: Routledge. 188–208.

Michaelowa, A., and K. Michaelowa. 2011. Coding Error or Statistical Embellishment? The Political Economy of Reporting Climate Aid. *World Development* 39 (11):2010–20.

Michaelowa, K., and A. Michaelowa. 2012. Development Cooperation and Climate Change: Political-Economic Determinants of Adaptation Aid. In *Carbon Markets or Climate Finance? Low Carbon and Adaptation Investment Choices for the Developing World.* London: Routledge. 39–52.

Miola, A., M. Marra, and B. Ciuffo. 2011. Designing a Climate Change Policy for the International Maritime Transport Sector: Market-Based Measures and Technological Options for Global and Regional Policy Actions. *Energy Policy* 39 (9):5490–98.

Moore, F.C. 2012a. Costing Adaptation: Revealing Tensions in the Normative Basis of Adaptation Policy in Adaptation Cost Estimates. *Science, Technology & Human Values* 37 (2):171–98.

Moore, F.C. 2012b. Negotiating Adaptation: Norm Selection and Hybridization in International Climate Negotiations. *Global Environmental Politics* 12 (4):30–48.

Müller, B. 2008. *To Earmark or Not to Earmark? A Far-Reaching Debate on the Use of Auction Revenue from (EU) Emissions Trading*. Oxford: Oxford Institute for Energy Studies. http://www.oxfordclimatepolicy.org/publications/documents/EV43.pdf.

Musgrave, P. 2008. Comments on Two Musgravian Concepts. *Journal of Economics and Finance* 32 (4):340–47.

Musgrave, R.A. 1956. A Multiple Theory of Budget Determination. *FinanzArchiv / Public Finance Analysis* 17:333–43.

Musgrave, R.A. 1959. *The Theory of Public Finance: A Study in Public Economy*. New York: McGraw-Hill.

Narain, U., S. Margulis, and T. Essam. 2011. Estimating Costs of Adaptation to Climate Change. *Climate Policy* 11 (3):1001–19.

OECD. 2011. *First-Ever Comprehensive Data on Aid for Climate Change Adaptation*. Paris: OECD. http://www.oecd.org/investment/aidstatistics/49187939.pdf.

OECD. 2013. *Statistics*. http://www.oecd.org/statistics/.

Olivier, J.G.J., G. Janssens-Maenhout, and J.A.H.W. Peters. 2012. *Trends in Global CO_2 Emissions: 2012 Report*. The Hague: PBL Netherlands Environmental Assessment Agency and the European Union.

Persson, Å. 2011. *Institutionalizing Climate Finance Adaptation under the UNFCCC and Beyond: Could an Adaptation 'Market' Emerge?* Stockholm: Stockholm Environment Institute.

Persson, Å., R.J.T. Klein, C.K. Siebert, A. Atteridge, B. Müller, J. Hoffmaister, M. Lazarus, and T. Takama. 2009. *Adaptation Finance under a Copenhagen Agreed Outcome*. Research Report. Stockholm: Stockholm Environment Institute.Pickering, J., and C. Barry. 2012. On the Concept of Climate Debt: Its Moral and Political Value. *Critical Review of International Social and Political Philosophy* 15 (5):667–85.

Pickering, J., and P. J. Wood. 2011. Climate Finance for Developing Countries. In K. Richardson, W. Steffen, D. Liverman, T. Barker, F. Jotzo, D.M. Kammen, R. Leemans, T.M. Lenton, M. Munasinghe, B. Osman-Elasha, H.J. Schellnhuber, N. Stern, C. Vogel and O. Wæver, *Climate Change: Global Risks, Challenges and Decisions*. Cambridge, UK: Cambridge University Press. 336–38 (Box 12.5).

Pickering, J., F. Jotzo and P. J. Wood. 2013. *Splitting the difference in global climate finance: are fragmentation and legitimacy mutually exclusive?* Centre for Climate Economics and Policy (CCEP) Working Paper 1308, November 2013. Canberra: Crawford School of Public Policy, The Australian National University.

Pielke, R., G. Prins, S. Rayner, and D. Sarewitz. 2007. Lifting the Taboo on Adaptation. *Nature* 445 (7128):597–98.

Pittel, K., and D. Rübbelke. 2013. International Climate Finance and Its Influence on Fairness and Policy. *The World Economy* 36 (4):419–36.

Rive, N., A. Torvanger, and J.S. Fuglestvedt. 2006. Climate Agreements Based on Responsibility for Global Warming: Periodic Updating, Policy Choices, and Regional Costs. *Global Environmental Change* 16 (2):182–94.

Roberts, J.T. 2009. The International Dimension of Climate Justice and the Need for International Adaptation Funding. *Environmental Justice* 2 (4):185–90.

Rübbelke, D.T.G. 2011. International Support of Climate Change Policies in Developing Countries: Strategic, Moral and Fairness Aspects. *Ecological Economics* 70 (8):1470–80.

Schalatek, L. 2011. *A Matter of Principle(s): A Normative Framework for a Global Compact on Public Climate Finance*. Publication series on ecology, Volume 13. Berlin: Heinrich Böll Stiftung.

Schipper, E.L.F. 2006. Conceptual History of Adaptation in the UNFCCC Process. *Review of European Community & International Environmental Law* 15 (1):82–92.

Schultz, K.H. 2011. Financing Climate Adaptation with a Credit Mechanism: Initial Considerations. *Climate Policy* 12 (2):187–97.

Scott, J., and L. Rajamani. 2012. EU Climate Change Unilateralism. *European Journal of International Law* 23 (2):469–94.

Seballos, F., and S. Kreft. 2011. Towards an Understanding of the Political Economy of the PPCR. *IDS Bulletin* 42 (3):33–41.

Seo, S.N. 2010. *A Theory of Adaptation to Climate Change as a Global Public Good*. Working Paper No. 10-041: United States Association for Energy Economics (USAEE) and International Association for Energy Economics (IAEE).

Shue, H. 1993. Subsistence Emissions and Luxury Emissions. *Law & Policy* 15 (1):39–60.

Stadelmann, M., J. Brown, and L. Hörnlein. 2012. Fast-Start Finance: Scattered Governance, Information and Programmes. In *Carbon Markets or Climate Finance? Low Carbon and Adaptation Investment Choices for the Developing World*, edited by A. Michaelowa. London: Routledge. 117–45.

Stadelmann, M., Å. Persson, I. Ratajczak-Juszko, and A. Michaelowa. 2013. Equity and Cost-Effectiveness of Multilateral Adaptation Finance: Are They Friends or Foes? *International Environmental Agreements: Politics, Law and Economics* (published online) 1–20.
Stadelmann, M., J.T. Roberts, and A. Michaelowa. 2011. New and Additional to What? Assessing Options for Baselines to Assess Climate Finance Pledges. *Climate and Development* 3 (3):175–92.
Stern, N. 2007. *The Economics of Climate Change: The Stern Review*. Cambridge, UK; New York: Cambridge University Press.
Stern, N. 2009. *A Blueprint for a Safer Planet: How to Manage Climate Change and Create a New Era of Progress and Prosperity*. London: Bodley Head.
Sterner, T. 2003. *Policy Instruments for Environmental and Natural Resource Management*. Washington, DC: Resources for the Future.
UNFCCC. 2011. The Cancún Agreements: Outcome of the Work of the Ad Hoc Working Group on Long-Term Cooperative Action under the Convention. FCCC/CP/2010/7/Add.1 (15 March 2011). http://unfccc.int/resource/docs/2010/cop16/eng/07a01.pdf.
UNFCCC. 2012. Launching the Green Climate Fund. Decision 3/CP.17 (Durban, 2011). 55–66. http://unfccc.int/resource/docs/2011/cop17/eng/09a01.pdf.
UNFCCC. 2013a. Amendment to the Kyoto Protocol Pursuant to Its Article 3, Paragraph 9 (the Doha Amendment). 1/CMP.8 (Doha, 2012). http://unfccc.int/resource/docs/2012/cmp8/eng/13a01.pdf.
UNFCCC. 2013b. Approaches to Address Loss and Damage Associated with Climate Change Impacts in Developing Countries That Are Particularly Vulnerable to the Adverse Effects of Climate Change to Enhance Adaptive Capacity. Decision 3/CP.18 (Doha, 2012). 21-24. http://unfccc.int/resource/docs/2012/cop18/eng/08a01.pdf.
Vanderheiden, S. 2008. *Atmospheric Justice: A Political Theory of Climate Change*. New York: Oxford University Press.
Verheyen, R. 2005. *Climate Change Damage and International Law: Prevention Duties and State Responsibility*. Leiden: Martinus Nijhoff Publishers.
Vieweg, M., A. Esch, L. Grießhaber, F. Fuller, F. Mersmann, F. Fallasch, and L. De Marez. 2012. *German Fast Start: Lessons Learned for Long-Term Finance*. Climate Analytics, Wuppertal Institute, Germanwatch.
Wheeler, D. 2011. *Quantifying Vulnerability to Climate Change: Implications for Adaptation Assistance*. Working Paper 240: January 2011. Washington, DC: Center for Global Development. http://www.cgdev.org/files/1424759_file_Wheeler_Quantifying_Vulnerability_FINAL.pdf.
World Bank. 2009. *Development and Climate Change*. World Development Report 2010. Washington, DC: World Bank. http://siteresources.worldbank.org/INTWDR2010/Resources/5287678-1226014527953/WDR10-Full-Text.pdf.
World Bank. 2010. *Economics of Adaptation to Climate Change: Synthesis Report*. Washington, DC: World Bank. http://beta.worldbank.org/sites/default/files/documents/EACCSynthesisReport.pdf.
World Bank. 2011. *Mobilizing Climate Finance: A Paper Prepared at the Request of G20 Finance Ministers*. October 6, 2011. http://www.imf.org/external/np/g20/pdf/110411c.pdf.
Yamin, F., and J. Depledge. 2004. *The International Climate Change Regime: A Guide to Rules, Institutions and Procedures*. Cambridge, UK: Cambridge University Press.

PART II

Uncertainty, equity, valuation and efficiency

4

SYNERGIES BETWEEN ADAPTATION AND MITIGATION AND THE COMPLEXITY OF REDD+

Asbjørn Aaheim and Jorge H. García[1]

[1] CICERO, CENTER FOR INTERNATIONAL CLIMATE AND ENVIRONMENTAL RESEARCH – OSLO

4.1 Introduction

Reduction of emissions by deforestation and forest degradation (REDD+) has been thought of as a way to combine mitigation with adaptation strategies: Preservation of forest stimulates biodiversity, which helps forests become more resilient towards climate change. Forests play an important role in limiting impacts of climate change, such as lowering the risk of threshold shifts in ecosystems, reducing problems of irrigation and limiting losses related to floods and landslides. At the same time, a huge amount of carbon is stored in forests, and more forests means that less carbon is concentrated in the atmosphere. As such, afforestation and reduced deforestation is also a way to mitigate climate change.

It is being pointed out, however, that the picture is more complicated (Barbier, 2012). Preservation implies that several current activities that are based on the utilization of the forests will have to be brought to an end. REDD+ initiatives cannot be evaluated without considering both the losses and alternative activities that are being stimulated as a result. Similarly, afforestation implies a change of land-use, meaning that there is a possibility of intensified utilization of other land, which has to be considered. In the end, an initiative within a REDD+ scheme may have indirect effects that reduce or even neutralize both the mitigation effect and the adaptation effect. REDD+ initiatives therefore need to be analyzed in a broader context with reference to specific cases, and the final result may be conditioned on the way the mechanism is implemented in each and every case.

This chapter addresses protection of selected forests in Tanzania, which has been proposed as a means to increase the storage of carbon, and possibly increase biodiversity or reduce degradation of biodiversity. Taken in isolation, the proposal will mitigate climate change by an increase of carbon storage. Increasing forested areas is likely to make people less vulnerable to some impacts of climate change, in particular by its effect on water irrigation. Thus, it enhances adaptive capacity.

However, the present harvesting of the forest aims primarily at producing charcoal, which provides an important source of added income to rural farmers, who sell the biofuel to the cities. A ban on this trade may limit income opportunities to farmers, and make them more vulnerable. Moreover, a lower supply of charcoal stimulates the use of alternative energy carriers. These are usually fossil fuels, and the impacts on emissions are therefore questionable. In other words, the proposal may as well have negative consequences for mitigation and moreover increase the vulnerability of the population. The objective of this chapter is to find under what conditions a REDD+ initiative such as the one suggested in Tanzania may lead to synergies between mitigation and adaptation.

The next section gives a short history behind the REDD+ mechanism, and explains the role it is meant to play as a policy instrument in the mitigation of global warming. We also discuss a number of implementation issues identified in the literature. A branch of the literature discusses socioeconomic aspects related to REDD+, both generally and with reference to specific cases. Another branch discusses whether, and to what extent replacement of fossil fuels by biofuels contributes to a reduction in the concentrations of greenhouse gases.

Section 4.3 addresses the impacts on emissions of a resulting switch from charcoal to alternative fossil fuels, and compares the resulting impacts on emissions with the impacts on carbon storage through forest protection. Then, we derive the resulting implications for the concentrations of carbon in the atmosphere. In particular, we examine the time frame of the concentration pathway for carbon in the atmosphere in order to make a distinction between long-term and short term impacts.

In Section 4.4, we discuss how small-holder farmers may respond to a ban on forest harvesting, which they use to produce and sell charcoal. The charcoal trade contributes a significant part of the monetary income to some of these farmers, while a part of the crop is used for consumption at the farm. Hence, the framing of the economic decisions to small-holders differs from that of the owners of large farms, where the economic problem better concurs with the standard assumption in economic theory. To the owner of a large farm, the monetary income is the only constraint on consumption, which is provided by market goods. The consumption to small-holders is partly constrained by what they produce. We analyze the consequences of a limitation to the potential income from selling charcoal in addition to farming on the consumption in small, medium and large farms, and ask to which extent forest protection changes the composite of farms subject to the different constraints. In Section 4.5, we ask what policies are needed to achieve both mitigation and adaptation with reference to the results of the previous sections.

4.2 The REDD+ mechanism

The idea of mitigating climate change by afforestation or avoiding deforestation is explained by the fact that the stock of carbon in the earth system is constant, while huge amounts are continuously being exchanged between the atmosphere, the biosphere, the oceans and the earth's ground. Deforestation implies that the uptake of carbon in the biosphere from the atmosphere is reduced, leaving less carbon in the biosphere and more in the atmosphere. This stimulates global warming.

REDD+ could play a prominent role as an effective climate change mitigation strategy for at least two reasons: 1) Deforestation and forest degradation, which mainly occurs in developing countries, accounts for about 20% of global emissions and 2) The costs of reducing emissions through reduction of deforestation rates have been estimated to be considerably lower as compared to other sectors such as industry and transport; see e.g., Kindermann et al. (2008) and van der Werf et al. (2009). REDD+ acknowledges that deforestation and forest degradation is driven by factors that make conservation particularly challenging for developing countries. This includes growing local and international demand for timber, meat, wood fuel and agricultural products. REDD+ thus seeks to compensate local governments and communities for giving up on certain local benefits associated with deforestation and forest degradation. At the same time, it is expected to improve forest governance and local welfare.

We will briefly review how REDD+ emerged within the context of the U.N. Framework Convention on Climate Change (UNFCCC) and where the scheme is currently at. Due to alleged difficulties associated with the measurement of carbon stocks and project additionality, forestry played a limited role under the Kyoto Protocol's Clean Development Mechanism (CDM) (IPCC, 2007), which is the only instrument that allows Annex I countries to meet reductions commitments through projects that reduce emissions in developing countries. Forest loss did not recede, and over the years it became clear that a mechanism that tackled a seemingly manageable and important source of carbon emissions would need to be created. The governments of Costa Rica and Papua New Guinea, with the support of other developing countries, requested to add the item "Reducing emissions from deforestation in developing countries: approaches to stimulate action" as part of the of the Conference of the Parties (COP) agenda at COP11 in Montreal in 2005. As a result, the secretariat invited the different countries to submit reviews and recommendations on this new item with a focus on scientific, technical and methodological issues. At COP13 in Bali in 2007, REDD+ became an integral part of the UNFCC negotiations. Decision 1/CP.13 (Bali Action Plan) calls for "Policy approaches and positive incentives on issues relating to reducing emissions from deforestation and forest degradation in developing countries; and the role of conservation, sustainable management of forests and enhancement of forest carbon stocks in developing countries." The semicolon initially implied that deforestation and forest degradation were to be given special status in terms of actions and policies. This was known as REDD (without the plus), and it underlined a difference between reduction of carbon emissions and enhancement of carbon stocks. Upon complaints from developing countries with relatively low deforestation rates, such as India, the enhancement of carbon stocks were to be given a similar status to that of reduced deforestation and forest degradation. Decision 2/CP.13 (Reducing emissions from deforestation in developing countries: approaches to stimulate action) spells out some of the details of the Bali Action Plan. It "Encourages all Parties, in a position to do so, to support capacity-building, provide technical assistance, facilitate the transfer of technology to improve, inter alia, data collection, estimation of emissions from deforestation and forest degradation, monitoring and reporting, and address the institutional needs of developing countries to estimate and reduce emissions from deforestation and forest degradation." At COP 16 in Cancun 2010, Decision 1/CP.16, it was agreed that REDD+ would be implemented in three phases: 1) development of national strategies or action plans, 2) implementation of policies and measures and 3) payment for performance on the basis of quantified forest emissions and removals. Countries were also encouraged to develop a national strategy or action plan, a national forest reference emission level and/or forest reference level or a national forest monitoring system for the monitoring and a system for providing information on a number of environmental and social safeguards.

The international community has mobilized readiness funds to lay down the foundations for REDD+ implementation. Donor contributions to the UNREDD Program, whose initial purpose is to develop country readiness for a REDD+ mechanism, has increased at an average annual rate of 46% since its creation in 2008 and total accumulated contributions amounted to 118 million dollars in 2011 (United Nations, 2011). The World Bank's Forest Carbon Partnership Facility (FCPF), which was also initiated in 2008, has a fund for REDD+ readiness (USD230') and a carbon fund (USD205') that is to be used for actual forest conservation and enhancement (FCPF, 2012). FCPF currently consists of 15 donor countries that have committed or pledged the aforementioned resources. Norway, a country that has strongly backed REDD+ since the beginning, has signed bilateral REDD+ agreements with Brazil and Indonesia for up to 1 billion dollars each.

With a Gross Domestic Product per capita of about US$1500, Tanzania is one of the poorest countries in the African continent. Forests and woodlands support about 87% of rural poor in the country, and the deforestation rate is around 1.16%. Agriculture and demand for fuel wood are among the main drivers of deforestation and forest degradation in the country. Tanzania is part of the FCPF and the UNREDD program, although a large share of REDD+ readiness resources have been channeled via bilateral agreements with other governments. In 2011, and consistent with phase 1 of REDD+ implementation, Tanzania submitted to the FCPF its final Readiness Preparation Proposal (R-PP). A Second Draft of the National REDD+ Strategy is currently being circulated for comments. Tanzania's R-PP states that The National Carbon Monitoring Center (NCMC) together with the National Carbon Accounting/Assessment System (NCAS) will be established to coordinate REDD+ matters and pave the way for the implementation of the country's REDD+ Strategy. Earmarked funds have been disbursed to civil society organizations to undertake pilot projects (REDD+ Readiness Factsheet: Tanzania 2012). The objectives of the demonstration projects vary widely and include the following: a) Improve knowledge and scientific understanding of forests, b) build village-level, local government and civil society organizational capacity towards understanding REDD mechanisms, c) build mechanisms for benefit sharing and empowerment of communities and d) Improve livelihoods and fuel wood availability to help address root causes of deforestation and degradation.

Although readiness activities related to REDD+ have begun to emerge around the developing world, key policy design issues remain unresolved. There exists large uncertainty regarding where the necessary funding is going to come from for a fully-fledged REDD+ scheme. While it is clear that the existing funds to build readiness can be instrumental in building local capacity, broader international involvement as well as financial resources are needed for scaling up REDD+ so it is an effective mitigation strategy to combat climate change (Angelsen et al., 2012; the Norwegian Agency for Development Cooperation, 2011). REDD+ funding may be supplied via two different channels, namely a donor-funded program or a compliance market. In the former case, countries participating in a carbon market will have the opportunity to buy REDD+ credits as offsets of their own emissions. In the latter, REDD+ is funded on a voluntary basis by donor countries. While potentially beneficial for forest conservation, the inclusion of REDD+ in carbon markets is yet to gain wide support in international negotiations. Some large developing countries, including some with significant forested areas such as Brazil, are skeptical of this approach. One of the main arguments against inclusion is that REDD+ credits would enable Annex I countries to buy their way out of one of the most important challenges posed by climate change, namely the decarbonization of industrialized economies (Eliasch, 2008).

While the flow of REDD+ funds may be propelled by differences in carbon abatement costs between developed and developing countries, these differences might have been overestimated. Major forestry governance challenges in developing countries shed doubt on the ability of REDD+ to deliver emissions reductions in certain countries. In the case of Tanzania, for instance, patrollers in charge of enforcing forest restrictions may sometimes take bribes rather than report violations (Robinson and Lokina, 2012). Similar and other inefficiencies at this and higher management levels cannot be discarded. Barbier (2012) explains that the transaction costs represented in measuring and verifying changes in deforestation in developing countries may be large. If reliable baseline deforestation scenarios cannot be constructed, it will not be possible to determine if certain REDD+ initiatives lead to emissions reductions. Romijn et al. (2012) show that there is a considerable capacity gap for reliable national forest monitoring in non-Annex I countries. On the other hand, it should also be noted that some large developing countries, such as Brazil and India, have comprehensive forest monitoring systems in place.

Forests provide a number of global ecological services such as biodiversity and carbon storage while supporting local livelihoods (Millennium Ecosystem Assessment, 2005). International climate negotiations recognized that forests provide both direct and ancillary benefits. At the same time, bundling several services in a single payment scheme such as REDD+ entails a number of difficulties. For instance a focus on forested areas with high carbon storage potential does not automatically lead to gains in biodiversity. It is not always possible to know ex-ante if a given REDD+ project will displace deforestation to particularly sensitive and highly diverse areas. Due to transaction costs including several ecological as well as socioeconomic goals under an overarching REDD+ scheme may hamper the chances for successful implementation of certain REDD+ projects.

The objective of this section has been to give the reader an overview of REDD+. In later sections we abstract away from institutional issues and focus on climate impacts and economic aspects of REDD+.

4.3 Mitigation by protection of forests

The long list of questions that can be raised regarding the functionality of REDD+ requires thorough documentation and examination to be accepted. In this respect, protection of a forested area can be considered a relatively straightforward suggestion. What is needed is a guarantee that the protection is actually carried through. If so, the development of the stock of the biomass can be predicted with an uncertainty related only to the knowledge of biophysical processes. To avoid issues related to the documentation of the projects, we therefore take protection of a forested area as our point of departure in this chapter. Instead, we focus on the effect that protection may have on concentrations of carbon in the atmosphere in the case considered in Tanzania, where most of the harvest is carried out by small-holder farmers who produce and sell charcoal.

It must be added that afforestation and deforestation have impacts on the climate beyond those related to the uptake and release of carbon focused in this chapter. First, land use changes affect the reflection of radiation from the sun. Reflection is usually increased by deforestation, with cooling as the result. Second, forests also store water, meaning that deforestation increases evaporation from land, which also has a cooling effect. These feed-backs are increasingly being analyzed, but there is huge uncertainty to what extent they moderate the climate effect of a change in the concentration of carbon in the atmosphere.

4.3.1 The carbon storage in forests

The development of forested land and implications of human interactions are complex processes that are best described by comprehensive biophysical models. In this chapter, we are searching mainly for general lessons of how forest management interacts with other economic activities in a developing country. We therefore simplify the representation of the biophysical processes to a relationship between the stock of the biomass, S, and the growth, $\dot{S}$, of a given forested area, the so-called Lotka-Volterra equation

$$\dot{S}_t = aS_t - bS_t^2 \tag{1}$$

where a and b are positive constants. This relationship captures a few central properties of the dynamics of biomasses. At sparsely covered forested areas with a low stock of biomass, the growth of the biomass is low, and may be negative if $S = 0$ is defined as the minimum stock needed to achieve a positive growth of the forested land without active planting. From this point, the growth increases along with increasing stock up to a certain point when $S = a/2b$, when the growth is at its highest level, S_{max}. For higher densities of the biomass, the growth of the forest declines until $\dot{S} = 0$, when $S = a/b$.

Below, we connect the carbon stored in forests to the stock of biomass: If the stock of biomass changes, the storage of carbon in forests changes proportionally. Hence, the development of carbon stored in a forest with biomass S_0, which is protected at $t = 0$, can be found by solving the differential equation (1), which is a simplified Bernoulli equation with solution:

$$S_t = \left(Ce^{-at} - \frac{b}{a} \right)^{-1} \tag{2}$$

where

$$C = \frac{1}{S_0} + \frac{b}{a} \tag{3}$$

and S_0 is the initial stock of biomass. According to (2) and (3), the potential for carbon storage by protecting forests depends on the initial biomass on the forested land. The initial stock is, of course, case specific. In some cases, protection may be motivated by a threat of deforestation, or the idea may be to increase the stock of biomass in forests that are being utilized in a sustainable manner. We shall concentrate on the economic motivations for utilizing the forest. Consider an agent with profit function

$$V(x_t) = \int_0^T v(x_t)e^{-\rho t}dt \tag{4}$$

who aims to maximize the surplus derived from extracting biomass over time. x_t is extracted biomass at time t, ρ is the discount rate, and T is the time horizon. The agent maximizes (4) under the biomass constraint:

$$\dot{S}_t = aS_t - bS_t^2 - x_t \tag{5}$$

The first-order condition for optimal harvesting is

$$a + 2bS_t = \mu \frac{\dot{x}}{x} + \rho \tag{6}$$

The left-hand side is the marginal increase in biomass growth and $\mu = -xv''_{xx}/v'_x$ is the forest owner's intertemporal elasticity of substitution. That is, from a given stock of biomass, the rate of change in harvest shall compensate the difference between the return on the biomass stock and the return on capital in alternative utilities (the discount rate) by the savings obtained in lower unit costs when smaller quantities are cut. These savings are expressed by the curvature of the profit function $v(x_t)$. Long-term stabilization of the harvest is achieved when

$$\dot{S}_t = aS_t - bS_t^2 \qquad (7)$$

Below, we will concentrate mainly on two cases, one where the dynamic equilibrium in (7) is sustained and one where forest exploitation is abandoned and the biomass develops in accordance with (2). The carbon storage, which we shall assume is linear in S_t is then fully determined, and protection at $t = 0$ means that the storage of carbon storage follows the path described by to (2) and (3) when S_0 is known.

The first lesson is that the potential for carbon storage by protection is critically dependent on the initial stock of the forest, which again depends on the return on capital in alternative utilities. Figure 4.1 shows examples of the development of carbon storage in a small number of trees in a forest that has been protected, under assumptions of 3, 5 and 7 percent discount rates. The parameters of the Lotka-Volterra equation were calibrated to fit the example with substitution between charcoal and kerosene below. The storage potential by protection, which is the difference between the maximum stock at approximately 43 t and the initial stock at $t = 0$, increases from 27.3 t at 3 percent discount rate to 35.0 t at 7 percent discount rate. Also, the time perspective for the realization of this potential differs. It takes 57 years to achieve

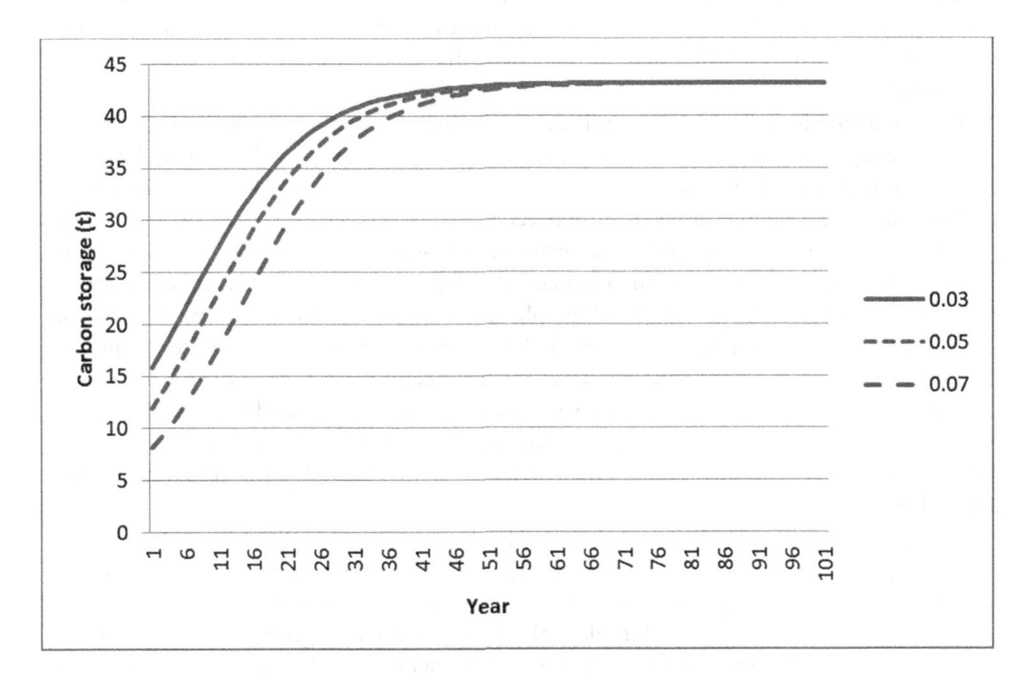

Figure 4.1 Carbon uptake paths with different discount rates

99.7 percent of the potential at a 3 percent discount rate, 60 years at 5 percent, and 65 years at 7 percent. The potential of 27.3 t at a 3 percent discount rate is, however, achieved after 27 years under a 7 percent discount rate.

Our point of reference is small-holder farming in Tanzania, and it may seem inadequate to compare their actions with the return on alternative investments and discount rates, as if their decisions could be compared with those of stock-holders in capital markets. We therefore emphasize that the use of the term discount rate in this context could be understood as a shadow price on the passing of time, which is reflected here by the initial size of the forest. A low initial biomass implies that the propensity to distribute income over time is low, which is interpreted as a high discount rate, and vice versa. To poor people, this may simply reflect an inability to save money. Correspondingly, deforestation may be the result of a shadow price on time that is higher than the return on the biomass at $S = 0$. Of course, it may as well be the result of an insufficient management of a common property resource ("the tragedy of the commons"). For now, we will, however, think of it as a well-managed resource subject to economic decisions, and come back to the alternative interpretations below.

4.3.2 The concentrations of carbon in the atmosphere

The primary aim of REDD+ is to contribute to the reduction of greenhouse gas concentrations in the atmosphere. At first sight, more carbon stored in biomass implies a corresponding reduction in the concentrations of greenhouse gases in the atmosphere. In this case, however, the increasing stock of biomass is followed by a reduction in the supply of bioenergy, primarily to the cities of Tanzania. To see the impact on the concentrations of greenhouse gases in the atmosphere, we need to know how the urban population responds to this change: to what extent do they reduce their demand for energy, and to what extent do they substitute charcoal with alternative fuels? Moreover, to the extent that they substitute, what is the preferred alternative?

To keep it simple, let us begin with a straightforward alternative: the protection of forests reduces the supply of energy to cities, and the urban population replaces all of it with kerosene, which is thought of as the most likely alternative to residential use of charcoal in cities. Such a substitution is not neutral in terms of impacts. It has been thought of as a way to improve public health, by enhancing air quality, in particular indoor air quality, which represents a major health problem to children and women in many developing countries such as Tanzania. After a review of the literature, Lam et al. (2012) emphasize, however, that the results from studies on health impacts of residential use of kerosene in low-income countries are inconsistent, and that the positive impacts of a substitution from charcoal to kerosene may have been exaggerated. Moreover, switching to kerosene also implies investments in new stoves, which is costly. Despite information campaigns and supporting programs to stimulate a replacement of charcoal ovens with kerosene or LPG, it turns out that the barriers to technology adoption are substantial but not well understood.

Although varying, we will assume that a sack of charcoal at 17–18 kilos can be replaced by 11 liters of kerosene, and give the same energy for the purposes considered here, which is mainly cooking. The energy budget for a family that does the switch would thereby increase by approximately 60 percent. It is thus likely that the total energy consumption will go down, but it is emphasized in other studies (IMF, 2008) that responses to changes in energy prices in developing countries are sometimes the opposite, if the availability of the alternative improves, which implicitly can be regarded as an increase in the supply of energy. Here, we assume a full replacement, but nothing more.

Comparisons of emissions of greenhouse gases from different sources depend on fuel qualities, production processes, transport, and user technologies. Hence, the change of emissions by a given switch of energy source is uncertain. We limit ourselves to an illustration of the different outcomes of protecting forests, and to address the dependencies on socioeconomic conditions. We therefore select burning values and emission factors within a reasonable range, without further discussing the variability of these factors. Table 4.1 shows the comparable costs and energy content of charcoal and kerosene. Prices refer to the approximate levels for buying energy in Dar es Salaam in 2011, and the quantities are based on an energy consumption of four bags of charcoal, or approximately 87 kilos, which we expect gives 9.52 PJ of energy. Note that the effect of 1 PJ may differ depending on fuel, which is not taken into account here.

Charcoal is, by far, the least costly energy type between the two. A switch to kerosene implies a nearly tripling of a household's energy budget, plus the possible need for new equipment. On the other hand, energy efficiency is likely to improve by a switch from charcoal to kerosene, although with expenses for the households. A further result may, therefore, be lower total energy consumption. For simplicity, we will disregard these potential effects, but it is important to note that replacement of charcoal by kerosene is likely to imply a loss of welfare to urban households, despite possible advantages.

As for the energy content, emission factors are also uncertain, and should in principle be considered in each and every case. Instead, we base our calculations on a combination of sources, which include Delmas et al. (1995) and Dualeh and Magan (2005). The emission factors used here are displayed in Table 4.2. For charcoal, it is essential to include both the production process and the burning of the charcoal, as the two processes generate approximately the same emissions of CO_2. For kerosene, emissions from the production process are very small, when compared with burning, and this explains most of the difference in greenhouse gas emissions between the two energy types. We have also included emissions of methane (CH_4). These are small, but methane is a much more potent greenhouse gas than CO_2 if calculated in terms of Global Warming Potentials (GWP), which is what UNFCCC applies to compare greenhouse gas emissions. The table shows that total greenhouse gas emissions will be reduced by nearly 50 percent by switching from charcoal to kerosene. This is due, first and foremost, to the lower emissions of CO_2 and CH_4 in the production process.

To compare the contributions to global warming from the two alternatives, we base ourselves on the instantaneous radiative forcing from concentrations, measured in CO_2 equivalents.

Table 4.1 Costs in Dar es Salaam (in Tanzanian Shillings, TZS) and quantities of charcoal and kerosene to obtain 9.52 PJ energy

Type	*Unit*	*TSH/unit (2011)*	*Consumption*		*kg*
			Units	*Costs (TSH)*	
Charcoal	Bag	32,500	4.0	130,000	280.0
Kerosene	Liter	1,370	273.3	374,458	221.4

Table 4.2 Emissions of 9.52 PJ energy from the production and burning of charcoal and kerosene (Kg)

Type	*Production process*		*Burning process*		*Total CO_2 equivalents*
	CO_2	CH_4	CO_2	CH_4	
Charcoal	455.3	11.1	667.8	1.5	1,437
Kerosene	20.9	1.8	652.5	0.1	720

For a precise measure, these will have to be derived from climate models. Here, we confine ourselves to the approximations of Hasselmann et al. (1997), who estimate the concentrations at t of a unit of CO_2 emitted at 0 as

$$C(t) = x_0 \sum_i \alpha_i e^{-\delta_i t} \tag{8}$$

where α_i is the share of emissions with decay rate δ_i, and i =1,.,5, such that $\sum \alpha_i = 1$. The five decay rates vary from 0.32 to 0, with larger rates used in earlier periods. A zero decay rate means that a share (0.178) of emitted CO_2 never disappears from the atmosphere. The pathway of concentrations from a constant emission of CO_2, x_0, over a period (0, t) becomes:

$$C(t) = x_0 \int_0^t \sum_i \alpha_i e^{-\delta_i \tau} d\tau = x_0 \sum_i \frac{\alpha_i e^{\delta_i t} - 1}{\delta_i e^{\delta_i t}} \tag{9}$$

For methane, there is only one, relatively high, rate of decay (0.083), such that i = 1 and $\alpha = 1$. However, methane is a much more potent greenhouse gas than CO_2 when emitted. Still, the contributions from the emissions of methane are small in this context.

Figure 4.2 shows the accumulated emissions of kerosene use and the resulting uptake of carbon from the protected forest, under the assumption that the forest is maintained at a stock of biomass that corresponds to the harvest at a discount rate, or real rate of return, at 3 percent. The net gain in protecting the forest can be read from the difference between the two curves. This gain grows and becomes notable over the first 50 years, but then, it declines as biomass of the forest achieves its maximum, and after 200 years, the climate effect of protecting the forest becomes negative when compared with continued harvest at the initial level.

The evaluation of the climate effect of protecting the forest thus depends on several factors. First, what is the alternative economic utilization of the forest? The example given here is based on a rather moderate discount rate, which means that the initial stock of biomass is relatively high. A higher rate of discount means that there is more to gain in protection. On the other hand, a high discount rate implies that the shadow price of protection is correspondingly high. It might also be that protection could be a way to avoid deforestation driven by the mismanagement of local commons. In that case, the management boils down to a question of "either protection or deforestation," while optimal management is considered impossible. We will come back to this later, but we will still concentrate on a case where sound management is considered possible.

Second, the evaluation of the climate effect of protection is a question of the time perspective considered appropriate. Two hundred years or even more may be regarded as extremely

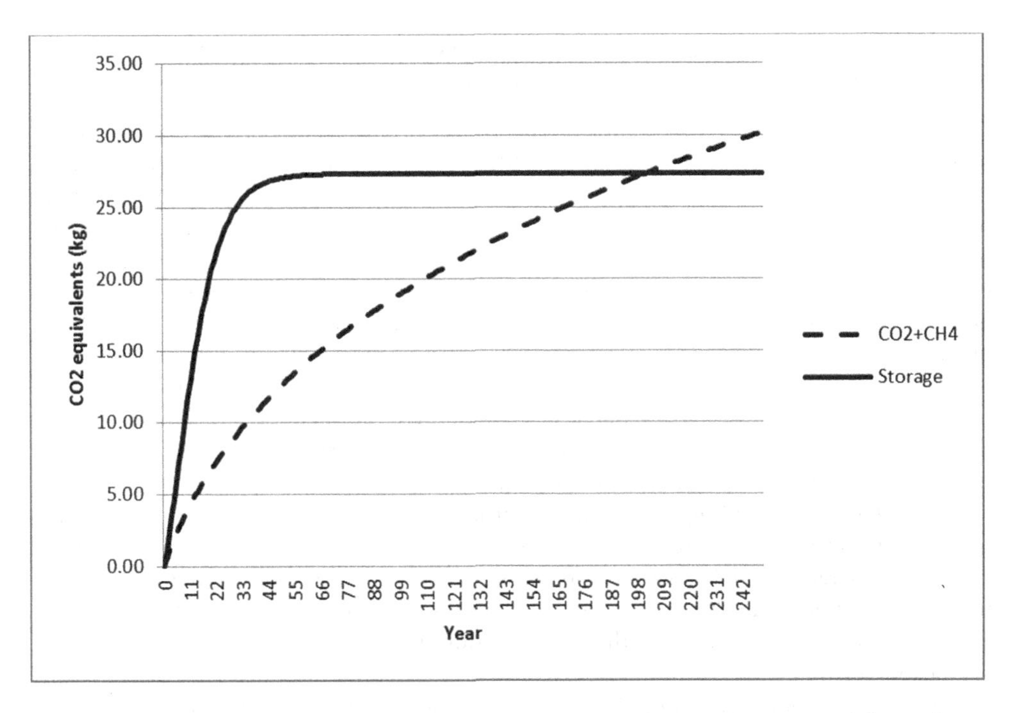

Figure 4.2 Concentrations of constant kerosene emissions from charcoal replacement, and the resulting added storage of carbon in forests

long, and is, in most cases, beyond interest. For climate change, however, it is not, because present emissions have partly irreplaceable impacts far beyond a 200-year horizon. Moreover, the main changes will occur in the far future, and protection may then turn out as a lost opportunity if initiated early. Pushing this argument further, one might consider protection of the forest as an attractive option to deal with shocks: if it is suddenly realized that climate change represents a more severe challenge than previously thought, protection of forests used as an energy source may slow down the speed of warming for a period of time, even if the biofuels are replaced by fossil fuels.

Recall, however, that the time frame considered here, that is 200 years, is due mainly to the emissions in the production process for charcoal, where all the energy produced in the production process is wasted. This is representative for the production that takes place in the case we address, but technologies exist where this energy can be utilized. In that case, the period over which a switch from charcoal to kerosene reduces concentrations would shorten significantly. There are also major uncertainties related to the energy efficiencies and emission coefficients that we have used in our illustration. A major reason for the positive impact on protection over the first and second century is that substitution to charcoal reduces emissions per unit of energy utilization by more than 50 percent. A substitution also has a range of other implications, both to the environment and to the welfare of people. In the next section, we concentrate on one of them, the loss of a source of income to small-holder farming.

4.4 Vulnerability and alternative income opportunities among poor farmers

There is limited statistical information about the contributions of charcoal trading to the budgets of small-holder farming in Tanzania. Most of what we know at the household level is derived from focus interviews of families in local communities. In some areas, charcoal represents a substantial contribution to some families while being quite insignificant in other communities. Fisher et al. (2011) present estimates of the opportunity cost of forest conservation at the district-level in Tanzania and report large differences across different districts. Quantities and prices depend largely on the access to a transport route and the distance to the large cities, where most of the charcoal is consumed. But income from charcoal trade varies also within smaller communities. Some families may earn half of their monetary income from charcoal while other families have no such income. What is clear is that selling charcoal is an important source of income to a notable number of families, and that this opportunity to earn an income is under threat if the forest is protected. In socio-economic terms, the resulting loss of income from these activities gives an estimate of the shadow price of protection. If the REDD+ transfer exceeds this shadow price, protection can be considered beneficial. Although there is insufficient information about the income from charcoal trading in the area, the amounts are likely relatively small, in at least certain areas (Fischer et al., 2011). Hence, the compensation needed to support protection as such is probably small.

An economic compensation need not imply that the vulnerability of people in the area is unaffected, however. First, a transfer implies that the income is redirected. While the income of charcoal trading is received directly by the people who produce it, a neutral transfer needs to be allocated correspondingly to those who suffer the losses. Transfer mechanisms are seldom perfect, and may moreover be considered a way to deliberately redistribute income. Even though this may be beneficial in socioeconomic terms, we can no longer talk about a Pareto improvement. Second, the loss of an income opportunity may affect the vulnerability of poor farmers by itself, as it represents a way for them to spread the risk connected to the outcome on the farm. Provision of easily accessible alternative income opportunities is, after all, a way to enhance the adaptive capacity among poor farmers.

To study the impacts to small-holder farmers who face a loss of income from trading charcoal, we shall apply a very simple extension of a standard general equilibrium model. Assume that the agricultural sector can be represented by an aggregate of farmers who possess farms of different sizes, which we denote by r. The aggregated "agricultural output," Y, is the sum of output from a total of N individual farms:

$$Y = \sum_{i=1}^{N} y_i = \sum_{i=1}^{N} f(n_i^f, r_i) \tag{10}$$

where n_i^f is the labor time spent on a farm of size r_i. We make no distinction between technologies at farms of different sizes, but assume that the production in the agricultural sector is independent on scale, which is the normal assumption in macroeconomic models.

Size matters, however, to the consumption pattern, as a part of the output is consumed by the households at the farm, while some of the output is sold to the market at a price $(p_x - t_x)$, where p_x is purchaser price of food in the market, and t_x is the difference between the price that farmers obtain when selling the food and what they receive as payment for the food. t_x can be associated with a transaction cost. In addition, the farmers may earn an income from other sources, either casual work or production of charcoal, which we will pay attention to here. Then, the budget constraint to a single farm is:

$$(p_x - t_x)(f(n_i r_i) - x_i^f) + w(\bar{n} - n_i^f) = (p_x - t_x)x_i^f p_x x_i^m + qz_i \tag{11}$$

where $\bar{n}$ is the total available working time on farms, and w is the average, or expected, rate of earnings in alternative work. x_i^f and x_i^m are food consumed from the output of the farm and food bought in market, respectively, while q and z are the price and quantity of other goods consumed at the farm, respectively.

The welfare of farmers is expressed as a function of the three consumption goods, $V = V(x_i^f, x_i^m, z_i)$. To maximize welfare, the farmer needs to balance the composite of consumption goods and decide where to spend time to get income. As we shall address small farms, we also impose a lower limit to the consumption of food, which we can call a nutrition constraint:

$$\bar{x} \leq x_i^f + x_i^m \tag{12}$$

The model differs from a standard equilibrium model by its linkage between what farmers produce and what they consume. However, the choice of consumption composite with a given output and monetary income is subject only to the relative prices, independent from how the goods are provided: from own farm or from the market. So far, there is no restriction in the model preventing farmers from demanding more from their own farm than they actually produce. To avoid such a situation, we therefore require that

$$y_i \geq x_i^f \tag{13}$$

Welfare maximization under the constraints (12) – (14) gives the following first-order conditions for the consumption composite:

$$\frac{V'_x f - \lambda^N - \lambda^O}{p_x - t_x} = \frac{V'_x m - \lambda^O}{p_x - t_x} = \frac{V'_Z}{q} \tag{14}$$

where λ^N and λ^O are the shadow prices imposed by the nutrition constraint (13) and the output constraint (14), respectively.

For the division of work, the first order condition is

$$\left(p_x - t_x - \frac{\lambda^O}{\lambda^B}\right) f'_n = w \tag{15}$$

where λ^B is the shadow price of the budget constraint, or marginal utility of money. In standard models, $\lambda^B = 1$ if all consumption is measured in a given currency. In the present model, the consumption of food from the households' own farm is, from the outset, a pure physical measure. To keep $\lambda^B = 1$ valid, we consider a separable welfare function that can be optimized in two stages. First, the composite of x_i^f and x_i^m is determined by minimizing the cost of consuming a given, total amount of food $x_i = x_i^f + x_i^m$. In the second step, market consumption is optimized given the amount of money available. The welfare function can then be written as

$$V(x_i^f, x_i^m, z_i) = v(x_i^f | x_i) + u(x_i^m, z_i) \tag{16}$$

Define the variables $\mu = \lambda^O p_z / u'_{x_i} m$ and $\eta = \lambda^N p_z / u'_{x_i} m$, which represent the direct amendments to the price of food when subject to the output constraint and the nutrition constraint, respectively. By use of (17), the conditions (15) give rise to demand functions for the two categories of food and for other market goods, while (16) and the production function (11) determine the demand for labor on the farm, and thereby available labor for work elsewhere, for people at a farm of a given size r_i. The demand functions can now be derived on the basis of standard cost-minimization and welfare maximization procedures (see e.g.,Varian, 1992).

As the production and welfare functions are assumed independent on scale, farm-specific variables follow as functions of prices and parameters for the production technology and preferences when the size of the farm, r_i, is given. The aggregates can then be derived from information about the distribution of farms. With this background, we can explore how a drop in the expected rate of earnings off farm, w, affects the welfare of people living at a farm of a given size. Below, we stick to an illustrative example, based on calibrated production and welfare functions taken from a case study in a district of Malawi (Chirwa and Matita, 2012). Tanzania and Malawi are neighboring countries, and there are similarities between the agricultural sectors, but also substantial differences. The point here is, thus, not to provide a quantified assessment of the impacts on agriculture of a drop in the rate of earnings, but rather to show how such a drop may affect the level of welfare in various households, and how the aggregates may turn out.

We base the illustration on calibrated CES functions, both for the production functions (11) and for the welfare functions (17). Figure 4.3 shows consumption composites and the yields at farms with increasing size (0.6 hectares to 4 hectares) under a base case with expected rate of earnings = 1 (black curves) and a 20 percent drop in the expected rates (grey curves). A reduction in the expected earnings is thus associated with vanishing opportunities for production and trading of charcoal.

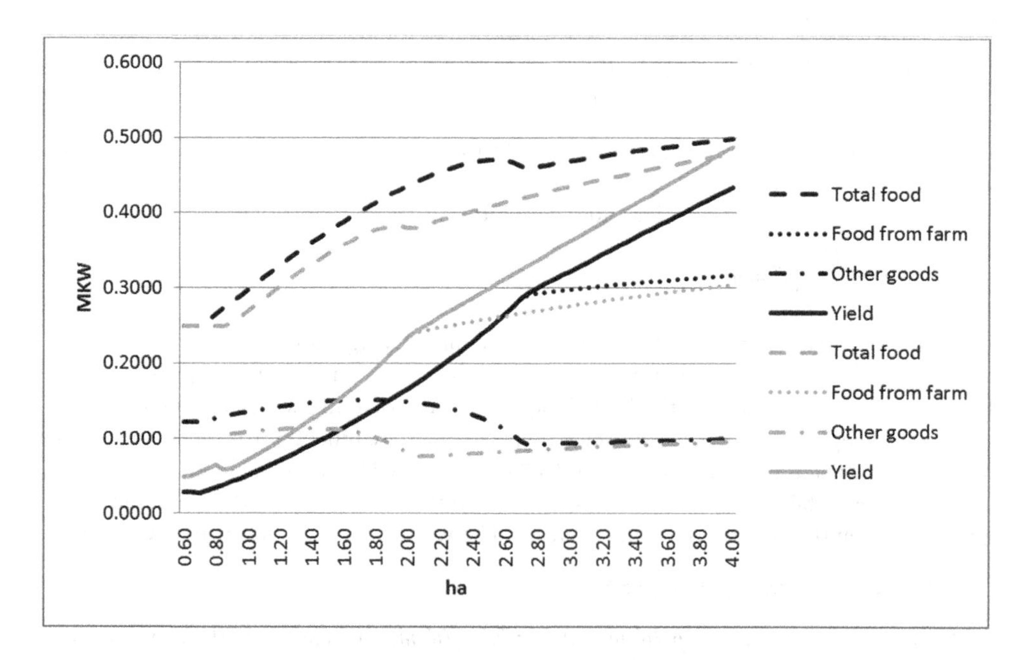

Figure 4.3 Consumption composites and yields at farms with increasing size under expected rate of earnings = 1 (black curves) and expected rate of earnings = 0.8 (grey curves)

As expected, consumption of both food and other goods declines when the earnings from outside the farm decline, but the effect is less the larger the farm , as larger farms depend less on earnings outside the farm. For some of the farms, notably those between 1.8 and 2.5 ha, the shift in consumption pattern is relatively large. This has to do with which farms encounter the output constraint, where the yield limits what farmers can consume from their own output. When external income is reduced, consumption is reduced in general, meaning that the output constraint is met at smaller farms, in this case from 2.65 ha to 1.90 ha. It is seen from Figure 4.3 that this enforces higher consumption of other goods at smaller farms.

Vanishing opportunities to earn an income from charcoal moreover implies that more farmers encounter the nutrition constraint. While the nutrition constraint was met at farms of size 0.6 ha, one needs a farm of 0.8 ha after the drop of the rate of earnings from external sources to avoid the nutrition constraint. In other words, if we look at the consumption opportunities created at farms, the drop of income opportunities from charcoal sales is therefore unequivocally negative: consumption declines in general, and more farmers will enter the category of poor people whose consumption is constrained by the need to survive.

If we turn to measures that oversea initiatives such as REDD+ are measured up against, the picture might change, however. Protection of forests is viewed as a means to enhance carbon sinks, but it should not put limits to the economic situation for regions or communities where the initiatives are put in place. Economic opportunities are usually measured in terms of value added (GDP), which is usually constrained to the part of the output which is brought to the market. In our model, the protection of the forest is clearly positive in this respect. The output from all farms, regardless of size, increases, and most notably the part brought to the market, which is indicated by the area between the total yield at the farm and the consumption from own farm. How much this increase means in terms of aggregated output from the agricultural sector depends, of course, on the distribution of farms. To give a sense of the impact, the increase of output is between 70 and 100 percent at the smallest farms, where the nutrition constraint is binding. Farms subject to the output constraint increase output between 20 and 40 percent, and most for the smallest farms.

For farms above the output constraint, the total yield increases by 12.3 percent. These are the farms that bring goods to the market. If the farms of sizes between 2 and 4 ha were distributed equally, the supply of food from these farms would increase by more than 130 percent. In other words, if regarded from the perspective of national governance, the economic impact of the suggested ban on charcoal harvesting may easily, but mistakenly, be considered positive.

4.5 Discussion

REDD+ has been thought of a possible means to achieve synergies between mitigation and adaptation. Reduced deforestation and forest degradation will reduce the decline of carbon stored in the biosphere and thereby mitigate anthropogenic contributions to climate change. An increase in the biomass of forests will reduce concentrations of greenhouse gases in the atmosphere. Forests moreover provide services to local communities, and reduce impacts of climate change, for example by limiting the impacts of extreme precipitation and flooding. Hence, mitigating forest degradation or increasing forested land can also be a way to adapt to climate change.

This chapter addresses a few questions that need to be answered before one can conclude whether an initiative within the REDD+ scheme contributes to mitigating climate change, and at the same time enhances the adaptive capacity of people affected by it. With reference to a suggested protection of forests in a district in Tanzania as a part of a REDD+ scheme, we explore the potential consequences for emissions of greenhouse gases to the atmosphere, and for the livelihood within small-holder farming. Our aim is to draw some general conclusions about the possibility of achieving synergies between mitigation and adaptation by means of REDD+. The examples are therefore meant as illustrations, and one should not put much confidence in the data used here. Below, we therefore discuss the results briefly on a rather general basis.

First, it was stated that social and economic drivers in the management of forests is essential in defining the potential of a REDD+ scheme: In this case, utilization of the forest for production of charcoal was thought to be the main objective of utilizing the forest prior to the protection. It was shown that alternative utilizations might imply that forest degradation is optimal, if sufficiently beneficial. However, depletion of a forest is also possible even though the social benefit is negligible. As long as the individual benefits are positive, the forest may degrade and become extinct if there is free access. On the other hand, full protection is seldom optimal, unless the social benefit of harvesting is negligible. Consequently, the challenge also in the context of REDD+ is to decide on the optimal management strategy for a forest, and to be able to carry the strategy through. There is reasonable confidence in the conclusion that protection does not belong to an optimal strategy.

In the case considered here, the alternative utilization of the forest is production of charcoal, which local farmers sell to households in cities. Protection of the forest therefore reduces the supply of energy in cities, and limits income opportunities for farmers in rural areas. A reduction in the supply of bioenergy stimulates use of alternatives. Here, we considered a replacement of the charcoal by kerosene. This assumption is a simplification, indeed, and raises several questions that were left aside. Such a switch implies a substantial increase in energy costs to urban dwellers, with resulting welfare losses. But even under an assumption of an economically welfare-compensating substitution, there are important aspects that were not dealt with, such as possible health effects. The general lesson is just that the implications of changing the alternative utilization of the forests need to be explored thoroughly before one can conclude about the impacts of REDD+.

Under the assumptions made in this study, the switch to kerosene contributed an intermediate reduction in the concentrations of greenhouse gases in the atmosphere. This is explained partly by the far lower emissions of producing kerosene than local production of charcoal, and partly by the quick decline in a part of the atmospheric concentrations from CO_2 emissions. Hence, a switch from charcoal to kerosene could contribute a reduction of greenhouse gas concentrations over nearly 200 years. From then on, a sustainable extraction of forests to produce charcoal is better. In other words, an evaluation of the REDD+ initiative depends on the time perspective. Two hundred years may seem extremely long. On the other hand, the challenges in keeping control of the climate system are expected to be much higher by then. In such a context, protection of forests may be an attractive means to respond quickly to unexpected negative news about climate change impacts. In that case, protection ought to be set aside as a means for use in case of negative surprises.

The other consequence of limiting the supply of charcoal is the reduction of off-site income to small-holder farmers. The effect is, as expected, negative to all farmers who are involved in the trading of charcoal, but the consequences increase as the farms become smaller and the farmers are poorer. Limitation of charcoal production implies that consumption in the farming households becomes more dependent on the farm yield. The total consumption is further constrained, which implies that more farming households can be categorized as poor. According to our numerical illustrations, the negative impacts increase significantly with smaller farms. It was also shown how the limitation of income from charcoal trading increases the output from the farm and most notably for food brought to the market. The significant increase in output combined with a lower level of welfare among farmers again point to the need to make careful evaluations of REDD+ initiatives.

The policy initially studied here, which is to protect forests in order to enhance the carbon stock in the biosphere and at the same time take advantage of ecosystem services from forests as a means to adapt to climate change therefore turns out to be questionable. First, the mitigation aspect is preliminary, as the charcoal produced from the forest will be replaced by alternative fuels, most likely kerosene. In the long run, estimated to 200 years here, concentrations from kerosene will exceed those of charcoal. This time span depends, however, on a range of factors, which include the biophysical properties of the forest and the value of alternative utilizations of forest services. Second, the impact of forest protection on the adaptive capacity among small-holder farmers is probably negative. We emphasize the term probably, because the potential benefits of higher ecosystem services for larger forests have not been studied here. However, we show that the welfare of poor people who take advantage of engaging in charcoal trade is unequivocally negative. We moreover argue that this negative effect is at danger of being ignored if REDD+ are evaluated exclusively from the perspective of impacts on sectoral GDP.

The relatively negative evaluation of the particular case focused on here does not imply, however, that there are no potentials for synergies between mitigation and adaptation if REDD+ were applied in Tanzania. In the end, the outcome depends crucially on the management of the forest. What we have shown is that protection is not a good management strategy. The challenge is rather to sustain a constant mass of forest, while at the same time ensure that the local farmers are given the opportunity to continue to add to their budgets from trading charcoal. To achieve this, a transfer of knowledge, monitoring experiences and management practices might be more important than transfers of money.

Acknowledgements

We owe great thanks to PhD student Meley Mekonen Araya and her supervisor Ole Hofstad at the University of Life Sciences in Norway for providing information about charcoal production and use in Tanzania, and for kind support and patience in answering silly questions. We also appreciate the financial support from the Norwegian Embassy in Dar es Salaam and the Norwegian Research Council.

References

Angelsen A., Wang Gierløff C., Mendoza Beltrán A., and den Elzen M. (2012). REDD credits in a global carbon market: Options and impacts. A report for NOAK – The Nordic Working Group on Global Climate Negotiations, the Nordic Council of Ministers.

Barbier, E.B. (2012). "Can global payments for ecosystem services work?" *World Economics* 13 (1): 157–172.

Chirwa, E.W., and Matita, M. (2012): "From subsistence to smallholder commercial farming in Malawi: A case of NASFAM commercialization initiatives," *Future Agricultures Working Paper* 037. www.future-agricultures.org.

Delmas, R., Lacaux, J.P., and Brocard, D. (1995). "Determination of biomass burning emission factors: methods and results," *Environmental Monitoring and Assessment* 38: 181–204.
Dualeh and Magan (2005). "Alternative energies and reduction of dependence on charcoal in Somaliland," Resource-based Conflicts Conference, Hargeisa, Somaliland, 2005. Candlelight for Health, Education and Environment.
Eliasch, J. (2008). Climate change: Financing global forests. The Eliasch Review. Office of Climate Change, London.
FCPF (2012). Forest Carbon Partnership Facility. http://www.forestcarbonpartnership.org/fcp/node/12.
Fisher, B., Lewis, S.L., Burgess, N.D., Malimbwi, R.I., Munishi, P. K, Swetnam, R.D., Turner R.K., Willcock S., and Balmford, A. (2011). "Implementation and opportunity costs of reducing deforestation and forest degradation in Tanzania." *Nature Climate Change*: 161–164.
Hasselmann, K., Hasselmann, S., Giering, R., Ocana, V., and van Storch, H. (1997). "Sensitivity study of optimal CO_2 emission paths using a structural integrated assessment model (SIAM)," *Climatic Change* 37: 345–386.
Intergovernmental Panel of Climate Change (2007). IPCC Fourth Assessment Report: Climate Change 2007. Cambridge: Cambridge University Press.
International Monetary Fund (2008). World Economic Outlook 2008. Washington DC.
Kindermann, G., Obersteiner, M., Sohngen, B., Sathaye, J., Andrasko, K., Rametsteiner, E., Schlamadinger, B., Wunder, S., and Beach, R. (2008). "Global cost estimates of reducing carbon emissions through avoided deforestation." *Proceedings of the National Academy of Sciences* 105 (30): 10302–10307.
Lam, N.L., Smith, K.R., Gauthier, A., and Bates, M.N. (2012). "Kerosene: A review of household uses and their hazards in low- and middle income countries," *Journal of Toxicology and Environmental Health, Part B* 15: 396–432.
Millennium Ecosystem Assessment (2005). Washington, DC: Island Press.
Norwegian Agency for Development Cooperation (2011). Real-Time Evaluation of Norway's International Climate and Forest Initiative Contributions to a Global REDD+ Regime 2007–2010. Evaluation Report, December, 2010.
REDD+ Readiness Factsheet: Tanzania (2012). Available at http://www.forestcarbonpartnership.org/fcp/sites/forestcarbonpartnership.org/files/Documents/PDF/June2012/REDD%20Tanzania%20Fact%20Sheet_June%202012_0.pdf.
Robinson E.J. Z, and Lokina R.B. (2012). "Efficiency, enforcement and revenue tradeoffs in participatory forest management: An example from Tanzania." *Environment and Development Economics* 17: 1–20.
Romijn, E., Herold, M., Kooistra, L., Murdiyarso, D., and Verchot, L. (2012). "Assessing capacities of non-Annex I countries for national forest monitoring in the context of REDD+." *Environmental Science and Policy* 19: 22–48.
United Nations (2011). UN-REDD Programme 2011 "Year in Review" Report.
van der Werf, G. R, Morton, D.C., DeFries, R, S, Olivier, J., Kasibhatla, P.S., Jackson, R.B., Collatz, G.J., and Randerson, J.T. (2009). "CO_2 emissions from forest loss." *Nature Geoscience* – Commentary (2009): 737–738.
Varian H. (1992). *Microeconomic analysis*, 3rd ed. New York: W.W. Norton & Company Inc.

5

INCORPORATING CLIMATE CHANGE INTO ADAPTATION PROGRAMMES AND PROJECT APPRAISAL

Strategies for uncertainty

Anil Markandya[1]

[1]BASQUE CENTRE FOR CLIMATE CHANGE (BC3), BILBAO (SPAIN)
IKERBASQUE, BASQUE FOUNDATION FOR SCIENCE, BILBAO

5.1 Introduction

This chapter reviews the role of uncertainty in the design of programmes and planning processes where climate change is an important factor. It provides a definition of uncertainty (Section 5.2), outlines the ways in which this key factor influences our understanding of the nature of climate impacts (Section 5.3), and goes on to consider its implications for the appraisal of projects and programmes, especially in a development context (Section 5.4). Section 5.5 discusses different methods for ranking options when uncertainty due to climate change is a key feature. Finally, Section 5.6 provides recommendations for best practice. The aim throughout is to provide practical guidance for mainstreaming climate into the broader development strategic objectives.

As background, we note that the issue of uncertainty has been discussed in a number of guidance documents, especially those by the UK government, including those that are not specifically related to climate change. Indeed, this factor has always been viewed as an important consideration in project appraisal, and methods for addressing it have been developed at different levels of sophistication. The H.M. Treasury Green Book (H.M. Treasury, 2007) the Orange Book (managing risk, H.M. Treasury, 2004), as well as the UK's Department for International Development's (DFID's own guidance on economic appraisal) all offer advice on how to handle uncertainty in project appraisal. In addition, there is supplementary guidance from DECC and DEFRA (H.M. Treasury, 2009) and UKCIP (2003) on how to address uncertainty in a context that includes climate change. There are also several international guidelines covering this issue (see, for example, Hallegatte, 2009; Dessai and Wilby, 2010). Given this extensive amount of information and guidance, the aims and content of this chapter need some clarification. We see the purpose as being to: (a) review the varied proposals from the different documents dealing with uncertainty in a project appraisal context and (b) provide a simple guide to the literature on uncertainty and adaptation to climate change.

5.1.1 Defining uncertainty

We adopt the definition of Klauer and Brown (2003) that a person is uncertain if s/he lacks confidence about the specific outcomes of an event. For most technical and natural science elements, uncertainty is primarily an objective matter, but this definition acknowledges that uncertainty includes subjective aspects, because the central focus is the degree of confidence that a person has about possible outcomes and/or probabilities of these outcomes.

Much in the spirit of this definition, the IPCC AR4 Synthesis Report (IPCC, 2007a – Annex II Glossary) defines uncertainty as:

> an expression of the degree to which a value (e.g., the future state of the *climate system*) is unknown. Uncertainty can result from lack of information or from disagreement about what is known or even knowable. It may have many types of sources, from quantifiable errors in the data to ambiguously defined concepts or terminology, or uncertain *projections* of human behaviour. Uncertainty can therefore be represented by quantitative measures, for example, a range of values calculated by various models, or by qualitative statements, for example, reflecting the judgement of a team of experts.

One can categorise all uncertainties covered under the definitions given above according to three dimensions: *nature, level and source* (Refsgaard et al., 2012a).

The *nature of uncertainty* can be categorised into *epistemic uncertainty, ontological uncertainty* and *ambiguity*. *Epistemic uncertainty* is the uncertainty due to imperfect knowledge, e.g., lack of observations and understanding of causalities, and is reducible by gaining more knowledge, e.g., through research, data collection and modelling. *Ontological uncertainty* (also termed aleatory or stochastic uncertainty) is the uncertainty due to inherent variability. It can be quantified, but is by nature stochastic and irreducible. *Ambiguity* results from the presence of multiple ways of understanding or interpreting a system. It can originate from differences in professional backgrounds, scientific disciplines, value systems and interests. In the context of climate change, Ranger et al. (2010) see ambiguity as a case of "incomplete probabilities, or multiple inconsistent probabilities".

The *level of uncertainty* characterises how well this phenomenon can be described within the range from determinism to total ignorance (Figure 5.1), where determinism is the ideal, non-achievable situation where everything is known exactly and with absolute certainty. In this range, *statistical uncertainty* can be described in statistical terms, e.g., measurement uncertainty due to sampling error, inaccuracy or imprecision. In contrast, *scenario uncertainty* in general cannot be described statistically. If probabilities are associated to scenarios, e.g., if the IPCC SRES scenarios are considered equally likely, they can be treated statistically; however, this is most often not the case. Scenarios are common in policy analysis to describe how a system may develop in the future as a function of known controls like changes in management, technology and price structure. They are used when the possible outcomes are known but not all probabilities of such outcomes are present (Brown, 2004). *Qualitative uncertainty* occurs when uncertainty cannot

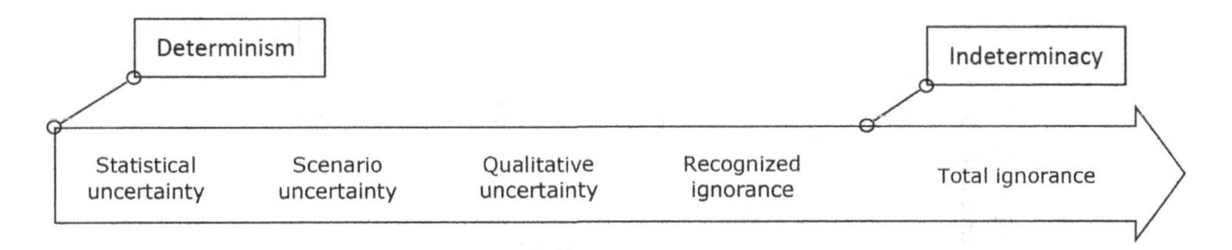

Figure 5.1 Levels of uncertainty according to van der Keur et al. (2008)

be characterised statistically, and where not even the possible outcomes are necessarily known (Brown, 2004). *Recognised ignorance* occurs when there is an awareness of lack of knowledge on a certain issue, but where it is not possible to categorise the uncertainty further. Finally, *total ignorance* denotes a state of complete lack of awareness about imperfect knowledge.

Another distinction that is made in this literature is between risk and uncertainty. There are different interpretations on these two terms, but a common one for economists is that attributed to Knight (1921), where risk is defined as a situation where the extent of lack of knowledge about the outcomes is measurable and uncertainty is one where it is not measureable. This is also interpreted as implying that a situation characterised by risk has probabilities associated with different outcomes. Not all scholars working in the field, however, take this interpretation, and sometimes risk is defined as the product of a probability of an outcome times the value attached to that outcome, or the expected value of that outcome.[1] In this chapter we use the terms in the Knightian sense.

The *sources of uncertainty* can be divided into uncertainties on *input data*; *model uncertainty*; *context uncertainty*, i.e., the boundaries of the systems to be modelled, such as future climate and regulatory conditions if these aspects are not explicitly included in the modelling study; and *uncertainty due to multiple knowledge frames*, reflecting that different persons may have different perceptions of what the main problems are, what is at stake, which goals should be achieved, etc. The simultaneous presence of multiple frames of reference to understand a certain phenomenon may cause ambiguity.

5.2 Uncertainty in a climate context

The uncertainty cascade in the climate modelling community includes uncertainties from emission scenarios, global models and regional models. If we use the term in a broader climate change adaptation context (Figure 5.2), we should also include uncertainties due to statistical downscaling, systems impacts and socioeconomic impacts.

The main uncertainties in the different steps are characterised in Table 5.1 with respect to the sources and nature of uncertainty following the typology outlined above (Refsgaard et al., 2012b). It is important to recognise that each step adds uncertainty to the step before, so that by the time we come to quantifying the impacts of the adaptation measures, we have to allow for the cumulative uncertainties through the different steps. Below we describe the nature of these uncertainties at each step.

The future greenhouse gas (GHG) emissions and socioeconomic scenarios cannot be known with certainty because they depend on future human decisions. They are typically characterised by scenario uncertainty based on IPCC scenarios (IPCC, 2007b) or variants thereof.

Table 5.1 Characterisation of key sources of uncertainty in the uncertainty cascade and their nature in relation to climate change adaptation. (X), X, XX is a general guide on the relative importance level of the sources, although it must be emphasised that the importance of the individual sources of uncertainty is context-specific.

Steps in climate change adaptation analyses (chain in uncertainty cascade, Figure 2)		*Sources of uncertainty*						*Nature of uncertainty*		
		Input data	*Model*			*Context*	*Multiple knowledge frames*	*Ambiguity*	*Epistemic uncertainty (reducible)*	*Ontological uncertainty (irreducible)*
			Parameter values	*Model technical aspects*	*Model structure*					
Greenhouse gas emissions						X	XX	XX	X	
Socio-economic scenarios		X			X	X	XX	XX	X	
Future climate (Climate models)	GCMs			X	XX				XX	
	RCMs			X	XX				XX	
	Initial conditions/ natural variability	X								XX
Downscaling/statistical correction			XX		X				X	X
Water system impacts (Hydro-ecological models)		(X)	XX	(X)	XX	X	(X)	(X)	XX	(X)
Socio-economic impacts (Socio-economic tools)		X			X	X	XX	XX	X	
Adaptation measures		X	XX	(X)	XX	X	XX	XX	XX	X

Source: Adapted from Refsgaard et al., 2012b

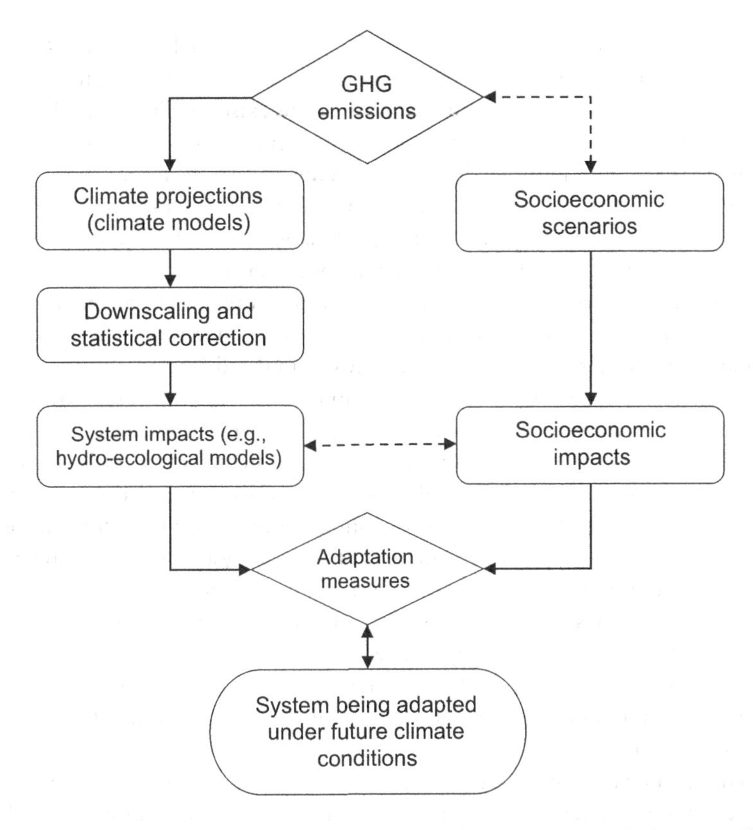

Figure 5.2 Structural elements in the assessment of climate change impacts and adaptation illustrating the uncertainty cascade
Source: Adapted from Refsgaard et al. (2012a,b)

The uncertainties related to global (GCM) and regional climate models (RCM) are mainly associated with lack of knowledge on physical processes at the spatial scale at which they are represented in the models. These uncertainties can be assessed by ensemble modelling where multiple climate models with different process equations are used for making probabilistic projections. In this context, it is important to include a wide range of models (as many as 10 if possible) and to cover wet and dry models in the range.

An important source of uncertainty is the lack of knowledge on the initial conditions, primarily of ocean characteristics, such as temperature and salinity distribution. Different plausible initial conditions may result in significantly different climate projection pathways reflecting the fact that climate systems have strong elements of natural variation, which for some type of climate simulations may dominate over the uncertainty due to choice of GHG emission scenarios and climate model structure.

It should be noted, however, that the ensemble-based probabilistic approach has some limitations that may result in underestimated (or overestimated) uncertainties. One problem is related to the lack of understanding of systematic model errors that could lead to misinterpretation of the probabilistic presentation of uncertainty (e.g., Christensen et al., 2008; Christensen and

Boberg, 2012). Another, more fundamental problem relates to the lack of knowledge on climate processes, where we are aware (recognised ignorance) that several of the natural feedbacks of the earth carbon balance are not included in most of the existing climate model projections. Although the development and application of earth system models will improve on this, the degree to which uncertainty in projections will be reduced remains to be seen.

In any case, once the data have been collected, it is important to provide the full set to the decision maker. This means that the whole range of values should be made available, not just the average. Furthermore it is useful to indicate what is better determined and what is less well determined (e.g., do we have a narrower range for temperature or for precipitation?).

For prediction of water systems (an important area for adaptation), climate variables are typically required at smaller spatial scales than those simulated by the climate models. Furthermore, it is generally recognised that climate models often do not represent the statistical properties of observed climate variables sufficiently precisely, not least precipitation, to enable modellers to use the outputs from these models directly in impact models. Therefore, data from climate models are typically downscaled and statistically corrected using some kind of statistical downscaling method. Different downscaling methods may give different results, and different methods should be applied depending on the water system impact being considered. For instance, in the assessment of critical water infrastructure where the properties of extreme precipitation are important, statistical downscaling methods that are especially tailored towards downscaling of extremes are more appropriate; whereas for general water balance studies, simpler bias-correction methods may be sufficient. In some cases, the uncertainty on downscaling methods may be just as important as the uncertainty on the climate models (e.g., Olesen et al., 2007; Sunyer et al., 2011).

Climate change impacts on human beings and on society more widely can be assessed with various economic models and analytical tools, including macroeconomic models, sectoral models, and cost-benefit analysis. These approaches rely on specific theoretical frameworks and assumptions reflecting concepts of welfare, aggregation over individuals' utility, preferences of present and future generations, and willingness to accept environmental impacts. The underlying theoretical issues have a character that reflects differences in framing of the issues and theoretical perceptions, which we refer to as ambiguity. Such different knowledge frames can have major implications on conclusions and results. In addition, there are also uncertainties related to specific assumptions that go into cost assessment, which often can be reduced by gaining more knowledge.

As a general rule, information that is passed on to decision makers from the different stages should include some guidance on trends in the key variables (predicted values increasing over time) and their qualitative implications. This will complement the more formal datasets and guide people in how much confidence they can have in the model outputs.

Finally, identification and assessment of possible climate change adaptation measures involves the use of basically the same tools as for assessing climate change effects, i.e., typically hydro-ecological models and socioeconomic tools, and hence all the uncertainties involved above arise again. As adaptation decisions often involve considerable uncertainties, aspects on stakeholders' and societal risk willingness becomes important, not least for situations where attitudes to risk may change. Furthermore, different stakeholders may often have different perceptions of the consequences of alternative adaptation measures, which also reflect their specific interests in relation to the area that is affected. This may lead to more ambiguity than in other elements of the uncertainty chain.

5.3 Climate uncertainty and adaptation decision-making in a development context

We start in this section by noting how uncertainty is generally addressed in project appraisal and then go on to consider how these tools for dealing with the problem need to be modified to handle climate change.

The H.M. Treasury Green Book identifies two issues in handling uncertainty. The first is to allow for risk and optimism to provide a Base Case, and the second is to consider the impacts of changes in key variables and of different future scenarios on the Base Case.

In a climate context, we can think of the Base Case as consisting of the impacts that would occur without the project or other form of intervention. The value attached to these impacts is generally uncertain and may be represented by anything ranging from a frequency distribution, to an array of possible outcomes, to a state of total ignorance (see Figure 5.1). If the range of impacts can be represented by a probability distribution, we can calculate an expected value. Moreover, we can also calculate a 'risk value' associated with the fact that there is indeed a range of possible outcomes. Methods for doing this are well established (see, for example, Ranger et al., 2010, Annex A). The problem, as noted in the previous section, is that such probability distributions are rarely available, and the calculation of the risk value or premium is frequently not possible.

When probabilities are not available, the Green Book proposes the use of sensitivity analysis, the consideration of scenarios and, in some cases, the use of Monte Carlo analysis. These tools are very much at the heart of project appraisal nowadays. A project document that did not report the sensitivity of the outcome indicator (e.g., the internal rate of return or net present value) to key parameters such as the costs or other uncertainties would be considered a poor piece of work. Likewise the analysis would be expected to identify the robustness of the chosen option to the dominant uncertainties. Indeed, software tools such as 'Tornado' or 'Crystal Ball' are commercially available to carry out the sensitivity analysis and supporting Monte Carlo exercises.

The other adjustment that the Green Book recommends for the Base Case is to allow for the fact that there is a tendency to underestimate the costs of the different intervention options. Allowance for optimism in estimating these costs is possible in the case of adaptation projects and one can use the standard methods of making such adjustments as outlined in the Green Book.

In this respect, climate change is not something qualitatively different, although one sometimes gets the impression it is. Where climate change is a major source of the uncertainties associated with the project or programme, however, the literature rightly draws attention to the following four key features:

a) The significant uncertainties about the impacts. Moreover, most of this is not statistical; some falls into the category of qualitative uncertainty or even of recognised ignorance. (See Figure 5.1.)
b) The changing knowledge base – we can expect to modify our views about the impacts of climate change over time.
c) The relatively long time periods over which the impacts will occur and over which the adaptation measures have to be evaluated.
d) The fact that there are multiple actors adapting to climate change, mostly in the private sector, and the actions of the public agencies have to take these account and indeed exploit all possible synergies with the actions of these actors.

These four aspects can make the evaluation of projects where climate change is a key feature, or indeed where it is the sole purpose of the intervention, somewhat different. In particular, the use of standard cost-benefit analysis indicators as measures to rank options may not be appropriate or indeed possible. In the next section, we look at whether this is in fact true and how the assessment of such projects needs to be adapted in a climate change context.

5.4 Ranking options when climate change is a feature

We take the view that it is useful to divide the kinds of projects of interest into those where climate change is only one of the issues that need to be addressed and those where climate change is the central feature and possibly the reason for undertaking the project. Examples of the first kind would be:

- Design of roads and other transport facilities
- Investments in the energy sector, such as construction of hydropower plants, thermal power stations, etc.
- Construction of buildings and similar infrastructure
- Projects to improve supply of potable water and sanitation

Examples of projects of the second kind would include:

- Design of defences to protect against sea level rise
- Design of stronger defences against alluvial flooding
- Measures to address expected changes in water availability
- Measures to increase resilience of agricultural activities to expected climate changes
- Measures to address increased risks from vector-borne diseases as a result of climate change
- Measures to address increased risks of heat-related health hazards

Clearly, the distinction between the two categories is not always apparent and there are projects, such as the selection and design of areas for urban settlement, where climate change is an important factor, but not the main one or the only one. Nevertheless, we believe the distinction is useful as the approaches taken in the two cases to addressing uncertainty could be quite different.

The distinction also cuts partly along the lines of decision-making under risk and decision-making under uncertainty. In the first category of projects, we believe that the uncertainty can be treated using risk management techniques because probabilities of the different outcomes under a given adaptation action can be defined. In the second category of projects, such probabilities cannot generally be obtained, although it may be possible to construct then in special cases.

5.4.1 Projects where climate change can be regarded as an overlay

A considerable amount of work has been done to provide guidance on how climate issues should be addressed in infrastructure projects of the kinds described above. These are projects that have long lifetimes, over which changes in climate will make an impact on them and therefore design changes may need to be incorporated now to account for them.[2] The World

Bank, for example, produced guidelines covering such projects some time ago and has updated them since (World Bank, 1997; World Bank, 2010a). The following is a simple presentation of the way in which the climate aspects can be overlaid into the standard project cycle (Figure 5.3).

The left-hand side of Figure 5.3 shows the broad phases in the lifecycle of a project, from conceptualisation through to divestment or decommissioning. The proposed climate risk management process for development projects is shown in the right-hand side of the figure. This process is based on standard approaches to characterising and managing risk (see for example, DETR, 2000 or CSA, 1997).

The proposed climate risk management process is linked to the project lifecycle through the "options appraisal" phase, which includes the range of options to be appraised. Instead of proceeding sequentially from 'preparation' to 'options appraisal' to 'decision to invest', and so on, the decision maker or adviser now works through the risk management process (beginning with "risk scoping") prior to 'options appraisal' and the remaining phases of the project lifecycle.

The framework provides both a systematic and flexible approach to climate risk management at the project level, where completion of one stage leads logically to the next until a decision is ultimately taken. The following key features are built into the framework:

- It considers climate risks during the earliest stages of project development, when the problem to be addressed is first defined and ideas for possible solutions developed into concept notes. The screening of projects at the early stages of the project lifecycle will allow due consideration to be given to material climate risks during the design of solutions to the problem in hand, and prior to post-feasibility evaluation.
- The risk management process is circular (as indicated by the thick bold black lines) to allow the problem, decision criteria, risk assessment (scoping, characterisation and evaluation, screening and quantification), options identification and options appraisal to be refined as a result of analysis undertaken in earlier steps, prior to any decision being taken.
- The project-life cycle is also circular (as indicated by the thick broken grey lines) to allow decisions to be monitored and revisited over time, as new information becomes available (e.g., updated climate scenarios) or when it is realised that the project is unlikely to meet its objective(s). It may also be the case that adaptation options, where justified, are best implemented sequentially. Moreover, for adaptation to be successful, it should be viewed as a dynamic process that evolves with changing information and circumstances over time. Monitoring and review procedures must therefore be in place to provide the necessary feedback.
- The risk management process is tiered, to allow decision makers to screen projects for their vulnerability to climate variability and climate change prior to undertaking more complex and detailed studies (quantitative risk assessments) and before identifying and appraising adaptation options. This will ensure that effort and resources are directed to only those cases where climate is a material risk, from the point of view of the decision maker and key stakeholders.

The whole approach is essentially based on the concept of acceptable risk. Climate change impacts have to be addressed in the design of the project to the extent that with the additional measures or changes in design the project meets the criteria of acceptable risk. The detail to which these risks are assessed will depend on a preliminary screening, the results of which are

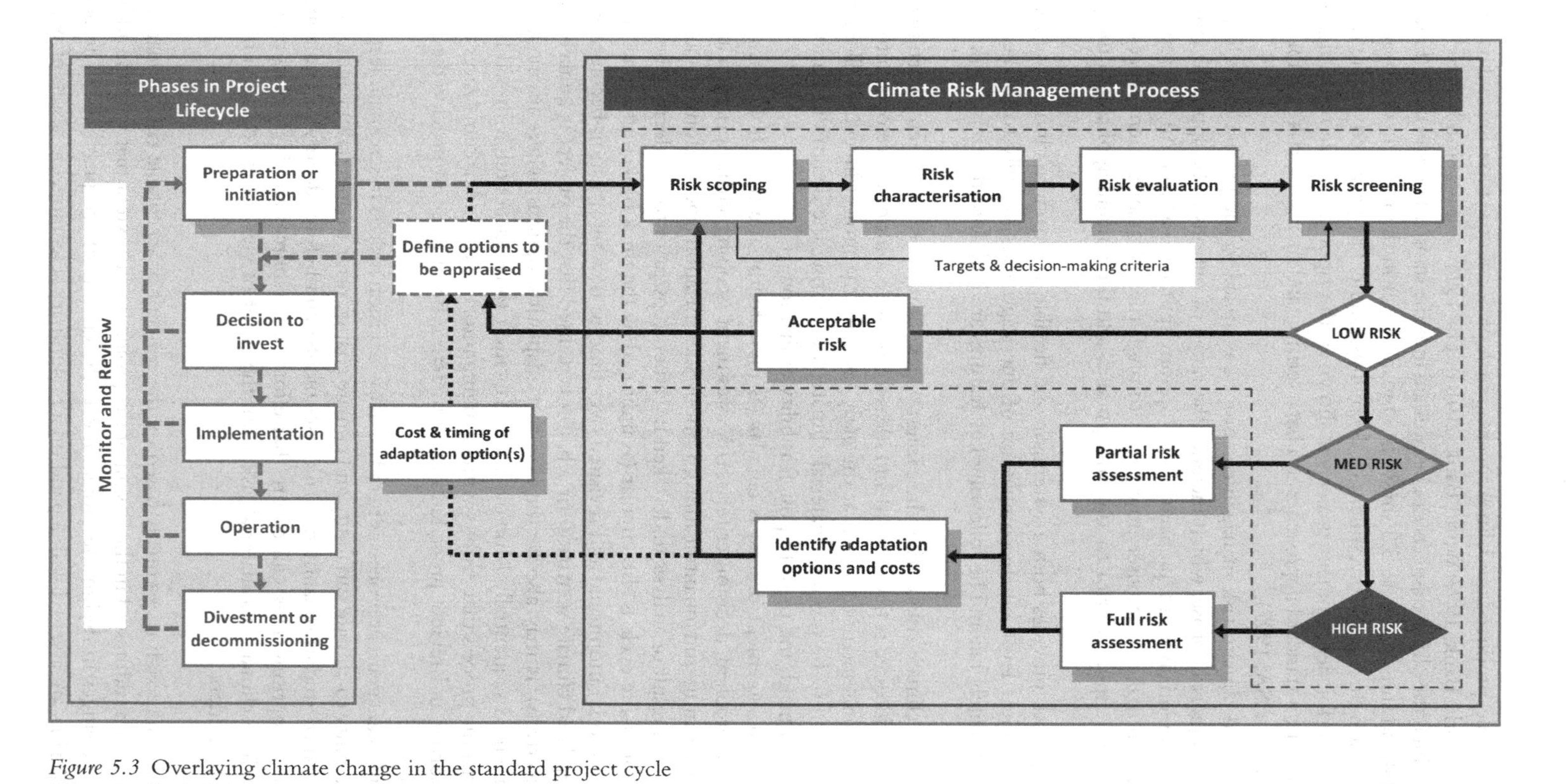

Figure 5.3 Overlaying climate change in the standard project cycle
Source: Author's design

classified as low, medium or high. In the case of low risk, no further action is taken. In the case of medium risk, a partial risk assessment is carried out, based on which adaptation options are identified and included in the project appraisal. In the case of high risk, the same is done, but with a full risk assessment.

Of course, any such risk assessment will face the difficulties in characterizing climate-related uncertainty in a sufficiently quantitative way. We have identified these problems in the previous section. Unfortunately, there is no easy way of resolving them; the possible impacts have to be taken into account when committing large resources, and where the information is poor or insufficient, one has to rely on what is available, combined with the expert judgment of the project appraisal team.

We would propose that such standard methods are applied in all cases where infrastructure-related development projects are liable to be impacted by climate change. Given the long duration of many infrastructure projects, it is suggested that all such projects be subjected to such a climate overlay procedure.

5.4.2 Projects where climate is the main reason

Where the areas of action are defined by climate change and the measures taken arise solely or mainly because of climate change, we would argue that a different approach is required. The literature that has reviewed such projects is generally sceptical about the possibilities of relying mainly on cost-benefit analysis to rank alternative options, largely because of the uncertainties associated with the climate impacts and the fact that most of them cannot be quantified (implying that the use of methods of risk management are difficult to apply); or are subject to multiple quantifications (reflecting ambiguity in the underlying knowledge). Alternative approaches have been proposed to address this, of which perhaps the most coherent is one based on robustness.

A robustness approach

A robustness approach is seen as having three components. The first is that it selects measures that are effective over the widest range of possible futures. One such set could be defined as 'no-regrets' measures, which are beneficial under any climate outcome. The problem with such measures is that they may not be enough to deal adequately with the consequences in the event of severe outcomes. For example, an early warning system for hurricanes would have a positive return under all future scenarios, but it would not by itself provide the desired level of protection in the event that there was a really severe hurricane. On the other hand, other 'robust' measures can be too costly.[3] For example, to design a protective wall that would operate effectively under the full range of future sea level rise would be prohibitively costly.

The second component of a robust approach is to build flexibility into the adaptation measure so that it can be adjusted in the future. An example would be to build a flood wall with deeper foundations than the structure put up initially demanded, so that it could be raised if necessary. In this case we are 'buying' robustness but then the question of whether it is good value for money inevitably arises. The approach used to make this evaluation is referred to as real options analysis (see Box 5.1).

Box 5.1 Real options analysis: A simple example

Consider the case where there are two options for protecting a coastal area against sea level rise. The first is to build a sea wall to protect against a one metre rise by 2030. The second is to build a wall that protects against a one-metre rise but also includes stronger foundations such that the wall could be raised to protect against a two-metre rise should that be discovered to be necessary in 2030. We assume that the higher wall will only be built in 2030 if it is found to be necessary in the second option and that we will know for certain whether it is required by 2030.

Table 5.2 lays out some costs and benefits. The trade-off is between incurring a higher cost now for the option of being able to protect against a higher risk in the future. The cost of a simple wall now is 100, while the more flexible wall is 130. If the simple wall is put up, the benefits with a one-metre rise are 200, but if the seal level rise is two metres, we will be forced into a retreat (it will be too late to put up a protective barrier when we get to know about the two-metre rise) and there will be a loss of 200. With the flexible wall, this second eventuality is avoided, but at an additional cost of 50.

The comparison hinges on what the probability is of a two-metre rise. With a small probability of 5%, the simple wall option has a higher expected value. With a probability of 10%, the two options have the same expected value and with any probability greater than 10% the flexible option is preferred. Of course, we may choose the flexible option even with a 5% probability of the two-metre rise if we place enough of a premium on protection against the risk or having to retreat.

Table 5.2 Real options analysis: Costs and benefits

	Period 1		*Period 2*			*Probability of 2m rise*	*Expected value*
	Costs	*Benefits*	*PV of costs*	*PV of benefits*			
				1m Rise	*2m Rise*		
Option 1	100	0	0	200	-150	5%	82.5
Option 2	130	0	50	200	200	5%	67.5
Option 1	100	0	0	200	-150	10%	65.0
Option 2	130	0	50	200	200	10%	65.0
Option 1	100	0	0	200	-150	15%	47.5
Option 2	130	0	50	200	200	15%	62.5
Option 1	100	0	0	200	-150	25%	12.5
Option 2	130	0	50	200	200	25%	57.5

As the above example demonstrates, a proper real options analysis does require the use of probabilities. This will often limit its application in a robustness assessment of proposed measures and may make it necessary to adopt more ad hoc approaches. These are discussed further below.

The third component of a robust approach is to build flexibility into the decision process itself, through waiting and learning. As Ranger et al. (2010) outline this, it would consist of undertaking the no-regrets options initially and delaying the more costly and inflexible options in anticipation of better information. Of course, if there is no loss of benefits in delaying action, then it is always better to do so as it increases the net present value of the benefits of any project. The decision on whether or not to wait depends on what kind of clarity of information is expected and what is the probability of a given impact. This approach is also referred to as a real options analysis.

While a lot has been made of the robustness approach, it follows from this discussion that it is not always a straightforward alternative. Indeed, as Ranger et al. (2010) note, "robustness-based strategies may not always be desirable as flexibility typically incurs additional costs or some productivity trade-off". Thus, we need some tools to be able to assess the benefits of flexibility, one of which is the real options analysis. Box 5.1 illustrates how such an analysis can help the decision maker. As the real options analysis example shows, however, the application of this analysis requires the definition of clear scenarios and the ability to define probabilities for outcomes, which are held with some confidence.

This raises the question of when is it possible to derive the required probability distributions. It is possible to devote resources to constructing probability distributions from limited data that can serve to allow the use of real options analysis or even cost-benefit methods? There is a difference of opinion on how useful are these 'ensemble based probabilistic approaches'. Of course these are not objective probabilities in the sense of representing long-run frequencies validated from data. Nevertheless, we take the view that when constructed with care, they are a reasonable method of bringing together a lot of the relevant information and can provide a good basis for at least a preliminary ranking of options. These can be supplemented by expert judgments and other ad hoc rules. The approach does, however, demand human and financial resources, which are not always available in developing countries.

Other approaches

Notwithstanding the position taken above, probabilities of different outcomes are not available or cannot be constructed in the required timeframe. Ranger et al. (2010) offer a range of decision tools such as maximin and minimax regret and info-gap analysis that could be used when probabilities are not obtainable. Table 5.3 summarises their proposals with annotated comments and Box 5.2 sets out some data illustrating the maximin and minimax criteria. Based on experience with actual studies, however, we doubt whether these tools could be used for anything other than providing illustrative examples of what the priorities might be.

One tool that could be used and that has been applied is multi-criteria analysis (MCA). Government advice on MCA can be found in the manual originally published by the Department for Communities and Local Government (DCLG) in the UK.[4] It has the advantage of being flexible enough to include partial information and to allow for uncertainty to be considered as a specific criterion. The range of possible outcomes for each option can then be 'scored' by the decision maker, with the weight used for deriving that score being based on expert opinions or wider participatory exercises. MCA has been proposed by the World Bank for use in adaptation projects where information on potential benefits is scarce and when "decision makers want to evaluate alternatives across a range of incommensurate criteria".[5] Its guidelines on climate change reference an example of the use of MCA in Uruguay where response options were identified and prioritised by local stakeholders. The weights

Table 5.3 Decision methods available when probabilities are not known (See Ranger et al., 2010 for more details.)

Method	*Decision criteria*	*Assumptions*	*Comments*
Maximin	Each option has a range of outcomes, one for each scenario. The score for each option is the minimum outcome value and the selected option is the one with the highest score.	This is an extreme risk aversion approach. In our example we would always build a sea wall as long as the cost is less than the maximum damage.	Such extreme risk aversion rarely reflects social preferences. It has the advantage that the outcomes need not be quantified on a cardinal scale.
Minimax regret	For each scenario, subtract the outcome of the option that does best in that scenario from the outcomes of other options in that scenario. This gives a regret table. For each option calculate the maximum regret and select the options with the smallest of these.	Rule is applicable when we care about missed opportunities. Generally less pessimistic than Maximin.	Ranking between two options can change if a third option is introduced. Does require outcomes to be quantified on a cardinal scale.
Info-gap decision theory for robustness	Construct a measure of the maximum amount of uncertainty we are exposed to and still ensure that losses do not exceed a given level. The decision maker then specifies an acceptable level of loss and chooses an option that has the lowest uncertainty subject to that loss limit.	Not a formal decision making process and can be expanded to include gains from different options as well (referred to a opportuneness).	Very much an ad hoc decision rule, but one that recognised the limits of simple robustness approaches.
Multi-criteria analysis	Each option is scored against a number of criteria and the option with the highest score is chosen. One of the criteria could be the robustness of outcomes in the face of uncertainty.	Agreement can be reached on the weight to be given to each criteria. This may be difficult.	Some sensitivity of the chosen option to different weights or scoring methods is normally required.

were based on stakeholder preferences between options. Furthermore, the scoring of the options was also derived from stakeholder opinions.[6] The criteria did not explicitly include robustness or uncertainty, and while a decision based on this method would have strong local support, its dependence on stakeholder knowledge, which must be regarded as incomplete at best, is questionable.

Box 5.2 Data illustrating the maximin and minimax regret options

Suppose we have two scenarios: no sea level rise and sea level rise of one metre; and two options: do nothing and build a sea defence. The cost of the sea defence is *C* and the worst outcome if sea level rises by one metre is *L*. Further suppose that the cost is less than the worst outcome ($C < L$).

For the maximin criteria, we note that the score for the do nothing option is *L*, while the score for the build sea wall option is *C*. Since $C < L$ the rule will always propose building a sea wall.

For the minimax criteria, we need to calculate a regret table, giving the difference between the outcome for a given option under a given scenario and the best outcome under that scenario. This produces the following regret table (Table 5.4).

The minimax regret rule says we should choose the option with the smallest maximum regret. The maximum regret in the case of the Build Sea Defence is C, while that of the Do Nothing Option is L – C. So we build the sea defence if C < L–C or L > 2C. In other words the rule says we should build the defence only if potential losses are twice as large as the costs of defence.

Table 5.4 Minimax regret table

	No sea level rise	*Sea level rise by 1m*
Build sea defence	C	0
Do nothing	0	L-C

Among these 'other approaches', MCA is the one that can be applied formally in a way that accounts for uncertainty. Actual applications do not in fact do that explicitly, although there can be an element of the value attached to uncertainty when choosing the weights and when selecting the criteria. There is room therefore to develop explicit methods of including uncertainty in the MCA approach.

5.4.3 Treatment of uncertainty in actual case studies

In practice, actual appraisals of climate adaptation projects or studies use a combination of an ad hoc robustness and a flexibility approach, along with rules based on thresholds for the values of key parameters that reflect vulnerabilities of the affected population and society to climate-related changes. These exploit a number of features that are present in many of the areas where we seek to make interventions and can be summed up as follows:

a) Action of a costly and irreversible nature often does not have to be taken immediately. Often low cost actions can serve to provide a cost-effective solution to the immediate risks while more information is gathered to reduce the epistemic uncertainty, and the more difficult decision can be postponed. Moreover, many of these actions, which also tend not to involve major investments and are classified as 'soft', serve to increase resource efficiency now, making them attractive from that perspective as well.

b) It can often make sense to preserve flexibility by taking measures now that leave open the option of a wider range of actions in the future. To be sure, such flexibility comes at a cost, but this cost can be modest and may be justifiable even when a probabilistic assessment of the flexible option using real options analysis is not possible.

In the rest of this section, we look at actual studies that have been conducted relating to adaptation.

Coastal zones and sea level rise

In coastal zones assessments, the focus is usually on a single estimate of projected sea level rise (usually one metre). Different options are then evaluated relative to that level. There are some that consider more than one level and one study (Nicholls, 2007), which examine the possible impacts of a major increase arising from the collapse of the West Antarctic Ice Sheet.

Perhaps one of the most sophisticated studies of sea level rise is that for the Thames Estuary (Haigh and Fisher, 2010). Details of that are given in the study itself, but we note here the main ways in which it handled uncertainty. The approach taken was one that used all available information in a practical way. The first thing was to lay out the adaptation options available for addressing the likely sea level rise to 2100 and, for each option, identify: (a) a threshold value in terms of sea level rise at which it would come into play and (b) the latest date at which a decision on its implementation would need to be taken. This resulted in identifying the actions that had to be taken urgently against those that could be postponed. Most of the urgent actions took the form of low-regrets measures, such as extending the lifetime of the existing flood management infrastructure. On current projections, the initial decision point for the irreversible decisions was found to be about 2050. Hence, the analysis showed that there was time for new information to be evaluated and decisions on what action to take to be revised if necessary. The choice between different options was then based on how each option performed under a range of indicators and under a range of scenarios, a methodology that combined multi-criteria analysis with a robustness assessment.

In summary, the uncertainty aspect of the problem was largely handled by 'buying time' – i.e., using the available data to identify how late key decisions could be left without compromising the response. Fortunately, the project allowed one to address short-term risks with small actions now and to leave the more costly decisions till later. It is likely that similar considerations apply in some other locations for sea level rise, but it would be risky to assume that this will always be the case. For example, a coastal area where there is heavy planned development and settlement will not be a case where all costly options can be postponed; someone will have to decide whether to curtail development now or leave a future decision maker with a much higher cost of adaptation.

Agriculture

In the agriculture sector, farmers make decisions based on the variability they face on a frequent basis. Hence, one can expect them to modify their behaviour to allow for the increased risks. This is something that can be modelled (Parry et al., 2009) and public programmes need to take account of this behaviour. Indeed, one of the clear benefits of such programmes is to reduce the variability of farm incomes in the face of climate change (the other being to reduce the loss of income). Hence, taking of risks is a key part of the design of the strategy.

To date, however, the reduction of variability of incomes is not directly accounted for. One approach has been to select adaptation measures that are low-cost and for which the cost-benefit ratios are likely to be high irrespective of the uncertainty dimensions. For example, Tan and Shibasaki (2003), using a GIS-based crop model and allowing for modelling of inter- and intra-regional bioclimatic differences, computed global adaptation benefits (measured in terms of the IRR) of low-cost adjustments in the range of 23% to 48%.

A similar approach is taken in the case study that looks at adaptation options in The Gambia, where climatic uncertainty is addressed by looking at two scenarios generated by two different climate models (Njie et al., 2006). These have significantly different predictions of rainfall, with the result that some adaptation options come up with very different outcomes depending on which scenario is taken. The final recommendations are based on selecting the option that is robust under the two scenarios.

The World Bank guidelines suggest another alternative method of assessment based on multi-criteria analysis (World Bank, 2010a) with heavy dependence on participatory methods. A range of options is selected through stakeholder consultation. The stakeholders then also decide on the criteria against which they are to be evaluated and the weights to be given to the criteria. The scores given to the options were then derived from a combination of stakeholder assessment and information provided by the local study team. Finally, a ranking was made based on these scores, and the weights attached to the different criteria. There was no criterion that specifically addressed uncertainty or robustness, although one criterion was "capacity to reduce damage caused by extreme events". Such an approach has the advantage of have a strong buy-in from local stakeholders, but it perhaps does not draw on the full range of information regarding robustness and flexibility of different options and on the advantage of analytical assessments of these important factors.

Water

The water sector is particularly difficult, given the huge uncertainties about precipitation and the frequency of extreme events. The case studies we looked at included one from a developing country (South Africa) and one from a developed country (the UK). The South African study used a range of scenarios to reflect uncertainties about climate change impacts and future water demand. It found one adaptation option to dominate under all scenarios, which is fortunate. It is also risky, as that option (setting up water markets) is politically difficult to implement. Having to go for second best would imply having to face choices where some options did better under one scenario and others did better under another.

In the UK case study, the uncertainty is treated in much greater detail. The Anglian water company conducted the analysis, which is based on providing a given level of service (water supply, water quality and waste water disposal) in its region. A 25-year water resource plan was drawn up, which outlines policies to enable the company to fulfil its obligations. Climate impacts on the supply of these services were estimated as changes from a baseline, arising from a number of factors: global temperature change, regional climate change, changes in precipitation, changes in water requirements as a result of changes in precipitation, etc. For each of these, a probability distribution was derived from which the company calculated the impact that each factor would have on the additional water required to meet demand with a 95% probability. Different plans were evaluated in terms of the cost of meeting the 95% probability requirement.

Such probabilities are of course subject to ambiguity. They are artificially constructed and are therefore not in the same category as those derived from long-run frequency data. Nevertheless, the analysis is a state of the art of how to build uncertainty into a framework that allows for its quantification in a credible way, to design appropriate responses.

Ranger et al. (2010) note that cost-effective options of meeting regulatory requirements under climate change, with all its uncertainties, should focus on measures such as demand management through metering and water transfer between water utilities in the first instance and then move on to resource based-solutions, such as increasing reservoir capacity, later. Not only does this reflect going for the lowest cost measures first, it also allows flexibility in terms of providing time for more information to be gathered before the design of the irreversible alternatives is undertaken.

Dessai and Wilby (2010) provide a case study for Yemen, in which the water sector problems are reviewed in the light of climate change. Their perspective is that the sector faces a crisis even without any climate change, and actions need to be taken to address that. While climate change may add to the problems of water scarcity (projections of changes in rainfall are so uncertain that they encompass both increases and decreases in annual levels), flash flooding and the like, these impacts will remain small compared to the large natural variability during the next few decades. Given the large uncertainty associated with regional rainfall, the authors propose that action be limited to those that perform well under a wide range of future conditions. These "low-regret" and robust measures include 12 elements for which the first five are: (a) development and implementation of Integrated Coastal Zone Programmes, (b) water conservation through reuse of treated waste water, (c) development and implementation of awareness-raising programmes on adaptation to potential impacts of climate change, (d) establishment and maintenance of a database for climate change adaptation and (e) planting and re-planting of mangroves and palms for adaptation to projected sea level rise.

This list, taken from the National Adaptation Programme for Action for Djibouti, consists of items that are sensible. There is a ranking, but it is unclear how this is derived. Furthermore, it is not clear how and to what extent the cost of the different measures have been taken into account in deriving the list. Some indication of cost effectiveness or value for money would normally be considered an important consideration, and that is missing.

Infrastructure and energy

As noted earlier, addressing climate change in infrastructure projects is best done though the traditional methods of risk assessment. Examples are mostly from developed countries and include investments in roads, railway lines (to account for subsidence from climate change), bridges (to account for sea level rise) and the like. An example from the developing world is the Qinghai-Tibet Railway. The railway crosses the Tibetan Plateau with about a thousand kilometres of the railway at least 13,000 feet (4,000 metres) above sea level. Five hundred kilometres of the railway rests on permafrost, with roughly half of it "high temperature permafrost" which is only 1°C–2°C below freezing. The railway line would affect the permafrost layer, which will also be impacted by thawing as a result of rising temperatures, thus in turn affecting the stability of the railway line. To reduce these risks to acceptable levels, design engineers have put in place a combination of insulation and cooling systems to minimise the amount of heat absorbed by the permafrost (IPCC, 2007c).

A similar approach is needed for the energy sector, where climate change will impact the design requirements, particularly for hydropower, thermal and nuclear facilities. For such investments, an acceptable risk approach is appropriate and one that can be used even when objective probabilities cannot be defined.

In addition, however, the energy sector also needs to look at wider issues; i.e., not only regarding the design of a particular facility, but also whether or not that facility is an appropriate part of the energy system. In other words, the whole energy system needs to be appraised. In this case, we need to go to the second method of assessment, where the system should be tested for robustness to climate change, and in particular on the changes in the availability of water and conflicting demand for water (World Bank, 2010b). Such assessments are yet to be undertaken anywhere.

5.5 Conclusions

Uncertainty is a major factor when selecting the best response to climate change through adaptation. Some of this uncertainty can be reduced through research, although this does not necessarily mean that we will have narrower projection ranges, merely that we will have greater confidence in them. Such uncertainty is referred to as epistemic. The second kind of uncertainty arises from the inherent variability of the phenomenon and cannot be reduced – it is referred to as ontological. The third kind is uncertainty arising from differences in the way different groups understand the problem and choose to model it. This is referred to as ambiguity and gives rise, in the climate context, to incomplete or multiple inconsistent probabilities associated with different outcomes. In the case of climate change, all three types are present but, at the present time epistemic uncertainty is pervasive at all steps of the decision-making frame. Ontological uncertainty is especially important in the climate models, in the downscaling measures and in characterising the adaptation measures. Finally, ambiguity is most serious in defining the socioeconomic scenarios and in characterising the socioeconomic impacts and adaptation measures.

When appraising any public project, not only one relating to climate, uncertainty is an important factor and standard tools have been developed to handle it. These tools can also be applied in a number of cases when climate change is the source of the uncertainty. For example, uncertainty arising from 'cost optimism' can be addressed for climate adaptation projects as much as it can for any other project. We also argue that tools for risk management in the design of projects through acceptable risk can be applied for some climate adaptation projects.

The literature suggests a number of reasons why climate change makes use of the main tool of project appraisal, namely cost-benefit analysis, inappropriate or impossible. These are:

a) The significant uncertainties about the impacts, most of which cannot be represented in statistical terms. Some fall into the category of qualitative uncertainty or even of recognised ignorance.
b) The changing knowledge base – we can expect to have a changed view about the impacts of climate change over time.
c) The relatively long time periods over which the impacts will occur and over which the adaptation measures have to be evaluated.
d) The fact that there are multiple actors adapting to climate change, mostly in the private sector, and the actions of the public agencies have to take these into account and indeed exploit all possible synergies with the actions of these actors.

We take the view that adaptation projects can be usefully divided into two categories. The first consists of projects where climate change is only one additional concern that needs to be addressed. It is of course critical that it be addressed, but this can be done through standard risk management techniques, involving the definition of acceptable risk and the search for cost effective solutions. Investments in infrastructure for transport, energy, industrial and commercial plants, housing, etc., all with long lives, fall into this category. In terms of the project appraisal cycle, these steps are part of the options appraisal phase.

The second consists of the areas of intervention where climate is the main driver of the action. Projects relating to flood protection, protection against sea level rise, adaptation to changes in water supply and to changes in agricultural yields fall into this category. The difficulties here arise from the factors just listed above. These make the use of cost-benefit methods that depend on estimating expected values and/or expected values with risk premiums very difficult to apply. Even modified criteria for ranking options such as robustness are difficult to apply formally without some estimate of credible probabilities for different outcomes. It is possible to use tools such as MCA for such projects, but care must be taken not to rely too heavily on rankings based on participatory approaches, when the stakeholders making these assessments are not fully informed about the impacts and consequences of different measures.

It is possible to devote resources to constructing probability distributions from limited data that can serve to allow the use of real options analysis or even cost-benefit methods. There is a difference of opinion on how useful these 'ensemble-based probabilistic approaches' are. We take the view that when constructed with care, they are a reasonable method of bringing together a lot of the relevant information and can provide a good basis for at least a preliminary ranking of options. These can be supplemented by expert judgments and other ad hoc rules. The approach does, however, demand human and financial resources, which are not always available in developing countries.

At the same time, practical tools have been developed that serve to provide rankings of options that are ad hoc but that make sense. These exploit a number of features that are present in many of the areas where we seek to make interventions.[7]

a) Action of a costly and irreversible nature does not have to be taken immediately. Often low-cost actions can serve to provide a cost-effective solution to the immediate risks while more information is gathered to reduce the epistemic uncertainty, and the more difficult decision can be postponed. Moreover, many of these actions also serve to increase resource efficiency now, making them attractive from that perspective as well.
b) It can often make sense to preserve flexibility by taking measures now that leave open the option of a wider range of actions in the future. To be sure, such flexibility comes at a cost, but this cost can be modest and may be justifiable even when a probabilistic assessment of the flexible option using real options analysis is not possible.

Examples of actions that have the characteristic of being low-cost (or even no-cost when one takes account of the co-benefits) include the following (Hallegatte, 2009):

For **coastal zones**, strategies to adapt to uncertain CC for coastal zones include 'easy-to-retrofit' defences, restrictive land use planning, insurance, warning and evacuation schemes, creation of risk analysis institution and long-term plans.

For **agriculture**, options that could form part of a low cost-strategy at the farm level include, e.g., diversification of production; especially to crops more resistant to climate variations; crop and farm income insurance; partial non-farm employment. And at the public level, options include, e.g., institutional support to diffuse information on CC and adaptation possibilities (like extension services, early warning systems); enhancing agricultural trade to spread the impact of regional supply shortages over the international market; etc.

For **water**, strategies especially noteworthy for their contribution to adapt to uncertain climate change include 'institutionalisation' of long-term perspectives, loss reduction (leakage control, etc.), demand control and water reuse.

For the **health** sector, strategies to adapt to uncertain climate change include R&D on vector control and vaccines, and improvements in public health systems.

While this appears to resolve the difficulties for the design of adaptation strategies, it would be a mistake to ignore the fact that certain hard decisions are emerging and need to be addressed relatively urgently. One relates to land use planning. Developing countries with their rapid rates of urbanization are expanding to areas where vulnerability to climate change could be high in the near future (i.e., in the next 20 years or so). Climate change considerations should be integrated into these plans now. The second relates to infrastructure where major investments are being made based on current availabilities of the supply of water and current frequencies of extreme weather events. Although the changes in these projections are not going to materialise immediately, investments made now have a lifetime of 50 years or more, and changes in the future will be very costly. Thus early action may be justified in these cases.

Acknowledgement

The review has been developed in the course of the BASE project (Bottom-up Climate Adaptation Strategies towards a Sustainable Europe), financed under FP7 program Grant agreement n.308337.

Notes

1 An assessment is made by Stafford Smith for adaptation projects by 'decision lifetime'. He notes that when this lifetime is 50 years or more and when the changes are relatively predictable, a precautionary risk management approach (similar to our acceptable risk approach) is the appropriate method. When the changes are unpredictable, he argues for a robust decision-making approach. We would argue that for many investments in the construction, energy and transport sectors, the former approach is feasible. Even if probabilities for different climate outcomes cannot be defined, subjective risk assessments can be derived that allow an acceptable risk to be calculated. The rule will be more difficult to apply when looking at system-wide changes.

2 Ranger et al. (2010) make the distinction as follows: "Economists and decision theorists have long distinguished between decisions under risk, in which the relative frequencies of states are known, and decisions under uncertainty, in which there is ambiguity about probabilities, or we are completely ignorant of which state of nature is likely to occur" (Annex, p. 14).

3 Dessai and Wilby (2010) further recognise that in practice even the 'no-regrets' options have costs associated with them and cannot be guaranteed to yield benefits irrespective of climate change. They make the good suggestion that we should refer to such options as 'low-regrets'.

4 http://www.communities.gov.uk/documents/corporate/pdf/1132618.pdf

5 http://siteresources.worldbank.org/EXTTOOLKIT3/Resources/3646250-250715327143/GN7.pdf

6 http://siteresources.worldbank.org/EXTLACREGTOPRURDEV/Resources/503766-1225476272295/PDF_Agricultue_Climate_change.pdf.

7 All of these have been identified earlier. See, e.g., H.M. Treasury (2009).

References

Brown, J.D. (2004). Knowledge, Uncertainty and Physical Geography: Towards the Development of Methodologies for Questioning Belief, *Transactions of the Institute of British Geographers*, 29(3), 367–381.

Callaway, J.M., Louw, D.B., Nkomo, J.C., Hellmuth, M.E., and Sparks, D.A. (2007). "The Berg River Dynamic Spatial Equilibrium Model: A New Tool for Assessing the Benefits and Costs of Alternatives for Coping with Water Demand Growth, Climate Variability, and Climate Change in the Western Cape," AIACC Working Paper 31, The AIACC Project Office, International START Secretariat, Washington, DC, p. 41.

Christensen, J.H., Boberg, F., Christensen, O.B., and Lucas-Picher, P. (2008). On the Need for Bias Correction of Regional Climate Change Projections of Temperature and Precipitation, *Geophys. Res. Lett.*, 35, L20709, doi:10.1029/2008GL035694.

Christensen J.H. and F. Boberg, 2012: Temperature dependent climate projection deficiencies in CMIP5 models, Geophys. Res. Lett., doi:10.1029/2012GL053650

CSA (1997). Risk Management Guidelines for Decision-makers, CAN / CSA-Q580-97, Canadian Standards Association.

Dessai, S., and R. Wilby. (2010). How Can Developing Country Decision Makers Incorporate Uncertainty about Climate Risks into Existing Planning and Policymaking Processes? World Resources Report, Washington DC. Available online at http://www.worldresourcesreport.org.

DETR (2000) Guidelines for Environmental Risk Assessment and Management – Revised Departmental Guidance, Institute for Environment and Health, HM Stationary Office, London.

Haigh, N., and Fisher, J. (2010). "Using a 'Real Options' Approach to Determine a Future Strategic Plan for Flood Risk Management in the Thames Estuary." Draft Government Economic Service Working Paper.

Hallegatte, S. (2009). Strategies to Adapt to an Uncertain Climate Change, *Global Environmental Change*, doi:10.1016/j.gloenvcha.2008.12.003.

H.M. Treasury (2004). *The Orange Book, Management of Risk – Principles and Concepts*, London: Her Majesty's Treasury.

H.M. Treasury (2007). *The Green Book: Appraisal and Evaluation in Central Government, Treasury Guidance*, London: Her Majesty's Treasury, TSO.

H.M. Treasury (2009). *Accounting for the Effects of Climate Change, June 2009*, Supplementary Green Book Guidance, http://www.defra.gov.uk/environment/climatechange/adapt/pdf/adaptation-guidance.pdf.

IPCC (2007a). Climate Change 2007: Synthesis Report, Contribution of Working Groups I, II and III to the Fourth Assessment Report of the Intergovernmental Panel on Climate Change [Core Writing Team: Pachauri, R.K. and Reisinger, A. (eds.)], Geneva, Switzerland: IPCC. 104 pp.

IPCC (2007b). Climate Change 2007: The Physical Science Basis, Contribution of Working Group I to the Fourth Assessment Report of the Intergovernmental Panel on Climate Change [Solomon, S., Qin, D., Manning, M., Chen, Z., Marquis, M., Averyt, K.B., Tignor, M., and Miller, H.L. (eds.)], Cambridge, United Kingdom and New York, NY, USA: Cambridge University Press, 996 pp.

IPCC (2007c). Climate Change 2007: Impacts, Adaptation and Vulnerability. Fourth Assessment Report (2007) [Parry Martin, Oswaldo Canziani, Jean Palutikof, Paul van der Linden, and Clair Hanson (eds.)], Cambridge University Press.

Klauer, B., and Brown, J.D. (2003). Conceptualising Imperfect Knowledge in Public Decision Making: Ignorance, Uncertainty, Error and 'Risk Situations', *Environmental Research, Engineering and Management*, 27(1), 124–128.

Knight, F.H. (1921). *Risk, Uncertainty and Profit*, Chicago: Houghton Mifflin Company

Nicholls, R.J. (2007). Adaptation Options for Coastal Areas and Infrastructure: An Analysis for 2030, Report to the UNFCC, Bonn.

Njie, M., Gomez, B.E., Hellmuth, M. E., Callaway, J.M., Jallow, B.P., and Droogers, P. (2006). "Making Economic Sense of Adaptation in Upland Cereal Production Systems in The Gambia." AIACC Working Paper No. 37, International START Secretariat, Washington, DC.

Olesen, J.E., Carter, T.R., Diaz-Ambrona, C.H., Frontzek, S., Heidmann, T., Hickler, T., Holt, T., Minguez, M.I., Morales, P., Paalutikof, J.P., Quemada, M., Ruiz-Ramos, M., Rubæk, G.H., Sau, F., Smith, B., and Sykes, M.T. (2007). Uncertainties in Projected Impacts of Climate Change on European Agriculture and Terrestrial Ecosystems Base Don Scenarios from Regional Climate Models, *Climatic Change*, 81, 123–143.

Parry, M., Arnell, N., Berry, P., Dodman, D., Fankhauser, S., Hope, C., Kovats, S., Nicholls, R., Satterthwaite, D., Tiffin, R., and Wheeler, T. (2009). Assessing the Costs of Adaptation to Climate Change: A Review of the UNFCC and Other Recent Estimates, International Institute for Environment and Development and Grantham Institute for Climate Change, London.

Ranger, N., Millner, A., Dietz, S., Fankhauser, S., Lopez, A., and Ruta, G. (2010). Adaptation in the UK: A Decision-Making Process, Policy Brief September 2010, Grantham Research Institute on Climate Change and the Environment & Centre for Climate Change Economics and Policy.

Refsgaard, J.C., Arnbjerg-Nielsen, K., Drews, M., Halsnæs, K., Jeppesen, E., Madsen, H., Markandya, A., Olesen, J.E., Porter, J.R., and Christensen, J.H. (2012a). Climate Change Adaptation Strategies: Water Management Options Under High Uncertainty – A Danish Example, *Mitig. Adapt. Strateg. Glob Change*, 18(3), 337–359.

Refsgaard, J.C., Arnbjerg-Nielsen, K., Drews, M., Halsnæs, K., Jeppesen, E., Madsen, H., Markandya, A., Olesen, J.E., Porter, J.R., and Christensen, J.H. (2012b). The Role of Uncertainty in Climate Change Adaptation Strategies – A Danish Water Management Example, *Mitigation and Adaptation Strategies, Global Change* doi: 10.1007/s11027-012-9366-6.

Sunyer, M.A., Madsen, H., and Ang, P.H. (2011). A Comparison of Different Regional Climate Models and Statistical Downscaling Methods for Extreme Rainfall Estimation Under Climate Change, *Atm. Res.*, In Press.

Tan, G., and Shibasaki, R. (2003). Global Estimation of Crop Productivity and the Impacts of Global Warming by GIS and EPIC Integration, *Ecological Modelling*, 168(3), 357–370.

UKCIP (2003). Climate Adaptation: Risk, Uncertainty and Decision-making [Willows, R. and Connell, R. (eds.)], Oxford: UKCIP Technical Report.

van der Keur, P., Henriksen, H.J., Refsgaard, J.C., Brugnach, M., Pahl-Wostl, C., DeWulf, A., and Buiteveld, H., (2008). Identification of Major Sources of Uncertainty in Current IWRM Practice, Illustrated for the Rhine Basin, *Water Resources Management*, 22, 1677–1708.

World Bank (1997). Guidelines for Climate Change Global Overlays. Global Environment Division, Paper No. 047.

World Bank (2010a). The Costs to Developing Countries of Adapting to Climate Change: New Methods and Estimates. Washington DC: World Bank.

World Bank (2010b). Climate Impacts on Energy Systems: Key Issues for Energy Sector Adaptation (with several authors). ESMAP, World Bank.

6

DISTRIBUTIONAL IMPACTS

Intra-national, international and inter-temporal aspects of equity in adaptation

Alistair Hunt[1] *and Julia Ferguson*[2]

[1]UNIVERSITY OF BATH (UK)

[2]CRANFIELD SCHOOL OF MANAGEMENT (UK)

6.1 Introduction

This chapter discusses the role of equity in the assessment of adaptation to climate change. In doing so, we consider equity in relation to the application of economic analytical concepts and metrics, thereby enabling us to explore the extent to which they are compatible. The development of climate change mitigation strategies has, of course, provided a lively decision context with which to explore the characterisation of equity alongside economic efficiency – and other criteria – in appraisal, primarily through the parameterisation of economy-climate models (Hope, 2008). It has also served to identify the role of equity in influencing decision outcomes – most obviously in the UNFCCC mitigation architecture adopted under the Kyoto Protocol that places the majority of the burden of reducing greenhouse gases in more affluent countries (known as Annex 1 countries in the Protocol). However, it is now well understood that no matter how successful global GHG mitigation efforts turn out to be in the next few decades, there will be residual risks (and opportunities) resulting from climate change that have impacts on human welfare and supporting ecosystem services (Khor, 2012).

Consequently, the need to consider responding directly to climate change risks on an individual and collective basis through adaptation implies the necessity for evaluating these responses against alternative criteria, including equity. Whilst the characterisation of the adaptation context parallels the mitigation context in a number of ways, including the relatively long appraisal periods and the broad range of geographical scales at which policies have to be implemented, the differences in context that result, for example, from its likely multi-sectoral application and a lack of a common metric, suggests that adaptation decisions may present unique challenges in the treatment of equity. In order to investigate whether – and to what extent – this is the case, this chapter reviews the literature that has sought to delineate the role of equity in a range of adaptation decision-making processes.

6.2 Principles of equity analysis: What do we understand equity to entail?

As our starting point, we understand equity to refer to the concept of fairness in the distribution of capital, goods or access to services, and the welfare associated with these. In this respect, equity is closely related to notions of justice and legitimacy, and these terms are used interchangeably in the body of adaptation literature. This chapter adopts "equity" as the principal term of reference since it may be judged to be less pejorative than "justice" and other similar terms. Indeed, it is defined in specific terms above in order to help promote a shared understanding of the concept of equity in the adaptation context that is currently absent. It is highlighted that the applicability of equity as a guiding principle in influencing decisions is independent of scale, and will operate at multiple scales at the same time. Consequently, in the extreme, climate risks and adaptation can be seen to have equity implications at an intra-household level, as well as a global level. For example, resistance measures may be introduced by a household in response to a climate change-induced flood risk, the financial burden of which is differentiated between household members perhaps in relation to their income. At the global scale, the financial burden of adaptation may also be determined on the basis of income; for example, the Adaptation Fund, operated by the United Nations Framework Convention for Climate Change (UNFCCC) is funded principally by higher income countries to support adaptation action in lower income countries, where vulnerability to climate change risks is projected to be greater.

In its application, this concept can then be further divided between horizontal equity and vertical equity (Musgrave, 1959). Horizontal equity is taken to mean that those benefiting (or being adversely affected) to a similar extent by a change in the distribution of these factors of production and goods and services should pay (or be compensated) the same amount; in other words, there should be equal treatment of equals. On the other hand, vertical equity expresses the principle that those with higher endowments of wealth, or higher income levels, who benefit should pay more than those with lower wealth endowments or income levels, reflecting a greater ability to pay, i.e., there may be un-equal treatment of un-equals. Thus, to take a simple example in the adaptation context, the principle of horizontal equity may be followed in the context of a sea defence system that reduces the impacts of climate change-induced sea-level rise may be paid for by the local community whose members all benefit to a similar extent and who therefore each contribute a similar amount. However, if the wealth endowments differ substantially across the local population who are potentially exposed to sea-level rise impacts, it may be agreed that those with higher wealth endowments should bear a higher burden of payment in absolute terms. In the latter instance, the notion of vertical equity as a guiding principle is therefore dominant.

Following the classification made by Paavola and Adger (2002), a further delineation of equity is relevant in the context of evaluating adaptation to climate change: the distinction between retributive, distributive and procedural equity.[1] Retributive equity is based on the idea that damages should be repaired by those who caused them; in the climate change context, it is likely to be most relevant to the determination of emissions mitigation responsibilities. Distributive equity is not concerned with the identification of responsibility, but rather is concerned with the equalization of resources and benefits according to needs and capacities, themselves determined by levels of vulnerability and adaptive capacity within society. The equalisation of resources and benefits in the context of climate change therefore needs to be conceived both geographically, across countries and world regions, and temporally, bearing in mind the intergenerational dimension of climate risks.

Whilst both retributive and distributive notions of equity are primarily concerned with the outcomes of climate risks and associated responses on human welfare, procedural equity refers to how, and by whom, decisions about adaptive responses are made. Thus, procedural equity refers to the degree to which parties (individuals, households, organisations, etc.,) who are likely to be affected by climate risks and adaptation are recognised and participate in formulating an adaptation response. Whilst these notions are well-established, their relative importance is disputed. Indeed, Hayek (1976), amongst others, suggests that as long as the process itself is fair, then that is sufficient since the process will determine an equitable outcome. Alternatively, Adger et al. (2006) argue that since the specific form of process referenced by Hayek, i.e., participation in markets, is less appropriate for many adaptation decisions that are made in the context of nonmarket welfare changes, distributive equity has to be addressed explicitly.

Thus, a further layer of complexity in the interpretation of the notion of equity arises as a result of the philosophical orientation adopted since this can profoundly influence how fairness is judged. For example, approaching equity from a utilitarian perspective would imply conceiving fairness as the achievement of the greatest good for the greatest numbers, (Varian, 1984). Thus, weights are attached to individuals, households or other appropriate unit on the basis of a distributional metric – most often income. In the simplest formulation of the measure, the change in utility resulting from the adaptation project may then be assumed to equate to the change in income resulting from the project, and the weighting given to each household is one. A more humanist perspective would lead to a view of equity that assesses and addresses differential vulnerability between households. In the extreme, this viewpoint has been characterised by Rawls (1972) who suggests that welfare change is dictated solely by how the poorest individual or household is impacted. Thus, the weighting given to this individual/household is one whilst the weighting given to all others is zero.

As has been noted previously (Markandya, 2011), whilst income is a metric that is often used to proxy for a measure of welfare in equity analysis, alternative metrics may be required to capture welfare effects of climate policy more fully. For example, the capabilities approach outlined first by Sen (1985) has, at its core, the idea that what people can do ('capabilities'), as opposed to what they actually do ('functionings'), should be the focus of well-being evaluations and government policy. It takes the view that society should maximise people's capability to achieve their potential, i.e., people's freedom to pursue valuable acts or reach valuable states of being, rather than maximise welfare or utility. Capability is defined as the ability to achieve different levels of functionings (beings and doings) and it refers to the *freedom* to enjoy various functionings (Sen, 1992). Nussbaum (1993) has provided a list of capabilities that she thinks are all essential for a good life including: normal life span; bodily health; bodily integrity; senses, imagination and thought; emotions; practical reason; affiliation; other species; play; and control over one's environment. It also includes security – e.g., of food or livelihood – likely to be important in the context of climate change risks. Anand (2005) notes that Nussbaum's list of capabilities is likely to be regarded as important albeit to varying degrees and argues that these capabilities would vary greatly with age, among people and across cultures.

From a policy point of view, it is necessary to select capabilities that are relevant to the particular context – in this case, the management of climate change risks. Alkire and Deneulin (2009) note that "the key questions to keep in mind when selecting capabilities are: (1) which capabilities do the people who will enjoy them *value* (and attach a high priority to) and (2) which capabilities are relevant to a given policy, project or institution?"[2] However a significant problem with operationalising the capabilities approach is that it is hard to secure agreement about the appropriate list of functionings or capabilities (Cookson, 2005). Thus, whilst frequently acknowledged as constituting a more holistic approach to the measurement of well-being than income, its practical implementation remains underdeveloped. The cross-sectoral, multi-risk nature of climate change, therefore, suggests that the adaptation context will present additional complexities to the application of the capabilities approach.

The complexity in evaluating equity arises partly through the necessarily subjective elements of the concept. As indicated above, the challenge of its evaluation is exacerbated by the lack of agreement that currently exists in the appropriateness of alternative metrics and their weighting. However, as highlighted by Adger et al. (2006), equity is a critical, though not the only, criterion against which actions are judged to be acceptable or not. As noted above, in the context of adaptation to climate change, equity issues are inextricably linked with the issue of justice that arise from the recognition that the spatial impacts of climate change – at a global scale at least – often fall disproportionately on those least able to bear them and who have reaped less economic benefits from historical emissions. As discussed below, it may also be the case that the impacts of climate change across future time periods are judged to be inequitable.

In the context of climate change adaptation, addressing the equity implications of taking action are likely to be important on multiple spatial and institutional scales if existing patterns of inequality are not to be exacerbated. However, they also need to be viewed alongside data that summarise effects on other criteria such as feasibility, economic efficiency, urgency, etc. An example, reported below, demonstrates the extent to which these criteria are evaluated relative to each other.

The remaining sections of this chapter demonstrate how – and to what extent – the principles identified as being relevant to equity analysis have been used in practical applications to date. Where relevant, an economic perspective is adopted in order to give an additional focus to the review.

6.3 Climate change risks, adaptation and equity – empirical evidence

6.3.1 The international context

In order to provide an impression of the spatial distribution of climate change risks, Figure 6.1 below, reproduced from the IPCC Fourth Assessment Report (Parry et. al., 2007) provides an overview of regionally disaggregated climate change impacts, projected across a range of mean annual temperature changes. The figure demonstrates that the impacts of climate change will be borne by all regions but that the human costs are likely to be greater in lower income countries in Africa and Asia.

Consequently, on the basis of the income metric, the findings shown in Figure 6.1 suggest that if the international community has a global equity objective such that existing inequalities are not to be exacerbated, there should be implicit weighting – of >1 – given to adaptation expenditure. Thus, lower income populations are favoured in allocation of adaptation funds over higher income populations at a global scale. In practice, such a weighting appears to exist: the international effort to respond to climate change risks through reduction of greenhouse gases (mitigation) and adaptation is represented by the United Nations Framework Convention

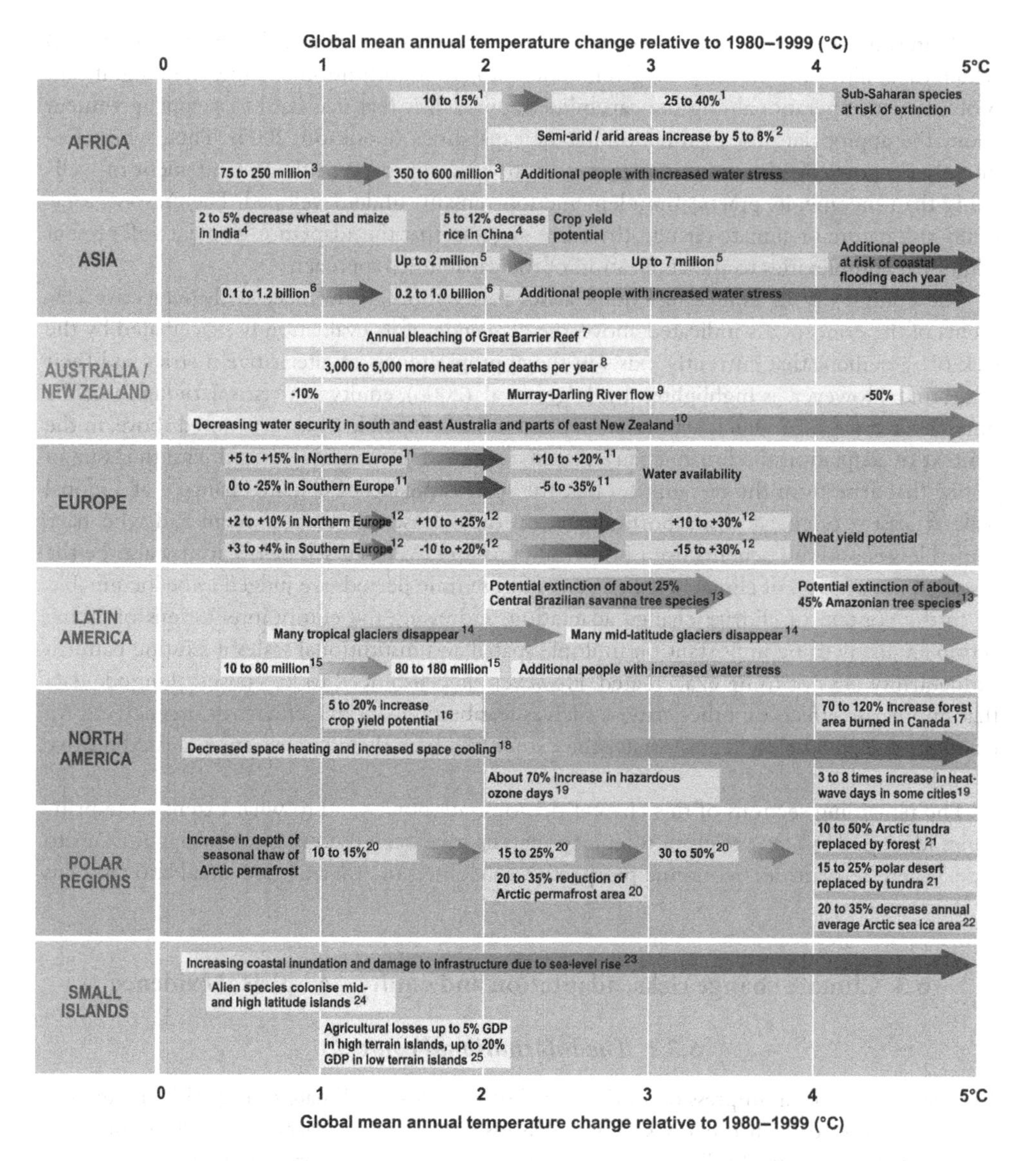

Figure 6.1 Examples of regional impacts of climate change in the twenty-first century
Source: Parry et al. [2007]. Figure TS.4. Page 67

on Climate Change (UNFCCC), and the UNFCCC uses differentiation among the parties to address equity in adaptation. Specifically, parties are distinguished along the development continuum (developed, developing and least developed); the vulnerability continuum (vulnerable and particularly vulnerable); the economic continuum (Annex I, Annex II and economies in transition) and on the basis of their physical characteristics (for example, small island countries and countries prone to flood, drought or desertification). The distributive justice aspect of equity is designed to be addressed through prioritization of needs according to vulnerability and through the elaboration of rules for accessing funding and technology transfer to meet the needs. The UNFCCC recognises that achieving equity is also at least partly dependent on procedural equity. Thus, procedural equity is designed to be addressed in the governance structures and operating mechanisms, key amongst them being the annual Conferences of the Parties (COPs) and the Global Environment Fund (GEF), in which effective access to, participation in – and operation of – both of these mechanisms by all parties is sought.

The effectiveness of the UNFCCC architecture in meeting equity objectives is, however, disputed. For example, Mace (2006) argues that whilst the UNFCCC, through some of its institutional arrangements (examples of which are the Least Developed Countries [LDC] work programme and the National Adaptation Programmes of Action [NAPA] process) attempts to address equity in adaptation, there remain a number of challenges to be overcome in achieving both procedural and distributive justice, and these fall into two broad interrelated sets of issues: access and 'voice', and funding.

In terms of access and voice, Mace argues that the most vulnerable of the developing countries are frequently the most disadvantaged in negotiations. They are often financially disadvantaged and may therefore only be able to put forward small delegations to the COPs, thereby reducing opportunities to participate in parallel sessions. Financial constraints may also result in reduced opportunities to attend preparatory regional meetings where the benefits of regional coalitions can be explored. Other challenges include lack of access to translation services in informal working groups and lack of institutional and technical capacity to prioritise their adaptation needs. It is also the case that the most vulnerable can be disadvantaged in some negotiating scenarios within GEF in particular. In this forum, a consensus is required, and the voting mechanism counts both participants and contributions to the fund. This means that GEF council members from countries that make the largest contributions to the fund carry the most weight, clearly constituting a challenge to distributive justice.

There are also significant funding challenges within the UNFCCC. The case for financing adaptation internationally arises from the differences in per capita income, per capita GHG emissions and energy intensities between countries. For example, in 2004, Annex 1 countries contained 20% of the world's population and produced 57% of global GDP and 46% of global greenhouse emissions. Average per capita emissions in Annex 2 countries were about a quarter of the Annex 1 levels. Figure 6.1 provides an indication that requirements for finance to support adaptation in Annex 2 countries will be considerable and additional to development finance. A range of estimates of the adaptation costs for these countries are summarised in Table 6.1. In evaluating the state of financing for adaptation, a paper in the UNEP Copenhagen Discussion Series (UNEP, 2009) identifies that there is an adaptation-funding deficit, i.e., funds available are not sufficient to meet adaptation funding requirements. The authors estimate that there is a current global total of $3 billion per annum in adaptation funding available in such funds as GEF-managed funds established under COP7 in 2001, the Least Developed Countries Fund, the Special Climate Change Fund and The Adaptation Fund, established under Article 12 of Kyoto Protocol. Given that current estimates of adaptation funding needs are of $1 billion per annum for Africa alone, rising to $7 billion by 2030, the global gap is significant.

In contrast to Mace, Thomas and Twyman (2005) present a view of global adaptation equity that seems to have significant commonalities with the Hayekian perspective outlined above. In assessing the extent to which the UNFCCC mechanisms facilitate equity at the global scale, they argue that whilst much of the discussion about equity in adaptation focuses on the distributional dimensions, it tends to be based on a deterministic view of people in the developing

Table 6.1 Adaptation costs, US$ billion per year

Assessment	*Developing countries*	*Benchmark year*
UNDP, 2007	86–109	2015
UNFCC, 2007	28–67	2030
World Bank, 2006	9–41	Present
Oxfam, 2007	>50	Present
Stern Review, 2006	4–37	Present

Source: Adapted from UNEP Copenhagen Discussion Series, Paper 3, July 2009

world as victims of global and climactic forces with little capacity for action or decision making and little resilience. They argue that these discourses undervalue and undermine the potential for such peoples to adapt under circumstances of climate change. Their view is therefore that equity in the context of climate change outcomes needs to be seen more broadly and should encompass issues including:

- Decision-making processes – who decides, who responds;
- Frameworks for taking and facilitating actions;
- Relationships between the developed and developing world and
- Relationships between climate change impacts and other factors that affect and disturb livelihoods.

These views are scrutinised by Huq and Khan (2006) in the context of implementing the Bangladeshi National Adaptation Program of Action (NAPA) – NAPAs constituting one of the key mechanisms for achieving a distribution of global resources to meet adaptation needs, with the process being approved and funded by the UNFCCC. The authors' focus is on procedural justice, based on the assumption that if there is fairness in the process used to prioritise adaptation activities, there would be equity in the outcomes achieved. On the basis of their experience in developing the Bangladeshi NAPA, which they judged as being partially effective, they suggest that procedural equity in the NAPA process needs to be addressed through consideration of the structure and functioning of the participatory process in terms of the composition of the NAPA team, including representation at various levels, and the nature of public participation in the process at both the plan development and review stages. Their analysis of the process in Bangladesh showed that while there are provisions in the NAPA guidelines to ensure a focus on process when developing plans, the effective application of such provisions depends on the nature of governance and institutional culture and traditions within a country. They conclude that in many developing countries, public consultation and review is a rapidly evolving norm that can serve the purpose of achieving equity, if careful consideration is given to it at every stage in the process. Huq and Khan (2006) also serve to highlight the importance of considering an indicator to identify adaptation needs broader than income alone. Specifically, they suggest that livelihood vulnerability is a more effective analytical concept since it captures security aspects of welfare that income measures aggregate out of any assessment of vulnerability. In so doing, it closely approximates the concept of functionings incorporated in Sen's capabilities approach.

Table 6.2 Alternative principles of distributive justice under resource scarcity

Allocation principle	*Interpretation*	*Countries favoured*
Horizontal equity	Same amount to all	Small countries
Proportionality	Proportional to damages	Countries with large (potential) damages
Vertical equity	Preserve the current distribution of climate-sensitive resources or risks	Countries where adaptation is relatively 'expensive'
Classical utilitarianism	Maximise adaptation benefits regardless of location	Countries where adaptation is relatively 'cheap'
Priority view	Maximise welfare of those worst off	Least developed countries

Source: Adapted from Füssel (2009)

A further rebalancing of these alternative positions is suggested by the findings of Tompkins and Adger (2004), in their study of adaptation processes in two Southern African natural-resource-dependent communities. The study highlights that local empowerment does not always lead to just and equitable decisions at the community and household levels. In doing so, it demonstrates that whilst empowerment is considered a cornerstone for equity in decision-making around adaptation (World Bank, 2000), it must be addressed along with considerations of distributive and procedural equity at all levels of decision-making.

The basis on which global distributive equity should be administered is, however, not easily established. For example, Füssel (2009), reproduced in Table 6.2, demonstrates that when the alternative allocation principles outlined above are adopted, the countries that benefit may differ significantly.

Furthermore, Füssel argues that – independently of the allocation principle that is adopted – whilst the use of national-level vulnerability indicators is necessary towards the allocation prioritization of adaptation assistance to the most vulnerable countries, it can lead to highly controversial results since it may implicitly reward poor governance. It is therefore likely to be controversial as to whether the country with poor governance – and thus higher vulnerability – should receive higher, equal or lower adaptation assistance than the other. The underlying – and unresolved – dispute then centres on whether the population in the country with poor governance should be regarded primarily as vulnerable, and warranting help, or as at least partly responsible for its vulnerability, thereby diminishing the equity rationale. This represents one of the biggest challenges to addressing adaptation equity at the international level.

6.3.2 The national context

On the basis of the brief overview in the preceding section, the justification for international support for adaptation in countries most vulnerable to climate change risks can be seen to be influenced by equity considerations, albeit without consensus. Most obviously, the UNFCCC architecture exists to facilitate the process of distributing financial support for adaptation. However, as highlighted above in Huq and Khan (2006) and Tompkins and Adger (2004), the concept and practicalities of that distribution process remain contentious. In order to illustrate how the challenges of equity analysis are currently met in practical terms at a national level,

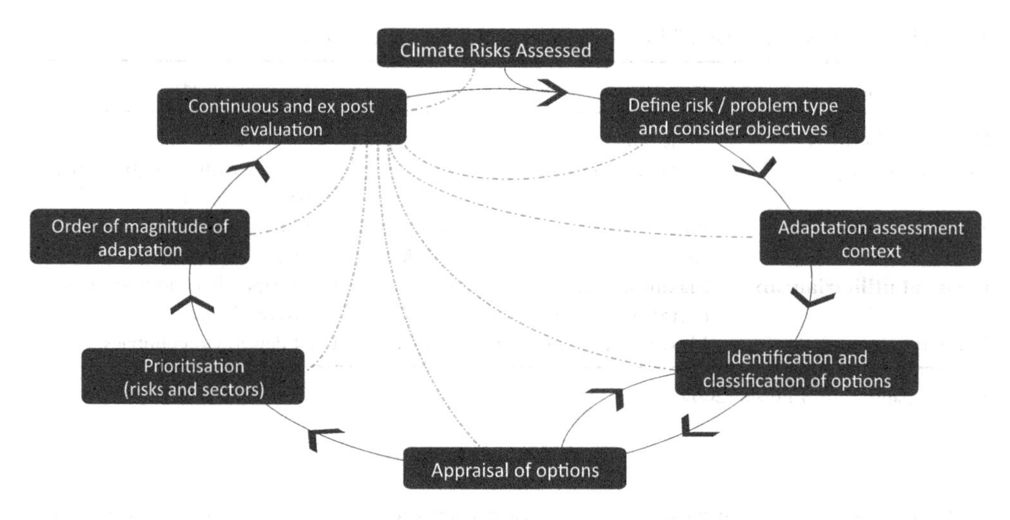

Figure 6.2 National Adaptation Assessment Framework
Source: Adapted from Watkiss and Hunt, 2010

we present two country-based case studies of adaptation assessment: one from the developing country context and the other from Western Europe – which can be seen to constitute best practice in these regions.

National-scale assessments are typically framed in cyclical terms, as sketched in Figure 6.2, where climate risks are identified, and adaptation objectives defined, prior to the adaptation assessment – option identification and appraisal – that leads to a prioritisation of adaptation actions and a scoping of the aggregate budgetary implications. An evaluation of the process then informs future assessments as new climate information becomes available and/or new assessment methods are developed.

Ethiopia

The context in which the experience of Ethiopia can be viewed is the vision that the national government has adopted – to build a Climate-Resilient Green Economy (CRGE) by 2025. The supporting strategy divides into two components: the green economy, focussed on GHG emission reductions on a development path compatible with becoming a middle-income economy by 2025, and the resilient economy that emphasises the need to adapt to climate risks (FDRE, 2011). The current climate resilience strategy – outlined in FDRE (2013) – is focussed solely on the agricultural sector, reflecting the fact that it is at present the dominant sector in the national economy, comprising 41% of GDP and 75% of total export commodity value, whilst employing 85% of the work-force; GDP per capita is currently \$380.[3] Current and projected future climate stresses are shown in Table 6.3.

The focus on the low per capita income agricultural sector in the initial stage of the climate resilient component suggests there to be a distributive aspect to the strategy design. Three further aspects of the strategy demonstrate the equity considerations being considered: adaptation option identification, option appraisal and option prioritisation. We outline each in turn.

Table 6.3 Climate stresses and key impacts

Climate stresses	*Key impacts*
High mean temperature	Shifting agro-ecological zones
Days with a max temperature above 35°C	Heat stress for some crops
Days with a max temperature above 40°C	Leads to heat stress on people and livestock
Lower mean rainfall	Shifts in agro-ecological zones plus drought regimes
Higher mean rainfall	Landslides, damage to crops and livestock
Large-scale floods	Damage to crops, livestock, infrastructure and people
Flash floods	Local damages to crops, livestock, infrastructure and people
High 1-hour rainfall intensity	Soil erosion and landslides, some local damages to crops
Heavy hail events	Crop damage at certain times in the growing season
Rainfall distribution (variability) within season	Significant impact on some crops
10-day dry spells	Significant impact on some crops
Seasonal droughts	Significant impact on most crops
Consecutive seasonal droughts	Significant impact on livelihoods and economic growth
Later onset of rainfall season	Shortens growing period – impacts on crops, fodder
Earlier end date of the rainfall season	Shortens growing period – impacts on crops, fodder
Decreased predictability of the rainfall season	Less reliable forecasts affects some enterprises
Increased uncertainty in rainfall distributions	Increases risk, important for some enterprises
Increases in cloudiness and humidity	Reduces radiation, increases thermal stress for people

Source: Adapted from FDRE (2013)

Adaptation option identification

Responses to the current and future climate risks listed in Table 6.3 were identified – through an expert elicitation process comprising a team including Ministry of Agriculture officials and representatives from the three main farming groups (pastoralists, agro-pastoralists and smallholder cropping) – in three discrete stages that successively reduced the number that could be regarded as most promising and so eligible for more detailed appraisal. The three stages comprised of: (a) compiling a long-list of options based on a stocktake of existing adaptation and development plans, programmes and literature in Ethiopia and internationally; (b) a clustering of options and (c) a filtering based on a number of criteria. The criteria were expressed in a series of questions:

1. Does the option pass an initial assessment of relevance and feasibility to be implemented in the local context?
2. Does the option provide a positive contribution to reaching the targets of the Ethiopian Growth and Transformation Plan (GTP)?
3. Does the option help to alleviate poverty, address distributional and equity issues (women, children, and people with disabilities) and ensure food security?
4. Does the option provide significant reductions to the current costs of climate variability and future climate change?

On the basis of this three-stage process, 41 distinct adaptation options were identified.

Option appraisal

In order to further inform resource allocation decisions regarding the identified adaptation options, a multi-attribute analysis (MAA) was undertaken: first, on the basis of a literature review and second, in a stakeholder workshop attended by farming representatives from across the country. The broad attribute groups included: institutional feasibility, effectiveness in climate risk reduction, synergies and co-benefits, economic costs and benefits and urgency. Within 'synergies and co-benefits' two specific attributes were 'poverty and equity' and 'gender'.

Each option was then scored on a scale of 1 to 4, with 4 representing a relatively high importance. However, it was not found to be possible to agree weightings to individual – or groups of – attributes. Implicitly, therefore, each was given equal weighting. The MAA also did not attempt to further rank the adaptation options. Rather, the MAA was interpreted as: (a) a validation of the option identification process; (b) a means by which implementing authorities could subsequently select options, given a particular policy focus (e.g., flood management) and (c) a sequencing of implementation according to their score against the 'urgency' attribute.

The process undertaken in both the Option Identification and Appraisal parts of the adaptation assessment relies on multi-attribute ratings, undertaken by principal stakeholder groups, which incorporate explicit consideration of equity. The strategy documentation states that the process was chosen to ensure that efficiency criteria do not dominate other decision criteria. It is notable, however, that the weightings given to the equity criteria are left to be made by the decision maker within the Ethiopian Government; the transparency in the process is diminished at this point.

Option prioritisation

In the Ethiopian climate resilience strategy, the process component subsequent to the option appraisal comprised a prioritisation of options based on a temporal criterion compatible with principles of iterative risk management (IRM) (IPCC, 2012). IRM seeks to identify how to implement options over time, based on current impacts and projected future climate change, against a background of future climate – and socioeconomic – uncertainty. In the present context, the IRM considered three time periods across which to sequence adaptation options based on the identified climate risks: the short term (now), the medium term (2020) and the long term (2050+). On the basis of this, short-term actions were identified to: develop capacity to more effectively implement options later (e.g., climate research and data gathering, institutional strengthening), exploit low-regret opportunities for social and economic benefits irrespective of the envelope of future climate change and undertake pro-active investigation of potential longer term options (e.g., development of drought resistant crop strains).

This management approach therefore effectively acts to facilitate protection against climate risks for agricultural producers and consumers in future time periods. Since Gollier (2001) shows that lower income groups within a population can be expected to be more risk averse – in absolute terms – than those with average and higher incomes, planning on this basis to reduce the risk of uncertainties in future income levels can be interpreted as implicitly accepting that relative welfare levels within the population should not be exacerbated by climate change risks.

This approach may therefore be seen to be equivalent to the notion of risk premia being attached to the weighting given to future climate risks in a cost-benefit analysis of climate change adaptation (Markandya, 2011). However, it also has a strong link with the concept of strong sustainability that is sometimes martialled in order to ensure that temporal aspects of resource allocation are more fully addressed than reliance on discounting allows. Specifically, whilst cost-benefit analysis – incorporating positive discount rates – may be justified in the short term, in the longer term, decisions about adaptation necessarily turn from consideration of specific options towards the provision of adaptive capacity. In turn, adaptive capacity – as with the notion of sustainability more generally – can be seen as the availability of capital of all forms.

Moreover, if capital substitutability cannot be assumed, it suggests that discounting in adaptation assessment may be best utilised in conjunction with the use of capital constraints similar to the notion of environmental stewardship suggested by Howarth (1995). This approach reflects the pluralistic ethical framework suggested by Norton and Toman (1997) to encompass alternative value systems concerning the future, combining utilitarian-based CBA decision rules with the rights-based rules outlined by e.g., Sen (1982). IRM adopted in the Ethiopian context therefore suggests a consistency with these two views of the appropriate treatment of equity. What is not so clear is whether the allowance – in terms of resource commitment – made to future time periods is sufficient to guarantee achievement of the desired degree of equity. This may be a principal practical challenge over the following decades for the implementing agencies.

UK

The policy context in which adaptation strategy is developed in the UK is shaped by the Climate Change Act 2008 which includes – inter alia – a legally binding commitment to build the country's ability to adapt. The evidence base with which to facilitate the National Adaptation Programme (NAP) has been compiled principally through the UK Climate Change Risk Assessment (CCRA) published in 2012 and related research. This related research specifically included a series of linked projects – the PREPARE Programme of research on preparedness, adaptation and risk – one of which was charged with understanding the equity effects of climate risks and adaptation options (Ricardo-AEA, 2013). This research, in particular, provides input into the adaptation assessments needed to inform the NAP – analysing the equity dimensions in the "Climate Risk Assessment" and "Option Appraisal" components of the adaptation assessment framework. These are outlined below.

Climate risk assessment

Ricardo-AEA (2013) identified key impact risks and vulnerable groups and conducted a detailed analysis of differential impacts for four key risks – flooding, heat-waves, drought and extreme cold – and associated vulnerable groups. These four risks were selected on the basis of an expectation, informed by the existing literature, that these risks would result in the most significant inequalities. Vulnerable groups identified were, the elderly, children and young people; those with compromised physical or mental health; people with mobility problems; people in poor quality housing or remote rural areas; the homeless; those with drug and alcohol misuse problems; transient communities; people on low incomes; and those for whom English is not their first language. Table 6.4 gives measures of the extent to which certain vulnerable groups may be impacted by the climate risks under current and future climatic conditions.

Table 6.4 Overview of differential risk estimates for climate risk–vulnerable group combinations

	Measurement of differential risk	*Scaling*		
		Current high risk	*Current enhanced risk*	*Future risk*
Flooding and economic deprivation	– Low income status – Number of households in deprived areas according to Index of Deprivation	– 66,000 (very low income in high flood risk areas) – 55,000 (properties in deprived areas at significant risk of flooding)	– 720,000 (very low income in areas with some flood risk) – 410,000 (properties in deprived areas at medium or low risk of flooding)	Increase in properties at risk in the most deprived areas of 1.5 to 2.8 times for the 2020s and 1.7 to 3.7 times for the 2050s.
Water scarcity and low-income users	– Water poverty (a threshold of 3% of disposable household income on water bills)	– Average of 46% of households dependent on benefits – Average of 40% of single adult and lone parent households	– 24% of all households	Estimate of 6.5M to 14M population in water poverty in areas of water deficits by the 2050s even without price rises
Heatwaves and ethnicity	– Numbers of ethnic groups in urban areas – Proficiency in English – New arrivals – low income	– 1.6% of England and Wales residents cannot speak English well or at all – About 1,836,000 new arrivals in England and Wales between 2007 and 2011 – About 40% of people from ethnic groups live in low – income households	– 99% of Black and Asian groups and 97% of Chinese live in urban areas compared to 90% of the general population	Not estimated
Cold-weather events and homelessness	– Statutory homeless and rough sleepers – No fixed-abode hospital admissions	– 2,300 (rough sleepers) – No data for hospital admissions during cold weather differential risk is likely to be significant	– 52,960 (statutory homeless) – No data for hospital admissions during cold weather differential risk is likely to be enhanced	Not estimated

Source: Adapted from Ricardo-AEA, 2013

Option appraisal

Adaptation appraisal of options included an economic analysis of the differential costs associated with the options. Taking one of the selected risk/vulnerability combinations as an example, the appraisal considered the possible differential cost impacts on households with varying ability to afford investment in flooding resilience and resistance measures, and restoration following flooding. Resistance measures are defined as those that prevent flood water from entering properties whilst resilience measures include those that reduce damage to properties when water has entered the property. Both resistance and resilience forms of adaptation are ex ante measures. In contrast, when these forms of measure are not implemented, households may be expected to rely on restoration of their property and its damaged contents, ex post of flood events.

For households living in high-flood-risk areas, five scenarios were developed to illustrate possible differential costs including:

1. Ex-post restoration measures undertaken following flooding,
2. Investment in a package of flood-resistance measures financed through loan finance,
3. Investment in a package of flood-resistance measures financed through savings,
4. Investment in a package of flood-resilience measures financed through loan finance and
5. Investment in a package of flood-resilience measures financed through savings.

Analysis of the above scenarios was undertaken in the case of a typical terrace house or flat. Key assumptions are as follows:

1. Cost of packages of adaptation measures: Scenarios have been calculated based on the cost of retrofitting measures. Indicative current costs were used for packages of resistance and resilience measures with 50-year lifetimes.
2. Restoration costs: The range of indicative current costs of restoration is based on estimates for flood depths of less than one metre. No account is taken of lost earnings or housing cost if the household temporarily relocates. Additionally, no account is taken of nonmarket costs such as the inconvenience and/or trauma of flooding to the householders. However, it is assumed that households – following a flood – restore the contents of the property to the state that it had before the flood. If it does not restore contents to the prior state, we would expect nonmarket costs to be higher.
3. Total financial costs to households: Total costs were estimated on the basis that one flooding event occurs during the lifetime of the adaptation measures (50 years). In other words, the property is located in an area that has a 1 in 50-year flood return rate.
4. Insurance availability. Results are estimated on the basis of the assumption that insurance is not available. Though the existence and availability of insurance would effectively act to smooth the size of payments over time, thereby avoiding the need to meet the cost in a lump-sum manner, the costs still have to be met by income over the time-period. Moreover, the assumption is increasingly realistic in the UK context as insurers are more often refusing to offer household insurance to those households identified as being at risk to flooding.

The results are presented in Table 6.5. The range of costs defined for each cost-type reflects the ranges in: (a) property size and (b) forms of resilience and resistance measures that were

Table 6.5 Indicative differential household costs of scenarios (£ per household)

Adaptation option scenarios: Terrace house or flat	*Indicative cost*		*Indicative cost of restoration per flood*		*Total financial cost*	
	Low	*High*	*Low*	*High*	*Low*	*High*
No adaptation	0	0	10,000	20,000	10,000	20,000
Resistance with loan	8,800	12,600	0	0	8,800	12,600
Resistance without loan	3,900	9,200	0	0	3900	9,200
Resilience with loan	3,700	40,000	2,000	10,000	5,700	50,000
Resilience without loan	2,300	28,400	2,000	10,000	4,300	38,400

Source: Adapted from Ricardo-AEA, 2013

considered in the modelling exercise. On the basis of a comparison of the cost estimates presented, a number of conclusions can be drawn with respect to the consequences for distributive equity. First, total costs are found to be greater to those households that do not introduce resistance or resilience measures. It does not seem unreasonable to assume that those who do not make these ex ante investments are not able to make them because they do not have access to sufficient funds.[4] Thus, low-income households are shown to bear higher costs – either in the form of adaptation costs or nonmarket impact costs – than higher-income households. A similar conclusion with regard to distributive equity can be drawn from comparison of costs borne by households that finance the adaptation cost through bank loans rather than solely from their own funds. Moreover, total costs to households that are able to finance resistance measures through banks loans were lower than for those where no ex ante adaptation was undertaken. Further, whilst these results pertain to the specific scenarios considered, further modelling suggests that our conclusions are robust to a wider range of assumptions relating to lifetime costs and frequency of floods.

In all of the four combinations of impacts and vulnerable groups considered by Ricardo-AEA (2013), it was found that impacts fell disproportionately on the vulnerable groups, and that in some cases the benefits of policy interventions were unevenly felt due to lowered adaptive capacity of some vulnerable groups. For example, in analysing the differential impacts of heat-wave policy interventions on ethnic minorities living in urban areas, findings suggested that adaptation measures and policy options can be less effective for some ethnic groups because economic poverty and lower educational attainment in such groups can lower their adaptive capacity and therefore their ability to cope with the effects of adaptation. However, it was also noted that the strength of family networks and social cohesion in such groups can augment and sometimes even circumvent the need for state health and social services. Similarly, the analysis of the differential impacts of adaptation policy to flooding on low-income households found that whilst flood-related policies can disadvantage these groups through weaker protection from floods and lower ability to recover assets lost through flooding, strong social networks can ameliorate the effects whilst poor language skills, disability and/or old age can exacerbate them. Clearly then, understanding social capital is therefore likely to have an integral role in addressing equity issues.

6.4 Summary and conclusions

This chapter explores the extent to which equity is considered in the determination of climate change adaptation resource allocation. It is clear that a number of the principal conceptualisations of equity – most notably distributive and procedural equity – are recognised as being both relevant and integral to the climate adaptation decision context. Thus, international adaptation initiatives – most obviously those that are implemented through the UNFCCC policy apparatus – have at their core a recognition that the current inequalities in both economic development and climate risks resulting from historical patterns of greenhouse gas emissions should be addressed in some way. Relative income level across world-regions is therefore the dominant metric by which inequality is measured. As the analysis of Füssel referred to in Section 6.3.1 indicates, however, there is a lack of consensus as to how these concepts should be applied in this context. Furthermore, his analysis, as summarised in Table 6.2, suggests that the allocation rule that is adopted is likely to make a material difference to which countries benefit from the resulting resource transfer from Annex 1 countries to Annex 2 countries. Moreover, even adopting the most generous interpretation of the concept, the discussion around Table 6.1 demonstrates that there remains a gap between acknowledging the principles of equity and their realisation of matching resource transfer commitments.

Recent experience of adaptation assessment at the national level also shows that equity is important in determining the acceptability of adaptation strategies. The example, presented above, of the Ethiopian adaptation strategy for the agricultural sector shows that – in including equity and gender as decision criteria within a formal multi-attribute assessment of adaptation priorities undertaken by stakeholder representatives – its development is at least influenced by this dimension. The example from the UK flood context, in which the cost of implementing alternative household-level flood management options is shown likely to be high for lower income households in both relative and absolute terms, is also likely to be an input to the development of a national adaptation strategy; it was commissioned by the Government Ministry responsible for coordinating the strategy. However, in both cases, whilst the decision-support processes recognise equity as a decision criterion, the final weightings that it receives in resource allocation are not yet clear. It therefore remains too soon to evaluate the strategies in terms of their effectiveness in achieving equity-based objectives.

Those country-level examples, and others referred to, also indicate that whilst implicit income weightings are the most common metric with which to address inequalities, other – broader – measures of human welfare – possibly of the type proposed by Sen and subsequent writers – can be used. Indeed, the UK assessment indicates that social capital – represented, for example, by family and community networks – may be highly important in determining levels of welfare. The discussion of temporal equity also highlights the potential importance of the preservation of capital stock – in this case to maintain economic options for future generations.

Overall, it seems that equity is established as a concept that has a role in the determination of adaptation resource allocation. Certainly, it is an intrinsic part of the design at the international level, where a retributive motive adds substantial weight to the case for resource transfer. In both the international and national contexts we have reviewed, however, the weight given to redistributive and procedural equity remains contentious and – in some cases – opaque. Thus, the practice of representing equity in adaptation processes and allocation is at an early stage. Ultimately, though, the extent to which equity is considered in decision processes may be expected to determine the acceptability and sustainability of the adaptation strategy. Future research therefore has an obvious role in developing metrics and ways with which to weight these metrics against other decision criteria.

Notes

1 Note that Paavola and Adger (2002) use the term *justice* rather than *equity* in their classification.
2 See http://www.idrc.ca/en/ev-146685-201-1-DO_TOPIC.html to find this quote
3 http://www.worldbank.org/en/country/ethiopia
4 Note, however, that available data on uptake of the national Flood Warning Service shows that around one-half of areas with significant and moderate flood risks are not yet registered with the service. If registration is considered a proxy of flood awareness, this suggested that a significant proportion of households may not be sufficiently aware of the opportunities for protecting themselves through resistance and resilience measures.

References

Adger, W.N., Paavola, J., and S. Huq (2006). Towards Justice in Adaptation to Climate Change. In Adger, W.N., Paavola, J., Huq, S. and M.J. Mace, (Eds.) *Fairness in Adaptation to Climate Change.* Massachusetts: The MIT Press.

Alkire, S., and Deneulin, S. (2009). The Human Development and Capability Approach. In Deneulin, S. and Shahani, L. (Eds.) *An Introduction to the Human Development and Capability Approach: Freedom and Agency.* London: Earthscan.

Anand, P. (2005). Capabilities and Health, *Journal of Medical Ethics*, 31: 299–303.

Cookson, R. (2005). QALYs and the Capability Approach, *Health Economics*, 14: 817–829.

FDRE (2011). *Ethiopia's Climate Resilient Green Economy: Green Economy Strategy.* Addis Ababa, Ethiopia: The Federal Democratic Republic of Ethiopia.

FDRE (2013). *Ethiopia's Climate Resilient Green Economy: Climate Resilient Strategy: Agriculture.* Addis Ababa, Ethiopia: The Federal Democratic Republic of Ethiopia.

Füssel, H.M. (2009). *Review and Quantitative Analysis of Indices of Climate Change Exposure, Adaptive Capacity, Sensitivity, and Impacts.* Background Note to the World Development Report 2010. Washington: World Bank.

Gollier, C. (2001). *The Economics of Risk and Time.* Cambridge MA: MIT Press.

Hayek, F.A. (1976). *Law, Legislation and Liberty: The Mirage of Social Justice.* London: Routledge and Kegan Paul.

Hope, C. (2008). Discount Rates, Equity Weights and the Social Cost of Carbon, *Energy Economics*, 30 (3), 1011–1019.

Howarth R.B. (1995). Sustainability Under Uncertainty: A Deontologist Approach, *Land Economics,* 71, 417–427.

Huq, S., and Khan, M.R. (2006). Equity in National Adaptation Programs of Action (NAPAs): The Case of Bangladesh. In Adger, W.N., Paavola, J., Huq, S. and Mace, M. J, (Eds.) *Fairness in Adaptation to Climate Change.* The MIT Press.

IPCC (2012). *Managing the Risks of Extreme Events and Disasters to Advance Climate Change Adaptation.* A Special Report of Working Groups I and II of the Intergovernmental Panel on Climate Change [C.B. Field, V. Barros, T.F. Stocker, D. Qin, D.J. Dokken, K.L. Ebi, M.D. Mastrandrea, K.J. Mach, G.D. K. Plattner, S.K. Allen, M. Tignor, and P.M. Midgley (Eds.)] Cambridge, UK and New York, NY, USA: Cambridge University Press, 582 pp.

Khor, M. (2012). Statement for UNFCCC AWG-LCA Workshop on Equity, South Centre, Bonn, 16 May 2012. At: http://unfccc.int/files/bodies/awg lca/application/pdf/20120517_south_centre_0922.pdf

Mace, M.J. (2006). Adaptation under the UNFCCC. In W.N. Adger, J. Paavola, S. Huq, and M.J. Mace (Eds.) *Fairness in Adaptation to Climate Change.* The MIT Press.

Markandya, A. (2011). Equity and Distributional Implications of Climate Change. *World Development*, 39 (6), 1051–1060.

Musgrave, R.A. (1959). *The Theory of Public Finance.* New York: McGraw-Hill.
Norton B.G., and Toman, M. (1997). Sustainability: Ecological and Economic Perspectives. *Land Economics,* 73 (4), 553–568.
Nussbaum, M. (1993). Non-Relative Virtues: An Aristotelian Approach. In M. Nussbaum and A.K. Sen (Eds.) *The Quality of Life.* Oxford: Oxford University Press.
Paavola, J., and Adger, W.N. (2002). *Justice and Adaptation to Climate Change,* Tyndall Centre for Climate Change Working Paper 23. University of East Anglia, Norwich, UK. In M.L. Parry, O.F. Canziani, J.P. Palutikof, P.J. van der Linden and C.E. Hanson (Eds.) Contribution of Working Group II to the Fourth Assessment Report of the Intergovernmental Panel on Climate Change, 2007. Cambridge, United Kingdom and New York, NY, USA: Cambridge University Press.
Rawls, J. (1972). *A Theory of Justice.* Oxford: Clarendon Press.
Ricardo-AEA (2013). *PREPARE – Understanding the Equity and Distributional Impacts of Climate Risks and Adaptation Options.* Report for Department for Environment, Food and Rural Affairs, London.
Sen, A.K. (1982). Approaches to the Choice of Discount for Social Benefit-Cost Analysis. In R.C. Lind (Ed.) *Discounting for Time and Risk in Energy Policy.* Washington DC: Resources for the Future, 325–353.
Sen, A.K. (1985). *Capabilities and Commodities.* Amsterdam: Elsevier.
Sen, A.K. (1992). *Inequality Re-examined.* Oxford: Clarendon Press.
Thomas, D.S.G., and Twyman, C. (2005). Equity and Justice in Climate Change Adaptation Amongst Natural-Resource-Dependent Societies, *Global Environmental Change,* 15, 115–124.
Tomkins, E.L., and Adger, W.N. (2004). Does Adaptive management of Natural Resources Enhance Resilience to Climate Change? *Ecology and Society,* 9 (2), 10.
UNEP (2009). Copenhagen Discussion Series. July 2009. At: http://www.unep.org/climatechange/Portals/5/documents/UNEP-DiscussionSeries_3.pdf
Varian, H. (1984). *Microeconomic Analysis.* New York, NY: W.W. Norton.
Watkiss, P., and Hunt, A. (2010). Method Report 3: Method for the Adaptation Economic Assessment. UK 2012 Climate Change Risk Assessment. Report for Defra, London.
World Bank (2000). *Can Africa Claim the 21st Century?* Washington, DC: World Bank.

7
DISCOUNTING

Ben Groom[1]

[1]DEPARTMENT OF GEOGRAPHY AND ENVIRONMENT,
LONDON SCHOOL OF ECONOMICS AND POLITICAL SCIENCE

7.1 Introduction

This thorny issue of social discounting generates a great deal of heated debate within economics. One need only scan through the numerous articles written on the subject in the aftermath of the Stern Review (Stern, 2007; Nordhaus, 2007) to gain some insight into just how different opinions can be. The reason for the heat generated on this topic is obvious. By necessity, governments and other public bodies now have to consider longer time horizons than they did in the past. An entire intergenerational policy arena encompassing biodiversity, nuclear power, public health and a range of other long-term policies has emerged over the past few decades. The outcome of traditional Cost Benefit Analysis (CBA) in such cases is highly sensitive to the level of the discount rate deployed, and policy recommendations can differ radically on the basis of the discounting assumptions. At the centre of disagreement on social discounting is the apparently paradoxical observation that high discount rates, such as those typically deployed for medium-term projects, imply that virtually no weight is placed on the costs and benefits that accrue to currently unborn future generations. Standard discounting approaches appear to 'tyrannise' future generations. Introspection on this issue often leads to the conclusion that, in the words of Weitzman (1999), '. . . something is wrong somewhere . . .' with the usual economic logic behind discounting.

Unsurprisingly, in recent years the natural battleground for the discounting debate has been in the context of climate change mitigation. The distinction between normative and positive approaches to discounting discussed in the first IPCC chapter on the economics of climate change (Arrow et al., 1996, Ch. 6) became extremely divisive post Stern Review. According to Nordhaus (2007) and others, the ostensibly normative view of Stern ignored the higher rates of return available in the market. Hence, Stern was in danger of simultaneously impoverishing future generations via low return investments in mitigation, and burdening current generations with unnecessarily onerous climate policy. Yet according to Stern, Nordhaus's positive perspective – that some or other market rate of interest could be inserted into the analysis of climate change – was 'a serious mistake' since doing so ignores, inter alia, the formidable ethical issues at stake with regard to intergenerational projects (Stern, 2008). Consequently, Stern recommended rapid and significant action on climate change, while Nordhaus recommended a 'policy ramp': a gradual ratcheting-up of mitigation measures over time.

By now the discounting debate post Stern, and the differences in opinion that it has engendered, is well-trodden ground. Opinions on discounting are informed by a long tradition of economic and philosophical theory since at least Sidgewick, via Ramsey, Pigou, Harrod, Sen, Arrow, and other luminaries of economic thought (Gollier, 2012). As the aftermath of the Stern Review illustrates, despite rumbling on for decades there appears to be little agreement among experts on the details of social discounting, particularly with regard to the normative-positive divide, but also within those schools of thought.

Yet in the past decade, something resembling a consensus has emerged in the theory of discounting long-term time horizons. There is now something of a consensus that the social discount rate should decline with the time horizon or maturity of the costs and benefits considered. Agreement on this aspect of social discounting is rooted in pioneering work by Martin Weitzman and Christian Gollier and a glut of follow-on work inspired by their contributions. Broadly speaking, their respective contributions focus on different sides of the Ramsey Rule: $r = \rho + \eta g$. Weitzman largely focussed on the production side: the rate of return to capital, r, while Gollier focussed on the consumption side: the social rate of time preference, $\rho + \eta g$, where ρ is the pure rate of time preference, η is the elasticity of marginal utility and g is the growth rate of consumption.[1] In each case, uncertainty and persistence in r or growth, g, drives a declining term structure of discount rates. While debate still continues with regard to the evaluation of the parameters of the Ramsey Rule – the pure rate of time preference ρ, and the elasticity of marginal utility η – there is now a general agreement that for risk-free projects, the term structure of discount rates should decline with the time horizon (Arrow et al., 2012; Gollier, 2012, Ch. 1).

The striking thing about this emerging consensus is that it has been extremely influential from a policy perspective. In 2003, the UK government responded to a report summarising the literature as it stood at that time on the rationales for declining discount rates (DDR) (Oxera, 2002).[2] Citing influential work by Weitzman and Gollier, the HM Treasury Guidelines on Cost Benefit Analysis, the so-called Green Book (HMT, 2003), was updated to include a declining schedule of discount rates. See Figure 7.1a. Some years later, the French government, strongly influenced by the prolific work of Christian Gollier, also inserted a declining term structure of discount rates for CBA into its guidelines.[3] See Figure 7.1b. The USEPA guidelines on discounting have also internalised this literature in their recommendations, albeit in a slightly different way. Citing empirical work by Newell and Pizer (2003), which operationalised Weitzman's theoretical contributions (Weitzman, 1998, 2001), a lower yet constant rate of 2.5% is recommended for projects with intergenerational consequences.[4] This should be compared to the rates of between 3% and 7% recommended generally for sensitivity analysis. Nevertheless, a recent panel of experts meeting organised by the USEPA and Resources For the Future recommended that the USEPA and Office of Management and Budgets (OMB) guidelines should be updated to reflect recent literature on DDRs. The outcome of these recommendations will be known soon. Lastly, and most recently, the Norwegian Government has also consulted on the issue of long-term discounting and recently concluded that a declining term structure for discount rates is also appropriate, as illustrated by Figure 7.1c.

This chapter reviews the theoretical and empirical underpinnings of the recent changes in government policy on social discounting towards discount rates that decline with the time horizon: DDRs. The review of the theory of DDRs is not exhaustive, but focussed on those aspects of the theory that have influenced the most recent policy changes in the UK, US,

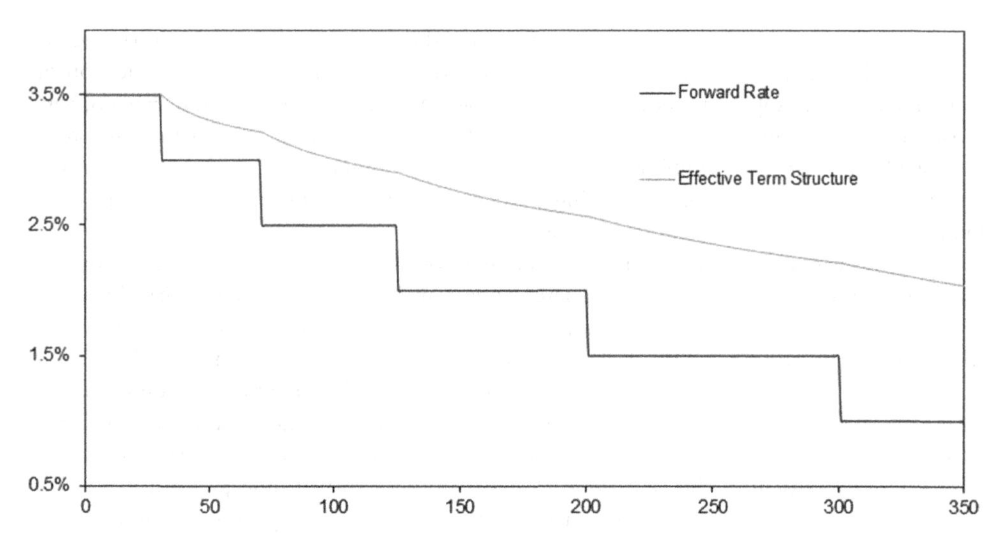

Figure 7.1a The UK government social discount rate term structure (HMT, 2003)

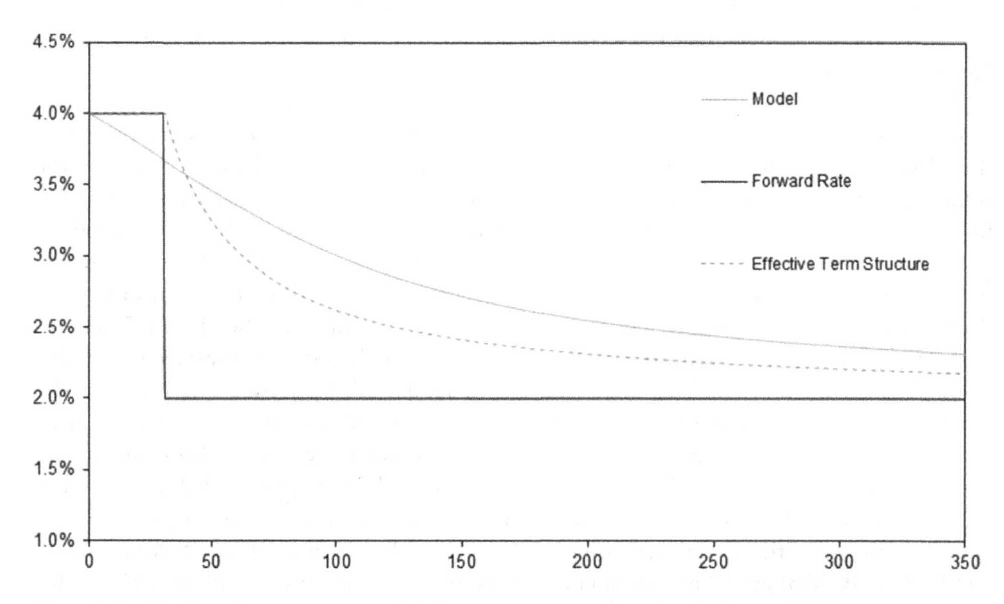

Figure 7.1b The French government social discount rate term structure (Lebegue, 2005)

France and Norway. The empirical contributions underpinning recent policy changes are also discussed so that a complete picture of the path from theory to practice is laid bare. The chapter concludes with a discussion of the implications for climate change adaptation, either planned or autonomous, of the recent discounting literature.

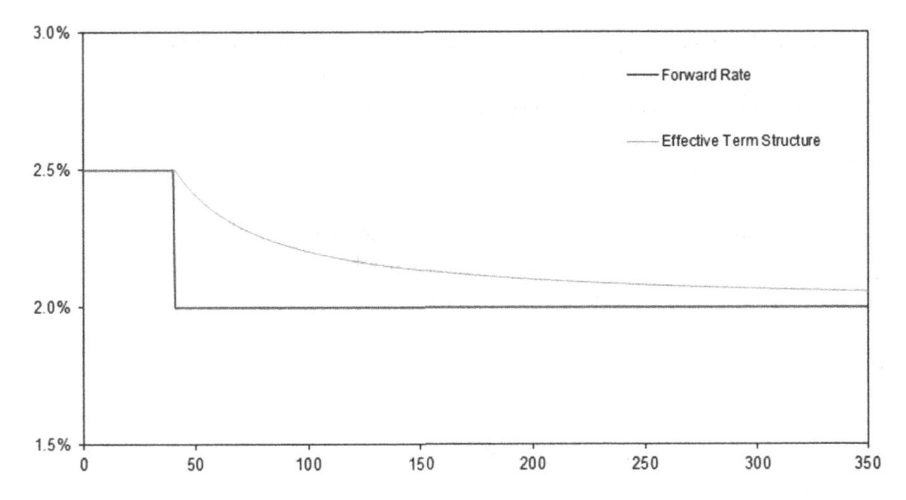

Figure 7.1c The Norwegian government social discount rate schedule (for risk-free projects) (MNOF, 2012)

7.2 Social discounting and the Ramsey Rule

7.2.1 The Ramsey Rule

The Ramsey Rule is typically the departure point for the analysis of the production and consumption side rationales for DDRs in the presence of uncertainty. We first present the deterministic case and then move on to some extensions.

The Social Discount Rate (SDR) is tied directly to the measure of inter-temporal welfare deployed. The workhorse approach to inter-temporal decision-making starts with a time separable discounted Utilitarian social welfare function (SWF). Here, inter-temporal welfare is measured by the sum of utilities, $u(c_t)$, over all time periods/generations from the present, $t = 0$, onwards discounted at the pure rate of time preference ρ:

$$W_0 = \sum_{t=0}^{\infty} \beta^t u(c_t) \tag{1}$$

where $\beta^t = (1+\rho)^{-t}$. Over and above its consequentialist underpinnings, this representation already embodies a number of crucial assumptions, including: (i) time-separability of utility; (ii) a constant exponential discounting of utility and (iii) addition of utilities as the appropriate metric of welfare. Separability rules out habit-forming and related behaviour, while constant discounting ensures time consistency. Each assumption simplifies the analysis of optimal and other consumption paths, $\{c_t\}$.

The social discount rate can be derived by considering the rate of return above which inter-temporal welfare would increase. Consider two time periods, period 0 and period t, so that the SWF can be written:

$$W_0 = u(c_0) + \beta^t u(c_t) \tag{2}$$

Now suppose that risk-free project i is available that costs one unit of consumption today, and yields $(1 + r_i)^t$ units at time t, where r_i is the internal rate of return (IRR) of the project. The change in welfare associated with project i is approximately:

$$\Delta W_0 = -u'(c_0) + \beta^t u'(c_t)(1 + r_i)^t \quad (3)$$

The social discount rate is the welfare preserving rate of return, that is, the return, δ, that sets the welfare change equal to zero. Rearranging (3) when $\Delta W_0 = 0$ and using a first order Taylor Series Expansion of the felicity function yields:

$$\delta = \rho - \frac{u''(c_t)}{u'(c_0)} c_0 g \quad (4)$$

where $g = (c_t - c_0)/tc_0$ is the annualised growth of consumption between time period zero and t. With iso-elastic preferences: $u(c) = c^{1-\eta}/(1 - \eta), - cu''/u' = \eta$ and (4) becomes:

$$\delta = \rho + \eta g \quad (5)$$

Equation (5) is often treated as a statement of the Ramsey Rule and measures what is known as the social rate of time preference (SRTP) in this framework. Equation (5) captures the essential message of Ramsey (1928). That is, it measures the minimum compensation required in the future to forego consumption today. In short, a project must compensate for any social preference for having utility now rather than in the future, and the fact that with diminishing marginal utility, additions to utility are worth less (more) in the future when there has been positive (negative) growth in the intervening period. Respectively, the project must compensate for impatience/different values placed on future generations' utility, ρ, and the wealth effect, ηg.

Strictly speaking, this is not quite the Ramsey Rule, which states that on the optimal consumption path, r will equal the social rate of return to capital in the economy. Where the aggregate production function is a function of aggregate capital stock k, $f(k)$, the Ramsey Rule states that:

$$r = \delta = \rho + \eta g \quad (6)$$

where $r = f_k(k)$. Furthermore, in a perfect decentralised economy with a perfectly functioning capital market, r will be equal to the market rate of interest, i. The Ramsey Rule shows how the production side, r, and the consumption side, SRTP, of the economy equate along the competitive or optimal consumption path. This differs from the interpretation in equation (5), which is a more general statement of the inter-temporal trade-off, and which does not require the consumption path in question to be necessarily an optimal one.

Despite the stylised nature of the theory, the Ramsey Rule forms the basis of most government policies on social discounting. For instance, the equivalence shown in equation (6) underpins the use of both the social opportunity cost of capital (production side) and the SRTP (consumption side) to inform CBA in different countries. Table 7.1 provides some international examples.

As indicated in Table 7.1, many countries have chosen to discount using an estimate of the SRTP. This choice is typically motivated by two assertions: (i) the Ramsey Rule (eq. 6) will not

Table 7.1 International experience with social discounting

Country	*SDR*	*Theoretical basis*
Australia	8%	SOC
China	8%	Weighted average approach. Lower for long term projects
France	4%	Social rate of time preference $\delta = 0$, $\eta = 2$ and $g = 2\%$
India	10%	SOC
Spain	6% (transport) 8% (water)	Social rate of time preference
UK	3.5% for intra-generational projects. Declining rates for inter-generational projects (See Figure 7.1a)	Social rate of time preference: $\delta = 1$, $\eta = 1$ and $g = 2.5\%$
	Office of Budget Management: 7%–3%: – 2.5%–3% for intra generational projects	SOC: pre-tax private rate of return. Different rates for cost effectiveness analysis
US	Congressional Budget Office and General Accounting Office	Rate of Treasury Debt with maturity comparable to project
	Environmental Protection Agency (US EPA): – 2.5%–3% for intra generational projects	Social rate of time preference

Source: ADB, 2007; IAWG, 2010

hold in a decentralised economy due to the presence of distortionary taxes and other market failures; (ii) public projects are predominantly funded from consumption side taxes and (iii) the market interest rate is not an appropriate SDR when (i) and (ii) hold.

When intergenerational projects are being evaluated a further motivation for using the SRTP arises: (iv) financial markets do not exist for such long time horizons, and hence no current market interest rates can price the relevant future costs and benefits. When conditions (i)–(iv) hold, the question becomes: how to estimate the *SRTP*?

7.2.2 Estimating the social rate of time preference

A great deal has been written concerning the parameters of the SRTP: ρ and η, and there are many possible strategies for estimating these parameters.

A standard approach to estimating the SRTP can be found in the UK Treasury Green Book (HMT, 2003). Here, the pure rate of time preference is estimated by looking not towards the ethical issues associated with pure time preference, but rather towards risk. A value of 1.5% is selected on the basis of an assessment of the risk of death within UK society (Pearce and Ulph, 1999). This is very much an intra-generational approach. Similarly, the elasticity of marginal utility, η, is estimated to be 1, based on microeconometric studies of inter-temporal savings and consumption behaviour witnessed in the UK. Coupled with an estimate of 2% long-term average growth, the SRTP becomes 3.5%. Similar approaches can be found in other countries. In France for example, the pure rate of time preference is set at 0% (Gollier, 2012; Lebegue, 2005), while the elasticity of marginal utility is set at 2 and growth is 2%.

When considering intergenerational investments, preference for one empirical strategy over another stems from whether one takes a positive view of discounting: that the discount rate should be informed by some observable market interest rate and the parameters of the SWF calibrated around this, or a normative approach: that the parameters should be based on consideration of intergenerational ethics. In fact, the disagreements on discounting witnessed in the aftermath of the Stern Review centred on this distinction, by and large.

The pure rate of time preference, ρ

In the social discounting literature, a normative view of the Ramsey Rule is typically taken. In this regard, the most heated debate has concerned the pure rate of time preference. A long Utilitarian tradition has argued, sometimes vociferously, for equal treatment of generations: $\rho = 0$, on ethical grounds. At this point, various luminaries of economics are routinely quoted using adjectives like 'indefensible', 'rapacious', 'myopic' and so on, to describe positive pure time preference.[5] Ultimately, however, this position entails treating future generations' utility equally to our own in the summation of the SWF in (1), no matter how far in the future they are. Obviously, some economists and ethicists disagree with this position. Arrow (1999) has discussed agent relative ethics and the trade-off between morality and self regard as an ethical argument for unequal treatment of generations utility: $\rho > 0$. Essentially, the argument here is that the imposition of equal treatment may tyrannise the present with onerous savings requirements. Individuals need not, it is argued, adhere to the morality of equal treatment if this comes at a great a cost to themselves. In particular, Arrow concludes that:

> . . . the strong ethical requirement that all generations be treated alike, itself reasonable, contradicts a very strong intuition that it is not morally acceptable to demand excessively high savings rates of any one generation, or even of every generation.
>
> *(Arrow, 1999, p. 7–8)*

So, with regard to the pure rate of time preference, there are strong and defendable arguments for zero and positive pure time preference. These are subjective differences, over which it is unlikely to get universal agreement, and this forms the basis of differing policies in different countries, and different perspectives among researchers. Now famously, the Stern Review took the normative view that, barring a small probability of catastrophe, each generations well-being should be treated equally: $\rho = 0$. This can be contrasted to the general UK guidelines, for instance.

Elasticity of marginal utility, η

Within the Ramsey framework, the elasticity of marginal utility reflects aversion to inequality of outcomes across possible states, among people and across time (Stern, 1977; Dasgupta, 2008). Empirical evidence suggests that individuals, even when considering intergenerational contexts, do not equate these concepts at all (Atkinson et al., 2009). While not necessarily descriptive of individual behaviour, the equality of these concepts in the Ramsey set up is taken as a normative stance on the treatment of difference across time, space and states (e.g., Gollier 2012). Doing so motivates radically different methods of estimating the parameter η: e.g., observed consumption behaviour at the aggregate or individual level (inter-temporal substitution), progressivity of the income tax schedules (inequality aversion), experiments involving risk (risk aversion), ethical introspection (inequality aversion), international transfers (international inequality aversion), and so on.

A complete review of the literature need not detain us here. Suffice it to say that Stern (1977), Oxera (2002) and more recently Groom and Maddison (2013) provide a review of the empirical evidence for the UK, which covers many of the methods typically deployed. Groom and Maddison's (2013) meta-analysis finds that the mean value across the different methods investigated supports a value of around 1.5–1.6, with a confidence interval between 0.9 and 2 in the UK.[6] There are, of course, some procedural issues to deal with here. For instance, Dasgupta (2008) was critical of Stern for taking an openly normative approach to the pure rate of time preference, but an obviously positive approach (econometric evidence on individual consumption) to inform η. Gollier (2012) recommends a value of 2 throughout his comprehensive review of discounting, based on normative considerations and personal introspection on inequality aversion, which coincides with the central value proposed by Dasgupta (2008), again for normative reasons relating to inequality aversion.

The difference perspectives on the parameters of the Ramsey Rule naturally lead to different recommendations for the social discount rate. With expected growth given as 2%, the UK's selection of $\rho = 1.5\%$ and $\eta = 1$ motivate the SRTP of 3.5%. In France, expected growth of 2% together with $\rho = 0\%$ and $\eta = 2$ lead to an SRTP of 4% following equation (5).

7.3 Declining discount rates: Consumption-side arguments

The Ramsey Rule provides a departure point for social discounting in the presence of uncertainty. Taking the parameters, ρ and η as given, uncertainty in growth affects how the social planner would 'price' future costs and benefits according to preferences over risk and the extent of the precautionary motive. In the intergenerational context, where the parameters of the iso-elastic Ramsey Rule typically have a normative interpretation, these two features of the utility function are closely related.

With uncertain growth, the inter-temporal SWF is typically augmented to measure the sum of expected utilities. In the two period case, the SWF becomes:

$$W_0 = u(c_0) + \beta^t E[u(\tilde{c}_t)] \tag{7}$$

Following the same procedures as before to find the welfare preserving rate of return, the expression for the SDR under uncertainty becomes:

$$\delta = \rho - \frac{1}{t}\ln\frac{E[u'(\tilde{c}_t)]}{u'(c_0)} \tag{8}$$

The nature of the SDR under uncertainty depends upon the nature of the uncertainty surrounding growth. Where growth is *i.i.d normal* and preferences are iso-elastic, equation (8) can be simplified to the following expression:[7]

$$\delta = \rho + R(c_0)\bar{g} - 0.5R(c_0)P(c_0)\,\text{var}(x)t^{-1} \tag{9}$$

where $\bar{g}$ is the expected growth rate and $\text{var}(x_t)t^{-1}$ is the annualised variance of growth. In this case, $\text{var}(x_t)t\sigma_c^2$, so the annualised variance remains constant over time. Now the SDR has an additional third term known as the 'prudence' effect (Gollier, 2012). $P(c)$ is a measure of 'relative prudence':

$P(c) = -cu'''/u''$, and $R(c)$ is the measure of relative risk aversion: $R(c) = -cu''/u'$. Both are positive when $u''' > 0$ and $u'' < 0$. The latter indicates diminishing marginal utility and leads to relative risk aversion. The former indicates that marginal utility is convex to the origin and indicates downside risk aversion, or aversion to differences in marginal utility. Where utility is iso-elastic, relative risk aversion is given by: $R(c) = \eta$ and relative prudence is given by: $P(c) = \eta + 1$, and (9) becomes:

$$\delta = \rho + \eta\bar{g} - 0.5\eta(\eta+1)\,\mathrm{var}(x)t^{-1} \tag{10}$$

So, with *i.i.d.* normal growth, the discount rate should be adjusted downwards where the social planner exhibits 'prudence' and therefore has a precautionary motive for saving in the presence of uncertainty. Such preferences are associated with a positive third derivative of the felicity function, which in turn reflects downside risk aversion (Kimball, 1990). In the normative context of the *SRTP* and where preferences are assumed to be iso-elastic, the selection of $\eta > 0$ ensures that the prudence effect reduces the social discount rate.

Of course, *i.i.d.* normal growth is potentially simplistic, and certainly does not lead to DDRs, the focus of this chapter. However, if growth exhibits persistence, that is, growth is correlated over time, the expansion of (8) leads to rather different recommendations for the SDR. Suppose, following Gollier (2012, Ch. 4), consumption growth follows the following process:

$$\begin{aligned} c_{t+1} &= c_t \exp(x_t) \\ x_t &= \mu + y_t + \varepsilon_{xt} \\ y_t &= \phi y_{t-1} + \varepsilon_{yt} \end{aligned} \tag{11}$$

That is, growth is dependent upon a state variable, y_t, which follows an Auto-Regressive process with one lag (AR(1)). The strength of the relationship of growth with past realisations is determined by the parameter ϕ. That is, ϕ measures the persistence of past shocks. Note that if $\phi = 0$ then the model returns to the *i.i.d.* model that we encountered above. Persistence increase as ϕ tends towards 1. In order to expand equation (8), in this case, one must define the mean and variance of the growth process, as undertaken in (9). Appendix 1 follows Gollier (2012, Ch. 4) and shows how the variance of growth now increases with the time horizon considered so that the SDR can be written as:

$$\begin{aligned} r_t = {} & \rho + \eta\mu - 0.5\eta^2\left[\sigma_x^2 + \frac{\sigma_y^2}{(1-\phi)^2}\right] \\ & + \left[\eta y_{-1}\phi\frac{1-\phi^t}{t(1-\phi)} - 0.5\eta^2\frac{\sigma_y^2}{(1-\phi)^2}\left[\phi^2\frac{\phi^{2t}-1}{t(\phi^2-1)} - 2\phi\frac{\phi^t-1}{t(\phi-1)}\right]\right] \end{aligned} \tag{12}$$

Note that the SDR is now indexed with a time subscript because it now depends upon the time horizon or maturity under consideration. In fact, r_t is declining with the time horizon since the last bracketed term shrinks as the time horizon gets larger. Precisely:

$$r_t = \begin{cases} \rho + \eta\mu - 0.5\eta^2[\sigma_x^2 + \sigma_y^2] & \text{for } t = 1 \\ \rho + \eta\mu - 0.5\eta^2\left[\sigma_x^2 + \dfrac{\sigma_y^2}{(1-\phi)^2}\right] & \text{for } t = \infty \end{cases} \tag{13}$$

where it is clear that the limiting SDR is less than the short run SDR.

There are two factors that determine the level of the SDR between these two extremes, that is, the term structure of the SDR. Firstly, the term involving the state variable y_{-1} reflects how expectations of future growth affect the SDR. If $y_{-1} > 0$ then current growth is above trend and the expectation is that growth will decline towards the long-term mean in the future. Secondly, and irrespective of growth expectations, the second term reflects the impact of the persistent growth uncertainty introduced by the auto-correlated state variable. With persistence, the variance of consumption growth is increasing with the time horizon. Given prudence, increasing uncertainty increases the precautionary savings motive and reduces the SDR with the time horizon.

This is but one rationale for DDRs. It is among the simplest theoretically and illustrates clearly the importance of persistence in determining the term structure of discount rates. Typically, the parameters of growth uncertainty are empirically estimated using historical data. Groom and Maddison (2013) estimate this simple model for the UK using historical data on growth for a period of 200 years.

Nevertheless, some empirical applications of the model above indicate that it is more appropriate for short-term term structures, with observed persistence insufficient to impact on long-term decision making. The French government's guidelines on social discounting were influenced by an extension of the model presented above, which introduces parameter uncertainty. That is, uncertainty with regard to the mean and variance of consumption growth. It seems realistic to assume that we are uncertain about the future mean and variance of growth, particularly over the long time horizons associated with the intergenerational investments in question. Gollier (2008, p. 178) shows that if the mean and variance of growth are themselves determined by a random parameter $\tilde{\theta}$, which reflects changes in growth regimes through, e.g. technological change, then equation (8) becomes:

$$r = \rho - \frac{1}{t}\ln E_{\theta}\frac{E[u'(\tilde{c}_t)\mid\theta]}{u'(c_0)} \tag{14}$$

By iterated expectations and using the Arrow-Pratt approximation this becomes:

$$r_t = \rho - \frac{1}{t}\ln E_{\theta}[\exp[-\eta t(\mu(\tilde{\theta}) - 0.5\eta\sigma(\tilde{\theta})^2)]] \tag{15}$$

The second term is increasing with the time horizon t. Defining the certainty equivalent M_t of $\mu(\tilde{\theta}) - 0.5\eta\sigma(\tilde{\theta})^2$ as follows:

$$\exp(-\eta t M_t) = E_{\theta}[\exp(-\eta t(\mu(\tilde{\theta}) - 0.5\eta\sigma(\tilde{\theta})^2))] \tag{16}$$

it can be seen that M_t is increasing in t since the exponential function becomes more concave in t. Therefore, the SDR is declining with the time horizon. The intuition follows from the previous example. In essence, uncertainty over the parameters of the growth distribution introduces further uncertainty about the future. For instance, where there is uncertainty about the mean of growth, even if the growth shocks are independent year on year, and in a sense rapidly mean reverting, uncertainty remains about which value of μ growth is reverting to. There are several growth scenarios that must be evaluated in calculating the certainty equivalent discount rate associated with (16). In doing so, future scenarios with high mean growth rates, and hence high SDRs and low NPVs, become unimportant in present value terms. Future scenarios with low mean discount rates therefore dominate the calculation of the certainty equivalent for long time horizons, and this drives the declining term structure of the SDR. This rationale for DDRs is offered as one motivation behind the French government's recommendations for discounting and the schedule shown in Figure 7.1b (Lebegue, 2005, p. 102).

There are numerous other consumption side rationales for DDRs. Catastrophic risks, as discussed by Weitzman (2009) in the context of climate change, and Barro (2006) in the context of economic depressions, have also been shown to motivate extra weight to be placed on future costs and benefits. The important issue in such cases is the presence of 'fat-tailed' risks, that is, an increased probability of adverse outcomes. Coupled with rapidly increasing marginal utility as consumption heads towards zero, as would be the case with an iso-elastic utility function of the type used above, a fat-tailed distribution means that the expected marginal utility in the presence of catastrophic outcomes is dominated by the enormous marginal utility associated with the bad states of the world. As one can see from equation (8), the larger the expected marginal utility at time t, the smaller the SDR. Weitzman's 'dismal theorem' (Weitzman, 2009) is essentially a formal statement of this in which the presence of catastrophic risks and fat tailed distributions (e.g., student–t rather than a normal distribution) indicates that an infinite weight be placed on the future![8] It can be argued that fat-tailed distributions underpin the arguments associated with parameter uncertainty also (Gollier, 2008).

7.4 Declining discount rates: Production-side arguments

Weitzman (1998, 1999) invoked a simple argument on the production side of the Ramsey Rule, focussing on uncertainty in the rate of return to capital, r. The original theoretical paper was rather ad hoc in that the decision rule under which the theoretical results were developed was an Expected Net Present Value (ENPV) rule. That is, in the presence of uncertainty in r the objective of the social planner is to choose investments to maximise ENPV:

$$ENPV = \int_{t=0}^{\infty} E[\exp(-\tilde{r}t)](B_t - C_t)dt \tag{17}$$

Where, for simplicity, it is assumed that the net benefits are measured in certainty equivalents.[9] The thought experiment that Weitzman (1998) invoked can be explained as follows. Imagine that a decision has to be made with regard to a particular project that costs £1 today and pays B_t at time t. The decision must be made under uncertainty about the opportunity cost of the project funds, r. As soon as a decision is made with regard to the project uncertainty is resolved and we find out which discount rate prevails for the duration of the project. The question is, what is the appropriate discount rate that should be applied in the moment of uncertainty?

The project described has the following ENPV:

$$ENPV = -1 + E\exp(-\tilde{r}t)B_t \tag{18}$$

and should be approved if $ENPV$ is greater than zero.[10] The equivalent decision criterion can be framed in terms of the certainty equivalent discount rate, that is, the certain discount rate that if applied over the time horizon t would yield the same $ENPV$. The certainty equivalent discount rate, $r_{CE}(t)$, can be defined as follows:

$$\begin{aligned} &\exp(-r_{CE}(t)t) = E\exp(-\tilde{r}t) \\ &\Rightarrow \\ &r_{CE}(t) = -\frac{1}{t}\ln(E\exp(-\tilde{r}t)) \end{aligned} \tag{19}$$

Due to the fact that the exponential function is convex, and more so with larger t, Jenson's inequality means that the certainty equivalent decreases with time. In fact, Weitzman (1998) shows that $r_{CE}(0) = E[\tilde{r}]$ and $\lim_{t\to\infty} r_{CE}(t) = r_{min}$; that is, the certainty equivalent discount rate should decline from its expected value to the lowest imaginable realisation of the return to capital. The essential insight here is that with the *ENPV* approach, one calculates the expected discount *factor* rather than the expected discount *rate*. The certainty equivalent discount rate is a DDR. A simple example suffices to complete the point.

Suppose that the discount rate is equally likely to be 0% or 5%. The expected discount factor is then: $0.5(\exp(-t * 0\%) + \exp(-t * 5\%))$. The certainty equivalent discount rate that yields the same discount factor is: $t^{-1} \ln(0.5(\exp(-t * 0\%) + \exp(-t * 5\%)))$. For $t = (10, 50, 100)$, $r_{CE}(t) = (2.2\%, 1.2\%, 0.7\%)$, which declines rapidly from the mean of 2.5% to the minimum of 0% as the time horizon extends.

As we shall see in the empirical section, this elegant yet powerful result can make an important difference to the evaluation of long-term projects for plausible characterisations of the uncertainty in the return to capital.

7.4.1 The Gollier-Weitzman puzzle

As subsequent literature showed though, the ENPV criterion is potentially problematic and certainly does not have a general theoretical justification. Gollier (2004) argued that the choice of ENPV is somewhat arbitrary, and illustrated his point by invoking an arguably equally valid objective, the Expected Net Future Value (ENFV), to evaluate projects when the rate of return to capital is uncertain:

$$ENFV = -E\exp(\tilde{r}t) + B_t \tag{20}$$

While at no point arguing for the use of the ENFV criterion, Gollier (2004) shows that the certainty equivalent rate associated with the expected *compound* factor, $E\exp(\tilde{r}t)$:

$$r_G(t) = \frac{1}{t}\ln[E\exp(\tilde{r}t)] \tag{20a}$$

has the opposite qualities to that under the ENPV criterion in that $\lim_{t\to\infty} r_{max}$. That is, if ENFV is the appropriate criterion, the appropriate discount rate should be *increasing* with the time horizon.

This apparent puzzle stems from the absence of a robust theoretical framework underpinning the *ENPV* criterion proposed by Weitzman (1998) and more famously in his Gamma Discounting paper (Weitzman, 2001). The puzzle framed by Gollier (2004) has become known as the Gollier-Weitzman puzzle, and has seen several resolutions in the literature, starting with Hepburn and Groom et al. (2007), and culminating with Gollier and Weitzman (2010) and Traeger (2012) via Buchholz and Schumacher (2008) and Freeman (2010).

Hepburn and Groom (2007) argued that the puzzle arose from a failure to specify the appraisal date, that is, the date at which the project is being appraised, τ. The present and the future date t of the project returns are two such dates, but the project could be evaluated at any intervening date, τ, from the perspective or a planner at that date. Introducing the appraisal date generates an Expected Net Value criterion of the following kind:

$$ENV(t,\tau) = -1 + E\exp(-\tilde{r}(t-\tau))B_t$$

and the associated certainty equivalent discount rate for Hepburn and Groom (2007) becomes:

$$r_{HG}(t,\tau) = -\frac{1}{t-\tau}\ln(E\exp(-\tilde{r}(t-\tau))) \quad (21)$$

This solution embodies both the *ENFV* and the *ENPV* criteria as special cases, respectively when $\tau = t$ and $\tau = 0$. Hepburn and Groom (2007) show that their certainty equivalent discount rate is increasing with τ but, for a given evaluation date, decreasing in t. In this sense, Hepburn and Groom claimed that Gollier and Weitzman were both right.

Whilst this solved the original puzzle as it was presented, what remained was the problem that the viability of a project depended crucially on the appraisal date, for which there was no clear selection criterion.[11] Ultimately, the puzzle was only properly solved by invoking more theoretical structure.

Gollier and Weitzman (2010) did exactly this. For a flavour of their more general result, take the standard Ramsey social welfare function:

$$W_0 = \Sigma_{t=0}^{\infty}\exp(-\rho t)u(c_t) \quad (22)$$

Suppose the economy is linear in production, but there is uncertainty in the rate of return capital, r_i. That is, we do not know which state i we will end up in. The optimality condition in state i for consumption is that a marginal reduction in consumption today should be offset exactly by the return delivered at time t:

$$u'(c_{0i}) = \exp(-\rho t)u'(c_{ti})\exp(r_i t) \quad (23)$$

For a project like the one above, which costs one unit of consumption now and pays off B_t at time t, the project will increase expected utility if:

$$\sum_i^n p_i u'(c_{0i}) \leq B_t \sum_i^n p_i u'(c_{ti})\exp(-\rho t) \quad (24)$$

As shown by Hepburn and Groom (2007), Weitzman and Gollier essentially use different evaluation dates. Weitzman places all costs and benefits in terms of the present. To do this, Gollier and Weitzman (2010) use (24) to purge (25) of the term c_{ti} to get, after rearrangement:

$$B_t \sum_i^n q_i^W \exp(-rt) \geq 1 \quad (25)$$

where $q_i^W = p_i u'(c_{0i}) / \sum^n p_i u'(c_{0i})$. Gollier essentially used a future evaluation date, and so used (24) to purge (25) of the term $u'(c_{0i})$ to get after rearrangement:

$$B_t \geq \sum_i^n q_i^G \exp(rt) \quad (26)$$

where $q_i^G = p_i u'(c_{ti}) / \sum^n p_i u'(c_{ti})$. Criteria (26) and (27) are the counterparts of (18) and (21), the *ENPV* and *ENFV* criteria, except that they have been adjusted in utility terms for risk. Respectively, they lead to the following certainty equivalent discount rates:

$$\begin{aligned} R^W(t) &= -\frac{1}{t}\ln E[q_i^W \exp(-r_i t)] \\ R^G(t) &= \frac{1}{t}\ln E[q_i^G \exp(r_i t)] \end{aligned} \quad (27)$$

While these discount rates are reminiscent of (20) and (21a), they have identical properties. In essence, in an optimising framework, adjusting for the risk preferences reflected in the utility function $u(c_t)$ makes the evaluation date irrelevant to the evaluation of the project, and apparently solves the Gollier-Weitzman puzzle.

There are caveats to this result, of course. For instance, the model relies on a rather particular thought experiment: uncertainty about the rate of return to capital for an instant, resolved forever thereafter. Similarly, it is assumed in all but one special case that current consumption, c_0, can adjust elastically when the uncertain return to capital is resolved. This too is a questionable assumption. From these perspectives, the model can be seen as instructive and illustrative, but perhaps ripe for extension.

Nevertheless, Gollier and Weitzman (2010) also provide a justification for the pure *ENPV* framework originally proposed by Weitzman (1998). It turns out that the special case in which current consumption is fixed irrespective of the resolution of the rate of return to capital is also the case in which the *ENPV* framework is theoretically valid. This occurs when the utility function is logarithmic. In that case, the weights in q_i^W all become simply p_i, and $R^W(t)$ in (28) converges to $r_{CE}(t)$ in (20).[12] Hence, Weitzman's original production side approach has a, albeit particular, theoretical justification.

7.4.2 Summary

The theoretical justification for the Weitzman approach is important and useful. The original contributions (Weitzman, 1998, 2001) have been extremely influential in policy circles, despite the puzzles that have occurred along the way. Both the UK Treasury Guidelines (HMT, 2003) and the US guidelines have been heavily influenced by this approach, and in some cases have based their discounting advice precisely on the concepts presented here (USEPA, 2010). The Norwegian government also refers to arguments about uncertainty of the rate of return to capital in its advice on the term structure of risk free discount rates (MNOF, 2012). So in some ways, in this thorny area of economic theory, it is something of a relief that there is a sound theoretical justification for the proposals of Weitzman.

It should be mentioned that there are other theoretical justifications for the ENPV approach in the literature. In the finance literature for instance, the ENPV approach is a manifestation of the local expectations hypothesis (Cox et al., 1981). In this case, uncertainty need only be realised one period ahead, rather than for all time, which is a more realistic formulation of the problem. Traeger (2012) has also shed further light on the issue, showing that, even in the Ramsey type framework, in the long run, the *ENPV* approach is valid.

We have now covered the theory relevant to the observed policies in the UK, US, France and Norway. In some cases the production side approach has been used, in others the consumption side has been favoured. We now turn to the empirics of the social discount rate and how the SDR was estimated in each case.

7.5 The empirics of the social discount rate

In this section, the empirics of the consumption side and production side rationales for DDRs are examined. In short, governments have deployed different approaches to this matter in developing their declining discounting schedules. We examine the full range of approaches here, providing both examples and insights into how the policies were developed.

7.5.1 The consumption side

Gollier (2012) has undertaken numerous estimates of the wealth and prudence effects for the social discount rate for different countries, taking as a starting point the choice of the ethical parameters of the SRTP as $\rho = 0$ and $\eta = 2$. In order to estimate the SRTP, one needs an estimate of the mean and the variance of growth for any given country. It is obvious from (9) that countries with high, stable growth will have a high discount rate. Those with low, volatile growth will have low, possibly negative discount rates. In the iso-elastic case, the precautionary effect is quite low in developed countries, and the wealth effect is moderate. In low growth developing countries, the wealth effect is typically outweighed by the precautionary effect, hence a lower discount rate.

The French government guidelines take a consumption side approach based on parameter uncertainty as described briefly above, but in more detail in Gollier (2007, 2008). The approach taken is a simplification and considers only uncertainty in the mean rate of growth, μ, which has n possible values with associated probabilities p_i, where $i = 1, \ldots, n$.

Ultimately, the guidelines use the following formula (Lebegue, 2005, p. 37):

$$r_t = -\frac{1}{t}\ln\left[\sum_{i=1}^{n} p_i \exp(-(\rho + \eta\mu_i)t)\right] \tag{28}$$

A normative approach is taken with respect to pure time preference and the elasticity of marginal utility. The French term structure could be motivated by the following parameter values: $\rho = 1\%$, and $\eta = 2$, reflecting aversion to income inequality, together with very simple characterisation of uncertainty with two possibilities for mean growth, with probability 1/3, $\mu_1 = 0.5\%$, with probability 2/3, $\mu_2 = 2\%$ (Lebegue, 2005, p. 102). Figure 7.1b shows the term structure associated with this characterisation of uncertainty and equation (29). The actual term structure used is an approximation, with 4% used for the first 30 years, and 2% used thereafter. There is no actual empirical motivation for the numerical example provided above. But the eventual term structure represents the result of a compromise of various opinions on the rationale for DDRs and the parameters underpinning the Ramsey Rule.

7.5.2 The production side

The empirical work on the production side has also been extensive and contributed to the policies now found in the US and UK guidelines, as well as informing the Norwegian government more recently. The key issue to the production side approach is that the uncertainty surrounding the rate of return to capital must be characterised in some way in order to calculate the certainty equivalent discount rate in equation (9). There are several approaches that have been taken in the literature.

Expert opinion

Perhaps the most widely cited empirical exercise in this field is Weitzman (2001) entitled 'Gamma Discounting'. Here, Martin Weitzman asked around 2,800 PhD-level economists for their proposed discount rate to evaluate climate mitigation investments. Specifically, he asked the following question:

> Taking all relevant considerations into account, what real interest rate do you think should be used to discount over time the (expected) benefits and (expected) costs of projects being proposed to mitigate the possible effects of global climate change?
>
> *(Weitzman, 2001, p. 266)*

The approximately 2,100 responses showed that the views on this issue were extremely heterogeneous, ranging from values of –3% to +27%. The sample frequency distribution of responses was approximately gamma distributed with a mean of 4% and a standard deviation of around 3%. The use of a gamma distribution is convenient and illustrative since it provides a closed form solution for the certainty equivalent discount rate. Defining the certainty equivalent discount rate as follows:

$$\exp(-r_{CE}(t)t) = \int_{r_{\min}}^{r_{\max}} \exp(-rt) f(r) \quad (29)$$

Weitzman used the gamma distribution to calculate this integral. That is: $f(r) = \Gamma(\alpha, \beta)$. Rearranging this to obtain the certainty equivalent discount rate when the uncertainty is gamma distributed gives the following closed form solution:

$$r_{\gamma}(t) = -\frac{\alpha}{t}\ln\left(\frac{\beta}{\beta + t}\right) \quad (30)$$

The associated term structure of the SDR is shown in Figure 7.2. It is sharply declining with the time horizon towards the sample minimum, which in the case of the gamma distribution is zero. The decline of the discount rate increases with the variance of the sample responses, or as Weitzman described it, the disagreement among experts. The policy implications of this term structure are potentially dramatic, and we return to this aspect below.

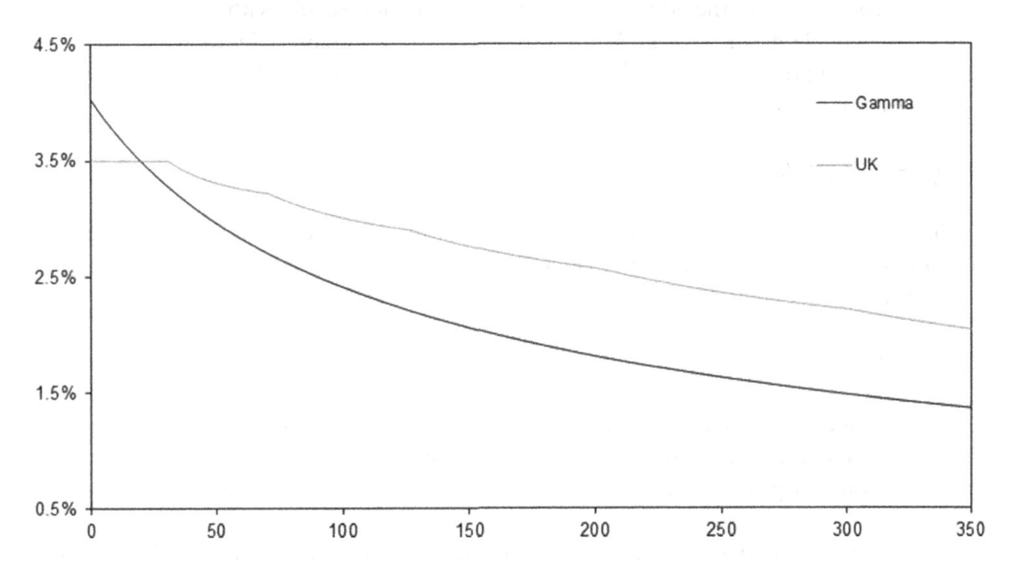

Figure 7.2 The Gamma Discounting and the UK term structures (HMT, 2003; Weitzman, 2001)

In order to understand the workings of the Gamma Discounting paper, it is helpful consider the non-parametric equivalent of (31):

$$r_Y^{NP} = -\frac{1}{t}\ln\left[\frac{1}{n}\Sigma_{i=1}^{n}\exp(r_i t)\right] \tag{31}$$

(32) shows that the certainty equivalent discount rate is calculated from the arithmetic average discount factor over the n experts sampled. That is, each expert's response is weighted equally in the calculation, no matter whether their response was –3% or 27%, each opinion carries equal weight, while the value of the response is weighted by its frequency in the sample.

In a recent paper, Freeman and Groom (2013) argue that, while the equal treatment of expert responses is an intuitive way to combine normative opinions, it is unlikely to be appropriate for positivist responses. It is unclear, Weitzman did not specify in the survey, whether responses come from the normative school or the positive school of thought on discounting. Indeed, the presence of responses in the order of 25%–27% indicates with a high degree of likelihood that at least some of the responses were of a positive nature.[13] Given that the context was intergenerational, it is also highly likely that many of the responses were normative. Freeman and Groom (2013) argue that this is an important distinction from an empirical perspective since different types of response should be combined in very different ways. A simple example illustrates their essential point.

Suppose that each expert response is a forecast of the future mean interest rate over the time horizon H. Suppose that their responses are contaminated with forecast error around the true value of the mean interest rate, $\bar{r}_H = H^{-1}\Sigma_{t=0}^{H} r_t$. The responses can be modelled as:

$$r_i = \bar{r}_H + \varepsilon_i \tag{32}$$

Suppose also that the experts are unbiased: $E(\varepsilon_i) = 0$, and independent such that the variance covariance matrix of the errors, Σ, is diagonal and homoskedastic with variance σ^2. Now consider the sample mean, $\bar{r}_H^n = n^{-1}\Sigma_{i=1}^{n} r_i$. With a large enough sample, by the Central Limit Theorem, the sample mean will be distributed normally about the population (true) mean, $\bar{r}_H$, and the *standard error* $n^{-1}\sigma^2$:

$$\bar{r}_H^n \sim N(\bar{r}_H, \sigma^2/n) \tag{33}$$

This is true irrespective of the initial distribution of the responses r_i. From this, the uncertainty in the true value of the discount rate can be characterised, as desired for the calculation of the certainty equivalent discount factor. Simple inversion yields:

$$\bar{r}_H \sim N(\bar{r}_H^n, \sigma^2 / n) \tag{34}$$

Conditional on the sample, the true mean is normally distributed around the sample mean with a variance equal to the standard error. Such an approach places much less weight on the extreme responses (e.g., 27%) since these are the likely to be the most erroneous when experts are unbiased.

It is important to notice that when one treats the expert responses as unbiased and independent forecasts, the uncertainty around distribution of the interest rate is dependent on the sample size. The more experts, so the argument goes, the more information and the less uncertainty about the future. In the limiting case considered here, the appropriate measure of spread is the standard error, not the standard deviation of the sample as was the case for Gamma Discounting. Since the decline in the discount rate in this framework increases with the uncertainty, the positive approach described here has important implications for the term structure of the discount rate. In fact, when one takes the positive approach to *ENPV* described above, the certainty equivalent discount rate can be calculated as follows. For time horizon *H* the certainty equivalent discount rate can be defined as:

$$\exp(-r_{CE}(H)H) = \int \exp(-\bar{r}_H H) f(\bar{r}_H) d\bar{r}_H \tag{35}$$

So the $r_{CE}(H)$ is a function of the $-Hth$ moment generating function of the normal distribution in (35), hence the certainty equivalent becomes:

$$r_{CE}(H) = \bar{r}_H^n - \frac{1}{2}\frac{\sigma^2 H}{n} \tag{36}$$

This is declining with the time horizon, *H*, but at a much slower rate than Weitzman's equally weighted approach.

Of course, there are a number of caveats to this result: what happens in small samples? What happens when experts are not independent? What happens if experts are biased? All of these questions are addressed by Freeman and Groom (2013), with similar outcomes. For instance, Figure 7.3 contrasts the term structures that emerge when smaller samples of experts are considered. A smaller sample has an equivalent effect to non-independence of experts.

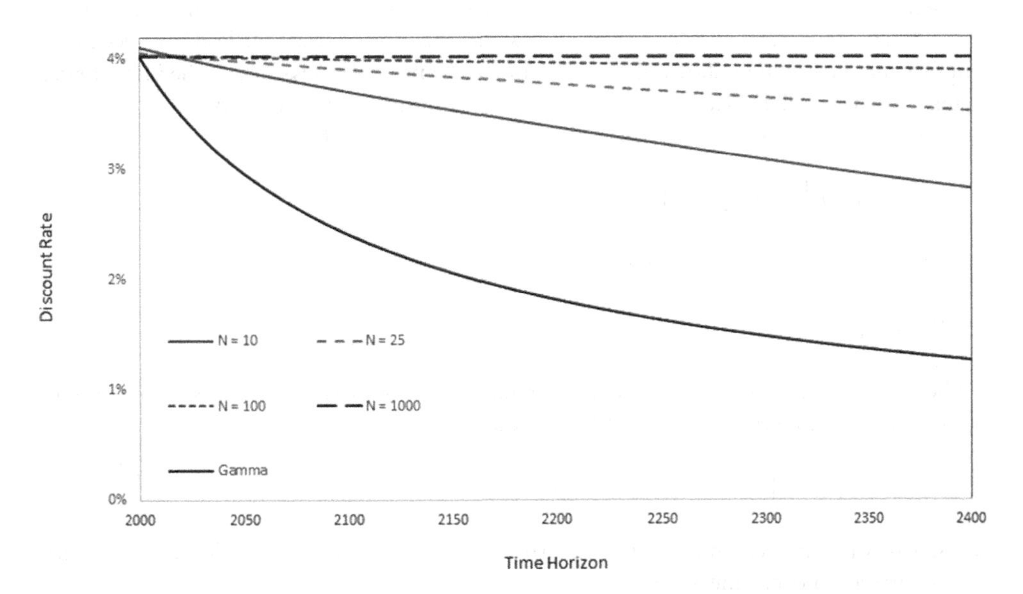

Figure 7.3 Positively Gamma Discounting (Freeman and Groom, 2013) vs. normative discounting (an interpretation of Weitzman, 2001)

The conclusion when it comes to expert opinions is not necessarily that the decline of the term structure will be radically slower when responses are positive rather than normative, but rather that the way in which the expert opinions should be combined differs radically. From Figure 7.3 it can be seen that as the sample size or, equivalently, the number of effectively independent experts declines, the term structure converges towards Weitzman's Gamma Discounting terms structure. The addition of bias into the positive response story adds another source of uncertainty, which may ultimately increase the decline compared to Gamma Discounting. Finally, one has to consider the likely range of normative and positive responses. In this author's experience, normative responses on the discount rate tend to be lower and less varied than positive recommendations, a consideration that could easily lead to a more rapidly declining positivist schedule than the one derived from normative responses. This is an empirical question, however.

The essential point of Freeman and Groom (2013) is that, for normative opinions on matters of pure time preference, ρ, and inequality aversion, η, it is likely that these subjective differences on ethical matters are irreducible. With positive responses, however, experts are error prone about some eventually observable value. These two types of responses can be thought of as reflecting heterogeneity on the one hand and uncertainty on the other, and deserve very different empirical treatments.

Historical data on interest rates

The alternative empirical approach, which has been highly influential in the discounting literature, has been to look to historical interest rates in order to characterise the uncertainty surrounding the discount rate. Many models of the interest rate find that persistence in the time series of growth manifests itself in persistence in the time series of interest rates. Implicitly underpinning the thought experiment of Weitzman (1998, 2001) and Gollier and Weitzman (2010) is the perfect correlation of interest rates over time. That is, once uncertainty is resolved, the interest rates do not deviate from that point onwards. We know that this extreme example leads to a declining term structure, yet it can be shown that at the other extreme, *i.i.d.* realisations of the interest rate, the term structure will be flat. Gollier et al. (2008) put this succinctly as follows:

In the *ENPV* framework, the social discount rate, $r(t)$, is given by:

$$\exp(r(t)t) = E\left[\exp\left(-\Sigma_{\tau=1}^{t} r_\tau\right)\right] \tag{37}$$

where the discount factor at time t is:

$$P_t = \exp\left(-\sum_{\tau=0}^{t} r_\tau\right) \tag{38}$$

The term structure of $r(t)$ depends on the random process driving the diffusion of r_t. If r_t is driven by an *i.i.d.* process, then the SDR turns out to be:

$$r = -\ln[\exp(-x_1)] \tag{39}$$

that is, it is constant over time: a flat term structure. If, on the other hand, the series exhibits perfect correlation over time, such that:

$$r(t) = -\frac{1}{t}\ln[E[\exp(-\tilde{x}_1 t)]] \tag{40}$$

which is essentially the Weitzman formula, albeit derived via consideration of the time series process of r rather than by the thought experiment about the resolution of uncertainty. The important point to take from this is that persistence in the time series of interest rates is an important determinant of the term structure of the SDR under uncertainty. It seems likely that there will be intervening cases of mean reversion of r in which DDRs emerge to varying degrees. This observation has led to a great deal of work looking at historical data on the interest rate.

The pioneers in this regard were Newell and Pizer (2003). They started with the definition of the certainty equivalent discount factor: $E(p_t)$, which is the expectation of (39). They then worked with the forward rate, r_t^f, that is, the period to period discount rate defined in discrete time:

$$r_t^f = \frac{E[P_t]}{E[P_{t+1}]} - 1 \tag{41}$$

This approach is closer to the Weitzman (1998) theoretical model. They the showed that in a simple AR(1) framework:

$$\begin{aligned} r_t &= \bar{r} + \varepsilon_t \\ \varepsilon_t &= \rho\varepsilon_{t-1} + u_t \qquad u_t \sim N(0,\sigma_u^2) \end{aligned} \tag{42}$$

the forward rate can be represented as a function of the parameters of this model: ρ, $\bar{r}$, σ_u^2, $\sigma_{\bar{r}}^2$. In particular, it was shown that with persistence, $\rho > 0$, the term structure is declining, and more so the greater the persistence. This provided an empirical framework with which to estimate the uncertainty surrounding the interest rate, and estimate the certainty equivalent discount rate, either the forward rate, or term structure of average rates that we have been considering to this point.

Newell and Pizer used annual market interest rates for long-term government bonds for the period 1798 to 1999. Starting in 1950, nominal interest rates are converted to real ones by subtracting a 10-year moving average of the expected inflation rate of the CPI, as measured by the Livingston Survey of professional economists. They estimated several models of the interest rate, including an AR(3): a simple extension of (43), and a random walk. In the end, the preferred model in Newell and Pizer (2003) was the AR(3) model, which showed high levels of persistence ($\Sigma_i \rho_i$ close to 1), and was statistically indistinguishable from a random walk ($\Sigma_i \rho_i = 1$).

In order to generate a usable term structure of discount rates, 10,000 forecasts of future interest rate paths were undertaken, the distribution of which at any point in time simulated the uncertainty in the interest rate. A simple numerical calculation of the expected discount factor resulted in a declining term structure of discount rates. Figure 7.4 shows the Newell and Pizer (2003) term structure of interest rates.

Subsequent work by Groom et al. (2007), Hepburn et al. (2008) and Freeman et al. (2013) have extended and tested the robustness of these results to, respectively, different models of interest rate diffusion and model selection, cross country analysis and treatment of the inflation. In each case, similar results emerge: there is sufficient persistence in the interest rate series for the empirical term structures to exhibit a significant and policy relevant decline.

For instance, Freeman et al. (2013) investigate the data series used by Newell and Pizer in greater depth. In particular, the observations that: (i) the pre-1950 data are nominal interest rates and (ii) logarithms of the series are used, which preclude negative rates and make interest rate volatility more sensitive to the level of interest rates (Newell and Pizer, 2001). As Figure 7.4

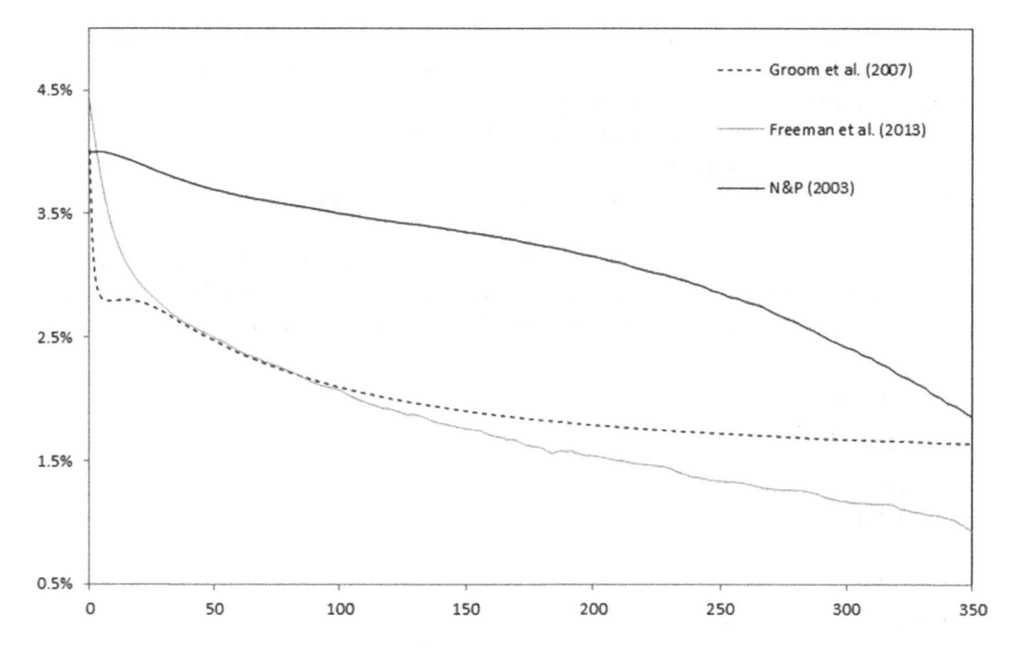

Figure 7.4 Empirical term structures for the certainty equivalent discount rate in the U.S. (Newell and Pizer [2003] and Freeman, Groom, Panopolou and Pantelidis [2013])

shows, these data issues were not behind the persistence estimated by Newell and Pizer (2003), since the Freeman et al. (2013) term structure (FGPP) declines in a similar manner. Nevertheless, the term structure is sensitive to these assumptions and is altogether lower once the relationship between nominal and real interest rates are modelled more appropriately.

Together with Weitzman (2001), the contributions of Newell and Pizer (2003), Groom et al. (2007) and Hepburn et al. (2008) have been highly influential in determining the US discounting policy. They are cited in the main policy documents on cost benefit analysis from the Office of the Management of Budgets (OMB). They are also cited in the Norwegian government guidelines (MNOF, 2012), and their preliminary work fed into the pioneering UK Green Book recommendations (HMT, 2003). However, at present the recommendations of the OMB are that, on the basis of the uncertainty arguments of Weitzman (1998, 2001) and the empirical work of Newell and Pizer (2003), a lower but flat rate of 2.5% should be applied to intergenerational projects. This, it is argued, reflects a halfway house between the random walk and the mean reverting models of Newell and Pizer (2003).[14]

The US, UK and Norwegian policies on discounting are yet further examples of how the somewhat stylised and abstract discounting theory has made headway in the policy world via some innovative empirical applications.

7.6 Policy implications of DDRs

Irrespective of the influence of the theory of declining discount rates on government policies in recent years, key questions to ask are: (1) will the DDRs enshrined in government policies make any difference? and (2) have they made any difference thus far? With respect to the latter question, it is early days, but certainly within the UK this is an area worthy of research. With respect to the potential for influencing the portfolio of investments undertaken a simple example serves to illustrate.

Following Newell and Pizer (2003) and Groom et al. (2007), we now present estimates of the social cost of carbon (SCC) under different discounting policies. The time profile of damages is shown in Figure 7.5 and is calculated using simulations of the DICE model. It represents the marginal damages resulting from an additional ton of carbon into the atmosphere. The profile is characteristic of a stock pollutant, with damages building over time, then declining over a lengthy time horizon as natural processes assimilate the initial pulse of carbon. It is precisely the profile of damages that is likely to be sensitive to discounting policy.

The SCC is simply the present value of this profile of damages. Table 7.2 illustrates the estimated SCC for a variety of different discounting procedures from theory and from the policy world, arranged in order of the dollar value of the SCC. At the top of the list is the SCC arising from the USEPA recommendations of a flat 2.5% rate, with a value of $21.8/tC. Next is Norway, with a value of $17.9/tC, whose recommended discount rate for risk-free projects begins at 2.5% and drops to 2% after 40 years. Gamma Discounting, using expert opinions to characterise interest rate uncertainty comes next at $15.9/tC. This schedule starts at 4% and declines rapidly. These are either side of the recommendations of Groom et al. (2007) and lead to an SCC of $14.4/tC, slightly higher than the French and UK policies, which lead to SCCs of $11.95/tC and $9.8/tC, respectively. Compared to the lower bound associated with the flat 2.5%, the empirical results of Newell and Pizer (2003) lead to more modest SCCs of $10.4/tC and $6.4t/C. The final theoretical contribution is from Freeman and Groom (2013), whose positivist interpretation of the expert opinions provides relatively low values of SCC for the reasons outlined above. Of course, while the work of Freeman and Groom (2013) makes a

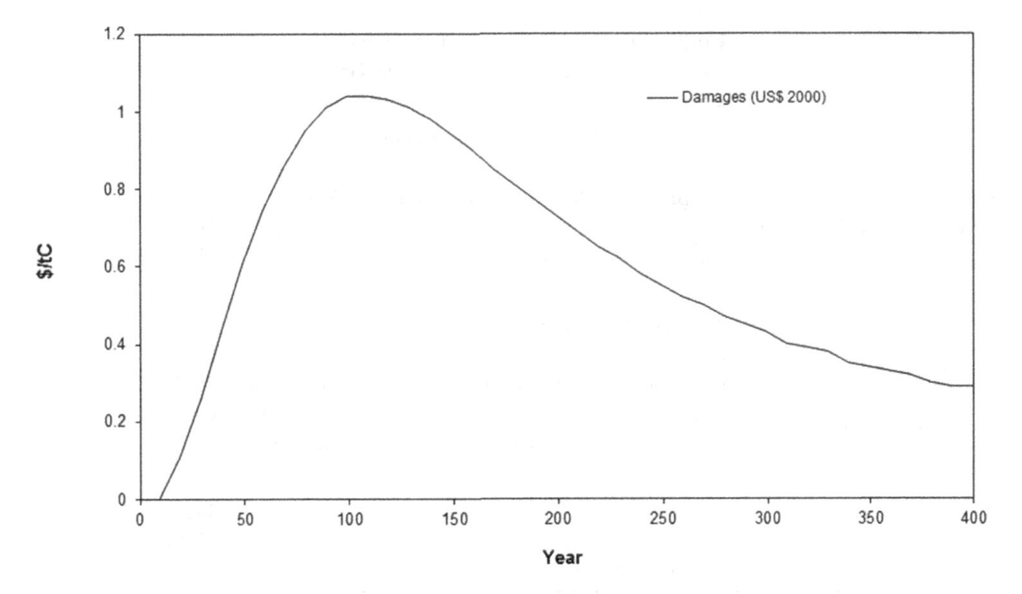

Figure 7.5 Damages from the marginal ton of carbon from the DICE model (Newell and Pizer, 2003)

Table 7.2 Declining discounting rates: policies, theories, empirics and SCC

Discounting policy and the social cost of carbon $ (2000)/tC		
Method	*Motivation*	*SCC*
Norwegian Government (NMOF, 2012)	**Theory:** Consumption side, and CAPM model. DDR motivated by uncertainty arguments (Weitzman, 1998; Gollier, 2008). **Empirical:** Motivated by time series arguments (Newell and Pizer, 2003; Groom et al., 2007, Hepburn et al., 2008)	**17.9**
Gamma Discounting	**Theory:** Production side, uncertainty in r. (Weitzman, 1998, 2001) **Empirical:** Expert opinions (Weitzman, 2001). Freeman and Groom (2013) argue this is a normative approach to expert opinions.	**15.9**
State Space Model (Groom et al., 2007)	**Theory:** Production side, uncertainty in r. (Weitzman, 1998, 2001) **Empirical:** Historical interest rates, interest rate theory, model selection.	**14.4**
French Government (Lebegue, 2005)	**Theory:** Consumption side, parameter uncertainty in growth process, g. (Gollier, 2007, 2008; Weitzman, 2007) **Empirical:** E.g., two possible states of mean growth: 2% with probability 2/3, 0.5% with probability 1/3.	**12.0**
UK Government (HMT, 2003)	**Theory:** Consumption and production side arguments taken on board. (Weitzman, 1998, 2001; Gollier, 2002a, 2002b; Oxera, 2002) **Empirical:** Nothing explicit, but influenced by time series models (Newell and Pizer, 2003; Groom et al., 2004) and Gamma discounting (Weitzman, 2001).	**9.8**
USEPA (IAWG, 2010; USEPA, 2010)	**Theory:** Production side, uncertainty in r. (Weitzman, 1998, 2001) **Empirical:** Influenced by work on historical interest rates (Newell and Pizer, 2003, Groom et al., 2007) and Gamma discounting (Weitzman, 2001).	Mean Reverting **6.4** Random Walk **10.4** Flat 2.5% **21.8**
Positive Gamma Discounting (Freeman and Groom, 2013)	**Theory:** Production side, uncertainty in r. (Weitzman, 1998, 2001) **Empirical:** Data from expert opinions, literature on combining forecasts and combining expert opinions (Winckler, 1981; Genest and Zidek, 1986; Bates and Granger 1969.	Independent **5.7** Correlated **6.3**
Flat 2.5% (Norway)	Pre-DDR policy	**14.8**
Flat 3.5% (UK)	Pre-DDR policy	**7.91**
Flat 4% (US, France)	Pre-DDR policy	**5.7**

powerful point with regard to the interpretation of expert opinions, and the different way normative and positive responses ought to be treated, the positive results rest on some important assumptions, chiefly independence and unbiasedness of experts. The comparison of the values labelled 'independent' and 'correlated' are worthy of mention here. With independence, the term structure is virtually flat and the SCC is the same as with a flat 4%: \$5.7.[15] With a constant correlation between expert opinions, the SCC rises somewhat to \$6.3. This is approximately a third of the value associated with the original Gamma Discounting schedule. In short, dependence of experts raises uncertainty surrounding the forecast of future interest rates, quickening the decline of the term structure and raising the estimated SCC.

The focus in this chapter is on the guidance that exists in the government policies of the UK, US, France and Norway. Table 7.3 summarises the impact of the new guidance on DDRs on the SCC in each case, by comparing the SCC with the original flat rate. Column 3 of Table 7.3 shows the percentage increase of the SCC in each case. Perhaps surprisingly, the biggest impact in percentage terms is found in the US, where discounting at the flat 2.5% rate increases the SCC by over 150% compared to the central 4% case. The French policy has the next largest impact on the SCC, raising it by around 110%, followed by the pioneering UK policy (+24%) and Norway (+21%). It should be recognised that these percentage changes arise from very different levels.

In conclusion, the inclusion of DDRs in government guidance on the SDR is likely to have a significant impact on the portfolio of investments that a government might choose. The precise effect is not clear, and will depend on the time profile of costs and benefits of the particular intergenerational project in question. It will naturally depend on the portfolio of projects available and the associated budget constraints. It seems likely that the balance will be tipped in favour of intergenerational projects and future generations.

It is also worth relating these implications of discounting guidance back to the present day economic situation. The guidance presented here supposes that a fixed term structure is applied to project appraisal going forward. Nevertheless, the consumption side theory, in many cases, makes a clear recommendation of how to proceed when times are bad: growth is well below trend. As shown in the theoretical section, when growth is below trend, reflected by the stock variable: $y^{-1} < 0$, an argument exists for deploying an increasing term structure that will attenuate and maybe outweigh the impact of the prudence effect (see equation 12). The effect of this might be to skew current investments towards those with more immediate payoffs.[16]

Table 7.3 Percentage change in SCC (\$US 1989/tC) in the UK, US, France and Norway due to DDRs

Country	*Pre DDR*	*Post DDR*	*% Increase*
UK[a]	**7.63**	**9.74**	27.8%
US[b]	**5.71**	**14.79**	158.9%
France[c]	**5.71**	**11.89**	108.2%
Norway[d]	**14.79**	**17.93**	21.3%

The comparisons are: (a) a 3.5% flat rate with the Green Book Schedule; (b) 4% flat rate with a 2.5% flat rate; (c) a flat 4% rate with the French guidelines in Lebegue (2005) and (d) a flat 2.5% rate with the new DDR guidelines in NMOF (2012).

7.7 Discounting and adaptation to climate change

In this final section, I make some initial tentative observations on potential impact of the latest theory and guidelines on social discounting on the economic analysis on climate adaptation.

The discounting debate has taken place in the arena of climate change mitigation, rather than adaptation. Adaptation and mitigation are often thought of as being opposed to one another when considering the climate issue, with those ostensibly less inclined to invest in mitigation citing our ability to adapt as a means of attenuating the need for mitigation investment. This view tends to ignore the potential for catastrophic risks associated with climate change, which would limit the extent of adaptation possible in catastrophic states of the world (e.g,. Weitzman, 2009; Gollier, 2008; Nordhaus, 2009). Also, a certain amount of adaptation is unavoidable by now, so the substitutability of mitigation and adaptation relates only to future adaptations rather than present. Over the long term, adaptation and mitigation will both play a part in an integrated response to climate change. Furthermore, on a global scale there are spatial issues to deal with. It is likely that the most radical changes in climate and hence the areas where adaptation is most imminently required will be in developing countries. In the welfare calculations associated with climate change, there are issues of spatial discounting and *intra*-generational equity that become important. Lastly, there is the issue of autonomous versus planned adaptation: e.g., decentralised adaptation by farmers to climatic changes vs. centralised plans for adaptive investments in flood defences, agricultural infrastructure, and so on, such as might be found in a National Adaptation Programme of Action (NAPA).

Given this, there are at least three interrelated ways to think about the relationship between discounting and adaptation:

- The impact of climate change adaptation on the discount rate,
- The impact of the discount rate on climate change adaptation, and
- The impact of income inequality on the discount rate.

We now briefly visit each of these in turn.

7.7.1 The impact of adaptation on the discount rate

Under uncertainty, consideration of economy-wide autonomous adaptation could enter into the discount rate via its impact on the uncertainty surrounding growth or the rate of return to capital. Taking the simple extended Ramsey Rule in equation (9), there are several possibilities, and the overall effect is likely to be ambiguous. Adaptation to climate change will attenuate the negative impacts of climate change compared to a "no-adaptation" scenario. However, adaptation comes at a cost to the economy and may reduce growth since such investments can offer lower returns than other investments available in the economy. Other effects are possible, but these seem like the most likely. Taken together adaptation could reduce mean growth while also reducing the spread, particularly on the downside. If this is the case then in principle the former decreases the discount rate via the wealth effect, and the latter may increase it by reducing the precautionary effect. The net effect is ambiguous.

7.7.2 The impact of the discount rate on adaptation

Looked at simply, this is a question of how the portfolio of investments changes in the context of climate change with the new discounting guidelines. For instance, Oxera (2002) considered the impact of the new UK guidelines on investments in flood defences. Such investments have short-term investment costs and long-term benefits, and are likely to become more viable under the new discounting guidance. In the specific UK case study identified, application of the UK Treasury DDR schedule has only a marginal effect on the benefit cost ratio, raising it from 1.2 with a flat 3.5% discount rate, to 1.3 with the DDR.

This seemingly minor impact is potentially specific to the case study, but ignores some important features of project-specific risk.[17] This chapter has ignored such issues but such adaptation measures, and a great deal of mitigation investments, are likely to have an insurance value: the payoffs are negatively correlated with the background growth level. At the project level, the inclusion of project risk may increase the value of such mitigation investments since their payoff is likely to be better in bad states of the world. Gollier (2013) has recently addressed the issue in the context of climate change mitigation. Here he argues that mitigation investments tend to pay off more in the high-growth states, since climate change is worse in those states of the world. This points to a positive, and in some cases increasing premium on these risky climate mitigation projects, which may outweigh the decline in the risk-free rate discussed above (Gollier, 2013). This study contradicts previous analyses of the issue that argued that climate mitigation rather has insurance type properties since it pays off in the low growth states of the world (Sandsmark and Vennemo, 2007), implying a negative premium for risk and a reduction in the required rate of return.

Recent theoretical work on adaptation looking at project risk through the lens of asset pricing theory argues that if adaptation reduces the impact of unexpected climate change, as asserted above, then there is an argument for using a lower discount rate for investments in adaptation to reflect this insurance effect (Aalbers, 2013). However, given that many adaptation strategies are essentially 'wait and see' strategies, it is likely that, from the perspective of project risk, adaptation is better placed within the real options framework rather than the standard discounting framework.

7.7.3 The impact of income inequalities on the discount rate

In the Ramsey model considered thus far, intra-generational income inequalities are usually captured by the parameter η. However, the actual distribution of income at a given point in time is ignored beyond its relationship with average consumption, c_t, which is the only aspect of interest to the social planner. An explicit treatment of income inequality at each point in time can affect the discount rate that one employs to the extent that society is averse to such inequalities.

Two recent papers have investigated the impact of income inequality on the discount rate: Gollier (2010) and Emmerling (2011). Emmerling (2011) disentangles preferences for intra-generational income inequality, γ, the risk preferences: the coefficient of relative risk aversion, φ and the inverse of the elasticity of intertemporal substitution, η (all of these are conceptually and numerically equal in the Ramsey framework). He then shows that in these circumstances the discount rate with *i.i.d.* growth when consumption differs spatially among *i* different individuals, countries or regions, can be written as follows:[18]

$$\begin{aligned} r_t = \rho + \eta\mu - 0.5\varphi(\eta+1)\frac{Var(E_t \ln(c_{it} \mid c_{i0}))}{t} \\ - 0.5\gamma(\eta+1)\frac{Var(E_t \ln(c_{it} \mid c_{i0})) - Var(\ln c_{i0})}{t} \end{aligned} \tag{43}$$

This shows that the term structure of the discount rate depends upon the usual features: the pure utility discount rate, the wealth effect, and the prudence effect, but also an additional effect that captures the evolution of income inequalities over time. Emmerling (2011) shows, and it is clear from (44) that if inequality among the i agents is increasing over time linearly, then there is a flat term structure in this case, and the SDR is reduced in the presence of inequality. If inequality increases more than linearly over time, the term structure will be declining. Emmerling's calibration to the US recommends close to a .4% reduction in the SDR based on the linear increase in income inequality witnessed over the past 50 years.

This is relevant for climate adaptation if we believe that inequalities over the relevant geographical scale are likely to increase as a consequence of climate change, and these inequalities are attenuated or exacerbated by autonomous and planned adaptations. The incorporation of these distributional aspects of climate change adaptation in the discount rate is a matter for future research. It may or may not be the best way to deal with inequalities in the CBA of climate change.

7.8 Conclusion

The literature on social discounting has covered a lot of ground over the past decade or so. It has also influenced government policy in a number of high profile cases. The UK in 2003, the US in 2010 and in ongoing revisions, France in 2005 and Norway in 2012, have all reviewed thoroughly the academic contributions in this field and updated or are in the process of updating their policies on long-term discounting to reflect this. In this chapter, we have reviewed the theory and empirical approaches that have underpinned these updated policies, and evaluated the potential impact in the context of the Social Cost of Carbon. In short, the updates clearly have the potential to shift the portfolio of public investments towards those with longer-term, intergenerational benefits. Whether they have done so in the case of the UK and France, or will do so elsewhere is difficult to say thus far.

The implications for climate change adaptation are tentatively argued to lie in three strands. First, adaptation may affect the uncertainty surrounding growth and the rate of return to capital and hence the term structure of the SDR. Second, the imposition of a declining term structure may affect the viability of adaptation projects and will most likely make them more attractive since their long-term benefits will have a higher present value. Here, though, the explicit consideration of project risk is likely to be crucial in determining the optimal portfolio of adaptation and mitigation investments, and there are some theoretical arguments to suggest that adaptation investments should be discounted at a lower rate due to their insurance like properties. Third, since income inequality can also affect the SDR, the impact of adaptation on income inequality across the relevant spatial scale should also be considered. Each of these avenues is worthy of future investigation.

Notes

1 The separation is not a perfect one, since a number of articles, including Weitzman (2007), are more general, and the correspondence between the two sides is readily seen in Gollier and Weitzman (2010).
2 A summary of the literature as it stood at that time can also be found in Pearce et al. (2003) and Groom et al. (2005). The Green Book (HMT, 2003) can be found here: www.hm-treasury.gov.uk/data_greenbook_index.htm
3 www.catalogue.polytechnique.fr/site.php?id=324&.leid=2389
4 For the USEPA See: yosemite.epa.gov/ee/epa/eed.nsf/pages/Guidelines.html/$.le/Guidelines.pdf. For the US Interagency Working Group on the Social Cost of Carbon see: www1.eere.energy.gov/buildings/appliance_standards/commercial/pdfs/sem_finalrule_appendix1.pdf

5 Respectively quoted are: Ramsey (1928), Harrod (1948) and Pigou (1920).
6 Groom and Maddison (2013) look at estimates of the elasticity of inter-temporal substitution via analysis of the consumption function, inequality aversion via progressive tax rates and the 'wants-independent' approach of Frisch.
7 That is $c_t = c_0 \exp(xt)$:, where $x \sim N(\mu, \sigma_c^2)$.
8 Millner (2012) has a clarifying discussion on this issue.
9 Weitzman (2001) describes this as a 'pragmatic decomposition'.
10 Strictly speaking, this expectation must use the Dirac delta.
11 Hepburn and Groom (2007) proposed testing a project at a variety of appraisal dates for sensitivity. The argument went that different future appraisal dates reflected how future generations would have made the decision with the same information at time $t = 0$.
12 Where the expectation and the summation are considered identical operations.
13 In the second IPCC report, Arrow et al. (1996) review the literature on candidate rates of interest that can be used to inform the positive or, in their words, descriptive approach to discounting. A value of 26% appears there stemming from actual interest rates paid on credit cards and other short-term loans.
14 The USEPA guidelines state: 'If a time-declining approach cannot be implemented, it is possible to capture part of its empirical effect by discounting at a constant rate somewhat lower than those used in the conventional case. For example, the current interagency guidance for valuing carbon dioxide emission reductions includes treatment with certainty-equivalent constant discount rates of 2.5, 3 and 5 percent' (USEPA, 2010).
15 The relevant comparator for Freeman and Groom (2013) is with Weitzman's Gamma Discounting Schedule, which starts at 4%.
16 One has to be careful here of course. Some intergenerational projects, such as nuclear power, which have non-unique internal rates of return, may well become more attractive with higher or increasing discount rates, since the long-term decommissioning costs become insignificant in present value terms.
17 The inclusion of project risk via a Capital Asset Pricing Model (CAPM) approach is explicit in the Norwegian guidance on social discounting, for example (MNOF, 2012).
18 See Emmerling 2011, p. 19, equation 21.

References

ADB (2007). Theory and Practice in the Choice of Social Discount Rate for Cost-Benefit Analysis: A Survey. Economics Working Papers. ERD Working Paper No. 94.

Arrow K.J. (1999). 'Discounting, Gaming and Morality'. In P. Portney and J. Weyant (Eds.), Discounting and Intergenerational Equity, Washington: Resources for the Future. Also found at: http://www-siepr.stanford.edu/workp/swp97004.pdf.

Arrow, K., Cline, W., Mäler, K.-G., Munasinghe, M., Squitieri, R. and Stiglitz, J. (1996). Intertemporal Equity, Discounting, and Economic Efficiency. In H.L.J.P. Bruce and E. Haites (Eds.), Climate Change 1996: Economic and Social Dimensions of Climate Change, Contribution of Working Group III to the Second Assessment Report of the Intergovernmental Panel on Climate Change. Cambridge University Press.

Arrow, K.J., Cropper, M.L., Gollier, C., Groom, B., Heal, G.M., Newell, R.G., Nordhaus, W.D., Pindyck, R.S., Pizer, W.A., Portney, P.R., Sterner, T., Tol, R.S.J. and Weitzman, M.L. (2012). How Should Benefits and Costs Be Discounted in an Intergenerational Context? Resources of the Future Discussion Paper, RFF DP 12-53. Washington: RFF.

Atkinson, G., Dietz, S., Helgeson, J., Hepburn, C. and Sælen, H. (2009). Siblings, Not Triplets: Social Preferences for Risk, Inequality and Time in Discounting Climate Change. Economics Discussion Papers, No. 2009-14, Kiel Institute for the World Economy. http://www.economics-ejournal.org/economics/discussionpapers/2009-14.

Barro, R.J. (2006). Rate Disasters and Asset Markets in the Twentieth Century. The Quarterly Journal of Economics, 121 (3), 823–866. doi: 10.1162/qjec.121.3.823

Bates, J.M. and Granger, C.W.J. (1969). The Combination of Forecasts. Operational Research Quarterly, 20, 451–468.

Buchholz, W. and Schumacher, J., (2008). Discounting the Long Distant Future: A Simple Explanation for The Weitzman–Gollier Puzzle. Working paper: University of Regensburg.
Cox, J.C., Ingersoll, J.E. and Ross, S.A. (1981). A Re-Examination of Traditional Hypotheses About the Term Structure of Interest Rates. Journal of Finance, 36, 769–799.
Dasgupta, P. (2008). Discounting Climate Change, Journal of Risk and Uncertainty, 37, 141–169.
Emmerling, J. (2011). Discounting and Intragenerational Equity. Mimeo: Toulouse University.
Freeman, Mark C. (2010). Yes, We Should Discount the Far-Distant Future at Its Lowest Possible Rate: A Resolution of the Weitzman-Gollier Puzzle, Economics, No. 2010-13, Economics: The Open-Access, Open-Assessment E-Journal, Vol. 4, Issue 2010-13, 1–21. http://dx.doi.org/10.5018/economics-ejournal.ja.2010-13.
Freeman, M.C. and Groom B. (2013). Gamma Discounting and the Combination of Forecasts. http://papers.ssrn.com/sol3/papers.cfm?abstract_id=1676793.
Freeman M.C., Groom B., Panopolou, E. and Pantelidis, T. (2013). Inflated Past, Discounted Future: Declining Discount Rates and the Fisher Effect. Centre for Climate Change Economics and Policy Working Paper No. 129, Grantham Research Institute on Climate Change and the Environment, Working Paper No. 109, London School of Economics and Political Science.
Genest, C. and Zidek, J.V. (1986). Combining Probability Distributions: A Critique and an Annotated Bibliography. Statistical Science, 1, 114–135.
Gollier, C. (2002a). Discounting an Uncertain Future. Journal of Public Economics, 85, 149–166.
Gollier, C. (2002b). Time Horizon and the Discount Rate. Journal of Economic Theory, 107, 463–473.
Gollier, C. (2004). Maximizing the Expected Net Future Value as an Alternative Strategy to Gamma Discounting. Finance Research Letters, 1, 85–89.
Gollier, C. (2007). The Consumption-Based Determinants of the Term Structure of Discount Rates. Mathematics and Financial Economics, 1 (2), 81–102.
Gollier, C. (2008). Discounting with Fat-Tailed Economic Growth. J Risk Uncertain, 37, 171–186. doi:10.1007/s11166-008-9050-0
Gollier, C. (2010). Discounting, Inequalities and Economic Convergence, http://idei.fr/display.php?a=22743.
Gollier, C. (2012). Pricing the Planet's Future: The Economics of Discounting in an Uncertain World. Princeton Press.
Gollier, C. (2013) Term Structures of Discount Rates for Risky Investments. http://idei.fr/doc/by/gollier/term_structure.pdf.
Gollier, C., Koundouri, P. and Pantelidis, T. (2008). Declining Discount Rates: Economic Justifications and Implications for Long-Run Policy. Economic Policy 23 (56), 757–795.
Gollier, C. and Weitzman, M. (2010), How Should the Distant Future be Discounted when Discount Rates are Uncertain? Economics Letters, 107 (3), 350–353.
Groom, B., Hepburn, C., Koundouri, P. and Pearce D. (2005). Discounting the Future: The Long and the Short of It. Environmental and Resource Economics (32), 445–493.
Groom, B., Koundouri, P., Panipoulou, K., and Pantelides, T. (2007). Discounting the Distant Future: How Much Does Model Selection Affect the Certainty Equivalent Rate? Journal of Applied Econometrics, 22, 641–656.
Groom, B., Koundouri, P., Panopoulou, E. and Pantelidis, T. (2004). Discounting the Distant Future: How Much Does Model Selection Affect the Certainty Equivalent Rate? Discussion Paper 04-02, Department of Economic, University College London.
Groom, B. and Maddison, D. (2013). Estimating the Elasticity of Marginal Utility: Revisions, Extensions and Problems. Grantham Research Institute on Climate Change and the Environment, Working Paper (Forthcoming).
Harrod R.F. (1948) Towards a Dynamic Economics: Some Recent Developments of Economic Theory and Their Application to Policy. London: Macmillan.
Hepburn, C. and Groom, B. (2007). Gamma Discounting and Expected Net Future Value. Journal of Environmental Economics and Management, 53 (1), 99–109.
Hepburn, C., Koundouri, P., Panopoulou, E. and Pantelidis, T. (2008), Social Discounting Under Uncertainty: A Cross-Country Comparison. Journal of Environmental Economics and Management, 57, 140–150.

HMT (2003). Guidelines on Cost Benefit Analysis. UK: HM Treasury.
IAWG (2010). Appendix 15a: Social Cost of Carbon for Regulatory Impact Analysis under Executive Order 12866. US Interagency Working Group on the Social Cost of Carbon of the United States of America.
Kimball, M.S. (1990). Precautionary Saving in the Large and in the Small, Econometrica, 58, 53–73.
Lebegue, D. (2005), Revision du taux d.actualisation des investissem-net publics. Rapport du groupe de experts, Commisariat Generale de Plan. http://catalogue.polytechnique.fr/site.php?id=324&.leid=2389.
Millner, A. (2012). On Welfare Frameworks and Catastrophic Climate Risks. Working paper, Berkeley. http://areweb.berkeley.edu/documents/seminar/WelfareClimateMillner.pdf.
NMOF (2012). Cost Benefit Analysis. Official Norwegian Reports NOU 2012: 16. Ministry of Finance, Norway.
Newell R. and Pizer, W. (2001), Discounting the Benefits of Climate Change Mitigation: How Much Do Uncertain Rates Increase Valuations? Discussion Paper 00-45, Washington DC: Resources for the Future.
Newell, R.G. and Pizer, W.A. (2003). Discounting the Distant Future: How Much Do Uncertain Rates Increase Valuations? Journal of Environmental Economics and Management 46, 52–71.
Nordhaus, W.D. (2007). A Review of the Stern Review on the Economics of Climate Change. Journal of Economic Literature 45, 686–702.
Nordhaus, W.D. (2009). An Analysis Of The Dismal Theorem. Cowles Foundation Discussion Paper, No. 1689.
Oxera (2002). The Selection of a Social Time Preference Rate to be Used in Long-Term Discounting. Oxford: Oxera Ltd. http://www.oxera.com/Publications/Reports/2002/.
Pearce, D.W. and Ulph, D. (1999) A Social Discount Rate for the United Kingdom. In D.W. Pearce (Ed.), Environmental Economics: Essays in Ecological Economics and Sustainable Development, Cheltenham: Edward Elgar.
Pearce, D., Groom, B., Hepburn, C. and Koundouri, P. (2003). Valuing the Future: Recent Advances in Social Discounting. World Economics 4: 121–141.
Pigou, Arthur C. (1920). The Economics of Welfare. London: Macmillan. (The particular chapter referenced can be found at: http://www.econlib.org/library/NPDBooks/Pigou/pgEW2.html.)
Ramsey F.P. (1928). A Mathematical Theory of Saving. Economic Journal, Vol. 38, 543–59.
Sandsmark, M. and Vennemo, H. (2007). A Portfolio Approach to Climate Investments: CAPM and Endogenous Risk. Environmental and Resource Economics, 37, 681–695.
Stern, N. (1977). Welfare Weights and the Elasticity of the Marginal Valuation of Income. In M. Artis and A.R. Nobay (Eds.), Studies in Modern Economic Analysis. Oxford: Basil Blackwell.
Stern, N. (2007). The Economics of Climate Change: The Stern Review. Cambridge University Press.
Stern, N. (2008). The Economics of Climate Change. American Economic Review: Papers & Proceedings 98, 1–37.
Traeger, C. (2012). What's the Rate? Disentangling the Weitzman and the Gollier Effect, CUDARE Working Paper 1121, University of California, Berkeley.
USEPA (2010). Guidelines for Preparing Economic Analyses. United States Environmental Protection Agency, Washington DC. EPA 240-R-10-001.
Weitzman, M.L. (1998). Why the Far-Distant Future Should Be Discounted At Its Lowest Possible Rate. Journal of Environmental Economics and Management, 36, 201–208.
Weitzman, M. (1999). Just Keep on Discounting, but .... In P. Portney and J. Weyant (Eds.), Discounting and Intergenerational Equity. Washington DC: Resources for the Future, 23–30.
Weitzman, M.L. (2001). Gamma discounting. American Economic Review, 91, 260–271.
Weitzman, M.L. (2007). A Response to the Stern Review. Journal of Economic Literature, 45 (3), 703–724.
Weitzman, M.L. (2009). On Modeling and Interpreting the Economics of Climate Change. The Review of Economics and Statistics, Vol. XVI, No. 1, 1–19.
Winkler, R.L. (1981). Combining Probability Distributions from Dependent Information Sources. Management Science, 27, 479–488.

Appendix 1

It is easy to show that the growth over the time period $0 - t$ is now given by:

$$\ln c_t - \ln c_0 = \mu t + \phi \frac{1-\phi^t}{1-\phi} y_{-1} + \sum_{\tau=0}^{t-1} \frac{1-\phi^{t-\tau}}{1-\phi} \varepsilon_{y\tau} + \sum_{\tau=0}^{t-1} \varepsilon_{x\tau}$$

where y_{-1} is the initial state. If the shocks ε_{yt} and ε_{xt} are mean zero and normal, the expected growth over this interval is given by the first two terms, the second of which reflects the effect of persistence on growth. The annualised variance of growth is given by:

$$t^{-1} \operatorname{var}(\ln c_t - \ln c_0) = \frac{\sigma_y^2}{(1-\phi)^2} \left[1 + 2\phi \frac{\phi^t - 1}{t(\phi - 1)} + \phi^2 \frac{\phi^{2t} - 1}{t(\phi^2 - 1)} \right] + \sigma_x^t$$

which contrasts to the *i.i.d.* due to the addition of the first term reflecting persistence. These terms can be used to expand equation (8) and define the SDR in the case of persistent growth shocks. See (Gollier, 2012, Ch. 4).

8

THE ROLE OF ECONOMIC MODELLING FOR CLIMATE CHANGE MITIGATION AND ADAPTATION STRATEGIES

Francesco Bosello[1]

[1]MILAN STATE UNIVERSITY, FONDAZIONE ENI ENRICO MATTEI, EURO-MEDITERRANEAN CENTER ON CLIMATE CHANGE

8.1 Introduction

Economics of climate change has been a field of research, but has also been a topic of fierce debate since, with the 1988 World Conference on the Changing Atmosphere in Toronto and the 1992 United Nations Framework Convention on Climate Change (UNFCCC), climate change was recognised as a problem to be tackled by policy measures.

The three well known "pillars" of a successful policy are effectiveness, efficiency, and equity. But getting the optimal blend of these often conflicting ingredients is particularly challenging when climate change is involved. It induces short and very long-term impacts and global and local effects, and it affects welfare of societies as a whole, but with important asymmetries across social groups. Moreover, the climatic, environmental and social dimensions which are affected are complexly intertwined; their evolution is characterised by non-linearity, irreversibility and inherent unpredictability. Climate change impacts are thus highly uncertain. On their turn, the cost of mitigation and adaptation policies may be better, but yet far from being fully known.

Against such a background, it may seem that economics and modelling are of very little help. On the contrary, economic models, when appropriately integrated with climatic and environmental knowledge, could capture important features of the social dimension of climate change while keeping its complexity manageable. In fact, since the beginning of the 90s, climate change became one research field where multidisciplinarity, the use of models, and of model coupling bridging the gap between climate, environmental and economic sciences, was pushed to a maximum. In more than 20 years of "integrated assessment", a continuously improving economic modelling contributed to clarify key aspects related to efficiency, effectiveness and equity of mitigation and adaptation policies and provided useful support to policy decisions.

How has this been done? What are the strengths and weaknesses of the different modelling approaches? What are the results? How should these be interpreted and what are the caveats in their use? What remains to be done?

All this is briefly touched on in what follows. The present work focuses in particular on the applied economic modelling literature addressing climate change impact, mitigation and adaptation. It does not cover theoretical works. Specifically: Section 8.2 summarises the challenges for the discipline; Section 8.3 presents major investigation approaches and modelling methodologies; Section 8.4 revises main results and Section 8.5 concludes.

8.2 Challenges in the economic modelling of climate change

The climate change issue is characterised by some peculiar aspects that need to be adequately captured by modelling tools if they aim to provide sound scientific insights and useful support to decision making. Their implementation in quantitative models is complex and challenging though. They are introduced below.

8.2.1 Uncertainty

Uncertainty on climate dynamics and the associated reaction of the environmental systems is the first background challenge for economic climate-change impact and policy assessment. Notwithstanding undeniable advancements, knowledge remains far from perfect, and in some key areas – like the link between a changing climate and the frequency and intensity of extreme meteo-climatic events – uncertainty is pervasive (IPCC, 2012). There is then a more social-economic uncertainty. It derives from the inherent unpredictability of many aspects of the individual and collective behaviour; from the difficulty in anticipating the evolution of societies over the very long term as often is required in climate change assessments; to the need to often evaluate goods and services without the support of any observable market transaction (non-market/existence values). All this makes it difficult to provide "objective" evaluations on the costs and benefits associated with environmental policy interventions. At the same time, it calls for the use of approaches able to appropriately handle and communicate uncertainty.

8.2.2 Discontinuities, tipping points, irreversibility

Natural phenomena do not generally follow linear evolutionary trends. They are characterised by radical changes and irreversibility that may dramatically modify living conditions. Economic discontinuities and irreversibility are also relevant. Technological breakthroughs are typical examples of the former. The latter are, for instance, associated to huge investments with long payback periods in anticipatory adaptation (e.g., coastal defence) or de-carbonization. Once implemented, both may be difficult to modify in the light of newly acquired knowledge. Or, if postponed too much, "waiting to learn before acting", they may produce the desired effects too late. Models need thus to capture these "jumps" in order to predict and avoid dangerous divergence from equilibrium paths.

8.2.3 Time scale

The long-term nature of climate change impacts and policy effects brings about the issue of discounting. This fuelled a heated debate never settled and recently reinvigorated by the Stern review (Stern, 2007; Nordhaus, 2007; Weitzmann, 2007; Tol, 2007). The problem in a nutshell boils down to deciding how to weight the future against the present. This choice is particularly relevant in the case of climate change as, typically, rather certain policy costs sustained today need to be compared with uncertain benefits in an often far off tomorrow.

Clearly, a higher consideration given to future outcomes, which numerically corresponds to low discount rates, implies a higher concern about potential future damages. This, on its turn, increases a society willingness to devote resources to avoid those damages (i.e., to mitigation or adaptation policies) in the present.

Unfortunately, there are not objective rules to "weight" time. This subjectivity is evident in the *prescriptive* approach to discounting which indeed proposes explicit ethical considerations of intergenerational equity to justify different values for the discount rates. But it is also embedded in the *descriptive* approach. Apparently, it is objective as the proposed discount rates are based on observed market or societal behaviour. In fact, even assuming that intertemporal preferences could be perfectly estimated, one has to characterised that these can then differ across societies or within the same society across different times (Arrow et al., 1996).

8.2.4 Different spatial scales

GHG emissions generate a global externality, however climate change impacts, mitigation, and adaptation effects, are highly differentiated locally. This imposes economic models to integrate different investigation scopes.

A global perspective is needed to assess the overall effectiveness of mitigation policies. Indeed, due to the public good nature of abatement, policies implemented by individual countries or groups can spill over economic and environmental effects (leakages) that need to be traced. A country or even sub-national perspective, ideally coupling a sectoral economic breakdown with spatially explicit detail, is necessary for the assessment of environmental impacts and cost effectiveness of adaptation policies.

Integrating the global, national and even sub-national dimensions is also needed to properly address the efficiency and equity trade off originated by mitigation policies. Policy cost minimization requires that the greatest efforts are produced where abatement is cheaper. Equity broadly requires that those who polluted and are most capable to pay, pay. Often, the three components do not coincide: typically low abatement options are available in developing countries, which are at a time less responsible historically for emissions and are less capable to pay.

Finding a "fair" balance between the three principles is crucial for policy viability and therefore is one of the answers that policymaking expects most from modelling.

8.2.5 Technical progress

Technical progress is a key variable influencing all the aspects of the complex relation linking human activity, climate and the environment. It affects energy and carbon intensity, thus it is a major determinant of the impact that anthropogenic activities exert on climate and eventually on environmental damages. Moreover, by endowing societies with new tools and processes, it influences their ability to implement mitigation policies, and to anticipate, bear and potentially recover from damages, i.e., their adaptive capacity. Accordingly, an appropriate modelling of technological progress is fundamental to produce an informative assessment of climate change impacts and policies. However, this is another area rich in uncertainty, potential existence of discontinuities and of non-linear phenomena.

8.3 Modelling the economic dimension of climate change

The interdependent multi-dimensionality of the climate change issue implies that the related economic assessments cannot be credibly conducted in isolation. Integration of economics with environmental and climatic disciplines is necessary. Operationally, this translated into the development of Integrated Assessment models (IAMs). IA can be defined as "a process aimed at combining, interpreting and communicating knowledge from diverse scientific fields in order to tackle an environmental problem comprehensively by stressing its cause-effect links in their entirety" (Rotmans and Dowlatabadi, 1998). IAMs propose an analysis of the "climate change issue" where its different dimensions – climatic, environmental, and socioeconomic – are all represented and connected. Developing since the beginning of the 90s, IA is now the leading approach in climate change assessments.

The treatment of the economic dimension inside IAMs is highly differentiated and the related literature very extended. Classifications are thus complex and prone to some subjectivity. This said, probably the two most important taxonomic criteria concern the way in which economics intertwines with the climatic and the environmental components, and the specific "representation" offered of the economic system. The former produces "hard" or "soft link" integration, the latter results in top-down or bottom-up models.

After revising these two aspects, this section describes how impacts, mitigation, adaptation, technology and uncertainty have been incorporated into the different modelling approaches.

8.3.1 *Hard vs. soft link*

In hard-linked IAMs, climate and economics are treated as a unified system represented by a consistent set of differential equations. The "loop" between economics, environment and climate is thus closed.

Typically, emissions from the economic activity build atmospheric CO_2 concentration, which leads to increases in global mean temperature. Then, a more or less refined climate-change damage function translates temperature increase into GDP losses. Therefore, environmental benefits are explicitly part of the model objective function. Mitigation and adaptation policies are also accounted for by specific relations linking dedicated expenditures to damage reduction. This happens indirectly, through lower emissions, in the case of mitigation, or directly, modifying the relevant parameters of the damage function, in the case of adaptation.

Strengths of hard-linked IAMs are the internal consistency and the rigorous mathematical foundations allowing a full characterization of the solutions found. These features are particularly appealing as they allow for multi-dimensional long-term policy optimization exercises. That is: models can be asked to compute the cost minimizing mix between mitigation cost, adaptation cost and residual damages. Hard-link models can guarantee by construction that the policy profiles produced are indeed cost efficient and time consistent.

Hard-linked IAMs have been thus extensively used to analyse optimal paths for abatement efforts, investment in R&D and green R&D and investment in renewable energy; to quantify the costs of given mitigation policies; to study the strategic incentives to participate to international environmental agreements and, more recently, to define the optimal balance between mitigation and adaptation.

The weakness of hard-linked IAMs is that mathematical and computational tractability require huge simplifications translating into an extended use of reduced form functions. This applies to the climate dimension where complex global circulation models are compressed into few equations of the carbon cycle. It applies to the economic system, which is usually represented with strong top-down, aggregated features, e.g., growth models providing no or extremely limited sectoral breakdown. Finally, it applies as said to climate change impacts' feedback on economic activity.

In soft-linking approaches climatic phenomena, environmental effects, and their economic evaluation belong to separate modelling exercises, which are connected in a sequential output/input exchange process. Inter-model consistency needs to be pursued, but just to allow a meaningful exchange of information across the different modules. This is usually less demanding than in hard-linked approaches, and provides enough flexibility to each single model in the chain to increase the complexity of the analysis performed. Soft-linked exercises thus take the form of huge interdisciplinary interconnected modelling frameworks that, compared to hard-link assessments, appear more comprehensive and more detailed at a time. This applies to the economic dimension as well that is accounted for by a mix of top-down models like applied or computable General Equilibrium Models (GEMs) presenting a high country and sectoral detail and bottom-up models offering richer description of available technological options.

Examples of soft-linked modelling frameworks are the MIT IGSM framework (Prinn et al., 1999), the IMAGE IA model (IMAGE, 2001) and the AIM Integrated Modeling framework (Kainuma et al., 2003). They have been extensively used during the IPCC SRES scenario building exercise (IPCC, 2000).

The major shortcoming of the soft-linked approach is that because of the often huge differences in geographical, time scales, and in the typology of variables considered across the different modules, the linked framework can show inconsistencies and non-converging solutions. This usually prevents the practical performance of fully intertemporal optimization exercises, as some feedback loops cannot be closed. The standard analysis performed is thus that of impact or policy simulation. That is: climatic impacts or policy targets are imposed and then the modelling framework computes the consequences for the environmental and the social economic system.

8.3.2 Top down vs. bottom up

Top-down models are usually defined as "economics-oriented": they aim to provide a *comprehensive* picture of the functioning of the economic system based on the behaviour of representative and rational agents maximizing objective functions.

In the economic analysis of climate change, different top-down modelling methodologies are applied: General Equilibrium Models (GEMs) "computable or applied", dynamic growth models, and macro-econometric models.

General equilibrium models

The view of GEMs is multi-market and multi-country providing explicit representation of domestic and international trade. All markets are linked as supply and demand of factors of production; goods and services are mobile between sectors. These flows are determined by changes in relative prices which signal to optimizing producers and consumers how to efficiently allocate their resources. All markets eventually clear. In this way, GEMs can capture and describe market adjustments induced by a localised shock onto the global context and the feedback of macroeconomic dynamics on each single market.

Computable General Equilibrium Models (CGE) have been extensively applied to the study of mitigation policies. These indeed work through taxes, subsidies and quotas, which are the typical policy variables focus of GEM analysis. The related literature is wide, starting from the first works by Whalley and Wigle (1991) and Burniaux et al. (1992) to end up with the recent assessment of the post Kyoto mitigation policies (e.g., Böhringer et al., 2009b, 2010; European Commission, 2008, 2010). Nowadays it is also common to see CGE as the last step of soft-linked integrated assessment exercises. Examples include the EPPA CGE model in the MIT IGSM framework (Prinn et al., 1999), the WORLDSCAN CGE model in the IMAGE IA framework (IMAGE, 2001) and the AIM-CGE model in the Asian-Pacific Integrated Modeling framework (Kainuma et al., 2003).

Since the end of the 90s, GEMs have been applied also to the study of climate change impacts while very few attempts have been devoted to the inclusion of planned adaptation.

CGE models present some structural limitations. Being equilibrium models, they are ill suited to represent market imperfections;[1] their adjustment to equilibrium is instantaneous, missing frictions/inertias; being calibrated to some specific year, they can offer reliable information when the economic context remains reasonably similar to the initial one; due to the large number of relations depicted and high data intensity, they present stylised dynamics and exogenous technological progress.

Dynamic growth models

Dynamic growth models operationalise the paradigm of neoclassical (Solow 1956, 1987; Swann, 1956; Ramsey, 1928; Cass, 1965; Koopman, 1965) or new growth theory (Romer, 1986, 1990; Grossmann-Helpman, 1994; Aghion and Howitt, 1998). They are the economic backbone of many hard-linked IA models like DICE (Nordhaus, 1991), RICE (Nordhaus and Yang, 1996; Nordhaus and Boyer, 2000), DICER (Ortiz et al., 2011), ENTICE (Popp, 2004), MERGE (Manne and Richels, 2004), FUND (Tol, 2002), WITCH (Bosetti et al., 2006), DEMETER (Gerlagh, 2007) and MIND (Edenhofer et al. 2005). They assume that perfectly informed global or regional planners decide how much to invest to maximise an intertemporal discounted utility function which depends on consumption. The rate of technological progress is exogenous and represented by a Hicks-neutral autonomous energy efficiency improvement parameter, or endogenous, driven by Learning by Doing or Learning by Researching processes. The last generation of growth model used in IA introduced interesting features like multiplicity of capital endowments or an increased detail of the energy sector, which add some bottom-up features to their top-down perspective. These models have been extensively applied to the study of optimal abatement policies and energy and carbon efficiency policies. Many of them can incorporate strategic behaviours of different countries/regions in an international mitigation effort as the solution algorithms can simulate full cooperation, non-cooperative games or mixed games where multiple coalitions of cooperating countries play non-cooperatively against other coalitions or singletons. They have thus been used to study those mechanisms fostering, enlarging and stabilizing the participation to international environmental agreements (Bosello et al., 2003; Heykmans and Tulkens, 2003; Bosetti et al., 2011).

The major limit of dynamic growth models is their "aggregated" perspective: they are typically one sector models, and as such, they miss endogenous price adjustments and the intersectoral international trade dimension.

Macro-econometric models

The main feature of macro-econometric models is to parameterize the relation between variables using time series analysis based on yearly or quarterly data. They consist in huge systems of dynamic equations mimicking demand and supply whose evolution ends up to be validated, and thus determined, by past empirical observations rather than economic theory. They are also called "neo-Keynesian" as they developed from Keynesian economics, where the "demand side drives the system". Their empirical derivation allows them to consider with some flexibility departures from perfect competition, e.g., in energy and labour markets accounting for market power and bargaining processes. Among the most used macro-econometric models in climate change analysis are the E3ME and E3MG models (Cambridge Econometrics, 2012) amply applied to assess the cost and structural implications of mitigation policies and at the EU and wider levels (Ekins et al., 2012; Pollitt and Thoung, 2009; Barker et al., 2008; Barker et al., 2005)

There are two main criticisms against macro-econometric models. One is shared with CGE models: being parameterized on the past, they may be weak in long-term projections. The second is structural: being tailored on macroeconomic empirical observation, they cannot capture the impacts of policy shocks on micro agents behaviour (also known as the "Lucas critique").

Bottom-up models

Bottom-up models are often labelled as engineering or partial equilibrium models. This characterization mainly defines those energy market models representing in detail a huge number of alternative technological options used for harnessing energy resources and convert them into energy services.[2] Most bottom-up models interconnect conversion and consumption of energy via energy carriers. These are on their turn disaggregated according to their involvement with primary supplies, conversion, processing and end-use demand for energy services. This is usually exogenous, and is specified by sector and functions within a sector. Key parameters and variables in a bottom-up model concern the costs and the initial values of installed capacities of technologies; their residual life; fuel and electricity costs and the potential rates and limits of alternative technology penetration. All these information are compounded in "conservation supply curves" (Stoft, 1995) which summarise the marginal cost of supplying energy.

Estimation of mitigation policy cost and the investment needed to meet emission reduction targets is the paramount field for the application of bottom-up models like MARKAL (Loulou et al., 2004), PRIMES (E3MLAB, 2010), POLES (JRC-IPTS, 2010) and TIMER (van Vuuren et al., 2004) (see European Commission 2008, 2010; Barker et al., 2007, Van Vuuren et al., 2009; Edenhofer et al., 2010).

8.3.3 Economic impact modelling

The IA literature proposes different methodologies for the economic modelling of climate change impacts.

In hard-linked IA models, it is accounted for by reduced-form functional relations. These Climate Change Damage Functions (CCDFs) translate temperature increases into GDP losses. The parameterization of these functions is derived from literature surveys, econometric estimations or expert opinions. Climate change impacts are calculated for given global mean temperature increases (e.g., 3 °C in RICE 96, for 3.5 °C and 6 °C in RICE 99), for a number of "impact areas" (e.g., sea-level rise, other market sectors, health, non-market amenity of the environment, human settlements and ecosystem and catastrophic events in RICE 99) and regions. Impacts are then aggregated to determine a total loss as a percent of GDP for each region. Temporal dynamics in damages are then originated by interpolating different "temperature-GDP loss" couples with linear or exponential functions.

The MERGE model (Manne and Richels, 2004) adopts a similar approach for the calibration of market damages, but adds the treatment of the non-market dimension. This, accounting for impacts on human health, species losses and environmental quality, is captured with a "corrected" willingness to pay (WTP) approach. Non-market damages are calibrated to match current national expenditure on environmental protection. That is MERGE implicitly assumes that (a) what is currently spent on environmental protection proxies the effective WTP to avoid non-market losses and (b) that WTP gives at least a lower bound to non-market damages.

A partly different methodology in calibrating economic losses is proposed by the FUND model (Tol, 2002). Climate change costs for different impact categories (agriculture, forestry, water, energy consumption, sea-level rise, ecosystems, vector-borne and heath/cold-stress related diseases) are calculated dynamically for distinct impact-specific CCDFs.

The Global Impact Model (GIM) (Mendelsohn et al., 2000) grounds its damage estimates on empirical studies of the U.S. economy to establish a relationship between annually averaged temperature, precipitation and market impacts for a set of climate-sensitive sectors: agriculture, forestry, energy, water, and coastal structures. Estimated sectoral response functions are then generalised to other countries on the basis of the different regional climatic and "geographical" specificities. The research relied upon both experimental method (controlled laboratory studies), and cross-sectional evidence. This last methodology in particular allows for the capturing of adaptation trends by measuring, for instance, the differences in farmers or households behaviour in cool versus warm locations. All this finally leads to calculations of money damages or benefits by sector and country.

Since the end of the 90s, CGE models have also been increasingly used for economic climate change impact assessments. The main difference with hard-linked exercises is the lack of a reduced-form CCDF. Rather, GDP losses are the direct outcome of economic pressures that a climate change induces in quantity/quality of endowments and consumers preferences in the model. For instance, sea-level rise is usually modelled as a loss of productive land and/or capital. Changes in the health status are implemented as decreases in labour productivity or in labour stock and as a change in demand for health care services, etc. In doing so, CGE offers damage estimates that do take into account price endogeneity and the associated relocation of resources.

There is an extended CGE body of literature related to climate change and agriculture (see, e.g., Tsigas et al., 1997; Darwin, 1999; Ronneberger et al., 2009), which is connected also to water scarcity (Calzadilla et al., 2009) and sea-level rise (see, e.g., Deke et al., 2002; Darwin and Tol, 2001; Bosello et al., 2007, 2012; Aaheim et al., 2010). Fewer CGE analyses address the issue of climate change impacts on health (Bosello et al., 2006), tourism (Berrittella et al., 2006; Bigano et al., 2008) or ecosystems (Bosello et al., 2011). CGE models have also been used to investigate the interactions of multiple impacts. For example, Bigano et al. (2008) analyzed the joint effect of adverse climatic impacts on sea-level rise and tourism activity. Eboli et al. (2010), Bosello et al. (2009), Aheim et al. (2011) and Ciscar et al. (2011) present combined assessments of an extended set of climate change impacts (health, tourism, agriculture, energy demand and sea-level rise, forestry, river floods and fishing).

8.3.4 Mitigation modelling

The issue of mitigation modelling strongly overlaps with that of technology modelling and partly with that of uncertainty modeling, which are addressed in detail by sections 8.3.6 and 8.3.7, respectively, and not developed here. The basic of mitigation in IA modelling is to implement emission reduction through price signals (carbon-energy taxes or subsidies to renewable) influencing emission quantities, or trough quotas affecting carbon-energy prices. In some models like RICE 96, the abatement rate is an explicit control variable; in the vast majority of models it is the outcome of adjustments in the production side, taking the form of a combined decrease in economic activity, fuel/input substitution, higher stimulus to green R&D. Abatement can be treated under a cost-benefit perspective (policy optimization) as is usually done by hard-linked IAMs or under a cost-effectiveness perspective (policy evaluation).

8.3.5 Adaptation modelling

The economic modelling aspects of adaptation have been amply revised in Agrawala et al. (2010, 2011) to which the interested reader is directly referred. This section summarises their main methodological insights.

The first IAM to include adaptation is the Policy Analysis of Greenhouse Effect (PAGE) model (Hope 1993; Plambeck et al., 1997, Hope 2006, 2009). In the model, a dedicated expenditure in adaptation can increase the macro-regional tolerable temperature rate of change, the tolerable maximum, and, finally, reduce the adverse impacts of climate change when temperature exceeds tolerability. The different levels of adaptation are however exogenously set by the user and the model typically produces cost effectiveness assessments.

A step to endogenize adaptation choices is provided by the FUND model (Tol, 2007) where adaptation against sea-level rise is an explicit continuous policy decision variable. Coastal protection is optimally chosen on the basis of a given cost-benefit criterion (Fankhauser, 1994). The costs and benefits of coastal protection are efficiently balanced in different scenarios with and without mitigation. Insights on adaptation dynamics are then derived.[3]

Further developments are provided by a group of hard-linked IAMs presenting strong methodological similarities. All share an intertemporal utility maximisation growth modelling core with climate dynamics inspired by the DICE/RICE model family; all disentangle adaptation costs rearranging the CCDF estimated by DICE or RICE and all model adaptation as an additional control variable taking the form of endogenous dedicated investment(s) whose effects are the reduction of the damage coefficient in the model CCDF. In this vein, Bosello (2008) introduces anticipatory adaptation in the FEEM-RICE model as a stock of defensive capital that is accrued over time by periodical endogenous investment. Adaptation investment reduces climate change damage, but must compete with investment in R&D, physical capital and mitigation expenditure in the planner's utility maximisation problem.

De Bruin et al. (2009) enriched the DICE model with reactive adaptation: a flow expenditure that needs to be adjusted period-by-period responding to current damages. This approach has also been implemented in a regionalised version of the model AD-RICE, based on RICE, and in the updated global version, DICE2007[4] (De Bruin, Dellink, and Agrawala 2009). Hof et al. (2009) re-propose the same methodology to model adaptation in the regional model FAIR, used to analyse the effectiveness of international financing of adaptation costs using the revenue from emission trading.

Bosello, Carraro, and De Cian (2010) and Agrawala et al. (2010) represent adaptation not as a single choice variable, but as a portfolio of alternative policy responses in a nested Constant Elasticity of Substitution (CES) functions. Anticipatory adaptation and reactive adaptation are assumed imperfect substitutes. In addition, Bosello et al. (2010) allowed for endogenous improvements in the effectiveness of reactive adaptation by investing in dedicated R&D. In Agrawala et al. (2010) a dedicated investment can be addressed to develop adaptive capacity before further allocating resources between reactive and anticipatory measures.

Bahn et al. (2010) developed a model in the spirit of DICE, but with anticipatory adaptation and distinguishing two macro sectors: the fossil-fuel-based economy and the carbon-free or clean economy. This feature allows exploring the effect of stock adaptation on the accumulation of clean capital.

Some attempts, finally, have been made to include planned adaptation in CGE models. This is usually captured as a redirection of resources, e.g., national investment, toward protection activities. These reduce adverse climatic impacts, but displace other investment or consumption. In this vein, Deke et al. (2001) and Darwin and Tol (2001) use respectively a recursive-dynamic and a static CGE model to estimate the economy-wide consequences of coastal protection. The related costs are subtracted from investment; this essentially reduces the capital stock, and hence economic output. This effect is particularly evident in Deke et al. (2001), as the dynamic nature of the model amplifies it over time. Bosello et al. (2006) use a similar approach in a static CGE, but forced investment in coastal protection displaces consumption rather than other investment forms.

8.3.6 Technology modelling

A realistic representation of technical progress is key to determine cost effectiveness of mitigation and adaptation polices. In fact, the standard approach still largely followed by both growth and CGE models is to consider technical progress as a free and exogenously increasing parameter ("manna from heaven") augmenting total factor productivity, and/or factor-specific productivity and/or activity-specific productivity (e.g., abatement).

A slightly richer characterization of technical progress is provided by those models assuming the existence of low or carbon free backstop technologies (e.g., Peck and Teisberg, 1992; Manne and Richels, 2004; Bosetti et al., 2006; Gerlagh et al., 2007). The typical assumption is that these are initially too costly to be economically viable. But as long as fossil fuel-based technologies became more expensive because of resources depletion or mitigation policies, they can become competitive and enter the market. In this way, the technology mix of the economic system changes under the pressures of endogenous energy prices.

A further way of modelling technological transition, rather common in CGE modelling, is that of capital vintages. This is for instance proposed by the ENVISAGE model (Van der Mensbrugghe, 2008), the SGM model (Edmonds et al., 2004), the EPPA model (Paltsev et al., 2005) and the ENVI-LINKAGE model (Burniaux and Chateau, 2008). The idea is that the process of capital accumulation progressively brings into existence new "vintages" of capital which substitute the old ones. New vintages are then characterised by a different (higher) rate of substitution with other factors of productions (typically energy) than older vintages. Accordingly, as capital accumulation is endogenous also the energy/carbon intensity of the production system and ultimately its production potential are endogenous.

There is then an increasing number of last generation hard-linked IA models (Buonanno, 2001; Popp, 2004; Kemfert, 2002; Gerlagh, 2007; Edenhofer et al., 2005; Bosetti et al., 2006) and CGE models (Goulder and Schneider, 1999; Sue Wing, 2003; Kemfert, 2005; Otto et al., 2007; Otto et al., 2008; Löschel and Otto, 2009; Schwark, 2010) representing learning by doing (LbD), learning by researching processes (LbR) or a combination of the two. In LbD, total factor productivity or that of abatement activities depends respectively on cumulated production or abatement. These last are typically endogenous, thus also the productivities evolve endogenously to eventually feed-back on the convenience to produce or abate.

In LbR a dedicated investment in R&D builds a stock of knowledge that then increases the productivity of those activities to which R&D is addressed (production, mitigation, adaptation). The main difference with LbD is the existence of an explicit control variable, R&D, and of its opportunity cost. Both LbD and LbR models can then consider spill-overs and the like.

Bottom-up models capture technology in the engineering sense. This means that different technological options with given performances and accessible at specific costs are compared and ranked in order of increasing net cost to form "packets of options" represented as marginal cost or curves, the abovementioned "Conservation Supply Curves" (CSCs). Technological change occurs through technology substitution whose penetration is given by costs and performance characteristics.

8.3.7 Uncertainty modelling

Methodological approaches to uncertainty implementation in IAMS and related results have been recently revised by Peterson (2006), Böhringer, (2009a) and Golub et al. (2011).

Standard to the literature is dividing uncertainty into "parametric", when it originates from the imperfect knowledge of some aspect of the problem, or "stochastic", when it originates from the inherently non deterministic nature of the investigated phenomenon.[5] Given the pervasive nature of both forms of uncertainty, and the difficulty to treat them analytically, different categories of IAMs models have been applied to provide their quantitative support to understand how uncertainty can affect the climate change policy decision processes (how much to mitigate or to invest in de-carbonizing technologies) and to estimate the value of information resolving that uncertainty.

Parametric uncertainty is the category most widely analysed by IAMs. This is done with a multiplicity of approaches. The most widely used is a more or less refined sensitivity analysis. This goes from the simple single-value deterministic sensitivity, to "uncertainty propagation" that consists in Monte Carlo sampling over joint probability distributions of input parameters. The parameters' focus of the investigations are usually climate sensitivity or the pure rate of time preference.

Another approach consists in building uncertainty directly inside agents' optimization problems, through expected utility functions. These researches usually investigate how optimal mitigation or adaptation policies respond when facing the probability of a catastrophic event (Gjerde et al., 1999). The probability can be exogenously determined or endogenously dependent upon emission paths.

A third one is the use of a special form of real option analysis (Golub et al., 2011), to evaluate, in the presence of uncertainty that could be resolved in the future, the option premium provided by flexibility in mitigation policy. This approach offers the advantage to deal with continuous distributions with asymmetric fat tails.

The study of stochasticity is at an early stage. It is being conducted with applied models given the impossibility to find analytical solutions. It consists in introducing randomness directly in the intertemporal constraints which characterise agents' optimization problem. At the end, what is affected is the temperature growth path that becomes random because of climatic stochasticity or because of stochasticity in the success (effectiveness) of decarbonization technologies.

8.4 Results

8.4.1 Impacts

The economic assessment of climate change impacts, or, differently said, the cost of policy inaction, is clearly the first information economic modelling is expected to provide. The literature in the field is extremely vast. It is well summarised by the periodical releases of the IPCC reports (Pearce et al., 1996; IPCC, 2001, 2007a), by the Stern Review (Stern, 2007) and by Tol (2008, 2009, 2011).

There are different ways to measure the cost of climate change.

One is to determine its overall impact on world or regional GDP for given scenarios of temperature increase and economic development. This is the typical format of many hard and soft linked IA models and exercises. Even in a context of "smooth" climate change, i.e., with extreme, but not catastrophic events, estimates vary a lot. Summarizing: damages are increasing in temperature with usually a quadratic trend; for temperature increases below 2 °C, the net loss at the world level is moderate (reaching at the maximum the 2% of GDP) or even slightly negative (Tol, 2002; Mendelsohn et al., 2000; Hope, 2006). It becomes unambiguously negative for higher temperature levels. However, the range of estimates remains large, spanning from the 5% to the 20% of GDP.

This global view, however, can be partly misleading. In particular, developing countries are more exposed as they are broadly located at low mid-latitudes experiencing higher temperature increases and impacts, more sensitive, show a higher reliance on climate-dependent sectors (primarily agriculture), and are less capable to adapt. There is thus a widespread agreement in the impact literature on huge damages hitting developing countries. This finding is robust also for those assessments according to which global losses are small, or even negative (see, e.g., Tol, 2002; Hope, 2006; Mendelsohn et al., 2000 and Maddison 2003 where negative impacts on world GDP of 2.3%, 0.9%, 0% and -0.4% translates into losses of the 4.1%, 2.6%, 3.6% and 14.6% in Africa, Asia, Africa, South America, respectively).

Another way to present the cost of climate change is to estimate how much an additional ton of carbon (or of CO_2) emitted can cost the society. This is referred to as "the social cost of carbon". Estimates are again highly differentiated (Tol, 2009, 2011): they range from negative figures (implying gains from climate change), to values higher than 800 $/t. The distribution of these values is not uniform though. The majority of available estimates concludes that the damage ranges between the 5 and the 25 $/t of carbon.

Interestingly, the main drivers of this diversity are not different background scientific fundamentals, but rather different assumptions on the components of the discount factor (primarily the pure rate of time preferences), on country weighting and on the length of the period used as reference for the assessment. A clear example is the debate originated by the Stern review (see, e.g., Tol and Yohe, 2006; Nordhaus, 2007; Weitzmann, 2007). Its high-damage estimates – "5% of global GDP each year, now and forever [...] to 20% of GDP or more" and the correspondent high marginal cost of emissions, 314 $ per tons of carbon – were indeed largely determined by the low pure rate of time preference (0.1%), the high weight on the "utility" of developing countries and on the long time horizon (2200) chosen for the assessment.

Two other factors finally appear to be important to determine the potential cost of carbon.

The first is the structural uncertainty in science coupled with the presence of low probability catastrophic events. In a series of papers, Weitzman, (2007, 2009a,b, 2010), showed that the two together "fatten" the tails of climate change damage distributions. On the one hand, it increases expected damages much more than a low discount rate; on the other hand, it can consequently increase the willingness to pay to avoid it (i.e., to mitigate) nominally to infinite. In this uncertain and catastrophic environment, the standard cost-benefit analyses performed by integrated assessment models seem inadequate.

The second is the very existence of a price system which reacts to scarcity. This intuition has been formalised by Sterner and Persson (2008). They argue that the relative price of a given environmental good or service, say ecosystems, increases as it becomes scarcer, e.g., because of disruption induced by climate change. Accordingly, its weight in the production of GDP/welfare also increases relative to other goods and services. As a conclusion, the economic loss that climate change keeps on exerting on that good also increases. Sterner and Person (2008) show that these relative price effects can produce higher damages than the use of a very low pure rate of time preference.

8.4.2 Mitigation

The modelling literature on the cost of mitigation policies is extremely extended and has been developed with both top-down and bottom-up approaches. The different IPCC reports (in particular IPCC 1996, 2001, 2007a) provide comprehensive surveys on the cost of the Kyoto protocol and of post Kyoto mitigation regimes. Other important summaries of information are the Stern Review (Stern, 2006), Edenhofer et al. (2010) and Luderer et al. (2012). Moreover, following the June 2009 adoption by the EU Council of the "20-20-20" climate-energy package, and the voluntary mitigation commitments taken by major emitters during COP 15 in Copenhagen, a renewed literature offers a set of assessments of the cost of these policies for the major players in the mitigation action (see, e.g., Böhringer et al. 2009b, 2012; European Commission, 2008, 2010; Peterson et al., 2011; Durand-Lasserve et al., 2010).

In general, there is considerable variability about the costs of climate change mitigation. For instance, the modelling estimates contained in the IPCC AR4 for stabilization at 550 ppm CO2eq range from small GDP gains to 4% GDP losses in 2050. The related carbon price ranges between 25 and 150 US$ per tons of CO_2 (Fisher et al., 2007). The Stern Review found for comparable abatement efforts (leading to a GHG concentration stabilization at 500–550 ppm CO2eq) a cost range between 2% to the 5% of GDP (Stern, 2006). Somehow more "optimistic" are Edenhofer et al. (2010). They found a cost of stabilizing at 550 ppm between 0.5% and 2.5%, and at 400 ppm between 1.2% and 4.5% of GDP. Along the same lines are Luderer et al. (2012) according to which stabilization at 450 ppm could entail a world consumption loss between 0.1% and the 1.4% all over the century with carbon prices between 100 and 2100 $/t CO2 in 2100.

Also, the cost estimates of meeting the Copenhagen Pledges varies quite substantively. For instance, meeting the 20% emission reduction target in 2020 compared to the 1990 level could imply for the EU27 small GDP gains (Peterson et al., 2011), or a loss reaching 2% of welfare (Böhringer et al., 2009b) and a carbon price ranging roughly between 30 €/tCO2 (Durand

Lasserve et al., 2010) and 70 €/tCO2 (Böhringer et al., 2009b). This assumes that emission reductions are implemented at the lowest possible cost. The results range widens considerably if different distortions and second best situations are considered.

This uncertainty in model-based assessments arises on the one hand from "structural" factors. These are: differences about the baseline development of socioeconomic drivers (population, GDP, prices of fossil fuels, international trade); assumptions about emissions from land-use change and non-CO_2 greenhouse gases; assumptions about the deployment of technical progress in its double influence on the cost and availability of low-carbon mitigation technologies and flexibility in substitutability between different inputs and assumptions about the nature of the decision process and formation of expectations. On the other hand, it is driven by different assumptions on the mitigation policy design itself. Mainly: the degree of where (sectoral/country sharing of the abatement burden) and when (timing of the abatement supported by banking/borrowing mechanisms) flexibility, on the emission permit allocation mechanism (grandfathering vs. auctioning) and on the recycling of carbon tax or auctioning revenues.[6]

Against this background, the following robust messages can, however, be summarised.

The greatest costs savings opportunities are offered by the development of a well-established and wide carbon market allowing for the exploitation of cheaper emission reduction at the country and sector levels. For instance, the rigidities imposed by the EU implementation of the 20-20-20 climate-energy policy – namely the presence of trading and non-trading sectors and the imposition of the renewable standard – can increase the policy cost by 100%–125% (Böhringer et al., 2009b).

The availability of proper technological options is crucial in determining not only the overall cost of mitigation, but the possibility itself to achieve the stricter stabilization targets. On average, models introducing endogenous technical change show GDP costs and carbon prices more than halved compared with the exogenous technical change assumption (Barker et al., 2007). Investments in decarbonisation technologies are not trivial though and could amount to about 0.2% to 1% of world GDP over the course of the century (Luderer et al., 2012). Further, the availability of negative emission technology like carbon capture and sequestration is fundamental to achieve a 450–400 ppm GHG stabilization target (Edenhofer et al., 2010; Luderer et al., 2012).

Reaching a 2 °C temperature increase stabilization by 2100 is a very ambitious goal requiring early action. If postponed after 2020, there will be a very high probability of missing the target; in addition, the delay will nearly double global mitigation costs (Luderer et al., 2012).

Allocating pollution permits through auctioning instead of grandfathering increases the costs for the energy intensive sectors; however, a proper recycling of the revenues to cut for instance labour costs can, especially in the presence of distortionary labour tax, entail GDP gains (European Commission, 2010).

One of the major concerns in the case of unilateral mitigation action is the issue of carbon leakage, strictly associated to the potential international competitiveness loss for those countries implementing cap and trade systems.

The literature typically quantifies the leakage effect at around 15–30%, even if higher values are not uncommon. In particular, the studies introducing oligopolistic competition in energy markets obtain a leakage effect higher than 100% (Böhringer et al., 2010, 2012). Both leakages and competitiveness losses could be decreased (up to 1/3) by adopting border tax adjustments; however, the global efficiency gain of these instruments is quite small as they shift toward non abating countries or sectors the burden of the environmental regulation. In addition, they require a complex regulatory framework with high administrative and transaction costs, casting doubts in many authors on their concrete viability (Demailly and Quirion, 2005; Mathiesen and Maestad, 2002; McKibbin and Wilcoxen, 2009; Böhringer et al., 2009b, 2012).

Finally, uncertainty is shown to play an important role in determining mitigation policies. Attempting to summarise a rich IA body of literature full of special cases and different methodologies, it emerges that the main drivers of outcomes is whether one of two forms of irreversibility prevails. The first is related to the difficulty of reverting temperature increase (environmental irreversibility), the second of recovering an investment in mitigation in case it proves to be unnecessary (economic irreversibility). In general in applied research, environmental irreversibility dominates and uncertainty on damages leads to a stronger abatement, especially when it is coupled with catastrophic events (Nordhaus and Popp, 1997; Gjerde et al., 1999; Castelnuovo et al., 2003; Baranzini et al., 2003, Bosetti et al., 2009, Lorentz et al., 2012). The effect on the timing of action is much more ambiguous with studies supporting both anticipated (Gjerde et al., 1999) and delayed action (Fisher and Narain, 2003; Baranzini et al., 2003) depending on a complex interplay between sunk costs and learning. The uncertainty on mitigation technology availability, costs or effectiveness is also a field in expansion (Böhringer et al., 2009a) so far producing contrasting results on the adoption rates, on policy costs and on final welfare.

8.4.3 Adaptation

The main insights from the applied literature on adaptation modelling are centred on its relation with mitigation, on its time profile and on its regional distribution. The first robust message conveyed is that adaptation and mitigation crowd out reciprocally. There are both financial and functional reasons for that. Financial, as they both compete for the same limited budget; therefore, one drains resources away from the other (Agrawala et al., 2010; Hof et al., 2009). Functional, as both ultimately reduce climate change damages; therefore, more adaptation reduces the need to mitigate and vice versa (Fankhauser, 1994; Bosello, 2008; De Bruin et al., 2009b). Moreover, adaptation may dynamically delay the transition towards a greener economy, leading to higher GHG concentrations at the end of the century (Bahn *et al.*, 2010). This crowding out is asymmetric (de Bruin et al., 2009a; Bosello et al., 2010). Due to climatic and partly economic inertias against which mitigation operates, it is effective in the long term. Accordingly, even aggressive emission reduction provides a small incentive to reduce adaptation in the short to medium term. On the contrary, adaptation is more rapidly effective in damage reduction. Therefore, it provides an almost immediate incentive to lower mitigation.

Nonetheless, a second important outcome, exactly because both adaptation and mitigation reduce climate change damages, net climate change policy costs can be minimised when the two are used in combination (de Bruin et al., 2009a,b; Bosello et al., 2010; Bahn et al., 2010). This expected outcome in a first best context is replicated also when adaptation is "added on top" of pre-existing CO_2 stabilisation policies (Agrawala et al., 2010; Hof et al., 2009; Bosello et al., 2010a), therefore in cost-effectiveness.

Addressing the composition of the climate change policy basket, adaptation emerges clearly as the preferred option in cost-benefit analyses, absorbing the majority of resources and being the major damage reducer (Bosello, 2008; de Bruin et al., 2009b; Bosello et al., 2010). To see a partial reversion of results, GHG stabilization policies need to be explicitly imposed (see Agrawala et al., 2010 and Hof et al., 2009 for a 550 ppm and 2 °C policy target, respectively).

Alternatively, to observe this as endogenous results of the models, the risk of catastrophic climatic events (Bosello and Chen, 2010) or more generally uncertainty on climatic damages (Felgenhauer and de Bruin, 2009) have to be incorporated.

Adaptation expenditure closely follows the dynamics of climate change damage, which in most models is a smoothly increasing, convex function. Therefore, it is low initially and relevant only after the first simulated decades. Mitigation on the contrary tends to be anticipated compared to adaptation (Bosello, 2008; Bosello et al., 2010, 2010a; Bahn et al., 2010, Hof et al., 2009; Agrawala et al., 2010).

Adaptation expenditures will be concentrated in developing countries which, depending on the scenarios and models' assumptions, are estimated to be 2 to 4 times larger than that of developed countries. It is also expected to sharply increase in the second half of the century, driven by growing climate change damages (Bosello et al., 2010; Bosello et al., 2010; Agrawala et al., 2010). Agrawala et al. (2010) pointed at the role of capacity building in developing countries, which should be prioritised. This suggests a specific direction for international cooperation on adaptation (Bosello et al., 2010; Hof et al., 2009).

To conclude, the adaptation modelling literature confirms the importance of discounting: all studies suggest that a lower discount rate increases the relative contribution of mitigation to damage reduction. However, when adaptation is modelled as a flow expenditure, abatement substitutes adaptation (de Bruin *et al.*, 2009a, 2009b) when adaptation builds a protection stock, this last also increases (Bosello, 2008; Bahn et al., 2010; Bosello et al., 2010, 2010a). A lower pure rate of time preference also induces agents to invest more in anticipatory adaptation and in adaptive capacity building rather than in reactive adaptation (Agrawala et al., 2010).

The sensitivity of the two strategies to damages increase is as expected: both increase. However, once again if adaptation is of the proactive type, then it tends to increase relatively more than mitigation (Bosello, 2008; Bosello et al., 2010, 2010a), whereas if it is reactive, it increases less (de Bruin et al., 2009a, 2009b).

8.5 Directions for future research and conclusions

Against all uncertainties and the incredible complexities of climate change economics, 20 years of applied Integrated Assessment Modelling provided useful and robust insights which advanced the understanding of the problem and supported the design of climate change strategies.

Nonetheless, there are still important gaps to cover.

In the field of impact assessment, one area where both consolidated methodologies and quantitative evidence are lacking is that of biodiversity-ecosystem losses. This is the joint effect of the still incomplete knowledge surrounding ecosystem and biodiversity supporting services, and of their inherently non-market nature. At the micro level, these are treated by revealed or stated preferences methods, but their inclusion in top-down economic modelling is still problematic. There is a need to increase the spatial and temporal detail of top-down economic models toward the resolution provided by bottom-up impact models. This will help at a time to incorporate richer inputs from impact models without losing information through the aggregation process, and to capture important sub-national economic interactions that play an important role in the determination of the final effects. The same holds for seasonal effects. These are crucial features of many climatic impacts – e.g. on agriculture, tourism or energy – but difficult to include in top-down assessments. The equity aspect of climate change is mostly confined to inter-country comparison. More effort could be devoted to capture asymmetries of impacts across different income classes within the same country.

A better top-down/bottom-up integration is also needed in adaptation research. This would require on the one hand the development of adequate methodologies to gather uniform and upscale bottom-up information, and on the other hand to expand the portfolio of adaptation options the latter can consider. In parallel, a constant effort should be placed in improving the knowledge on adaptation costs and benefits in those areas where they are largely missing, like, e.g., health or developing countries. Another field of promising research is that of the strategic interaction of adaptation with mitigation in climate change agreements. While a stream of theoretical literature started, applied exercises are still very few.

Specific to technical progress modelling, there is a still limited representation of the diffusion and the innovation processes within and across sectors, including the role of Foreign Direct Investment (FDI) and trade. Engineering elements can enrich learning curves, providing better insights on this phenomenon. Agents are usually homogeneous while the role of institutions is not modelled.

Common to all fields, there is a still unsatisfactory treatment and communication of uncertainty (Swart et al., 2009; Saltelli and D'Hombres, 2010). The role of active learning, especially on technology cost-effectiveness and, more in general, on inertias and irreversibility, is not well characterised and would require more investigation. Findings on the effects of stochasticity are not yet conclusive.

Finally, as pointed out by Nordhaus (2012), fostered by software development and computational power increase, IAMs are rapidly growing in complexity. Especially when uncertainty is introduced through a set of complicated functions of random variables, there is an increasing probability of code mistakes and uncontrolled changes of the structural properties of the models. This imposes modellers to devote an increasing attention not only to the theoretical foundations of their models, but also to software engineering.

Obviously, the use of models requires caution. Even remaining in the field of economics, no one-fits-all-purposes model can exist. On the contrary, different questions need to be addressed by different tools. Given overarching uncertainty, models should be used to highlight trends, orders of magnitude and non-trivial interrelations among variables, rather than to provide one-point certain estimates. Models finally, should support decisions, and not replace the scientists or the decision makers' judgment. Modellers in particular, should be the first to understand when models can or cannot be used.

Notes

1 However, market unbalances can be introduced see Roson (2006).
2 Currently, in IA, "bottom-up" is loosely associated to all that assessments, with or without economic content, which taking as exogenous macro social and economic drivers, provide a detailed descriptions of specific markets, activities and impacts with a focus on engineering technological aspects. In this broader sense, crop models like, e.g., CLIMATECROP (Iglesias et al., 2009), sea-level rise engineering models like DIVA (Vafeidis et al., 2008) and flood models like LISFLOOD (Feyen and Dankers, 2009) provide huge spatial details and focus on specific impacts of climate change are often referred to as "bottom-up".
3 In some applications of FUND, general adaptive capacity is indirectly considered through international aid supporting health policies (Tol and Dowlatabadi, 2001), or, more in general, GDP growth (Tol, 2005).
4 http://nordhaus.econ.yale.edu/dice_mss_091107_public.pdf

5 A third form of uncertainty, but related to the modeler or model user, is that arising from subjective decisions on intertemporal or interpersonal weighting.
6 Interestingly, in the first assessments of the cost of mitigation, a general difference emerged between bottom-up and top-down estimates, with the former usually presenting lower costs than the latter (Bosello and Lanza, 2004). This was justified by a number of reasons: top-down models tend to represent a stylised technological progress, thus missing part of the emission reduction potential of the energy sector; they do include "rebound effects" where lower prices for cleaner technology imply both an increase in demand for that technology or savings that lead to a lower decrease if not an increase in energy consumption, and emissions. More recently, due also to the increasing refinement of top-down models that can include more realistic features of technological transitions, there seems to be no systematic difference in the reduction potential and costs reported by top-down and bottom-up approaches at the global scale (Barker et al. 2007; van Vuuren et al., 2009). Notable differences between the two approaches remain however in identifying emission reduction potentials at the sectoral level stressing their still complementary role.

References

Aaheim, A., Dokken, T., Hochrainer, S., Hof, A., Jochem, E., Mechler, R., and van Vuuren, D.P. (2010). "National Responsibilities for Adaptation Strategies: Lessons from Four Modelling Frameworks". In M. Hulme and H. Neufeld (Eds.), Making Climate Change Work for Us: European Perspectives on Adaptation and Mitigation Strategies. Cambridge: Cambridge University Press.

Aalbers, R (2013). Optimal discount rates for investments in mitigation and adaptation. CBP dicussion paper No. 257.

Aghion, P. and Howitt, P. (1998). Endogenous Growth Theory. Cambridge, Massachusetts: MIT Press.

Agrawala, S., Bosello, F., Carraro, C., De Cian, E., and Lanzi, E. (2011). "Adapting to Climate Change: Costs, Benefits, and Modeling Approaches", *International Review of Environmental and Resource Economics* 5, pp. 245–284.

Agrawala, S., Bosello, F., Carraro, C., De Cian, E., Lanzi, E., de Bruin, K., and Dellink, R. (2010). "PLAN or REACT? Analysis of Adaptation Costs and Benefits Using Integrated Assessment Models". OECD Environment Working Papers, No. 23, OECD Publishing.

Arrow, K.J., Cline, W.R., Maler, K.G., Munasinghe, M., Squitieri, R., and Stiglitz, J.E. (1996). "Intertemporal Equity, Discounting, and Economic Efficiency". In J.P. Bruce, H. Lee., and E.F. Haites, (Eds.), Climate Change 1995. Economic and Social Dimensions of Climate Change, Contribution of Working Group III to the Second Assessment Report of the Intergovernmental Panel on Climate Change. Cambridge, UK: Cambridge University Press.

Bahn, O., Chesney, M., and Gheyssens, J. (2010). "The Effect of Adaptation Measures on the Adoption of Clean Technologies". Paper presented at the WCERE Congress, Montreal, 2010.

Baranzini, A., Chesney, M., and Morisset, J. (2003). "The Impact of Possible Climate Catastrophes on Global Warming Policies", *Energy Policy,* 31, pp. 691–701.

Baker, E., Clarke, L., and J. Weyant (2006). "Optimal Technology R&D in the Face of Climate Uncertainty", *Climatic Change,* 78 (1), 157–179.

Barker, Terry, Junankar, Sudhir, Pollitt, Hector, and Summerton, Philip (2007). "Carbon Leakage from Unilateral Environmental Tax Reforms in Europe, 1995–2005", *Energy Policy,* 35, pp. 6281–6292.

Barker, T., Foxon, T., and Scrieciu, S.S. (2008). "Achieving the G8 50% Target: Modelling Induced and Accelerated Technological Change Using the Macro-Econometric Model E3MG", *Climate Policy* 8, S30–S45.

Barker, T., Bashmakov, I., Alharthi, A., Amann, M., Cifuentes, L., Drexhage, J., Duan, M., Edenhofer, O., Flannery, B., Grubb, M., Hoogwijk, M., Ibitoye, F.I., Jepma, C.J., Pizer, W.A., and Yamaji, K. (2007). Mitigation from a Cross-sectoral Perspective. In B. Metz, O.R. Davidson, P.R. Bosch, R. Dave, and L.A. Meyer (Eds.), Climate Change 2007: Mitigation. Contribution of Working Group III to the Fourth Assessment Report of the Intergovernmental Panel on Climate Change. Cambridge, United Kingdom and New York, NY, USA: Cambridge University Press.

Berrittella, M., Bigano, A., Roson, R., and Tol, R.S.J. (2006). "A General Equilibrium Analysis of Climate Change Impacts on Tourism", *Tourism Management,* 27 (5), pp. 913–924.

Bigano, A., Bosello, F., Roson, R., and Tol, R.S.J. (2008). "Economy-wide Impacts of Climate Change: A Joint Analysis for Sea Level Rise and Tourism", *Mitigation and Adaptation Strategies for Global Change,* Vol. 13, No. 8.

Böhringer, C., Balistreri, E.J., and Rutherford, T.F. (2012). "The Role Of Border Carbon Adjustment In Unilateral Climate Policy: Overview of an Energy Modeling Forum Study (EMF 29)", *Energy Economics,* 34, S97–S110.

Böhringer, C., Fisher, C., and Rosendahl, K.E. (2010). "The Global Effects of Subglobal Climate Policies", Resources for the Future Discussion Paper, pp. 10–48.

Böhringer, C., Mennel, T.P., and Rutherford, T.F. (2009a). "Technological Change and Uncertainty in Environmental Economics", *Energy Economics,* 31, S1–S3.

Böhringer, C., Rutherford, T., and Tol, R. (2009b). "The EU 20/20/20 Targets: An Overview of the EMF22 Assessment", *Energy Economics,* 31, 268–273.

Bosello, F. (2008). "Adaptation, Mitigation and Green R&D to Combat Global Climate Change. Insights from an Empirical Integrated Assessment Exercise", CMCC Research Paper, No. 20.

Bosello, F., Buchner, B., and Carraro, C. (2003). "Equity, Development and Climate Change Control", *Journal of the European Economic Association,* Vol. 1.

Bosello, F. Carraro, C. and De Cian, E. (2010a), "Climate Policy and the Optimal Balance Between Mitigation, Adaptation and Unavoided Damage", *Climate Change Economics,* Vol. 1 N. 2, 71–92.

Bosello, F. and Chen, C. (2010). "Adapting and Mitigating to Climate Change, Balancing the Choices Under Uncertainty", FEEM Note di Lavoro 2010.59.

Bosello, F., Eboli, F., Parrado, R., Nunes, P.A.L.D., Ding, H., and Rosa, R. (2011). "The Economic Assessment of Changes in Ecosystem Services: An Application of the CGE Methodology", *Economía Agraria y Recursos Naturales,* 11–1, 161–190.

Bosello, F. and Lanza, A. (2004). "Modeling Energy Supply and Demand: A Comparison of Approaches", *Encyclopedia of Energy,* Vol. 4, pp. 55–64.

Bosello, F., Nicholls, R.J., Richards, J., Roson, R., and Tol, R.S.J. (2012). "Economic Impacts of Climate Change in Europe: Sea-Level Rise", *Climatic Change,* 112, pp. 63–81.

Bosello, F., Roson, R., and Tol, R.S.J. (2006). "Economy-wide Estimates of the Implications of Climate Change: Human Health", *Ecological Economics,* 58, pp. 579–591.

Bosello, F., Roson, R., and Tol, R.S.J. (2007). "Economy-Wide Estimates of the Implications of Climate Change: Sea-Level Rise", *Environmental and Resource Economics,* 37 (3), pp. 549–571.

Bosetti, V., Carraro, C., De Cian, E., Massetti, E., and Tavoni, M. (2011). "Incentives and Stability of Environmental Coalitions: An Integrated Assessment", FEEM Note di Lavoro, 97.2011.

Bosetti, V., Carraro, C., and Galeotti, M. (2006). "The Dynamics of Carbon and Energy Intensity in a Model of Endogenous Technical Change", *The Energy Journal,* Special issue: Endogenous Technological Change and the Economics of Atmospheric Stabilisation, pp. 191–206.

Bosetti, V., Carraro, C., Sgobbi, A., and Tavoni, M. (2009). "Delayed Action and Uncertain Stabilization Targets. How Much Will Mitigation Policy Cost?", *Climatic Change,* 96, pp. 299–312.

Brechet, Gerard and Tulkens, H. (2011). "Efficiency vs Stability in Climate Coalitions: A Conceptual and Computational Appraisal", *The Energy Journal,* Vol. 32, No. 1.

Buonanno, P., Carraro, C., Castelnuovo, E., and Galeotti, M. (2001). "Emission Trading Restrictions with Endogenous Technological Change", *International Environmental Agreements: Politics, Law and Economics,* 1, pp. 379–395.

Burniaux, J.M. and Chateau, J. (2008). "An Overview of the OECD ENV-LINKAGE Model", OECD Economic Department Working Paper No. 653.

Burniaux, J-M., Martin, J.P., and Martins, J.O. (1992). "GREEN: A Global Model for Quantifying the Cost of Policies to Curb CO_2 Emissions", OECD Economic Studies, No. 19.

Calzadilla, A., Rehdanz, K., and Tol, R.S.J. (2011). "Trade Liberalization and Climate Change: A CGE Analysis of the Impacts on Global Agriculture", *Water,* 2 (3), 526–550.

Cambridge Econometrics (2012). "E3ME an Economy, Energy, Environment Model for Europe. Technical Manual Version 5.5". Available at: http://www.camecon.com/Libraries/Downloadable_Files/E3ME_Manual.sflb.ashx

Carraro, C., Eykmans, J., and Finus, M. (2006). "Optimal Transfers and Participation Decisions in International Environmental Agreements", *The Review of International Organization,* Vol. 1.

Cass, David (1965). "Optimum Growth in an Aggregate Model of Capital Accumulation", *Review of Economic Studies,* 32, p. 233–240.
Castelnuovo, E., Moretto M., and Vergalli, S. (2003). "Global Warming, Uncertainty and Endogenous Technical Change", *Environmental Modeling and Assessment* 8, pp. 291–301.
Ciscar, J-., Iglesias, A., Feyen, L., Szabó, L., Van Regemorter, D., Amelung, B., Nicholls, R., Watkiss, P., Christensen, O.B., Dankers, R., Garrote, L., Goodess, C.M., Hunt, A., Moreno, A., Richards, J., and Soria, A. (2011). "Physical and Economic Consequences of Climate Change in Europe". *Proceedings of the National Academy of Sciences of the United States of America,* 108 (7), pp. 2678–2683.
Darwin, R.F. and Tol, R.S.J. (2001). "Estimates of the Economic Effects of Sea Level Rise", *Environmental and Resource Economics,* 19, pp. 113–129.
de Bruin, K.C., Dellink, R.B., and Agrawala, S. (2009b). "Economic Aspects of Adaptation to Climate Change: Integrated Assessment Modeling of Adaptation Costs and Benefits", OECD Environment Working Paper, No. 6.
de Bruin, K.C., Dellink, R.B., and Tol, R.S.J. (2009a). "AD-DICE: An Implementation of Adaptation in the DICE Model", *Climatic Change,* 95, pp. 63–81.
Deke, O., Hooss, K.G., Kasten, C., Klepper, G., and Springer, K. (2001). "Economic Impact of Climate Change: Simluations with a Regionalized Climate-Economy Model?, Kiel Institute of World Economics, 2001: 1065.
Demailly D. and Quirion P. (2005). "Leakage from Climate Policies and Border Tax Adjustments: Lessons From a Geographic Model of the Cement Industry", CESifo Venice, Summer Institute, http://ideas.repec.org/p/hal/wpaper/halshs-00009337.html
Durand-Lasserve, O., Pierreu, A., and Smeers, Y. (2010). "Uncertain Long-Run Emissions Targets, CO_2 Price and Global Energy Transition: A General Equilibrium Approach". CORE Discussion Paper No. 2010.27.
Eboli F., Parrado R., and Roson R. (2010). "Climate Change Feedback on Economic Growth: Explorations with a Dynamic General Equilibrium Model", *Environment and Development Economics,* 15 (5), pp. 515–533.
Edenhofer, O., Bauer, N., and Kriegler, E. (2005). "The Impact of Technological Change on Climate Protection and Welfare: Insights from the Model MIND", *Ecological Economics,* 54, pp. 277–292.
Edenhofer, O., Knopf, B., Barker, T., Baumstark, L., Bellevrat, E., Chateau, B., Criqui, P., Isaac, M., Kitous, A., Kypreos, S., Leimbach, M., Lessmann, K., Magne, B., Scrieciu, S., Turton, H., and van Vuuren, D.P. (2010). "The Economics of Low Stabilization: Model Comparison of Mitigation Strategies and Costs", *The Energy Journal,* Vol. 31 (Special Issue 1).
Edmonds, J., Pitcher, H., and Sands, R. (2004). "Second Generation Model 2004: An Overview", Pacific Northwest National Laboratory.
Ekins, Paul, Pollitt, Hector, Summerton, Philip, and Chewpreecha, Unnada (2012). "Increasing Carbon and Material Productivity through Environmental Tax Reform", *Energy Policy,* Vol. 42, pp. 365–376, Elsevier.
European Commission (2008). "Package of Implementation Measures for the EU's Objectives on Climate Change and Renewable Energy for 2020" Commission Staff working document SEC (2008) 85 II.
European Commission (2010). "Analysis of Options to Move Beyond 20% Greenhouse Gas Emission Reductions and Assessing the Risk of Carbon Leakage" Commission Staff working document SEC (2010) 650.
Eyckmans, J. and Tulkens, H. (2003). "Simulating Coalitionally Stable Burden Sharing Agreements for the Climate Change Problem", *Resource and Energy Economics,* 25, pp. 299–327.
E3MLAB (2010). "The PRIMES Model", Available at: http://www.e3mlab.ntua.gr/e3mlab/index.php?option=com_content&view=category&id=35&Itemid=80&lang=en
Fankhauser, S. (1994). "Protection vs. Retreat – The Economic Costs of Sea Level Rise", *Environmental Planning,* 27 (2), pp. 299–319.
Feyen, L. and Dankers, R. (2009). Impact of Global Warming on Streamflow Drought in Europe. *Journal of Geophysical Research,* 114(D17116), doi:10.1029/2008JD011438.
Fisher A.C. and Narain, U. (2003). "Global Warming, Endogenous Risk, and Irreversibility", *Environmental and Resource Economics,* 25, pp. 395–416.
Fisher, B.S., Nakicenovic, N., Alfsen, K., Corfee Morlot, J., de la Chesnaye, F., Hourcade, J.-C., Jiang, K., Kainuma, M., La Rovere, E., Matysek, A., Rana, A., Riahi, K., Richels, R., Rose, S., van Vuuren, D.,

and Warren, R. (2007). "Issues Related to Mitigation in the Long Term Context". In B. Metz, O.R. Davidson, P.R. Bosch, R. Dave, and L.A. Meyer (Eds.), Climate Change 2007: Mitigation. Contribution of Working Group III to the Fourth Assessment Report of the Inter-governmental Panel on Climate Change. Cambridge: Cambridge University Press.

Gerlagh, R. (2007). "Measuring the Value of Induced Technological Change", *Energy Policy*, 35, pp. 5287–5297.

Gjerde, J. Grepperud, S., and Kverndokk, S. (1999). "Optimal Climate Policy Under the Possibility of a Catastrophe", *Resource & Energy Economics*, 21 (3–4), pp. 289–317.

Golub, A., Narita, D., and Schmidt, M.G.W. (2011). "Uncertainty in Integrated Assessment Models of Climate Change: Alternative Analytical Approaches", Fondazione Eni Enrico Mattei Note di Lavoro N. 02.2011.

Goulder, L.H. and Schneider, S.H. (1999). "Induced Technological Change and the Attractiveness of CO_2 Abatement Policies", *Resource and Energy Economics* 21, pp. 211–253.

Grossman, G.M. and Helpman, E. (1994), "Endogenous Innovation in the Theory of Growth", *Journal of Economic Perspectives*, 8 (1), pp. 23–44.

Hof Andries, F., de Bruin, K.C., Dellink, R.B., den Elzen, M.G.J., and van Vuuren, D.P. (2009). "The Effect of Different Mitigation Strategies on International Financing of Adaptation", *Environmental Science and Policy*, 12, pp. 832–843.

Hope, C. (2006). "The Marginal Impact of CO_2 from PAGE2002: An Integrated Assessment Model Incorporating the IPCC's Five Reasons for Concern." *The Integrated Assessment Journal*, 6 (1), pp. 19–56.

Hope, C. (2009). "The Costs and Benefits of Adaptation". In Assessing the Costs of Adaptation to Climate Change: A Review of the UNFCCC and Other Recent Estimates, IIED, London, pp. 100–110.

Hope, C., Anderson, J., and Wenman, P. (1993). "Policy Analysis of the Greenhouse Effect: An Application of the PAGE Model." *Energy Policy*, 21 (3), pp. 327–338.

Iglesias, A., Garrote, L., Quiroga, S., and Moneo, M. (2009). "Impacts of Climate Change in Agriculture in Europe. PESETA-Agriculture Study", European Commission, Joint Research Centre Seville Spain.

IMAGE (2001). The IMAGE 2.2 Implementation of the SRES Scenarios, RIVM CD-ROM Publication 481508018, The Netherlands: Bilthoven.

IPCC (2000). "Special Report on Emission Scenarios. A Special Report of Working Groups III of the Intergovernmental Panel on Climate Change", N. Nakicenovic and R. Swart (Eds.), Cambridge, UK: Cambridge University Press, 570 pp.

IPCC (2001). "Climate Change 2001: Impacts, Adaptation, and Vulnerability. Contribution of Working Group II to the Third Assessment Report of the Intergovernmental Panel on Climate Change", J. McCarthy, O.F. Canziani, N.A. Leary, D.J. Dokken, K.S. White, (Eds.), Cambridge, UK: Cambridge University Press.

IPCC (2007a). "Climate Change 2007: Impacts, Adaptation, and Vulnerability. Contribution of Working Group II to the Fourth Assessment Report of the Intergovernmental Panel on Climate Change", M.L. Parry, O.F. Canziani, J.P. Palutikof, P.J. van der Linden and C.E. Hanson (Eds.), Cambridge, UK: Cambridge University Press.

IPCC (2007b). "Climate Change 2007: Mitigation. Contribution of Working Group III to the Fourth Assessment Report of the Intergovernmental Panel on Climate Change", B. Metz, O.R. Davidson, P.R. Bosch, R. Dave and L.A. Meyer (Eds.), Cambridge, United Kingdom and New York, NY, USA: Cambridge University Press.

IPCC (2012). "Managing the Risks of Extreme Events and Disasters to Advance Climate Change Adaptation. A Special Report of Working Groups I and II of the Intergovernmental Panel on Climate Change", C.B. Field, V. Barros, T.F. Stocker, D. Qin, D.J. Dokken, K.L. Ebi, M.D. Mastrandrea, K.J. Mach, G.-K. Plattner, S.K. Allen, M. Tignor and P.M. Midgley (Eds.), Cambridge, United Kingdom and New York, NY, USA: Cambridge University Press, 582 pp.

JRC-IPTS (2010). "Prospective Outlook on Long-Term Energy Systems – POLES Manual – Version 6.1". Available at: http://ipts.jrc.ec.europa.eu/activities/energy-and-transport/documents/POLES description.pdf

Kainuma, M., Matsuoka, Y., and Morita, T. (2003). (Eds.) Climate Policy Assessment Asia-Pacific Integrated Modeling, Springer-Verlag.

Kemfert, C. (2002). "An Integrated Assessment Model of Economy-Energy-Climate – The Model WIAGEM", *Integrated Assessment,* 3 (4), pp. 281–298.
Kemfert, C. (2005). "Induced Technological Change in a Multi-Regional, Multi-Sectoral, Integrated Assessment Model (WIAGEM): Impact Assessment of Climate Policy Strategies." *Ecological Economics,* 54 (2–3), pp. 293–305.
Koopmans, T.C. (1965). "On The Concept of Optimal Economic Growth", The Econometric Approach to Development Planning, Amsterdam: North-Holland, pp. 225–195.
Loeschel, A. (2002), "Technological Change in Economic Models of Environmental Policy: A Survey", *Ecological Economics,* 43, pp. 105–126.
Lorenz, A., Schmidt, M.G.W, Kriegler, E., and Held, H. (2012). "Anticipating Climate Threshold Damages", *Environmental Modelling and Assessment,* 17, pp. 163–175.
Löschel, A. and Otto, V. (2009). "Technological Uncertainty and Cost-Effectiveness of CO_2 Emission Reduction", *Energy Economics,* 31.
Loulou. R., Goldstein, G., and Noble, K. (2004). "Documentation for the Markal Family of Models", ETSAP.
Luderer, G., Bosetti, V., Jakob, M., Leimbach, M., Steckel, J., Waisman, H., and Edenhofer, O. (2012). "On the Economics of Decarbonization – Results and Insights from the RECIPE Project", *Climatic Change,* 114 (1), pp. 9–37.
Maddison, D.J. (2003). "The Amenity Value of the Climate: The Household Production Function Approach", *Resource and Energy Economics,* 25 (2), pp. 155–175.
Manne, A. and Richels, R. (2004). "MERGE an Integrated Assessment Model for Global Climate Change". Available at: http://www.stanford.edu/group/MERGE/GERAD1.pdf
Mathiesen, L. and Maestad, O. (2002). "Climate Policy and Steel Industry: Achieving Global Emission Reduction by an Incomplete Climate Agreement", Norwegian School of Economics and Business Administration Discussion Paper 20/02, Bergen.
McKibbin W.J. and Wilcoxen, P. (2009). "The Economic and Environmental Effects of Border Adjustments for Climate Policy". In L. Brainard and I. Sorkin (Eds.), Climate Change Trade and Competitiveness: Is a Collision Inevitable? Brookings Trade Forum 2008/09, pp. 1–35.
Mendelsohn, R.O., Morrison, W.N., Schlesinger, M.E., and Andronova, N.G. (2000). "Country-specific Market Impacts of Climate Change", *Climatic Change,* 45, (3–4), pp. 553–569.
Messner S. and Schrattenholzer, L. (2000). "MESSAGE-MACRO: Linking an Energy Supply Model with a Macro-Economic Module and Solving it Iteratively", *Energy International Journal,* 25 (3), pp. 267–282.
Nordhaus, W.D. (2007). "A Review of the Stern Review on the Economics of Climate Change", *Journal of Economic Literature,* 45, pp. 686–702.
Nordhaus, W.D. (2012). "Integrated Economic and Climate Modeling"; Keynote Address 19th Annual Conference of EAERE, Prague, June 29, 2012.
Nordhaus, W.D. and Boyer, J. (2000). Warming the World: Economic Models of Global Warming. Cambridge, MA: MIT Press.
Nordhaus, W.D. and Yang, Z. (1996). "A Regional Dynamic General-Equilibrium Model of Alternative Climate-Change Strategies", *American Economic Review,* 86 (4), pp. 741–765.
Ortiz, R.A., Golub, A, Lugovoy, O, Markandya, A., and Wang, J. (2011). "DICER: A Tool for Analyzing Climate Policies", *Energy Economics,* 33, S41–S49.
Otto, V.M., Loeschel, A., and Dellink, R. (2007). "Energy Biased Technical Change: A CGE Analysis", *Resource and Energy Economics,* 29 (2), p.p 137–158.
Otto, V.M., Loeschel, A., and Reilly, J. (2008). "Directed Technical Change and Differentiation of Climate Policy", *Energy Economics,* 30 (6), pp. 2855–2878.
Paltsev, S., Reilly, J., Jacoby, H.D., Eckaus, R.S., McFarland, J., Sarofim, M., Asadoorian, M., and M. Babiker (2005). "The MIT Emissions Prediction and Policy Analysis (EPPA) Model: Version 4", MIT Joint Program on the Science and Policy of Global Change, Report No. 125.
Parry, M., Canziani, O., Palutikof, J., van der Linden, P., and Hanson, C. eds. (2007). Climate Change 2007. Impacts, Adaptation and Vulnerability, Contribution of Working Group II to the Fourth Assessment Report on Climate Change, Cambridge University Press.
Pearce, D.W., Cline, W.R., Achanta, A.N., Fankhauser, S., Pachauri, R.K., Tol, R.S.J., and Vellinga, P. (1996). "The Social Costs of Climate Change: Greenhouse Damage and the Benefits of Control". In

J.P. Bruce, H. Lee., and E.F. Haites (Eds.), Climate Change 1995. Economic and Social Dimensions of Climate Change, Contribution of Working Group III to the Second Assessment Report of the Intergovernmental Panel on Climate Change, Cambridge, UK: Cambridge University Press.

Peck, S. and Teisberg, T. (1992). "CETA: A Model for Carbon Emission Trajectory Assessment" *The Energy Journal*, 13 (1), pp. 55–77.

Peterson, E.B., Schleich J., and Duscha, V. (2011). "Environmental and Economic Effects of the Copenhagen Pledges and More Ambitious Emission Reduction Targets", *Energy Policy*, 39, pp. 3697–3708.

Peterson, S. (2006). "Uncertainty and Economic Analysis of Climate Change: A Survey of Approaches and Findings", *Environmental Modelling and Assessment*, 11 (1), pp. 1–17.

Plambeck, E.L., Chris, H., and John, A. (1997). "The PAGE95 Model: Integrating the Science and Economics of Climate Change", *Energy Economics*, 19 (1), pp. 77–101.

Pollitt, Hector and Thoung, Chris (2009). "Modelling a UK 80% Greenhouse Gas Emissions Reduction by 2050", *New Scientist*, December 4, 2009.

Popp, D. (2004). "ENTICE: Endogenous Technological Change in the DICE Model of Global Warming", *Journal of Environmental Economics and Management*, 48, pp. 742–768.

Prinn, R., Jacoby, H., Sokolov, A., Wang, C., Xiao, X., Yang, Z., Eckaus, R., Stone, P., Ellerman, D., Melillo, J., Fitzmaurice, J., Kicklighter, D., Holian, G., and Liu, Y. (1999). "Integrated Global System Model for Climate Policy Assessment: Feedbacks and Sensitivity Studies", *Climatic Change*, 41 (3/4), pp. 469–546.

Ramsey, F.P. (1928). "A Mathematical Theory of Saving", *Economic Journal*, 38 (152), pp. 543–559.

Romer, P. (1986). "Increasing Returns and Long-run Growth", *Journal of Political Economy*, 94 (5), pp. 1002–1037.

Romer, P. (1990). "Endogenous Technological Change", *Journal of Political Economy*, 98 (5), S71–102.

Ronneberger, K., Berrittella, M., Bosello, F., and Tol, R.S.J. (2009). KLUM@GTAP: Introducing Biophysical Aspects of Land Use Decisions into a General Equilibrium Model. A Coupling Experiment", *Environmental Modelling and Assessment*, Vol. 14, No. 2.

Roson, R. (2006). "Introducing Imperfect Competition in CGE Models: Technical Aspects and Implications", *Computational Economics*, Vol. 28, pp. 29–49.

Rotmans, J. and Dowlatabadi, H. (1998). "Integrated Assessment of Climate Change: Evaluation of Methods and Strategies". In S. Rayner and E.L. Malone, (Eds.), Human Choices and Climate Change: A State of the Art Report. Washington, D.C.: Batelle Pacific Northwest Laboratories.

Saltelli, A. and D'Hombres, B. (2010). "Sensitivity Analysis Didn't Help. A Practitioner's Critique of the Stern Review", *Global Environmental Change*, 20 (2), pp. 298–302.

Schwark, F. (2010). "Economics of Endogenous Technical Change in CGE Models – The Role of Gains from Specialization", CER-ETH Working Paper 10/130 Zurich.

Solow, R. (1957). "Technical Change and the Aggregate Production Function", *Review of Economics and Statistics*, 39 (3), pp. 312–320.

Stern, N. (2007). "The Economics of Climate Change: The Stern Review". Cambridge and New York: Cambridge University Press.

Sterner, T., and Persson, U.M. (2008). "An Even Sterner Review: Introducing Relative Prices into the Discounting Debate', *Review of Environmental Economics and Policy*, 2 (1), pp. 61–76

Swan, T. (1956). "Economic Growth and Capital Accumulation", *Economic Record*, 32 (November), pp. 334–361.

Swart R., Bernstein L., Ha-Duong M., and Petersen, A. (2009). "Agreeing to Disagree: Uncertainty Management in Assessing Climate Change, Impacts and Responses by the IPCC", *Climatic Change*, 92, pp. 1–29.

Tol, R.S.J. and Dowlatabadi, H. (2001), "Vector Borne Diseases, Development and Climate Change", *Integrated Assessment*, 2, 173–181.

Tol, R.S.J. (2002). "Estimates of the Damage Costs of Climate Change – Part 1: Benchmark Estimates", *Environmental and Resource Economics*, 21 (1), pp. 47–73.

Tol, R.S.J. (2007). "The Double Trade-Off between Adaptation and Mitigation for Sea Level Rise: An Application of FUND", *Mitigation and Adaptation Strategies for Global Change*, 12, pp. 741–753.

Tol, R.S.J. (2008). "The Social Cost of Carbon: Trends, Outliers and Catastrophes", *Economics: The Open-Access, Open-Assessment E-Journal*, 2, pp. 2008–2025.

Tol, R.S.J. (2009). "An Analysis of Mitigation as a Response to Climate Change", Paper prepared for the Copenhagen Consensus on Climate Center.
Tol, R.S.J. (2011). "The Uncertainty about the Social Cost of Carbon: A Decomposition Analysis Using FUND." ESRI Working Paper 404.
Tol, R.S.J., and Yohe, G.W. (2006). "A Review of the Stern Review", *World Economics*, Vol. 7, No. 4.
Tsigas, M.E., Frisvold, G.B., and Kuhn, B. (1997). "Global Climate Change in Agriculture". In Thomas W. Hertel (Ed.), Global Trade Analysis: Modeling and Applications. Cambridge University Press.
Vafeidis, A.T., Nicholls, R.J., McFadden, L., Tol, R.S.J., Spencer, T., Grashoff, P.S., Boot, G., Klein, R.J.T. (2008). "A New Global Coastal Database for Impact And Vulnerability Analysis to Sea-Level Rise", *Journal of Coastal Research*, 24 (4), pp. 917–924.
Van der Mensbrugghe, D. (2008). "The Environmental Impact and Sustainability Applied General Equilibrium (ENVISAGE) Model". The World Bank.
van Vuuren, D.P., de Vries, B., Eickhout, B., and Kram, T. (2004). "Responses to technology and taxes in a simulated world", *Energy Economics*, 26 (4), pp. 579–601.
van Vuuren, D.P., Hoogwijk, M., Barker, T., Riahi, K., Boeters, S., Chateau, J., Scrieciu, S., van Vliet, J., Masui, T., Blok. K., Blomen, E., and Krama, T. (2009). "Comparison of Top-Down and Bottom-Up Estimates of Sectoral and Regional Greenhouse Gas Emission Reduction Potentials", *Energy Policy* 37, pp. 5125–5139.
Webster, M. (2009). "Uncertainty and the IPCC: A Comment", *Climatic Change*, 92, pp. 37-40.
Weitzman, M.L. (2007). "A Review of the Stern Review on the Economics of Climate Change", *Journal of Economic Literature*, 45 (3), pp. 703–724.
Weitzman, M. (2009a). "Additive Damages, Fat-Tailed Climate Dynamics, and Uncertain Discounting." *Economics: The Open-Access, Open-Assessment E-Journal*, Vol. 3, 2009-39. http://www.economics-ejournal.org/economics/journalarticles/2009-39
Weitzman, M. (2009b). "On Modeling and Interpreting the Economics of Catastrophic Climate Change", *The Review of Economics and Statistics*, Vol. XCI, No. 1.
Weitzman, M. (2010). "What is the 'Damages Function' for Global Warming – and What Difference Might It Make?" *Climate Change Economics*, 1 (1), pp. 57–69.
Whalley, J. and Wigle, R. (1991). "The International Incidence of Carbon Taxes". In R. Dornbush and J.M. Poterba (Eds.), Economic Policy Responses to Global Warming. Cambridge: MIT Press.
Wing, I.S. (2006). "The Synthesis of Bottom-Up and Top-Down Approaches to Climate Policy Modeling: Electric Power Technologies and the Cost of Limiting US CO_2 Emissions", *Energy Policy*, 34, pp. 3847–3869.

9

ECOSYSTEM-BASED ADAPTATION

Elena Ojea[1,2]

[1]BASQUE CENTRE FOR CLIMATE CHANGE (BC3)
[2]BREN SCHOOL OF ENVIRONMENTAL SCIENCE & MANAGEMENT, UCSB

9.1 Introduction: Adaptation in ecosystems

Ecosystems have complex interactions with climate change, being directly affected and impacted but also playing an important role in both mitigation and adaptation. This chapter introduces basic notions to understand the recent and growing concept of ecosystem-based adaptation to climate change. As an introduction, we begin with the evidence of climate impacts in ecosystems, then we justify the need for adaptation and finally we introduce and explore the notion of ecosystem-based adaptation.

9.1.1 The impacts of climate change in biodiversity and ecosystem services

Climate change is already having a strong impact in ecosystems, affecting their functions and the many benefits and services they provide to people (Parry *et al.*, 2007). These impacts are well documented in the literature and supported by the Intergovernmental Panel on Climate Change (IPCC) Fourth Assessment Report.

Apart from the impacts that have already been observed, even greater impacts in biodiversity and ecosystem services are expected in the future (Pimm and Raven, 2000; Thomas *et al.* 2004; Araujo *et al.*, 2006). By the end of the 21st century, climate change impacts are expected to be the primary cause for biodiversity loss and changes in ecosystem services on a global scale (MEA, 2005). Biodiversity is a key component of ecosystems, influencing their capacity to respond to external stresses such as climatic impacts, but biodiversity loss is expected to accelerate in the near future stressed by climate change (Pimm and Raven, 2000; Thomas *et al.*, 2004). Global losses of biodiversity are of key relevance since they are frequently irreversible. It is also estimated that a 3 °C warming increase would transform one fifth of the world's ecosystems (Fischlin *et al.*, 2007). As ecosystems change and their services are eroded, the implications of impacts will reach people, communities and economies around the world. Under a better case scenario of 1.5–2.5 °C temperature increase by the end of the century, an increase in the risk of extinction to approximately 20% to 30% of the plant and animal species known will occur (Parry *et al.*, 2007).

Climate change and other anthropogenic causes are affecting ecosystem resilience, reducing their capacity to recover from external stresses and compromising the production of ecosystem services. This loss of services supposes a significant barrier to the achievement of the millennium development goals (UNFCCC, 2011).

Climate change impacts on biodiversity and ecosystems are likely to occur even if strong mitigation efforts are made (Parry *et al.*, 2007). Because of this, adaptation strategies are needed in order to adapt to the imminent climatic effects, in a way that adaptation as well as mitigation strategies are developed simultaneously. However, their effects vary over time and place. The IPCC expects the effects of mitigation policies on biophysical systems to be globally noticeable from the mid 21st century (Parry *et al.*, 2007). In contrast, the benefits of adaptation will be on a regional and local scale and can be immediate. In spite of the differences in adaptation and mitigation strategies, climate change policies are not about choosing between adaptation and mitigation. Adaptation is necessary because climate change impacts cannot be avoided in the next decades by any mitigation strategy, and mitigation is necessary in order to reduce future costs of adaptation measures (Parry *et al.*, 2007). These complementarities make both strategies important. However, many times policy makers face trade-offs due to financial constraints, leading to disequilibria. In fact, climate change policies have, in general, tended to pay disproportionately more attention to mitigation than to adaptation, although adaptation is receiving increasing political attention (Parry *et al.*, 2007; Munang *et al.*, 2013).

9.1.2 The need for adaptation

The need for immediate and significant reduction in greenhouse gas emissions (GHG) to tackle the impacts of climate change and avoid catastrophic consequences in the long term is now indisputable. But even as countries work to mitigate GHG emissions, climate change-related impacts will continue and increase over the short to medium term, making adaptation an urgent need (TNC, 2011).

Adaptation is defined by the IPCC as the "adjustments of natural or human systems in response to actual or expected stimuli, or its effects to moderate the harm or exploit beneficial opportunities" (Parry *et al.*, 2007; Andrade *et al.*, 2011). Adaptation is among the biggest challenges humanity faces in the next century (Jones *et al.*, 2012), and therefore it is becoming an increasingly important part of the development agenda, especially in developing countries most at risk from climate change (World Bank, 2010; Eakin and Lemos, 2010). Adaptation in general, is also at the forefront of scientific inquiry and policy negotiations today (Andrade *et al.*, 2011).

In the biodiversity and natural resource management sectors, there is a strong need to advance in the development of adaptation strategies (Campbell *et al.*, 2009; Heller and Zavaleta, 2009: Munang *et al.*, 2013). This is critical, not only for achieving biodiversity conservation goals, but also for maintaining the contribution that biodiversity and ecosystem services provide for social adaptation (Campbell *et al.*, 2009). Current adaptation work is challenged by this latter effect, to understand and demonstrate how adaptation works and what the implications of adaptation for ecosystem resilience are (Tschakert and Dierich, 2010). This dynamic notion of adaptation allows promoting resilience of both ecosystems and human societies, beyond mere technological options that mainly focus on building hard infrastructure and other similar measures (Andrade *et al.*, 2011). Indeed, adaptation strategies that incorporate natural resource management can result in positive feedbacks for both people and biodiversity (Campbell *et al.*, 2009), as well as for climate mitigation (Munang *et al.*, 2013).

The vulnerability of a system measures the degree to which that system is susceptible or unable to cope with adverse effects of climate change, including climate variability and extremes (Andrade *et al.*, 2011). Natural responses of biodiversity and ecosystem services to changes resulting from climate change are known as "autonomous adjustments" (CBD, 2009). These include properties such as resilience, recovering capacity, vulnerability and sensitivity. In this line, the adaptive capacity of a system is the ability to adjust to climate change to moderate potential damages, to take advantage of opportunities or to cope with the consequences (Parry *et al.*, 2007; Andrade *et al.*, 2011).

Autonomous adaptation alone is not considered sufficient to halt biodiversity loss and ecosystem services (Andrade *et al.*, 2011). The development of activities proposed by societies known as "planned adaptation" are required (CBD, 2009). Adaptive management provides criteria from the Convention of Biological Diversity on how "planned adaptation" should be addressed, prioritising actions based on the maintenance of natural infrastructures and the ecological integrity of ecosystems (Andrade *et al.*, 2011). Adaptation strategies involve a range of actions, including behavioural change, technical or hard engineered solutions such as the construction of sea defences or risk management, and reduction strategies such as the establishment of early warning systems (UNFCCC, 2011).

Three main categories of adaptation are being used in theory and practice: soft policy approaches; hard engineered approaches and ecosystem-based approaches. Soft policy approaches are normally the first actions that are taken and encompass policy and behavioural shifts (e.g., early warning systems, national plans, strategies, etc.). Hard or engineered adaptation consists of the creation or maintenance of man-made infrastructures, such as seawalls, dams, river canalization, irrigation systems, etc. Ecosystem-based adaptation is directed to the conservation and enhancement of biodiversity and ecosystem services to increase ecosystems resilience and reduce people's vulnerability to climate change. In this chapter, we will focus on the third adaptation category.

9.1.3 Ecosystem-based adaptation

There is a growing recognition of the role that healthy ecosystems can play in increasing resilience and helping people to adapt to climate change through the delivery of ecosystem services that play a significant role in maintaining human well-being (UNFCCC, 2011). Maintaining and increasing resilience in ecosystems reduces climate-related risk and vulnerability. Ecosystem-based management offers a valuable yet unexploited approach for climate change adaptation, complementing traditional actions such as infrastructure development. Such ecosystem-based approaches are not simply about saving ecosystems, but rather about using them to help people and conserve the resources on which they depend (Burgiel and Muir, 2010). Building on this concept, a recent approach known as "ecosystem-based adaptation" uses biodiversity and ecosystem services as part of an overall adaptation strategy to help people and communities adapt to the negative effects of climate change at local, national, regional and global levels (Burgiel and Muir, 2010; UNEP, 2012; TNC, 2011; Travers *et al.*, 2012; Jones *et al.*, 2012; Munang *et al.*, 2013). In this context, ecosystem services can be defined as the benefits people obtain from ecosystems, including provisioning, regulating, cultural and supporting, according to the millennium ecosystem assessment framework (MEA, 2005).

The second CBD Ad Hoc Technical Expert Group (AHTEG) on biodiversity and climate change, convened in 2008–2009 to provide scientific and technical advice and assessment on the integration of the conservation and sustainable use of biodiversity into climate change mitigation and adaptation activities. Ecosystem-Based Adaptation (EBA onwards) has been defined in the CBD as the use of "biodiversity and ecosystem services in an overall adaptation strategy. It includes the sustainable management, conservation and restoration of ecosystems to provide services that help people adapt to the adverse effects of climate change". Through considering the ecosystem services on which people depend to adapt to climate change, EBA integrates sustainable use of biodiversity and ecosystem services in a comprehensive adaptation strategy (CBD, 2009). EBA has been since then increasingly used as a strategy to adapt to climate change (Munang *et al.*, 2013), with some new particularities respect to previous approaches:

- Social perspective: EBA explicitly includes both people and biodiversity, recognising the potential for well-managed, resilient ecosystems to provide services that enable people to adapt to the impacts of climate change.
- System resilience: EBA includes a range of actions for the management, conservation and restoration of ecosystems in order to increase the resilience of the system (Andrade *et al.*, 2011).
- Vulnerability: Vulnerability assessments are the foundation for any EBA strategy and are instruments fundamental to understand where climate change will have impacts and which ecosystems are more susceptible to change (Parry *et al.*, 2007).
- Reducing threats: EBA aims at reducing other major threats to ecosystems, which when compounded with the effects of climate change would push a system beyond its ability to function properly (Burgiel and Muir, 2010).
- Applicability: EBA are widely applicable at different spatial and temporal scales. EBA approaches have been shown to be effective for adaptation across sectors, contributing to livelihood sustenance and food security, sustainable water management, disaster risk reduction and biodiversity conservation (UNFCCC, 2011). Additionally, EBA approaches consider that both natural and managed ecosystems can reduce vulnerability to climate-related hazards and gradual climatic changes (Andrade *et al.*, 2011).
- Co-benefits: In addition to protection from climate change impacts, EBA also provides many other benefits to communities, for example through the maintenance and enhancement of crucial ecosystem services, such as clean water and food provision (TNC, 2011; Paterson *et al.*, 2008), or through mitigation co-effects (Munang *et al.*, 2013).

The concept of using ecosystems as a basis to adapt to the impacts of climate changes has gained momentum in recent years and has now emerged as an important technology in the adaptation 'toolbox' (Travers *et al.*, 2012: Munang *et al.*, 2013). EBA approaches are therefore increasingly being implemented and recommended in the international policy agenda, although their potential is not yet fully recognised by national governments (Munang *et al.*, 2013).

Munang *et al.* (2013) define three forms of EBA: targeted management, conservation, and restoration activities. Examples of EBA[1] approaches include flood defence through the maintenance and/or restoration of wetlands, or the conservation of agricultural biodiversity in order to support crop and livestock adaptation to climate change (UNFCCC, 2011). At present, EBA is seen as a complement to other hard adaptation actions, and both research and implementation of EBA initiatives are necessary to mainstream EBA into climate adaptation and ecosystems sustainability.

This chapter continues as follows: Section 9.2 describes the adoption of EBA by the main international environmental and development players and provides some specific examples of implementation; Section 9.3 reviews the economics of EBA compiling data on the finance and costs of EBA, and discussing its efficiency. Section 9.4 analyses the current gaps and research directions, and Section 9.5 concludes.

9.2 Emergence of ecosystem-based adaptation

Ecosystem-based adaptation is growing in importance in the international adaptation debate, with increasingly more case studies and projects being applied, and more contributions from research and international non-profit organizations defending its advantages over other more traditional adaptation options. This section introduces the evolution of EBA in international environmental policy, reviews its basis in ecosystem-based management and community-based adaptation, and presents some examples of implementation.

9.2.1 Adoption of EBA in the international policy agenda

Major international environmental policy is beginning to recognise the connection between healthy ecosystems and resilience to climate change. The Convention on Biological Diversity (CBD) is prioritising attention to the management of the major drivers of global change, as critical for decreasing the present rate of biodiversity loss. Similarly, discussions under the United Nations Framework Convention on Climate Change (UNFCCC) are also highlighting the role of mitigating and adapting to the effects of climate change by protecting the natural systems around (Burgiel and Muir, 2010).

International organizations and institutions from both the environmental and development sectors have engaged in research and implementation of EBA in the last decade. These include, among others: the UNFCCC, United Nations Environment and Development Programs (UNEP, UNDP) or the CBD as relevant institutions, and Conservation International (CI), the World Wide Fund for Nature (WWF), the International Union for the Conservation of Nature (IUCN), the Nature Conservancy (TNC), Birdlife International (BI), the World Resources Institute (WRI), the World Bank (WB) and the Global Environment Facility (GEF) as international organizations.

The UNFCCC set up the Nairobi work program in 2005 to assist all countries, in particular developing countries, to improve their understanding and assessment of the impacts of climate change and to make informed decisions on practical adaptation actions and measures (UNFCCC, 2011). The Nairobi work program disseminates knowledge and information on adaptation, including EBA, providing several reports and a database on the evidence of EBA worldwide from the work done by partner institutions.

The Nairobi Program involves many agencies and organizations working in adaptation. UNEP has teamed up with the UNDP and the IUCN to establish the Ecosystem-Based

Adaptation Flagship Program, which includes among its activities the development of pilot EBA approaches and comparisons of the costs and cost-effectiveness of EBA with other adaptation strategies. The UNEP works to develop effective EBA approaches, and helps vulnerable communities adapt to climate change through good ecosystem management practices and their integration into global, regional, national and local climate change strategies and action plans. UNEP's EBA Program is being implemented in diverse ecosystem settings, including mountains, river basins, dry-lands and low-lying coasts (UNEP, 2012). The work is delivered through three main components: (i) assessments and knowledge support, (ii) capacity building and demonstration and (iii) integration of EBA options into national development and adaptation plans (UNEP, 2012). The program has an emphasis on Least Developed Countries (LDC), Small Island Developing States (SIDS), and Africa (Travers *et al.*, 2012).

EBA is increasingly being used as a new approach to adaptation in national and international reports, discussion papers and policy documents and strategies related to climate change (Munang *et al.*, 2013). A recent UNDP and GEF report on adaptation states that EBA will play an increasingly important role as part of the UNDPs integrated adaptation investment strategy (UNDP, 2010). The World Bank is among the organizations that have seen the number of projects and programs that emphasise the linkages between ecosystems and climate change increase in the last decade. Since 2009, many members of IUCN also have promoted the use of EBA as a practical tool for climate change adaptation (Andrade *et al.*, 2011). Among other initiatives, international conservation organizations have supported and implemented EBA initiatives around the world (TNC, 2011; CI, 2012, among others). Despite of this effort, EBA is still far from capturing the main attention in adaptation policy and remains under-utilised by policy makers and associated stakeholders (Murang *et al.*, 2013).

9.2.2 Ecosystem-based management and community-based adaptation

Adaptation is important in all countries, but particularly in least developed countries (LDCs), in small island developing countries (SIDS) and in those countries that have economies that depend on climate-vulnerable sectors such as agriculture, tourism and fisheries (Parry *et al.*, 2007). Additionally, the IPCC states that climate change affects poor human communities disproportionately (Parry *et al.*, 2007). These communities are often marginalised and receive only limited services or support from governments. This condition has given rise to the concept of Community-Based Adaptation, which describes a set of activities aimed at climate change adaptation by local communities and the poorest people (Kotiola, 2009; Andrade *et al.*, 2011).

We have seen that EBA approaches to adaptation are receiving increasing attention in a policy context. Despite the fact that some initiatives did not start out as adaptation projects, there is evidence of the application of such approaches as a part of national and local adaptation portfolios (UNFCCC, 2011). Also, as the theoretical concept of EBA is fairly recent, practical approaches to adaptation that utilise the services of healthy ecosystems have been implemented in various forms by different communities for some time (UNFCCC, 2011). EBA complements these community-based adaptation activities by adding to and supporting these practices as EBA unifies approaches to ecosystem management in terms of adaptation (Andrade *et al.*, 2011).

It is increasingly recognised that the thinking that is embedded in both current ecosystem-based management and community-based adaptation practice is to consider a holistic approach to support an optimal management response. Ecosystem-based management is an approach that recognises ecological systems as a rich mix of elements that interact with each other in important ways, going beyond examining single issues, species or ecosystem functions in isolation (UNEP, 2011).

EBA draws on multiple streams of knowledge from ecosystem-based management, community-based adaptation and climate change adaptation. It is worth noting that broad stakeholder participation can be a measure of successful implementation, since many ecosystems and ecosystem services are managed by local users (CBD, 2009). Effective EBA should encompass a participatory process with a systems-based view that considers all relevant drivers and responses to change, including climate driven change, disaster risk response, climate variability, and broader long-term socioeconomic change (Travers *et al.*, 2012). Lessons learnt from community-based adaptation and ecosystem-based management are important to start building the basis for a strong EBA.

9.2.3 Implementation and performance of EBA

Nowadays, it is possible to access a number of reports and publications with information on EBA initiatives on the ground and learn how they are being implemented worldwide. Andrade *et al.* (2011), for example, present case studies from different parts of the world covering a variety of ecosystems. The case studies cover a range of adaptation interventions, some focused on adaptation for conservation purposes, and some focusing on supporting people to adapt to climate change through EBA. The IUCN also recommends best practices from an analysis of several case studies in coastal systems (IUCN, 2010). The UNFCCC collects examples from various types of ecosystem-based measures and how they have contributed to several sectors, including livelihood sustenance and food security, sustainable water management, disaster risk reduction and biodiversity conservation (UNFCCC, 2012). The OCDE has a database as well on adaptation projects from which several EBA projects are collected.

Evidence on the performance of EBA points to a series of direct benefits the approach offers. These include the sustenance of livelihoods and food security, the sustenance of irrigation and potable water for the future, disaster risk reduction as healthy ecosystems act as buffers for extreme events, and biodiversity conservation (CBD, 2009). Additionally, EBA has possible co-benefits that can occur in the areas of social and cultural benefits, economic, biodiversity and also mitigation (Munang *et al.*, 2013). Appropriately designed ecosystem management initiatives can also contribute to climate change mitigation by reducing emissions from ecosystem loss and degradation, and enhancing carbon sequestration helping to mitigate climate (UNEP, 2012). For example, forest conservation or restoration of degraded wetlands can also contribute to climate change mitigation measures, and land use can be determining fresh water provision (Garmendia *et al.*, 2012). Such win-win outcomes could also help to avoid maladaptation (UNFCCC, 2011; Paterson *et al.*, 2008). Sustainable forest management, for example, has the adaptive option of maintenance of nutrient and water flow, and the prevention of landslides. It also has social and cultural benefits such as opportunities for recreation, preservation of local culture and protection of indigenous peoples and communities as well as economic benefits with the potential generation of income through ecotourism, recreation, sustainable logging, etc.

From a review of the existing evidence of EBA implementation, studies from marine and coastal ecosystems have been found more frequently. A detailed description of the benefits and co-benefits, approaches and effectiveness of coastal and marine EBA is available in Box 9.1.

Box 9.1 Coastal EBA

Coastal EBA includes a range of actions for the management, conservation and restoration of ecosystems such as mangroves, coastal wetlands, coral reefs, shellfish reefs and sea grass beds in order to help reduce the vulnerability and increase the resilience of coastal human communities in the face of climate change (TNC, 2011). This is particularly important for sustaining natural resources on which vulnerable communities depend for their subsistence and of providing alternative livelihoods in the face of climatic uncertainty (Andrade *et al.*, 2011). As in many sectors, EBA is extremely linked to nature conservation in coastal ecosystems. EBA does not only appear in Marine Protected Areas (MPAs), but these can be important for adaptation in coastal and marine ecosystems. A debate exists however on whether MPA networks can contribute to the overall objectives of ecosystem-based management (EBM) or whether approaches beyond the protected areas are necessary to implement (Granek *et al.*, 2010). Halpern *et al.* (2010) provide scientific guidance regarding this debate and the use of MPAs within the EBM approach and evaluate the potential for marine reserves to achieve specific EBM goals at global and regional scales. They conclude that MPAs will almost always be a necessary component of EBM, playing an important role in promoting EBM, but that MPAs alone will rarely, if ever, be sufficient to achieve the range of goals inherent in comprehensive ecosystem-based management. Furthermore, planners and managers have acknowledged the importance of looking at areas beyond protected area boundaries (such as the buffer zones around the MPA) as any impacts on those will likely have a profound influence on what occurs at the MPA. For example, Shepard *et al.* (2011) found in a meta-analysis that salt marshes have value for coastal hazard mitigation and climate change adaptation, although the magnitude of this value is uncertain. Based on that uncertainty, Shepard *et al.* (2011) recommend decision makers to use natural systems to maximise the benefits these may have and to be cautious when making decisions that erode these services.

Marine ecosystem-based management efforts approach the management of marine resources from an ecosystem perspective, aimed at achieving multiple objectives. If interactions between activities can be understood in an ecosystem context through ecosystem service trade-offs, spatially explicit valuation of potential ecosystem services changes could guide marine spatial planning, informing whether particular trade-offs are worth it (Chan and Ruckelshaus, 2010).

Coastal EBA has the potential to be a cost-effective and accessible means of adaptation that can help address multiple threats and local priorities (TNC, 2012). It has been suggested that wetlands, mangroves, oyster reefs, barrier beaches, coral reefs and sand dunes can protect coastlines from flooding and storm surge in a more cost effective manner than traditional engineered infrastructures (Adger *et al.*, 1995).

EBA can be both a substitute and a complement to other adaptation options, such as hard infrastructure or soft measures. Table 9.1 presents some examples of adaptation measures for different climate change impacts, indicating the ecosystem services being altered and the potential soft, hard and ecosystem-based adaptation strategies.

Table 9.1 Examples of soft, hard and ecosystem-based adaptation measures to climate change impacts

Adaptation measure	*ES maintained*	*Soft measures*	*Hard measures*	*EBA*
Coastal erosion control	Storm protection, erosion control, recreation and aesthetic, nutrient cycling, production of atmospheric oxygen, soil formation and retention, water cycling	Beach drainage restoration	Seawalls Breakwaters Heighten dykes and floodwalls	Beach nourishment, Artificial sand dunes Brush mattressing Revegetation Wetland restoration Mangrove forestation and conservation Coral reef conservation
Coastal biodiversity and ecosystem conservation	Food, fibre and fuel, recreation and aesthetic, nutrient cycling, storm protection, erosion control, genetic resources, invasion resistance, seed dispersal	Reef restoration Slow-forming terraces	Artificial reef construction	Marine protected areas
Agricultural soil conservation and management	Erosion control, food, fibre and fuel, water cycling, nutrient cycling, soil formation and retention	Conservation tillage		Slow-forming terraces Integrated nutrient management
Agricultural sustainable crops & farming and agrobiodiversity conservation	Food, fibre and fuel, pest and disease control, genetic resources, seed dispersal, pollination, nutrient cycling, soil formation and retention, production of atmospheric oxygen, climate regulation, water cycling		Chemical pest management Inorganic fertilizers Micro-irrigation Drip-irrigation	Crop diversification Ecological pest management Mixed farming Agro-forestry
Water resources improvement	Freshwater provision, water cycling, soil formation and retention, food fibre and fuel		Bores/tube wells for domestic water supply during droughts Desalination Protecting wells from flooding with bunds Rainwater harvesting from rooftops	Rainwater collection from micro catchments, small reservoirs Catchment thinning

Source: Expanded from Travers *et al.*, 2012; World Bank, 2010; CBD, 2009; Jones *et al.*, 2012

9.3 Economics of ecosystem-based adaptation

The implementation of adaptation policies in ecosystems is demanding more and more information of how economically successful these practices are. Few studies attempt to compare the cost-effectiveness of EBA with hard adaptation infrastructures and more evidence is needed to encourage implementation of EBA on the ground. This section illustrates the financial flows in adaptation for ecosystems, the costs of these adaptation measures and discusses the efficiency of EBA measures.

9.3.1 Financial flows in adaptation

From an economic perspective, investment in adaptation policies is widely recognised to be necessary and cost-effective. The Stern review emphasised that climate change is a reality and that reducing its impacts through adaptation and mitigation policies makes good economic sense (Stern 2007). The Economics of Ecosystems and Biodiversity consortium (TEEB) recommends significant investment in protecting ecosystem services and biodiversity and supporting the development of ecological infrastructure as a contribution to both climate change mitigation and adaptation (TEEB, 2009). The TEEB climate report even presents investment in nature-based adaptation as an attractive area for high-return investment, where the natural capital can be a source of growth in a time of recession, a provider of jobs and a solution to persistent poverty (TEEB, 2009).

Under this opportunity and need, countries have started to commit important financial flows to adaptation programs through the globe. According to Schalatek *et al.* (2012), countries have contributed with $2.73 billion pledged to adaptation funds. The United Kingdom, the United States, Germany, the European Union, Canada and Japan represent more than 70% of this quantity. From these, about $2.23 billion has been deposited for adaptation and only $1.22 billion of this finance has been approved to support projects and programs (Schalatek *et al.*, 2012). These amounts cover all adaptation and not only ecosystem-based. Estimates for ecosystem-based adaptation are hard to find yet.

This trend in climate finance is being increasingly directed through Official Development Assistance (ODA). The OECD[2] estimated that in 2010, 15% of ODA was counted as climate finance (UNFCCC, 2012). There is much debate within the UNFCCC and in other international forums over whether international public finance under a future climate agreement should be separate from ODA or at least additional to the 0.7% target (as a number of developing countries argue), or whether ODA has a legitimate role to play in meeting future climate finance commitments (as argued by some developed countries) (Parker *et al.*, 2009).

9.3.2 Costs of adaptation and EBA

Several international reports assess the gaps in knowledge on climate change research, pointing out the lack of information about the costs of adaptation measures, specially for the impacts occuring on biodiversity and ecosystems (TEEB, 2010; Parry *et al.*, 2007; Markandya and Mishra, 2010; Martin-Ortega, 2011). Whilst there is an urgent need for large-scale climate financing to allow developing countries to mitigate and adapt to climate change, there is a large gap in the current scale of climate finance (Parker *et al.*, 2009). The scale of financing required for climate change adaptation in developing countries is of the order of hundreds of billions of dollars (Parker *et al.*, 2009). The study by Parker *et al.* (2009) estimates a range of costs in developing countries of between USD 10–70 billion per year to adapt to the impacts of climate change. Global adaptation needs range in between $49 and $171 billion per year (Jones *et al.*, 2012). Other recent figures were given by UNFCCC, with an estimate of between $28–67 billion per year for developing countries by 2030 (Montes, 2012). These figures refer to all adaptation and not only EBA approaches.

When looking specifically at EBA, cost estimates are hard to find in the literature, and it is very difficult to accurately measure adaptation costs. However, estimates point to larger costs than adaptation to other sectors. This may be due to the links already explained between ecosystem services and socioeconomic activity. Montes (2012) provided an estimate of costs needed in this sector to be larger than the figures above, with a range of $65–300 billion a year communicated in 2012 to the UNFCCC (Montes, 2012; Schalatek *et al.*, 2012).

9.3.3 Economic efficiency in EBA

In general terms, reviews of existing projects and initiatives coincide in the advantages of integrating ecosystem-based approaches within adaptation and development strategies, by defending that EBA delivers a range of co-benefits, provides cost-effective opportunities to achieve multiple objectives relating to climate change and helps development and biodiversity conservation (UNFCCC, 2011; Munang *et al.*, 2013). Effective EBA has been defined as when ecosystem-based approaches replace or augment conventional adaptation approaches to deliver superior outcomes for people and the community (Travers *et al.*, 2012).

Although the importance of EBA in the adaptation toolbox is being increasingly recognised, robust information on specific benefits of EBA and the conditions under which those benefits are likely to be received is generally lacking (World Bank, 2010; UNFCCC, 2011; Travers *et al.*, 2012). Collecting data on the costs and benefits of EBA approaches in the initiatives will provide the evidence base for sustained investment. It is therefore important to consider how the cost-benefits of ecosystem-based technologies will be evaluated during the life of the intervention and beyond (Travers *et al.* 2012).

EBA delivers important co-benefits, but the evidence base that demonstrates contribution to co-benefits is weak. These win-win outcomes also help to avoid maladaptation (Paterson *et al.*, 2008). Although there is no universal definition of an effective EBA (Travers *et al.*, 2012), and evidence is still scarce, ecosystem-based adaptation is gaining attention as a cost-effective means of protecting human and ecological communities against the impacts of climate change (Heller and Zavaleta, 2009; Mooney *et al.*, 2009, World Bank, 2010; Burgiel and Muir, 2010; Munang *et al.*, 2013). But more evidence is needed on costs and benefits of EBA case studies.

Estimating the economic value of ecosystem services can play an important role in conservation planning and ecosystem-based management (Plummer 2009; Stenger *et al.*, 2009). In contrast, a lack of economic valuation can reduce the real importance of such resources. There is increasing consensus about the importance of incorporating ecosystem services into resource management decisions, but quantifying the levels and values of these services has proven difficult (Nelson *et al.*, 2009). This is the main reason for this lack of evidence also in ecosystem-based adaptation projects. Biodiversity and ecosystem services, although having recognised values for society, are still difficult to value in economic terms. Biodiversity and ecosystem services are valued by society for a wide range of reasons; from the functional to the aesthetic (MEA, 2005). Healthy ecosystems provide a wide range of ecosystem services that serve as a basis for the provision of food, fibre, building materials, or potable water. Additionally, many cultural practices and traditions were developed and depend on particular ecological elements or functions, and this dependence has been maintained over the generations (Burgiel and Muir, 2010). Economic valuation of ecosystem services requires up-to-date and reliable information and considerably better understanding of the landscapes that provide such services (Troy and Wilson, 2006; Baral *et al.*, 2009). Efforts have been made to value ecosystem services (TEEB, 2010), but still important improvements are needed. The consideration of trade-offs between ecosystem services, non-linearities, or the estimation of physical flows of services are areas of research that need further development. Additionally, problems such as the scope effect (Ojea and Loureiro, 2011) or double counting (Ojea *et al.*, 2012) need to be avoided in future economic assessments.

In addition to the broader value of ecosystem services for society, there is also an inextricable link between poverty and the loss of ecosystem services. Subsistence livelihoods, particularly those relating to farming, animal husbandry, fishing and forestry, are the most immediate benefits of healthy ecosystems and their services (TEEB, 2010; Burgiel and Muir, 2010). EBA is particularly relevant given that the poor are often the most directly dependent on the services ecosystems provide (UNFCCC, 2011; CBD, 2009), with three quarters of the world's poorest citizens dependent on the environment for their livelihoods (WRI, 2008). Indeed, as the UNFCCC shows from a review of several case studies, EBA is widely applicable to and particularly accessible to the most vulnerable communities (UNFCCC, 2011). These approaches may be more cost-effective for rural or poor communities than measures based on hard infrastructure and engineering (CBD, 2009). Coastal EBA for example has the potential to be a cost-effective and accessible means of adaptation that can help address multiple threats and can be particularly accessible to the most vulnerable communities (TNC, 2012; UNFCCC, 2011).

Hagerman *et al.* (2010) pointed out that policy adaptation in conservation should be based on existing scientific information and value-based commitments (Andrade *et al.*, 2011). However, uncertainty about cost effectiveness is still a common denominator in current EBA approaches, and meanwhile urgent action is needed. Therefore some studies defend that it is wise to apply the precautionary principle and adopt a strategy which is sustainable and aims at protecting both social and natural systems. This means investing in ecological infrastructure, as a way to both mitigate and adapt. In this line, Travers *et al.* (2012) present a series of recommendations on good EBA practice that are built upon the lessons learnt from the climate change and ecosystem management literature. These recommendations include the use of participatory approaches and stakeholder engagement; the adoption of a systems perspective (by not only looking at impacts but also to drivers that influence service delivery); the inclusion of people's perceptions on ecosystem services; and the adoption of a flexible approach to evaluation and the setting of flexible baselines, among others.

9.4 Current gaps and future research directions

From the review conducted in this chapter, we have observed that EBA is an emerging field in adaptation with a great political interest and also defended and promoted by international institutions and NGOs. Important investments are being pledged although there is still a big gap between investment and expected adaptation costs in general. Despite these efforts, EBA is still seen as a complement to hard and soft adaptation measures and not as a substitute policy choice (UNFCCC,[3] 2011). This situation may be influenced by the short history of EBA in climate change (although EBA-like approaches have been implemented for conservation and development for decades), together with the lack of scientific evidence on the efficiency and effectiveness of EBA interventions.

In this section, we synthesise the existing information on EBA performance to identify the main challenges for a potential mainstreaming of EBA as an important pillar of adaptation policy.

- Technology: There is a lack of information on EBA technologies in comparison to more 'traditional' adaptation technologies. There are limits to what is possible when undertaking an ecosystem-based approach, where some circumstances may require a more engineered or technical solution (CBD, 2009). While an increasing number of EBA resources are becoming available, information has not been yet collated to allow easy access for those who need to inform the decision-making process (Travers *et al.*, 2012).
- Effectiveness: On-the-ground data availability to assess EBA effectiveness is still very limited and not enough to generalise findings. Establishing an evidence base for decision-making with respect to EBA remains a challenge (Travers *et al.*, 2012). In EBA, while the theoretical qualities or principles that underlie effectiveness are well defined, there remains limited robust evaluation in practice (Travers *et al.*, 2012).
- Uncertainty: EBA in a warmer world will require new approaches and increasing intensive management of ecosystems to maintain ecosystem services. There are also constraints to implementing EBA that have to do with lack of information, the uncertainty of how ecological processes will react to both climate change and management, the tipping points of socio-ecosystems, a lack of adequate institutions, technology and funding, and the need to deal with extreme events that affect local communities and sectors (Colls *et al.*, 2009; Andrade *et al.*, 2011; Shepard *et al.*, 2011).
- Long-term solutions: Hard adaptation measures are often implemented when looking for fast solutions in the short term. In contrast, EBA involves solutions that are not immediate but that assure long-term effects and resistance to climate stressors. Recent studies have shown a negative impact of many adaptation strategies on biodiversity, especially in the case of 'hard defences built to prevent coastal and inland flooding' (Campbell *et al.*, 2009). This could result in so called 'mal-adaptation' in the long term if the ecological attributes that regulate the modified ecosystems are disturbed (Andrade *et al.*, 2011).
- Effective finance: Countries have pledge important amounts for adaptation from which only a small portion has been already disbursed. In the case of EBA approaches, information on invested quantities is not easy to access, and further analysis will be needed in the future. Also, assuring a sustainable and long-term finance mechanism for EBA is fundamental.
- Science of ecosystem services: EBA relies on a good understanding of ecosystem services and their relative importance. Ecosystem services have been many times overlooked, misunderstood or ignored in adaptation planning (UNFCCC, 2011). A lack of practical guidance on how to build resilience/incorporate ecosystem-based approaches to adaptation in strategies exists. Further developing the evidence base for EBA would help to enhance understanding of ecosystem interactions and the economics of EBA. Outcomes would need to be monitored and evaluated (UNFCCC, 2011; CBD, 2009).

- Trade-offs: Many ecosystems are vulnerable to the adverse effects of climate change, so care must be taken to ensure that measures are climate-proofed. EBA relies on a good understanding of ecosystem services and the relative importance of different ecosystem services. Managing ecosystems for adaptation may require prioritization of certain services that ecosystems provide at the expense of others (UNFCCC, 2011). It is therefore important that decisions to implement ecosystem-based adaptation are subject to risk assessment, scenario planning and adaptive management approaches that recognise and incorporate these potential trade-offs (CBD, 2009).
- Enforce conservation: The enhancement and maintenance of biodiversity conservation and ecosystem services delivery are climate non-regret actions that should be enforced. EBA should not conflict with conservation policies and investment as both actions share a common background.

9.5 Conclusions

The information reviewed for this chapter demonstrates that ecosystem-based adaptation is still a relatively new scientific and policy field, but that it is deeply rooted in long-standing approaches applied by communities locally in response to episodic and/or long-term climate change, as well as on lessons learnt from nature conservation, especially community-based and ecosystem-based management.

This systematic review of the current status of EBA suggests that ecosystem-based adaptation is not without complexity, uncertainty and risk, but existing examples and discussions defend the cost-effectiveness and the non-regrets character of this adaptation strategy. International institutions and conservation agencies have been fast in adopting the new terminology and the expectance of field examples, and implementations of EBA are greater than ever before. The main challenges allowing EBA to perform successfully have been discussed here. For a full EBA policy, there is a need to involve research on the complexities of delivering ecosystem services and conservation, assuring sustainable and effective finance to adaptation and looking at the benefits for the longer term. These challenges may need more field implementation together with a strong commitment from international organizations and donor countries that have already started to work in this way.

In the way forward to mainstream EBA as a basic approach to climate change adaptation, measures can and should be integrated within the broader adaptation context to avoid maladaptation, and at the same time, EBA should complement other approaches (including engineered or technical solutions) through realising multiple co-benefits, including mitigation. The experience in mitigation efforts taking place now such as Reduced Emissions from Deforestation and Forest Degradation (REDD+) can also be the basis for learning successful and unsuccessful practices, as many of the problems mentioned (trade-offs between services, sustainable finance, efficiency, etc.) are common to ecosystem-based adaptation.

Notes

1 See Section 9.2.3 for more examples on EBA approaches.
2 Organisation for Economic Co-operation and Development.
3 A UNFCCC report states that "Ecosystem Based approaches to adaptation are found to be most appropriately integrated into broader adaptation and development strategies, complementing, rather than being an alternative to, other approaches" (UNFCCC, 2011).

References

Adger, W.N., Brown, K., Cervigni, R. and Mora, D., 1995. Total economic value of forests in Mexico. Ambio 25(5), 286–296.

Andrade, A., Cordoba, R., Dave, R., Girot, P., Herrera Fernandez, B., Munroe, R., Oglethorpe, J., Pramova, E., Watson, J. and Vergara, W., 2011. Draft principles and guidelines for integrating ecosystem-based approaches to adaptation in project and policy design: A discussion document. CEM/IUCN, CATIE.

Araujo, M.B., Thuiller, W. and Pearson, R.G., 2006. Climate warming and the decline of amphibians and reptiles in Europe. Journal of Biogeography 33, 1712–1728.

Baral, H., Kasel, S., Keenan, R., Fox, J. and Stork, N., 2009. GIS-based classification, mapping and valuation of ecosystem services in production landscapes: A case study of the Green Triangle region of south-eastern Australia. In: Forestry: A climate of change, Thistlethwaite, R., Lamb, D. and Haines, R. (Eds). pp. 64–71. Proc. IFA Conf. Caloundra, Queensland, Australia, 6–10 September 2009.

Burgiel, S. and Muir, A. 2010. Invasive species, climate change and ecosystem-based adaptation: Addressing multiple drivers of global change. Global Invasive Species Program. 55pp.

Campbell, A., Kapos, V., Scharlemann, J.P.W., Bubb, P., Chenery, A., Coad, L., Dickson, B., Doswald, N., Khan, M.S.I., Kershaw F. and Rashid, M., 2009. Review of the literature on the links between biodiversity and climate change: Impacts, adaptation and mitigation. Technical Series No. 42. Secretariat of the Convention on Biological Diversity (CBD). Montreal, Canada. 124 pp.

CBD, Convention on Biological Diversity, 2009. Connecting biodiversity and climate change mitigation and adaptation: Report of the second ad hoc technical expert group on biodiversity and climate change. Technical Series No. 41. Secretariat of the Convention on Biological Diversity (CBD). Montreal, Canada. 126 pp.

Chan, K.M. and Ruckelshaus, M.H., 2010. Characterizing changes in marine ecosystem services. F1000 Biology, 2(54).

CI, Conservation International, 2012. Climate Solution: Ecosystem-based adaptation. Available at: http://www.conservation.org/Documents/CI_Climate_Solutions_Adaptation.pdf

Colls, A., Ash, N. and Ikkala, N., 2009. Ecosystem-based adaptation: A natural response to climate change. International Union for the Conservation of Nature (IUCN). Gland, Switzerland. 16 pp.

Eakin, H. and Lemos, M.C. (Eds.), 2010. Adaptive capacity to global change in Latin America. Special Issue of Global Environmental Change 20(1), 1–210.

Fischlin, A., Midgley, G.F., Price, J.T., Leemans, R., Gopal, B., Turley, C., Rounsevell, M.D.A., Dube, O.P., Tarazona, J. and Velichko, A.A., 2007. Ecosystems, their properties, goods, and services. In: Climate Change 2007: Impacts, Adaptation and Vulnerability. Contribution of Working Group II to the Fourth Assessment Report of the Intergovernmental Panel on Climate Change, M.L. Parry, O.F. Canziani, J.P. Palutikof, P.J. van der Linden and C.E. Hanson (Eds.), pp. 211–272. Cambridge University Press, Cambridge.

Garmendia, E., Mariel, P., Tamayo, I., i Aizpuru, I. and Zabaleta, A., 2012. Assessing the effect of alternative land uses in the provision of water resources: Evidence and policy implications from southern Europe. Land Use Policy 29, 761–770.

Granek, E.F., Polasky, S., Kappel, C.V., Reed, D.J., Stoms, D.M., Koch, E.W., Kennedy, C.J., Cramer, L.A., Hacker, S.D., Barbier, E.B., Aswani, S., Ruckelshaus. M., Perillo, G.M.E., Silliman, B.R., Muthiga, N., Bael, D. and Wolanski, E., 2010. Ecosystem services as a common language for coastal ecosystem-based management. Conservation Biology 24, 207–216.

Hagerman, S., Dowlatabadi, H., Satterfield, T. and McDaniels. T., 2010. Expert views on biodiversity conservation in an era of climate change. Global Environmental Change 20, 192–207.

Halpern, B.S., Lester, S.E. and McLeod, K.L., 2010. Placing marine protected areas onto the ecosystem-based management seascape. PNAS 107(43), 18312–18317.

Heller, N.E. and Zavaleta, E.S., 2009. Biodiversity management in the face of climate change: A review of 22 years of recommendations. Biological Conservation 142, 14–32.

IUCN, International Union for the Conservation of Nature, 2010. Building resilience to climate change: Ecosystem-based adaptation and lessons from the field. Available at: Data.iucn.org/dbtw-wpd/edocs/2010-050.pdf

Jones, Holly P., Hole, David G. and Zavaleta, Erika S., 2012. Harnessing nature to help people adapt to climate change. Nature Climate Change 2, 504–509.

Kotiola, P. (Ed.), 2009. Adaptation of forests and people to climate change. World Series Vol. 22. International Union of Forest Research Organizations (IUFRO). Vienna, Austria.

Markandya, A. and Mishra, A. (Eds.), 2010. Costing adaptation: Preparing for climate change in India. The Energy and Resources Institute, TERI Press, New Delhi, India.

Martin-Ortega, J., 2011. Costs of adaptation to climate change impacts on freshwater systems: Existing estimates and research gaps. Revista Española de Economía Agraria y de los Recursos Naturales 11(1), 5–28.

MEA, Millennium Ecosystem Assessment, 2005. Ecosystems and human well-being: Current state and trends. Island Press, Washington, DC.

Montes, M.F., 2012. Understanding long-term finance needs of developing countries Presentation to UNFCCC-South Centre. Available at: http://unfccc.int/files/cooperation_support/financial_mechanism/longterm_finance/application/pdf/montes_9_july_2012.pdf

Mooney, H., Larigauderie, A., Cesario, M., Elmquist, T., Hoegh-Guldberg, O., Lavorel, S., Mace, G.M., Palmer, M., Scholes R. and Yahara, T., 2009. Biodiversity, climate change and ecosystem services. Current Opinion in Environmental Sustainability 1(1), 46–54.

Munang, R., Thiaw, I., Alverson, K., Mumba, M., Liu, J. and Rivington, M., 2013. Climate change and ecosystem-based adaptation: A new pragmatic approach to buffering climate change impacts. Current Opinion in Environmental Sustainability 5, 1–5.

Nelson, E., Mondoza, G., Regetz, J., Polasky, S., Tallis, J., Cameron, D.R., Chan, K.M.A., Daily, G.C., Goldstein, J., Kareiva, P.M., Londsdorf, E., Naidoo, R., Ricketts, T.H. and Shaw, M.R., 2009. Modeling multiple ecosystems services, biodiversity conservation, commodity production, and tradeoffs at landscape scale. Frontiers in Ecology and the Environment 7(1), 4–11.

Ojea, E. and Loureiro, M.L., 2011. Identifying the scope effect on a meta-analysis of biodiversity valuation studies. Resource and Energy Economics 33, 706–724.

Ojea, E., Martin Ortega, J. and Chiabai, A., 2012. Defining and classifying ecosystem services for economic valuation: The case of forest water services. Environmental Science & Policy, 19–20, 1–15.

Parker, C., Brown, J., Pickering, J., Roynestad, E., Mardas, N. and Mitchell, A.W., 2009. The little climate finance book. Global Canopy Programme, Oxford (UK).

Parry, M.L., Canziani, O.F., Palutikof, J.P., van der Linden, P.J. and Hanson, C.E. (Eds.), 2007. Climate Change 2007: Impacts, adaptation and vulnerability. Contribution of Working Group II to the Fourth Assessment Report of the Intergovernmental Panel on Climate Change. Cambridge University Press, Cambridge, UK. 976 pp.

Paterson, J.S., Araujo, M.B., Berry, P.M., Piper, J.M. and Rounsevell, M.D.A., 2008. Mitigation, adaptation and the threat to biodiversity. Conservation Biology 22(5), 1352–1355.

Pimm, S.L. and Raven, P., 2000. Extinction by numbers. Nature 403, 843–845.

Plummer, M.L., 2009. Assessing benefit transfer for the valuation of ecosystem services. Frontiers in Ecology and the Environment 7(1), 38–45.

Schalatek, L., Stiftung, H.B., Nakhooda, S., Barnard, S. and Caravani, A., 2012. Climate finance thematic briefing: Adaptation finance. Climate Finance Fundamentals 3, 2012.

Shepard, C.C., Crain, C.M. and Beck, M.W., 2011. The protective role of coastal marshes: A systematic review and meta-analysis. PLoS ONE 6(11): e27374.

Stenger, A., Harou, P. and Navrud, S., 2009. Valuing environmental goods and services derived from the forests. Journal of Forest Economics 1(1), 1–14.

Stern, N., Peters, S., Bakhshi, V., Bowen, A., Cameron, C., Catovsky, S., Crane, D., Cruickshank, S. , Dietz, S., Edmonon, N., Garbett, S. L., Hamid, L., Hoffman, G., Ingram, D., Hones, B., Patmore, N., Radcliffe, H., Sathiyarajah, R., Stock, N., Taylor, C., Vernon, T., Wanjie, H. and Zenghelis, D., 2006. Stern Review: The economics of climate change. HM Treasury, London.

TEEB, The Economics of Ecosystem Services and Biodiversity, 2009. TEEB climate issues update. September, 2009.

TEEB, The Economics of Ecosystems and Biodiversity, 2010. Ecological and economic foundations. Pushpam Kumar (Ed.). Earthscan, London and Washington.

Thomas, C., Alison, D., Cameron, A., Green, R.E., Bakkenes, M., Beaumont, L.J., Collingham, Y.C., Erasmus, B.F.N., Ferreira de Siqueira, M., Grainger, A., Hannah, L., Hughes, L., Huntley, B., van Jaarsveld, A.S., Midgley, G.F., Miles, L., Ortega-Huerta, M., Peterson, A.T., Phillips, O.L. and Williams, S.E., 2004. Extinction risk from climate change. Nature 427, 145–148.

Thompson, I., Mackey, B., McNulty, S. and Mosseler, A., 2009. Forest Resilience, Biodiversity, and Climate Change. A synthesis of the biodiversity/resilience/stability relationship in forest ecosystems. Secretariat of the Convention on Biological Diversity, Montreal. Technical Series no. 43. 67 pp.

TNC, The Nature Conservancy, 2011. Ecosystem-based adaptation: Bridging science and real-world decision making. Second international workshop on biodiversity and climate change in China. Anne Wallace Thomas, Global Climate Change Adaptation Program.

Travers, A., Elrick, C., Kay, R. and Vestegaard, O., 2012. Ecosystem-based adaptation guidance. Moving principles to practice. Working document, April 2012.

Troy, A. and Wilson, M.A., 2006. Mapping ecosystem services: Practical challenges and opportunities in linking GIS and value transfer. Ecological Economics 60, 435–49.

Tschakert, P. and Dietrich, K., 2010. Anticipatory learning for climate change adaptation and resilience. Ecology and Society 15(2).

UNDP, United Nations Development Programme, 2010. Adapting to climate change UNDP-GEF initiatives financed by the least developed countries fund, special climate change fund and strategic priority on adaptation. New York, 2010.

UNEP, United Nations Development Programme, 2012. Building resilience of ecosystems for adaptation. Available at: http://www.unep.org/climatechange/adaptation/EcosystemBasedAdaptation/tabid/29583/Default.aspx

UNFCCC, United Nations Framework Convention on Climate Change, 2011. Ecosystem-based approaches to adaptation: Compilation of information. Durban, 2011.

UNFCCC, United Nations Framework Convention on Climate Change, 2012. Database on ecosystem-based approaches to adaptation. Available at: http://unfccc.int/adaptation/nairobi_work_programme/knowledge_resources_and_publications/items/6227.php

World Bank, 2010. Convenient solutions to an inconvenient truth: Ecosystem-based approaches to climate change. Environment Department. The World Bank, Washington DC, US.

WRI, World Resources Institute, 2008. Roots of resilience – growing the wealth of the poor. WRI, Washington DC.

PART III

Adaptation in activity sectors

10

CLIMATE CHANGE AND THE ENERGY SECTOR

Impacts and adaptation

Alberto Ansuategi[1,2]

[1]UNIVERSITY OF THE BASQUE COUNTRY (UPV/EHU)

[2]PARTS OF THIS CHAPTER ARE BASED ON A BACKGROUND PAPER PREPARED BY THE AUTHOR FOR THE WORLD BANK WHITE PAPER ENTITLED "CLIMATE IMPACTS ON ENERGY SYSTEMS: KEY ISSUES FOR ENERGY SECTOR ADAPTATION" EDITED BY JANE EBINGER AND WALTER VERGARA (2011).

10.1 Introduction

The rise in the presence of the so-called greenhouse gases (GHG) in the Earth's atmosphere experienced in the past decades has caused a rise in the amount of heat from the sun withheld in it. This greenhouse effect has resulted in climate change, which is expected to increase average global temperature (global warming) and produce effects such as changes in cloud cover and precipitation, a rise in sea levels, melting of ice caps and glaciers and more frequent and severe extreme weather events. Thus, the rate of warming averaged over the last 50 years (0.13 °C ± 0.03 °C per decade) is nearly twice that for the last 100 years (IPCC, 2007). According to recent studies, some extreme weather events have changed in frequency and/or intensity over the last 50 years (Durack et al., 2012). There are suggestions of increased intense tropical cyclone activity in some regions (Webster et al., 2005) and emerging evidence of increased variability of climate parameters such as temperature and precipitations (Seager et al., 2012). There is also high confidence that hydrological systems are being affected due to increased runoff and earlier spring peak discharge in many glacier- and snow-fed rivers (Diffenbaugh et al., 2012). Melting of ice sheets, glaciers and ice caps has accelerated (Joughin et al., 2012) and, globally, sea levels have raised an average of 18 cm since the late 19th century at an accelerating rate of rise (Cazenave and Llovel, 2010).

Most of the GHG emissions that have triggered the changes described above are produced by the combustion of fossil fuels, with the global energy sector being a major producer of emissions. Nowadays, nearly 70% of global GHG emissions come from fossil fuel combustion for electricity generation, transport, industrial activity and heat in buildings or cooking in homes (IPCC, 2007). However, the energy sector is not only contributing to climate change, it is also vulnerable to climate impacts. Although the impacts on energy supply and demand are the most intuitive, climate change can also have direct effects on energy endowment, infrastructure and transportation, and indirect effects through other economic sectors. The potential economic damage of such impacts is by no means negligible. Ackerman and Stanton (2008) estimate that annual U.S. energy expenditures (excluding transportation) will be $141 billion higher in 2100 – an increase equal to 0.14% of GDP – in the business-as-usual case than they would be if today's climate conditions continued throughout the century.

This chapter provides an overview on climate impacts on energy systems and adaptation of energy systems to climate change. It is structured in four sections. After this brief introduction, Section 10.2 identifies climate vulnerabilities in the energy sector. Section 10.3 discusses main issues regarding adaptation to climate change in the energy sector. Finally, Section 10.4 draws some conclusions and suggests topics where more research would be valuable.

10.2 Impacts

A necessary condition for adaptation to climate change to be effective is to understand and be aware of how climate change impacts the energy sector. Most of the studies of the relationship between the energy sector and climate change have focused on both the climate impact of GHG emissions of the energy sector and the impact of climate mitigation policies on the energy sector. Only recently have some reviews been produced on climate change impacts on the energy sector. Some of these reviews have focused on sub-sectors such as renewable energy (Lucena et al., 2009), wind energy (Pryor and Barthelmie, 2010) and nuclear energy (Kopytko and Perkins, 2011) or the electricity market in general (Mideksa and Kallbekken, 2010). Other studies have reviewed climate vulnerability of the energy sector for specific regions of the world such as the United States (CCSP, 2007), Sub-Saharan African countries (Williamson et al., 2009) or Nordic countries (Fenger, 2007). Finally, two very recent studies (Ebinger and Vergara, 2011 and Schaeffer et al., 2012) provide a comprehensive assessment of the impact of climate change in energy systems. In this section, we summarize the contribution of all these studies.

A changing climate can lead to changes in (1) the amount of primary energy available, (2) the capacity to supply energy to consumers and (3) energy consumption patterns. In what follows, we will discuss them separately.

10.2.1 Changes in primary energy availability

Hydropower potential

According to Euroelectric (1997) the Gross Hydropower Potential (GHP) is defined as "the annual energy that is potentially available if all natural runoff at all locations were to be harnessed down to the sea level (or to the border line of a country) without any energy losses". As it is pointed out by Schaeffer et al. (2012), the evolution of GHP is an indicative measure of possible trends related to climate change, but it is not enough to draw conclusions about the actual impacts of changes in climate variables.[1] Thus, a complementary measure for such assessment would be the so-called Developed Hydropower Potential (DHP), which measures the actual potential of all existing hydropower stations. Hydropower potential depends directly on the hydrological cycle, that is, on the excess water that turns into runoff and on the seasonal pattern of the hydrological cycle. Therefore, the potential of hydrogeneration is impossible to assess without additional locally specific study. Lehner et al. (2005) take a model-based approach for analysing the possible effects of global change on Europe's hydropower potential at a country scale and conclude that for the whole of Europe the GHP is estimated to decline by about 6% by the 2070s, while the DHP shows a decrease of 7%–12%.

Wind power potential

Climate change may alter the geographical distribution and the variability of the wind and, depending on local conditions, we could have either positive or negative impacts of climate change on wind power potential. To quantify these effects, it is required the application of downscaling methodologies designed to extract projections with higher resolution of certain climate parameters from Global Climate Models. Some studies show that it is unlikely that mean wind speeds and energy density will change by more than the current inter-annual variability (± 15%) over most of Europe (Pryor et al., 2005; Bloom et al., 2008) and North America (Sailor et al., 2008) during the present century, but some studies suggest that changes over South America may be larger (Lucena et al., 2010).

Solar power potential

The solar energy reaching the surface of the planet in direct and scattered radiation form is determined by astronomical factors such as the length of the day or the solar declination and by other factors related to actual atmospheric conditions such as cloudiness and the presence of aerosols and water vapour. According to Salby (1996) cloudiness is the most important determinant of the solar radiation flux of the Earth-Atmosphere system. Again, the impact of climate change on solar power potential will not be the same in every region of the planet and there will be "gainers" and "losers". For instance, Bartok (2010) reports increase in solar radiation of 5.8% compared to the 1992–1996 time average in south-eastern Europe, whereas Cutforth and Judiesch (2007) report a decrease of incoming solar energy on the Canadian Prairie.

Wave energy potential

Waves are created by the transfer of energy from wind flowing over water bodies. Therefore, we can assert that wave energy, in common with wind energy and other renewables, will be sensitive to changes in climate. In some regions, impacts are expected to be positive. This is the case of the coast of mid-Norway (Vikebo et al., 2003). In other regions, such as the southern Californian coast (Cayan et al., 2009), impacts are expected to be negative.

Availability of biofuels

Changes in temperature, rainfall and levels of carbon dioxide may place new stresses in agricultural production in general and crops to produce biofuels in particular (Hatfield, 2010). The overall effect of climate change in biofuel availability is hard to predict since many key factors in agriculture production are affected. Higher levels of carbon dioxide can improve photosynthesis in certain crops (Bernacchi et al., 2007); temperature changes can affect either directly the rate of plant development (Hatfield et al., 2008) or indirectly modify factors such as soil conditions or incidence of pests. Extreme climate conditions, such as droughts, frosts and storms can also affect crops (Mishra and Cherkauer, 2010; Zhao and Running, 2010).

Availability of fossil fuels

Climate change will not directly affect the actual amount of fossil fuels, but it can have an impact on the access to and exploration for reserves of fossil fuels. Burkett (2011) discusses extensively how increased ocean temperature, changes in precipitation patterns and runoff, sea level rise, more intense storms, changes in wave regime and increased carbon dioxide levels and ocean acidity have the potential to independently and cumulatively affect coastal and offshore oil and gas exploration, production and transportation. Harsem et al. (2011) discuss the role played by climate change as one of the main factors influencing future oil and gas prospects in the Artic.

10.2.2 Changes in the capacity to supply energy to consumers

Apart from the effects on the energy generating potential, climate change may also have an impact on the capacity of the system to convert this potential into final energy to be supplied to consumers to meet different energy services. Impacts on energy supply can be classified into two groups: impacts on energy-transforming technologies and impacts on transmission, distribution and transfer of energy.

Impacts on energy-transforming technologies

The main focus here would be the impacts on long life-span facilities that will still be in operation when the new climate conditions occur.[2] Hydro-power plants with built reservoirs that are not designed to manage earlier increased flows due to seasonal shift (Vicuña et al., 2007), thermal power plants whose output and efficiency will be affected by variation of ambient temperature and humidity (Arrieta and Lora, 2005) or by variation in the quantity and/or quality of water resources for competing uses (Feeley et al., 2008; Durmayaz and Sogut, 2006), and energy facilities sited in coastal low-lying lands subject to more severe storm surges and coastal erosion (Neumann and Price, 2009) constitute examples of impacts on energy-transforming technologies.

Impacts on transmission, distribution and transfers

Extreme weather events induced by climate change may affect the transmission of energy through disruption of infrastructure. Landslides, flooding, permafrost thawing, extreme wind and ice loads and other extreme meteorological events can affect both transmission power lines and gas transmission systems (Kiessling et al., 2003; Vlasova and Rakitina, 2010). Energy distribution may also be affected by meteorologically induced factors such as fires or falling trees and heat waves that may induce power transformer failures and losses in substation capacity (Sathaye et al., 2011). Note also that some of the impacts could also be "positive", as it is the case of the opening of new shipping routes as Arctic sea ice melts (Valsson Trausti and Ulfarsson, 2011).

10.2.3 Changes in energy consumption patterns

One of the most obvious effects of climate change on energy consumption patterns is that higher temperatures will reduce demand for heating and will increase demand for cooling. Isaac and Van Vuuren (2009) have estimated global residential sector energy demand for heating and air conditioning in the context of climate change and have concluded that heating energy demand will decrease by 34% worldwide by 2100 and air conditioning demand will increase by 72%. Climate change will also likely affect energy sectors in other sectors such as transportation (increased air conditioning in private cars and refrigerated vans)[3] and agriculture (energy requirements for irrigation).[4]

10.3 Adaptation

Adaptation to climate change and its impacts is receiving increasing attention as a complementary response strategy to reducing net emissions of GHG (termed "mitigation" in the literature). While the main objective of adaptation solutions is to ensure the security of people and assets, in the case of the energy system, the primary objective is to guarantee the supply of energy, balancing production and consumption throughout time and space.

The process of adapting to climate change is complex and consists of a multitude of behavioural, structural and technological adjustments. Several typologies of adaptation measures have been proposed.[5] Here we describe and differentiate adaptation measures based on a set of attributes used in the studies mentioned above:

- Based on the *timing* of the action, adaptation measures may be **proactive** or **reactive**. A proactive approach aims to reduce exposure to future risks, for instance by new coastal power plant siting rules to minimize flood risk or installing solar photovoltaic technology to reduce effects of peak demand. A purely reactive approach aims only to alleviate impacts once they have occurred, for instance reinforcing existing energy infrastructure with more robust control solutions that can better respond to extreme weather-related service interruption.
- Based on the *nature of agents involved in the decision-making*, adaptation measures can be **private** or **public**. Some adaptation measures, such as protection of coastal areas from sea level rise, provide public benefits and therefore it is governments who provide this form of adaptation as a public good. In many other cases, however, adaptation measures offer private benefits that accrue to individuals or firms, and actions do not have to be directed centrally by a public authority. Note that this distinction can also be referred as autonomous or "market driven" versus planned or "policy-driven" adaptation.
- Based on the *spatial scope*, adaptation measures can be **localized** or **widespread**. Adaptation is primarily local, since the direct impacts of climate change are felt locally and responses have to address local circumstances. However, for these measures to be implemented, most often they must also be supported by national or even international policies and strategies. Thus, it can be said that a successful adaptation measure has to proceed at several levels simultaneously.
- Based on the *temporal scope*, adaptation measures can be **short-term** or **longer term**. This distinction can also be referred to as tactical versus strategic, or as instantaneous versus cumulative (Smit et al., 1999). In the natural hazards field it is referred to as adjustment versus adaptation (Smit et al., 2000). The distinction between short-run and long-run adaptation has to do with the pace and flexibility of adaptation measures.
- Based on the *form*, adaptation measures can be **infrastructural**, **behavioural**, **institutional**, **regulatory**, **financial** and **informational**. To be effective, adaptation measures have to work through a wide range of interrelated channels. Sectors that could

face significant climate risks are those with long-term planning and investment horizons and dependent on extensive infrastructure and supply chains. This means that some measures will aim at reducing the vulnerability of energy infrastructure to environmental change (**infrastructural measures**). Adaptation will in part occur autonomously, with individuals and societies switching to new technologies and new practices. This implies that another group of adaptation options will target the behaviour of economic and social agents (**behavioural measures**). Climate risk management requires high-level coordination. All levels of governments should ensure that policies and programmes take account of climate change and adaptation strategies. Stakeholders also need to be organized in civic bodies that are able to contribute to decision-making processes (**institutional measures**). As the impacts of climate change become more direct and critical economic sectors are affected, governments are more likely to resort to prescriptive regulation and controls to ensure that critical actors take appropriate action on adaptation (**regulatory measures**). The Stern Review (Stern, 2007) identified financial constraints as one of the main barriers to adaptation. Thus, there is scope for the uptake of adaptation action targeting better the use of available financial resources and instruments (**financial measures**). Last, but not least, an improved informational and knowledge base is a necessary step with a view to defining scientifically sound measures of adaptation to climate change (**informational measures**).

- Based on their *ability to face associated uncertainties and/or to address other social, environmental or economic benefits*, measures can be **no-regrets** options, **low-regrets** options or **win-win** options. **No-regrets** adaptation measures are those whose socioeconomic benefits exceed their costs whatever the extent of future climate change. **Low-regrets** adaptation measures are those for which the associated costs are relatively low and for which the benefits under projected future climate change may be relatively large. **Win-win** adaptation measures are those that minimize social risk and/or exploit potential opportunities but also have other social, environmental or economic benefits. This distinction is clearly related to the debate on "hard" versus "soft" adaptation options and the irreversibility that they imply. A key feature of "soft" adaptation measures, involving policies and instruments that are designed to change behaviour, is that they imply less inertia than "hard" engineering measures. Thus, in the face of uncertainties with regard to climate projections, the risk of "sunk-costs" is much lower for soft adaptation measures than for hard adaptation measures.

In what follows, we will use this typology of adaptation options to review the most important adaptation measures and strategies that could be found in the energy sector. The description will be structured in two main blocks of measures: measures aiming at **building adaptation capacity** and measures aiming at **delivering adaptation actions**.

10.3.1 Building adaptation capacity

A system's ability to undertake specific adaptation actions is largely a function of its adaptive capacity. Broadly, adaptive capacity reflects fundamental conditions such as access to information (research, data collecting and monitoring and awareness raising), supportive social structures (organizational development and institutions) and supportive governance (regulations, legislations and guidance).

Access to information

The energy sector is critically exposed to weather and climate events in one way or another. Thus, the potential for improving the performance of the energy sector by using the best weather and climate information is apparent (Troccoli, 2010). To make optimal adaptation decisions, decision makers require detailed information about the impacts of climate change in space and time. At present there are several areas where knowledge is inadequate. Hence, the complexity of climate information calls for cooperation in order to undertake basic research into future changes. Article 5 of the United Nations Framework Conference on Climate Change (UNFCCC) refers to the need for the international community to support and further develop climate research and systematic observation systems, taking into account the concerns and needs of developing countries. Reliable, systematic climate data helps countries determine their current climate variability and model future changes. The development of higher resolution regional models for developing countries is important for improved predictions as well as analysing the disparity between the model outcomes. This would help enhance capacity for reaching informed decision-making.

Climate adaptation measures in the energy sector are critically dependent on observations. There is a need to continue to provide reliable and timely observations as required by weather forecast models, to supplement them for high-resolution models and to verify them for the energy sector.

Experts also stress the importance of assuring consistency of data (Troccoli et al., 2010) in order to be used in energy demand and energy production models. Small errors might be amplified by the transfer models to unacceptable levels.

Ready and reliable access to data and forecasts of some weather services should also be facilitated using grid computing technology. This would be particularly useful for small companies in the energy sector and would also serve in the regulatory and scientific communities to carry out climate/energy research activities.

Research is also a central activity in building adaptive capacity. The development of effective policies to face climate change relies on greatly improved scientific understanding of global environmental processes and their interaction with socioeconomic systems. This requires an unprecedented interdisciplinary effort to generate the knowledge needed by decision makers in governments and vulnerable sectors, such as the energy sector, to manage the risks of climate change impacts. The proliferation of climate change research centres and programs both at the international and the national (or even regional) scale is a clear response of the scientific community to the challenge of building our capacity to respond to climate change.

All the above reinforces the view that generating data and knowledge is a necessary condition for effective action. However, it is also important to succeed in persuading businesses, communities and individuals to adjust their behaviour in ways that promote adaptation and limit emissions (UNEP, 2006).

Supportive social structures

Mainstreaming climate change adaptation at the strategic level requires that clear policies on adaptation are developed in broad consultation and participation of staff and supported by senior management. Adaptation to climate change is becoming increasingly important from the perspective of corporate governance, strategic risk assessment and community planning.

Regulation and pressure from informed investors and rating agencies as well as consumers lead to an increased demand for disclosure of environmental and climate-related risks. Initiatives such as the Carbon Disclosure Project[6] encourage industry to better identify and manage risks, including those posed by climate change, to support the investor in making decisions.

Local public institutions (local governments and agencies), civil society institutions (producer organizations, cooperatives, savings and loan groups, etc.) and private institutions (NGOs and private businesses that provide insurance or loans) have an important operational significance in the context of climate change adaptation (Agrawal et al., 2008). Given that adaptation is inevitably mainly local, the involvement of local institutions is critical to the planning and implementation of adaptation policies and projects.

Connor et al. (2005) have reported some recent efforts made by European countries legislating and creating councils of energy users, which work side-by-side with their national energy boards or regulatory bodies.

Multi-sectoral partnerships between governmental, private and non-governmental actors are also an important part of any adaptation strategy. The extensive list of platforms/networks on adaptation practices maintained by the UNFCCC secretariat and the Nairobi Work Programme partner organizations[7] constitutes an example of efforts made to offer supportive social structures to adaptation.

Supportive governance

There are many ways in which societies and economic sectors can adapt to climate change. However, such adaptation has to be supported by governments in a variety of ways. Governance for adaptation to climate change requires effective administrative executive bodies, and enabling legal and regulatory frameworks. Uncertainty and imperfect information, missing markets and financial constraints constitute reasons that explain the necessity of government support in helping to promote effective adaptation.

Governments have an important role in providing a clear policy framework to guide effective adaptation by social and economic agents in the medium and longer term. In particular, governments are responsible for contributing to the provision of high-quality climate information, establishing land use plans and performance standards, defining long-term policies for climate-sensitive public goods such as coastal protection or emergency preparedness, and providing a safety net for those least able to afford protection and/or insurance. In developing countries, governments also have an important potential role in building adaptive capacity through good development practice.

Integrated planning within the energy sector and with others such as the water sector is highly important (Haas et al., 2008). Energy and water systems are closely linked. On the one hand, the production/consumption of one resource cannot be achieved without making use of the other. On the other hand, climate change affects the supply of both resources. Therefore, policy makers cannot provide a good adaptation plan without integrating both sectors as parts of a single strategy.

International governance also plays an important role in building capacity for adaptation. Given that the most vulnerable countries are often among the poorest, international assistance for adaptation is critical. The international community has managed to create a range of funding streams to support adaptation in developing countries. Thus, we have The Global Environment Facility, which manages two separate, adaptation-focused Funds under the UNFCCC:

the Least Developed Countries Fund (LDCF) and the Special Climate Change Fund (SCCF), which mobilize funding specifically earmarked for activities related to adaptation, and the latter also to technology transfer. More recently set up has been the so-called "Adaptation Fund", established by the Parties to the Kyoto Protocol of the UNFCCC to finance concrete adaptation projects and programmes in developing countries that are Parties to the Kyoto Protocol. The Fund is financed with the 2% of the Certified Emission Reductions issued for projects of the Clean Development Mechanism and other sources of funding. According to the World Development Report 2010 (World Bank, 2010) current levels of finance for developing countries fall far short of estimated needs. Total climate finance for developing countries is $10 billion a year today, compared with projected annual requirements by 2030 of $30 to $100 billion for adaptation.

10.3.2 Delivering adaptation actions

In the previous section, we described the framework within which adaptation actions can be delivered. Here, we will discuss effective responses made by stakeholders to the threats and opportunities of a changing climate.

Preventing effects or reducing risks

As indicated by the IPCC in its Fourth Assessment Report, certain short- and medium-term effects of climate change will be almost unavoidable. We have already seen in the previous section that climate change can have potential impacts on the energy production, energy transmission and supply and energy requirements. Therefore, some adaptive actions should try to alleviate or minimize these negative effects. Now we are going to offer several examples of "hard" and "soft" adaptation measures in the energy sector intended to minimize negative impacts due to long-term changes in meteorological variables and extreme events.

In many cases, the high vulnerability of energy infrastructure to environmental change is due to the fact that these infrastructures have a long lifespan and the risk of climate change related impacts was not factored into their design. A "hard" adaptation strategy is to invest in protective infrastructures to physically protect the energy infrastructure from the damages and loss of function that may be caused by climate change extreme events. Measures involve improving the robustness of offshore installations that are vulnerable to storms, building dikes and desilting gates, increasing dam heights, enlarging floodgates, improving the design of turbines to withstand higher wind speeds, installing mobile ventilation and refrigeration, burying or re-rating the cable of the power grid, etc.

There are also four types of "soft" adaptation strategies. A first option for adapting energy infrastructure to climate change is to reconsider the location of investments. For instance, Neumann and Price (2009) state that a key vulnerability of the U.S. energy infrastructure is that much of it is concentrated along the Gulf Coast, where hurricanes are fairly common during the summer and the fall. It is pointed out that if climate change leads to more frequent and intense storm events in the Gulf region, this concentration of energy infrastructures along the Gulf Coast could be particularly costly, and it could be in the interest of energy producers to shift their productive capacity to safer areas. Note that, as Paskal (2009) mentions, substantial investment in new emerging infrastructure is likely to take place in the next decades as a result of scheduled decommissioning, revised environmental standards, stimulus spending and new development. Location decisions of these new investments should take into account the impact of a changing environment in the infrastructure.

A second "soft" measure for minimizing the impact of climate change to energy systems consists in anticipating the arrival of a climate hazard through the development of meteorological forecasting tools inside the energy companies or improving the communication with meteorological services. These measures will require complementary actions such as the support of emergency harvesting of biomass in the case of an alert for rainfall or temperature anomalies. Here we would also include measures intended to hedge costs of protecting energy infrastructure if a disaster does strike. An example of such type of measures in the energy sector is the Deepwater Gulf of Mexico Pipelines Induced Damage Characteristics and Repair Options (DW RUPE) project (Stress Subsea, Inc., 2005). DW RUPE is a Joint Industry Study, including the U.S. Minerals Management Service and eight operating companies to address deepwater pipeline repairs. The implementation of repair plans to ensure functioning of distributed solar systems after extreme events would also constitute an example of anticipatory measures to minimize losses.

A third group of "soft" measures for minimizing the impact of climate change to energy systems comprises all the changes in the operation and maintenance of existing infrastructures. The management of on-site drainage and run-off of mined resources, changes in coal handling due to increased moisture content and the adaptation of plant operation to changes in river flow patterns constitute examples of this group of measures.

Finally, the fourth group of "soft" measures comprises technological changes and improved design of infrastructures. Examples include the improved design of wind and gas turbines in order to cope with changing climate conditions or the introduction of new biofuel crops with higher tolerance to high temperatures and water stress.

It is worth noting here that all these measures imply integrating future climate risks into every decision-making process. Thus, adaptation would be mainstreamed into all relevant policy interventions and planning and management decisions. This means that decision makers must consider future climate projections when deciding on issues such as coastal land-use planning, hazard management or emergency preparedness, and that these policies and plans should be regularly updated and upgraded.

Sharing responsibilities for losses and risks

Preventing losses/risks is not the only way the energy sector can adapt to climate change. It can also try to share responsibilities for losses and risks.

Insurance is an important tool to deal with risk. However, weather-related insurance has always posed a challenge to the insurance sector. These difficulties may be exacerbated by the increasing risk and unpredictability of extreme weather associated with climate change.

Even in developed countries, the climate insurance market is limited by poor information and understanding of risks by both insurance companies and potential clients. Nevertheless, even in these countries, insurance must be a key element in any climate change adaptation strategy.

It should be evaluated whether certain private actors/sectors that provide public services such as the energy sector need to be covered by compulsory standard weather-related insurance. In cases where insurance is not available, for example for infrastructures located in flood plains, publicly supported insurance schemes may be required. Due to the cross-border effects of climate change, there may be benefits in promoting international insurance as opposed to national or regional schemes.

When considering insurance as an adaptation strategy to deal with climate change, the Weather Risk Management Facility (WRMF) should be mentioned. The WRMF is a joint International Fund for Agricultural Development (IFAD) and World Food Programme (WFP) initiative to support the development of weather risk management instruments in developing countries. The WRMF has recently produced an overview of the issue of weather index insurance (WRMF, 2010). Index insurance, as defined in the report, "is a financial product linked to an index highly correlated to local yields". In contrast to traditional crop insurance, index insurance covers the risk of adverse environmental conditions (e.g., rainfall deficit) as opposed to suboptimal yields or production. As such, there's no threat of moral hazard as those who realize poor yields under favourable conditions cannot benefit from an insurance payout. Furthermore, as index insurance is based upon a verifiable indicator, it is eligible for reinsurance, which further spreads the risk.

The report highlights the potential benefits of index insurance for agricultural risk management at a range of scales (e.g., individual farmers to government agencies or relief organizations), but also notes some of the challenges. These include the complexity of establishing an index insurance market, which is dependent upon access to reliable environmental monitoring data and the ability to cultivate and maintain consistent market demand. Nevertheless, the report showcases a number of case studies where index insurance markets have been developed, often with success.

Energy diversification can be seen as an adaptation measure to increase resilience within the energy sector in responding to anticipated impacts of climate change. One approach would be to further expand the portfolio of energy sector (adoption on new forms of energy production such as solar, wind and hydro power).

Exploiting opportunities

Fortunately, as will be illustrated below, some opportunities exist to decrease the vulnerability of the energy sector to weather extremes and climate variability. For instance, ageing of existing infrastructures may open a new window of opportunity to build a more decentralized energy structure, preferably based on locally available renewable energy sources situated in secure locations. This would reduce the probability of suffering large-scale outages that result when centralized power systems are compromised. This sort of regional, network-based system might also prove more flexible and adaptive, and therefore more able to cope with the increasing variability and unpredictability caused by environmental change.

Another opportunity arises from urban design and land use planning. More than half of the world's population now lives in cities. According to the United Nations' estimates, the population living in urban areas is projected to pass from 3.49 billion in 2010 to 6.29 in 2050. This implies that cities are important and growing consumers of energy. Thus, urban policy and land use planning will play an important role in improving resilience of the energy system. In most cases, this strategy will take place through demand side management: building design (insulation, orientation), codes and standards (efficiency standards for appliances) and change of consumption patterns (district heating/cooling, flexible working hours, etc.). There is a wide range of examples of urban initiatives to reduce energy consumption and improve resilience (ETAP, 2006).

But there are also supply-side opportunities to be exploited from increasing urbanization. The electricity industry (Acclimatise, 2009) recognizes that it will face major challenges in providing new generation capacity and supply reliability within urban areas and that in the future they will need to develop a new supply and demand system where consumers can also be suppliers with a variety of home generators.

10.4 Conclusions

In the past decades, most studies on the relationship between climate change and the energy sector have focused on emissions from or mitigation by the energy sector. However, climate change is also expected to affect both energy supply and demand, and research on the adaptation options of the sector to the effects of climate change is surprisingly scant. In this chapter, we have offered a broad review of the main issues involved in the adaptation of the energy sector to climate change.

Based on this review, it is possible to draw some tentative conclusions about what the potential impacts of climate change will imply for the energy sector:

1. The major current risk for both supply and use is from episodic disruptions related to extreme weather events.
2. In many cases, "soft" and/or "hard" adaptation measures can reduce risks and prospects of negative consequences for energy supply and use.
3. Successful adaptation activities require the cooperation of a wide range of organizations and individuals.
4. Improving knowledge about vulnerabilities and possible risk management strategies is essential for effective climate change risk management in the energy sector.
5. "Climate-proofing" current and future energy systems should be mainstreamed among decision makers.
6. Climate change will very likely have significant effects on the potential power for many renewable energy sources, such as wind-power and hydro-power, and these potential changes should be considered in both siting and design decisions.
7. In regions where increases in average temperatures and temperature extremes are expected to increase the demand for electricity for cooling, measures to increase supply in general and peak load supply in particular should be considered.
8. Energy demand management should also be implemented as an adaptation measure.

Notes

1 Note that GHP does not take into account technical and economical feasibility of harnessing that energy.

2 Short life-span facilities will have much more margin for technological advances and/or relocation and, therefore, climate impacts will be much lower.

3 Parker (2005) estimates that the use of air conditioning reduces the efficiency of vehicles by around 12% at highway speeds.

4 Burt et al. (2003) studied current and future energy requirements for irrigation in California, including the projected loss of water from the state's reservoir system due to changed timing of snowmelt and surface water runoff. Assuming that the lost capacity would be replaced with groundwater, they calculated an increase in groundwater pumping energy of 163 GWh.

5 Burton et al. (1993), Stakhiv (1993), Carter et al. (1994), Smit et al. (1999, 2000), UKCIP (2007), and OECD (2008) provide some useful distinctions and discuss the nature of adaptation processes and forms.

6 http://www.cdproject.net

7 Available at http://unfccc.int/adaptation/nairobi_work_programme/

References

Acclimatise (2009), "Building Business Resilience to Inevitable Climate Change", *Carbon Disclosure Project Report*. Global Electric Utilities, Oxford.

Ackerman, F. and E.A. Stanton (2008), "The Cost of Climate Change: What We'll Pay if Global Warming Continues Unchecked", Natural Resources Defence Council, New York.

Agrawal, A., C. McSweeney and N. Perrin (2008), "Local Institutions and Climate Change Adaptation", Social Development Notes No 113, World Bank, Washington.

Arrieta, F.R.P. and E.E.S. Lora (2005), "Influence of Ambient Temperature on Combined-Cycle Power-Plant Performance", *Applied Energy* 80(3): 261–272.

Bartok, B. (2010), "Changes in Solar Energy Availability for South-Eastern Europe with Respect to Global Warming", *Physics and Chemistry of the Earth* 35: 63–69.

Bernacchi, C.J., B.A. Kimball, D.R. Quarles, S.P. Long and D.R. Ort (2007), "Decreases in Stomatal Conductance of Soybean under Open-Air Elevation of CO_2 are Closely Coupled with Decreases in Ecosystem Evapotranspiration". *Plant Physiology* 143: 134–144.

Bloom, A., V. Kotroni and K. Lagouvardos (2008), "Climate Change Impact of Wind Energy Availability in the Eastern Mediterranean using the Regional Climate Model PRECIS", *Natural Hazard and Earth System Sciences* 8(6): 1249–1257.

Burkett, V. (2011), "Global Climate Change Implications for Coastal and Offshore Oil and Gas Development", *Energy Policy* 39: 7719–7725.

Burt, C.M., D.J. Howes and G. Wilson (2003), "California Agricultural Water Electrical Energy Requirements", ITRC Report No R03-006. Prepared for the California Energy Commission. http://www.itrc.org/reports/reports.htm

Burton, I., R.W. Kates and G.F. White (1993), *The Environment as Hazard,* Guildford Press, New York.

Carter, T.P., M.L. Parry, H. Harasawa and N. Nishioka (1994), *IPCC Technical Guidelines for Assessing Climate Change Impacts and Adaptations,* University College London, London.

Cayan, D., M. Tyree, M. Dettinger, H. Hidalgo, T. Das and E. Maurer (2009), *Climate Change Scenarios and Sea Level Rise Estimates for the California 2009 Climate Change Scenarios Assessment*. California Climate Change Center; 2009. CEC-500-2009-014-D.

Cazenave, A. and W. Llovel (2010), "Contemporary Sea Level Rise", *Annual Review of Marine Science* 2: 145–173.

CCSP (2007), *Effects of Climate Change on Energy Production and Use in the United States. A Report by the U.S. Climate Change Science Program and the Subcommittee on Global Change Research*. Thomas J. Wilbanks, Vatsal Bhatt, Daniel E. Bilello, Stanley R. Bull, James Ekmann, William C. Horak, Y. Joe Huang, Mark D. Levine, Michael J. Sale, David K. Schmalzer, and Michael J. Scott (Eds.). Department of Energy, Office of Biological & Environmental Research, Washington, DC, USA. 160 pp.

Connor, H., R. Gould, R. Janssen and C. Rynikiewicz (2005), "New Governance Imperatives for Energy Planning in Liberalised European Markets?", Proceedings of the ECEEE 2005 Summer Study, panel 1, Volume I, pp. 223–230, Mandelieu, France, May 31–June 3, 2005.

Cutforth, H.W. and D. Judiesch (2007), "Long Term Changes to Incoming Solar Energy on the Canadian Prairie", *Agricultural and Forest Meteorology* 145: 167–175.

Diffenbaugh, N., S.M. Scherer and M. Ashfaq (2012), "Response of Snow-Dependent Hydrologic Extremes to Continued Global Warming", *Nature Climate Change* doi:10.1038/nclimate1732.

Durack, P.J., S.E. Wijffels and R.J. Matear (2012), "Ocean Salinities Reveal Strong Global Water Cycle Intensification During 1950 to 2000", *Science* 336(6080): 455–458.

Durmayaz, A. and O.S. Sogut (2006), "Influence of Cooling Water Temperature on the Efficiency of a Pressurized-Water Reactor Nuclear Power Plant", *International Journal of Energy Research* 30(10): 799–810.

Ebinger, J. and W. Vergara (Eds.) (2011), *Climate Impact on Energy Systems: Key Issues for Energy Sector Adaptation*, ESMAP, The World Bank, Washington.

ETAP (2006), "Development of Eco-Cities in the World", Environmental Technologies Action Plan. http://ec.europa.eu/environment/etap/pdfs/june06_ecocities.pdf

Euroelectric (1997), *Study on the Importance of Harnessing the Hydropower Resources of the World*, Union of the Electric Industry (Euroelectric), Hydro Power and other Renewable Energies Study Committee, Brussels.

Feeley, T J., T.J. Skone, G.J. Stiegel, A. McNemar, M. Nemeth, B. Schimmoller, J.T. Murphy and L. Manfredo (2008), "Water: A Critical Resource in the Thermoelectric Power Industry", *Energy* 33(1): 1–11.

Fenger, J. (Ed.) (2007), *Impacts of Climate Change on Renewable Energy Sources: Their role in the Nordic Energy System*, Nordic Council of Ministers, Copenhagen.

Haas, L.J.M., K. Schumann and R. Taylor (2008), *Climate Adaptation: Aligning Water and Energy Development Perspectives*, International Hydropower Association.

Harsem, O., A. Eide and K. Heen (2011), "Factors Influencing Future Oil and Gas Prospects in the Arctic", *Energy Policy* 39: 8037–8045.

Hatfield, J.L. (2010), "Climate Impacts on Agriculture in the United States: The Value of Past Observations", Chapter 10 in D. Hillel and C. Rosenzweig (Eds.), "*Handbook of Climate Change and Agroecosystems: Impact, Adaptation and Mitigation*". Imperial College Press, London UK.

Hatfield, J.L., K.J. Boote, P. Fay, L. Hahn, C. Izaurralde, B.A. Kimball, T. Mader, J. Morgan, D. Ort, W. Polley, A. Thomson and D. Wolfe (2008), Agriculture. In *The Effects of Climate Change on Agriculture, Land Resources, Water Resources, and Biodiversity in the United States.* A report by the U.S. Climate Change Science Program and the Subcommittee on Global Change Research, Washington, DC. 362 pp.

IPCC (2007), *Climate Change 2007: Synthesis Report.* Contribution of Working Groups I, II and III to the Fourth Assessment Report of the Intergovernmental Panel on Climate Change. R. K. Pachauri and A. Reisinger (Eds./Core Writing Team). IPCC, Geneva, Switzerland.

Isaac, M. and D.P. van Vuuren (2009), "Modeling Global Residential Sector Energy Demand for Heating and Air Conditioning in the Context of Climate Change", *Energy Policy* 37(2): 507–521.

Joughin, I, R.B. Alley and D.M. Holland (2012), "Ice Sheet Response to Oceanic Forcing", *Science* 338(6111), 1172–1176.

Kiessling, F., P. Nefzger, J.F. Nolasco, and U. Kaintzyk (2003), *Overhead Power Lines: Planning, Design, Construction*. Springer.

Kopytko, N. and J. Perkins (2011), "Climate Change, Nuclear Power, and the Adaptation-Mitigation Dilemma", *Energy Policy* 39: 318–333.

Lehner, B., G. Czisch and S. Vassolo (2005), "The Impact of Global Change on the Hydropower Potential of Europe: A Model-Based Analysis", *Energy Policy* 33: 839–855.

Lucena, A.F.P., A.S. Szklo and R. Schaeffer (2009), "Renewable Energy in an Unpredictable and Changing Climate", *Modern Energy Review* 1: 22–25.

Lucena, A.F.P., A.S. Szklo, R. Schaeffer and R.M. Dutra (2010), "The Vulnerability of Wind Power to Climate Change in Brazil", *Renewable Energy* 35: 904–912.

Mideksa, T.T. and S. Kallbekken (2010), "The Impact of Climate Change on the Electricity Market: A Review", *Energy Policy* 38: 3579–3585.

Mishra, V. and K.A. Cherkauer (2010), "Retrospective Droughts in the Crop Growing Season: Implications to Corn and Soybean Yield in the Midwestern United States". *Agricultural and Forest Meteorology* 150(7–8): 1030–1045.

Neumann, J.E. and J.C. Price (2009), *Adapting to Climate Change: The Public Policy Response: Public Infrastructure*, RFF Report, Washington, DC.

Organisation for Economic Co-operation and Development (2008), *Economic Aspects of Adaptation to Climate Change: Costs, Benefits and Policy Instruments,* OECD, Paris.

Parker, D.S. (2005), *Energy Efficient Transportation for Florida,* Energy Note FSEC-EN-19, Cocoa, Florida: Florida Solar Energy Center, University of Central Florida. Available at: http://www.fsec.ucf.edu/Pubs/energynotes/en-19.htm

Paskal, C. (2009), *The Vulnerability of Energy Infrastructure to Environmental Change*, Briefing Paper, Chatham House, The Royal Institute of International Affairs.

Pryor, S.C. and R.J. Barthelmie (2010), "Climate Change Impacts on Wind Energy: a Review", *Renewable and Sustainable Energy Reviews* 14: 430–437.

Pryor, S.C., R.J. Barthelmie and E. Kjëllstrom (2005), "Analysis of the Potential Climate Change Impact on Wind Energy Resources in Northern Europe using Output from a Regional Climate Model", *Climate Dynamics* 25: 815–835.

Sailor, D.J., M. Smith and M. Hart (2008), "Climate Change Implications for Wind Power Resources in the Northwest United States", *Renewable Energy* 33(11): 2393–2406.

Salby, M.L. (1996), *Fundamentals of Atmospheric Physics,* Academic, San Diego, CA.

Sathaye, J., L. Dale, G. Fitts, P. Larsen, K. Koy, S. Lewis, and A. Lucena (2011), *Estimating Risk to California Energy Infrastructure from Projected Climate Change.* California Energy Commission. Publication number: CEC-500-2011-xxx.

Seager, R., N. Naik and L. Vogel (2012), "Does Global Warming Cause Intensified Interannual Hydro-climate Variability?" *Journal of Climate* 25(9): 3355–3372.

Schaeffer, R., A.S. Szklo, A.F.P. de Lucena, B.S.M.C. Borba, L.P.P. Nogueira, F.P. Fleming, A. Troccoli, M. Harrison and M.S. Boulahya (2012), "Energy Sector Vulnerability to Climate Change: A Review", *Energy* 38: 1–12.

Smit, B., I. Burton, R.J.T. Klein and R. Street (1999), "The Science of Adaptation: A Framework for Assessment", *Mitigation and Adaptation Strategies for Global Change* 4: 199–213.

Smit, B., I. Burton, R.J.T. Klein and J. Wandel (2000), "An Anatomy of Adaptation to Climate Change and Variability", *Climatic Change* 45: 223–251.

Stakhiv, E. (1993), *"Evaluation of IPCC Evaluation Strategies"*, Institute for Water Resources, U.S. Army Corps of Engineers, Fort Belvoir, VA, draft report.

Stern, N. (2007), *The Economics of Climate Change: The Stern Review,* Cambridge University Press, Cambridge.

Stress Subsea, Inc. (2005), *Deep Water Response to Undersea Pipeline Emergencies,* Final Report, Document N° 221006-PL-TR-0001, Houston, TX.

Troccoli, A. (Ed.) (2010), *Management of Weather and Climate Risk in the Energy Industry*, NATO Science Series, Springer, Dordrecht.

Troccoli A., M.S. Boulahya, J.A. Dutton, J. Furlow, R.J. Gurney and M. Harrison (2010), "Weather and Climate Risk Management in the Energy Sector", *Bulletin of the American Meteorological Society* 91(6): 785–788.

United Kingdom Climate Impacts Programme (2007), *Identifying Adaptation Options*, UKCIP Technical Report, UKCIP, Oxford.

United Nations Environment Programme (2006), *Raising Awareness of Climate Change,* UNEP, Nairobi.

Valsson Trausti, T. and G.F. Ulfarsson (2011), "Future Changes in Activity Structures of the Globe under a Receding Arctic Ice Scenario", *Futures* 43: 450–459.

Vicuña, S., E.P. Maurer, B. Joyce, J.A. Dracup and D. Purkey (2007), "The Sensitivity of California Water Resources to Climate Change Scenarios", *Journal of the American Water Resources Association* 43(2): 482–498.

Vikebo, F., T. Furevik, G. Furnes, N.G. Kvamsto and M. Reistad (2003), "Wave Height Variations in the North Sea and On the Norwegian Continental Shelf, 1881–1999", *Continental Shelf Research* 23: 251–263.

Vlasova, L. and G.S. Rakitina (2010), "Natural Risk Management in the Gas Transmission System of Russia and Contribution of Climate Services Under Global Climate Change". In A. Troccoli (Ed.), *Management of Weather and Climate Risk in the Energy Industry*, NATO Science Series, Springer.

Weather Risk Management Facility (2010), *The Potential for Scale and Sustainability in Weather Index Insurance for Agriculture and Rural Livelihoods*, International Fund for Agricultural Development and World Food Programme, Rome.

Webster, P.J., G.J. Holland, J.A. Curry and H.R. Chang (2005), "Changes in Tropical Cyclone Number, Duration, and Intensity in a Warming Environment", *Science* 309: 1844–1846.

Williamson, L.E., H. Connor and M. Moezzi (2009), *"Climate-Proofing Energy Systems"*, Helio-International.

World Bank (2010), *World Development Report: Development and Climate Change*, The World Bank, Washington DC.

Zhao, M. and S.W. Running (2010), "Drought-Induced Reduction in Global Terrestrial Net Primary Production from 2000 through 2009", *Science* 329: 940–943.

11

WATER FOR AGRICULTURE

Some thoughts on adaptation to climate change from a policy perspective

Luis Garrote,[1] *Ana Iglesias*[2] *and Alberto Garrido*[3]

[1]DEPARTMENT OF HYDRAULIC AND ENERGY ENGINEERING, UNIVERSIDAD POLITÉCNICA DE MADRID (UPM), SPAIN

[2]DEPARTMENT OF AGRICULTURAL ECONOMICS AND SOCIAL SCIENCES, UNIVERSIDAD POLITÉCNICA DE MADRID (UPM), SPAIN

[3]RESEARCH CENTRE FOR THE MANAGEMENT OF AGRICULTURAL AND ENVIRONMENTAL RISKS (CEIGRAM), SPAIN AND DEPARTMENT OF AGRICULTURAL ECONOMICS AND SOCIAL SCIENCES, UNIVERSIDAD POLITÉCNICA DE MADRID (UPM), SPAIN

11.1 Introduction

In the beginning of the 21st century, water to support food production seems to stand at a crucial juncture. Projections of water availability remain complex and uncertain (Vorosmarty et al., 2010), not least due to changes in population, consumption patterns and environmental policy (Iglesias et al., 2011a). Research and technology have been unusually vigorous and have shed light on many possible innovations (Sunding and Zilberman, 2001: Nordhaus, 2002; Smit and Skinner, 2002). Part of this transformation is due to the pressure of agricultural practices and policies to face climate change, especially in relation to water for agricultural production.

Even though agriculture is an established economic activity in many areas, a string of worries about negative externalities and water conflicts makes it hard to predict how water and agriculture will come to terms with each other. The projections of future food supply and population shed doubts on the sustainability and productive capacity of the current agricultural model (Kearney, 2010). In the last decades the concern over agricultural sustainability has come to include environmental worries about the proper use of natural resources.

Exploring policy choices for future agricultural water 10 years ago would produce a picture full of uncertainties. In part, projections were erratic because of global changes in agricultural trade and policy.The role of emerging centres of power in agricultural trade was only just becoming clear, and environmental policy was defined and yet not implemented in Europe, lagging behind the United States in important aspects. Climate change science did not influence adaptation policy until the end of the last century.Therefore, policies need to be successfully balanced to achieve sustainability of water resources management (Vorosmarty et al., 2010), even if pressures to intensify agricultural production do not subside due to world's growing food demand.

In this chapter, we explore some aspects of water management through a series of assertions designed to stimulate discussion and debate. We present three assertions organised around the idea that understanding (1) vulnerability, (2) policy trade-offs and (3) flexible mechanisms may assist in formulating adaptation strategies for water for food production. The following sections provide evidence in support of these assertions. Following this introduction, Section 11.2 explores the vulnerability of the global agricultural system to climate change. Section 11.3 provides examples of trade-offs of water policies and the role of regional disparities in the outcome. Section 11.4 provides a discussion of the role of flexible allocation mechanisms for irrigation water. Section 11.5 outlines the main conclusions.

11.2 Planning for future water demand for agriculture

Assertion 1. Understanding and reducing vulnerability does not demand accurate predictions of the impacts of climate change.

Planning for a future that is uncertain and radically different from the present is a major concern for planners in most sectors. In the case of water for agriculture, the sector's profound linkages to the natural regime imply that understanding how different the world will be in coming years may help water management adapt to climate change.

The climate change challenge has forced research and policy to look at the interactions of agriculture and water with a longer-term perspective. Climate change science has dedicated a major effort to dealing with an uncertain future; the idea is to prepare society today so that it can deal with changes tomorrow in such a way as to make robust strategic decisions now but embed flexibility and adaptability mechanisms in the core of decision-making processes.

Here we look into the patterns of additional irrigation demand under a range of scenarios and to the adaptive capacity to define the regions where water management needs to be adapted to climate change. We conclude the section with some policy implications.

11.2.1 Projections of additional water for irrigation

Iglesias et al. (2011b) evaluated global irrigation water demand scenarios taking into account changes in the physical variables of the scenario (precipitation and temperature), changes in socioeconomic conditions (management at the farm level, markets and trade, and policy), and changes in technology (agricultural and hydraulic). The results show that, when policy and technology remain constant, agricultural water demand increases in all scenarios in the region. Here we present a summary of the results of the study by Iglesias et al. (2011b), expanding on the discussion and implications for future policy. In the study, the main drivers of this irrigation demand increase are the decrease in effective rainfall and the increase in potential evapotranspiration (due to higher temperature and changes of other meteorological variables). The calculation of changes in irrigation water requirements aim to reach demand satisfaction according to assumptions on technological capacity of the country, limited by environmental water needs such as the country environmental flow requirements.

Here we evaluate the need for additional irrigation as an adaptation strategy, drawing from the database of irrigation needs published by Iglesias et al. (2011b). The uncertainty of the climate scenario is characterised by selecting three emission scenarios (A1B, E1 and RCP8) and several global climate models, some of them downscaled across Europe (scenarios summarised in Table 11.1).

Table 11.1 Summary of the climate change scenarios used to drive changes in irrigation demand

Greenhouse gas emissions in 2080 (ppm CO_2)	*Assumptions on energy and emissions policy*	*Number of climate change scenarios*	*Average global climate signal in 2080*
A1B (712)	A balanced emphasis on all energy sources	Average of 12 GCMs	About 4 °C global temperature increase and large intensification of the hydrological cycle with large area extensions of more pronounced drought, especially in Africa and the Mediterranean regions.
E1 (489)	The so-called global 2 °C-stabilisation scenario, characterised by an emphasis in mitigation policy	Average of 14 GCMs	About 2 °C global temperature increase with a defined pattern of precipitation increases in areas of the northern hemisphere and small decreases in Africa and the Mediterranean regions.
RCP8	The so-called rising scenario, without mitigation policy	1 GCMs	About 5 °C global temperature increase and large intensification of the hydrological regime. Pronounced drought in the Mediterranean region.

Source: Iglesias et al., 2011b

The additional demand for water aims to achieve a level of production independent of the market (here irrigation demand is not a demand in economic terms that takes into account the market).

Global scenarios of irrigation demand change for the 2080s (Figure 11.1) were developed based on scenarios of changes in the environmental and socioeconomic variables by using the ClimateCrop model (Iglesias et al., 2011b). The study assumed that: (1) irrigated areas do not increase significantly, and (2) demand satisfaction was according to the technological capacity of each country. The first assumptions reflect current global irrigation in most agricultural areas; in Europe and the United States due to societal environmental concerns and in Asia due to limitations in irrigated areas. Potential irrigation expansion in Africa and some regions of South America may not be accounted for in our analysis. The second assumption reflects empirical data on global irrigation development. These assumptions imply that there is an optimisation of the environmental water requirements.

The results demonstrate variability for the irrigation demand changes but also demonstrate the clear linkages between needs for additional irrigation in regions that have water shortages. Most clearly, the results show that the global pattern of irrigation demand under climate change is very similar even under the less extreme E1 scenario. Of course, climate change is generally considered as a negative threat for irrigation – due, for instance, to changes in temperatures and precipitation or an increased likelihood of extreme events. The results presented in Figure 11.1 show a marked heterogeneous nature of climate change impacts. Furthermore, the results suggest that: (1) not all regions will be equally affected, and (2) there is a regional pattern of vulnerability that does not depend on the scenario. The most vulnerable areas in all cases are in the Indian sub-continent, Southeast Asia, East Africa and Mexico. The vulnerability in Europe

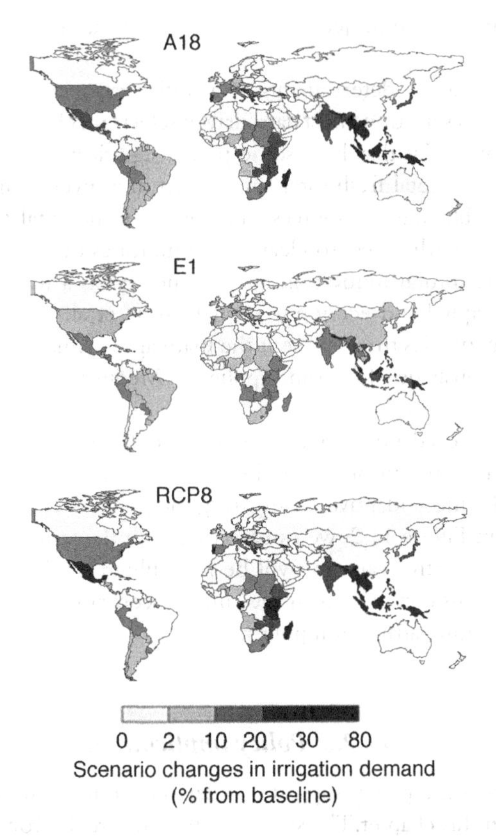

Figure 11.1 Projected water demand changes aggregated at the country level for the average A1B, E1 and RCP 8 scenarios for the 2080s

and the United States indicates a limitation in the potential expansion of litigation areas due to compliance with increasing environmental concerns.

11.2.2 Adaptive capacity

Understanding resiliency or adaptive capacity is a starting point for understanding the future. Adaptive capacity gives an idea of the ability of a society to respond to and recover from an impact or perturbation. Understanding the determinants of adaptive capacity helps formulate future strategies that may be required to dampen the negative effects of climate change (Yohe et al., 2006; Iglesias et al., 2011b; Iglesias, 2013). Because of this focus, research on adaptive capacity is ideally situated for pointing out possible opportunities for improving adaptation policy (Fankhauser et al., 1999).

A number of indices of adaptive capacity have been developed (Yohe et al., 2006; Ionescu et al., 2009; Iglesias et al., 2011b; Iglesias, 2013) to capture different elements of social and economic vulnerability to climate change. The uncertainties in terms of estimating adaptive capacity are high; those uncertainties are a consequence of the difficulty in defining the components of adaptive capacity and how to project those components into the future (Adger and Vincent, 2005).

Defining and selecting the determinants of adaptive capacity is essential; there are three options: a concept-driven selection and a data-driven selection. Here we take a mixed approach, selecting data-driven variables that have some policy significance and concept-driven determinants. For example, the total freshwater withdrawn in a given year is an indication of the pressure on the renewable water resources, and the environmental policies implemented in the area and population with access to clean water indicates development and health limitations. We consider agricultural innovation and technology as a main determinant as well as natural capital, social capital and economic capital (Yohe et al., 2006). We calculate adaptive capacity for different countries that represent the main agro-climatic regions of the world. The methodology was previously developed and applied to Mediterranean countries in Iglesias et al. (2011c).

Figure 11.2 shows the adaptive capacity components of countries that represent major agricultural regions where irrigation demand is projected to increase. Adaptive capacity ranges from 0 to 1, 1 being the most adaptive. Comparing the projected climate change impacts with the adaptive capacity in Figure 11.2, we can see that with the same level of potential impacts, countries with higher adaptive capacity will be less vulnerable. Here we show that adaptive capacity varies more across countries than the climate scenario, and therefore could be a main determinant for formulating adaptation policy.

11.2.3 Policy implications

Table 11.2 summarises some policy suggestions derived from the scientific indicators as they have been presented in this chapter. The suggested policies are far from comprising an exhaustive list, and should not be taken as a set menu of policies. Rather, they are meant to reflect the kind of policies that may be appropriate for redressing weaknesses in adaptive capacity where water scarcity is a problem. For instance, in countries with considerable social and economic inequality and where water scarcity is not a pressing issue, water management policies should focus on ensuring equitable access for disadvantaged populations to guarantee health and economic benefits. These types of measures are primarily concerned with managing the supply of water in a way that maximises equality. However, in the same country, if water scarcity becomes a serious problem, it is likely that emergency actions will have to be implemented to prioritise certain water users over others. In such a situation, it is important to ensure that these emergency actions do not aggravate pre-existing socioeconomic inequalities, a circumstance that is likely to occur if water markets are poorly regulated. This requires enforcing water regulations and transfers to guarantee the equitability of supply as well as ensuring that demands do not become excessive. The suggested policy mechanisms are in line with key priorities in European and international water policy and are based on the concept of integrated water management, which is concerned with ensuring that all of a region's water resources are efficiently managed while making room for social and environmental concerns to ensure that water use is sustainable. The appropriate policy mechanisms follow from the kinds of policy interventions that are required as determined by a combined analysis of water scarcity levels and weaknesses in adaptive capacity. The policy implications are outlined in Table 11.2.

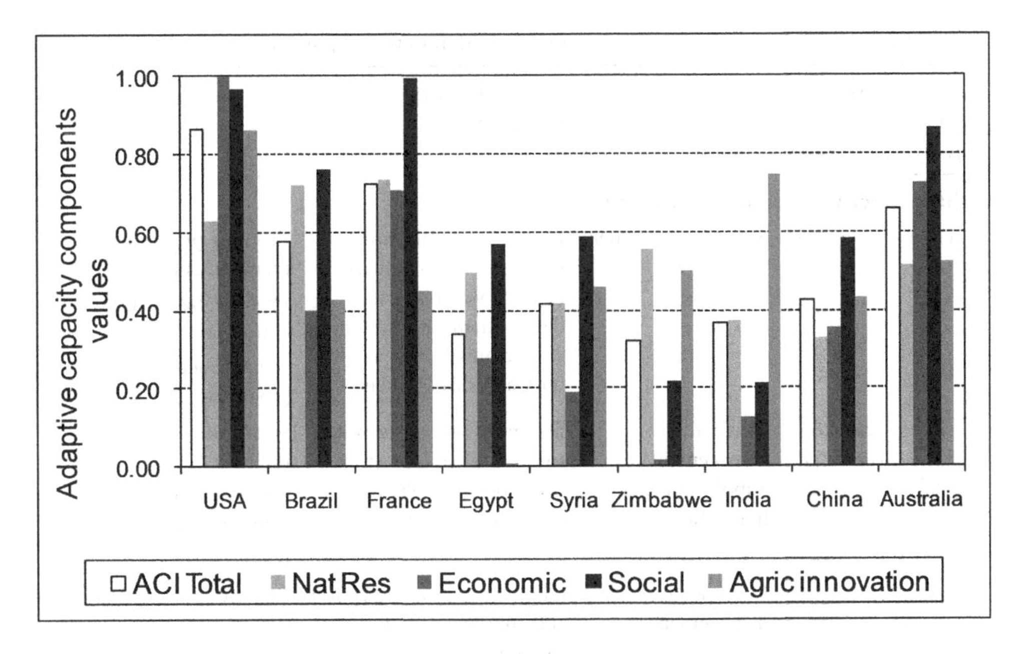

Figure 11.2 Components of the adaptive capacity for selected agro-climatic regions

11.3 Trade-offs in water policies

Assertion 2. Policy-based choices for adapting to climate change are the major drivers of future water availability.

Public policies are far more uncertain and difficult to project than private-local adaptation, since they do not respond only to optimising climate-water availability interactions. Therefore an essential component of public policies on adaptation may be to trigger adaptive capacity (i.e., the ability to adjust to climate change).

Here we explore some choices to thinking about water policy for agriculture in the future. We use the term "water availability" to describe the volume of water that could be used for irrigation after subtracting the domestic use and the environmental flow requirements.

Our analysis bridges the gap between traditional impact assessment and policy formulation by directing attention to the causes of the water management challenges for agriculture and evaluation of alternatives. This evaluation helps define the sensitivity of a system to external shocks and to identify the most relevant aspects that can decrease the level of risk posed by climate change.

Water policies are strongly constrained by available natural resources, especially in regions where water is scarce. Technology or management can partly compensate for a deficit of natural resources, but they are always constrained by economic limitations. One key factor in this context is the amount of water allocated for agriculture. Water use by agriculture must be compatible with other uses of higher priority, like environmental flows, urban and industrial

Table 11.2 Summary of some policy suggestions derived from the adaptive capacity indicators

Weakest component of the adaptive capacity	*Type of policy intervention required*	*Appropriate policy mechanisms*	*Policy recommendations*
Water scarcity level: Low to medium			
Social and economic factors	Supply management	Infrastructure investments, ensure access to sanitary and safe water sources, development of institutional and organisational capacity for water management	Promote pro-poor management Promote health and education Improve access to water for production and sanitation
Technological eco-efficiency & natural capital	Demand management, Supply management (regulation)	Design of system for water rights trading, appropriate water pricing policies and efficiency incentives	Focus on environmental mitigation Promote more efficient technologies Reduce water use
Climate	Extreme event management	Consultation on perception and appropriate response to potential climate impacts – design of policies and priorities of water provisioning in emergencies	Promote flexible water storage options Invest in physical infrastructure Develop safety net programmes
Water scarcity level: Serious to very serious			
Social and economic factors	Supply management (transfers & regulation) Demand management	Guarantee that water-pricing and efficiency policies do not disproportionately affect poor	Develop micro-irrigation technologies Integrate ground and surface water management
Technological eco-efficiency & natural capital	Demand management	Mandatory minimum efficiency standards, ensure balance between economic sectors and environmental flow	Reform conflict prevention institutions Promote information sharing and cooperative management
Climate	Extreme events management Demand management (transfers & regulation)	Implementation of policies for extreme events, water rights trading and reallocation to priority sectors	Promote policies that help create a paradigm shift Design adaptation policies Develop new alternative sectors

water demand and cooling needs of nuclear or thermal energy plants. It must also result in a global benefit, considering not only economic factors, but also social and environmental implications of agricultural activity. Most climate change scenarios present a picture where water availability is significantly reduced in the future in many regions of the world. Since agriculture is one of the largest water consumers, it is an obvious target for demand management policies that tend to reduce agricultural water use. However, the amount of water that can be allocated to agriculture depends on policy decisions. Establishing water policies frequently implies deciding on trade-offs between alternative actions and opportunities and costs of different water uses. Understanding these trade-offs is essential to respond to regional disparities in water use.

11.3.1 Water availability under climate change

One of the aspects of this analysis is the evaluation of water availability for agriculture under future climate change scenarios. Water availability was computed for 400 sub-basins in the European Mediterranean region using the Water Availability and Adaptation Policy Analysis (WAAPA) model (Garrote, 2013). The methodology of analysis is based on the demand-reliability curve and is shown in Figure 11.3. We consider a simplified system with only two types of demands (α and β, where α represents urban demand and β represents irrigation demand). If the value of the irrigation demand, β, is allowed to change while urban demand, α, is fixed, the supply reliabilities of both demands change accordingly. In the supply-reliability curve, reliability values for demand components α (p^{α}_{β}) and β (p^{β}_{β}) are represented as a function of demand value d_{β}. The minimum required reliabilities for every type of demand are established by water policy according to the socioeconomic context. If required reliabilities for urban supply and irrigation are, respectively, p^{α}_{req} and p^{β}_{req}, the maximum availability for

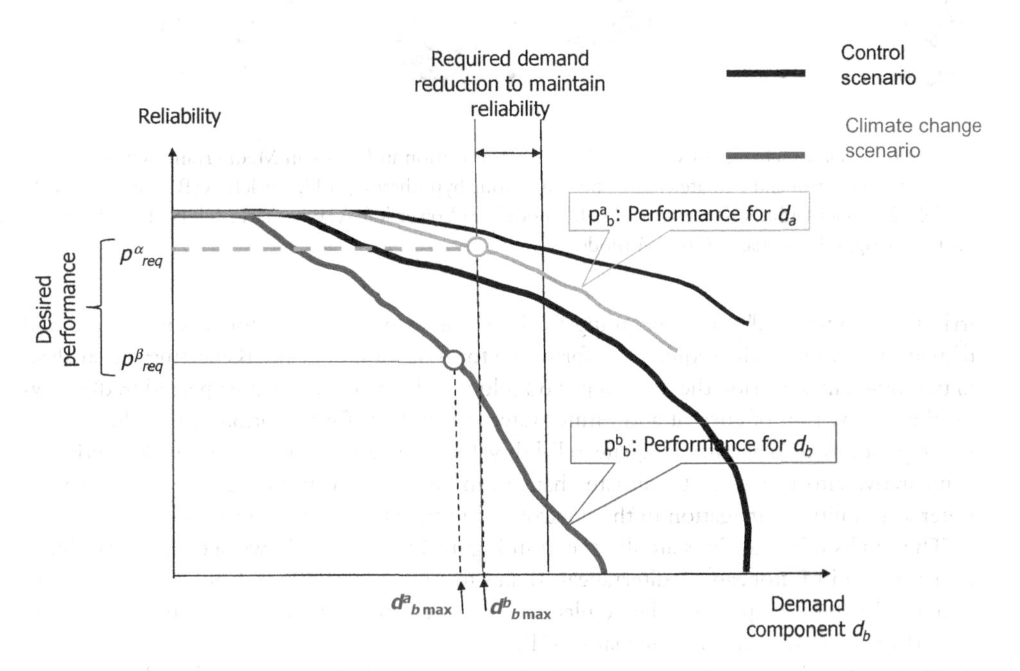

Figure 11.3 Schematic of demand reliability analysis in the control and climate change scenarios. Vertical lines represent water availability for irrigation satisfying minimum required reliabilities for urban and irrigation demands.

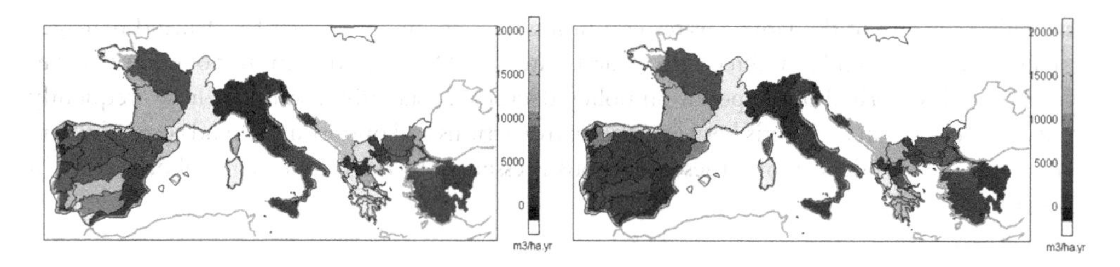

Figure 11.4 Specific values of water availability for irrigation (in m^3/ha.yr) in European Mediterranean basins for the control period (left) and the climate change period (right)

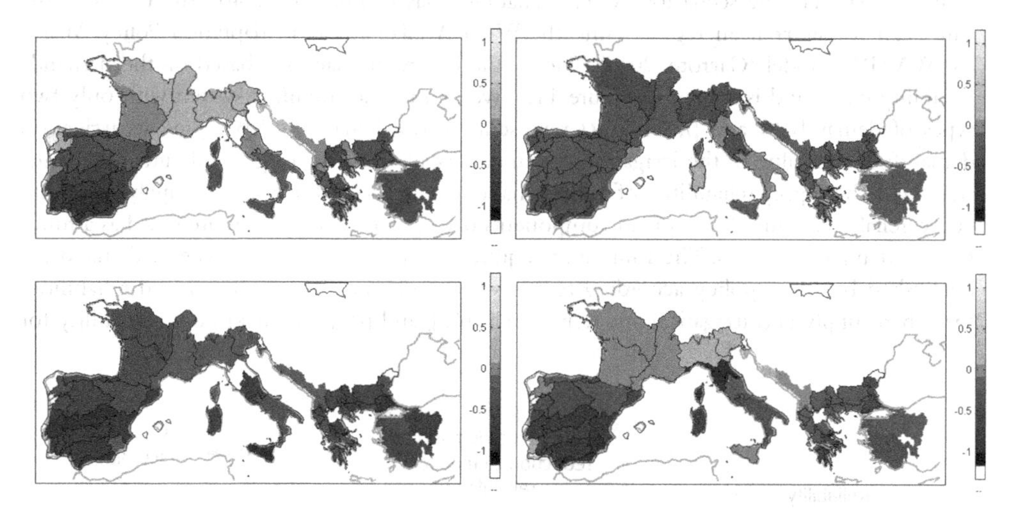

Figure 11.5 Relative change of water availability for irrigation in European Mediterranean basins between the control and climate change period in four hypotheses: (1) Upper left: A1B scenario KNMI model; (2) Upper right: A1B scenario ETHZ model; (3) Lower left: A1B scenario CRNM model; and (4) Lower right: E1 scenario CRNM model

irrigation would be $d^{\alpha}{}_{\beta}max$, according to the required performance for urban demand and $d^{\beta}{}_{\beta}max$, according to the required performance for irrigation demand. Repeating this analysis in two different scenarios, the control period (blue) and the climate change period (red) allows for the comparison of current and future water availabilities for irrigation. If the objective of water policy is to maintain adequate reliability for both urban and irrigation demands, the comparative effort to adapt to climate change can be estimated from the difference between water availability for irrigation in the control and in the climate change period.

The results of the analysis are illustrated in Figure 11.4, which shows net water availability for irrigation in European Mediterranean Basins for the control (1960–1990) and the climate change (2070–2100) period. The results are expressed in m^3/ha.yr, and correspond to the KNMI model under emissions scenario A1B.

The results obtained are sensitive to the emission scenario and the model applied. A comparative summary is presented in Figure 11.5 where the reduction of water availability for irrigation is shown in four modelling hypotheses. Although model uncertainty is high, results show a clear tendency to a reduction of water availability. However, there are strong regional disparities in the changes of water availability. The basins located on the South-East and South-West of Europe are affected the most. If we assume that water availability in the control period is similar to current irrigation demand, these basins would have to reduce water consumption by agriculture by more than 50% by the end of the century. This poses a serious threat to the sustainability of irrigated agriculture in the region, and calls for immediate action for adaptation.

11.3.2 Policy choices

Irrigation demand management is the main alternative for adaptation. However, it will be applied in combination with other public policies that may reduce or enhance its effectiveness. To illustrate this effect, we present the changes of the estimates of water availability for irrigation in the climate change scenario under four different public policies that support the provision of water for agriculture: (1) Improvement of water resources systems management; (2) Allocation of hydropower reservoirs for regulation; (3) Reduction of environmental flow requirements and (4) Improvement of urban and industrial water use efficiency. These policies have been introduced in WAAPA simulations by adopting suitable model parameters. The improvement of water resources system management is represented by changing from local to global perspective in water resources management. Under local management, each reservoir is managed to supply water to local demands, while under global management, all reservoirs in the basin are operated jointly to satisfy all demands in the basin. Moving from local to global management requires significant investment in water transportation and distribution infrastructure to enable water supply from any reservoir to any demand. The allocation of hydropower reservoirs is represented by adding the storage in these reservoirs to improve water regulation. The reduction of environmental requirements is introduced by changing the specification of environmental flows. In the reference case, environmental flows are computed as the 10% percentile of the marginal monthly distribution of naturalised flows. The reduction is represented by changing the percentile to 5%. Finally, the improvement of urban water use efficiency is introduced by assuming a maximum per capita consumption in cities of 200 l/p.day, compared to 300 l/p.day used in the reference case.

Figure 11.6 shows the change in water availability for the future period if each of these policies is applied. The results correspond to the KNMI model under the A1B emission scenario.

The disparities in the effectiveness of these policies in improving water availability for irrigation are apparent. Factors like variability of hydrologic series, configuration of hydraulic infrastructure or population density are some of the principle determinants of variations in policy effectiveness across European basins. These factors are local and present great variability. Some of the most significant disparities exist between basins with urban and industrial development on the one hand and basins with extensive irrigation on the other. Unlike urban and industrial areas, agricultural communities have a limited base for economic growth. Their residents are facing disproportionate exposure to climate change that has to be compensated by public policies. Extreme care must be taken in understanding these disparities while designing climate change adaptation policies.

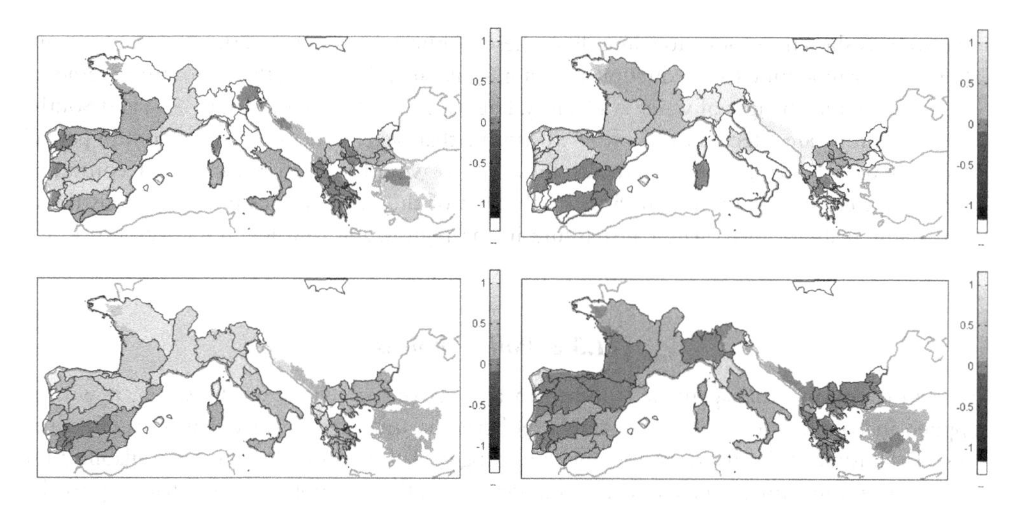

Figure 11.6 Relative change of water availability for irrigation in European Mediterranean basins with respect to the climate change period (2070–2100) in four policy hypotheses: (1) Upper left: Improvement of water resources systems management; (2) Upper right: Allocation of hydropower reservoirs for regulation; (3) Lower left: Reduction of environmental flow requirements and (4) Lower right: Improvement of urban water use efficiency. Results are computed for the A1B emission scenario and the KNMI model

11.4 Flexible mechanisms

Assertion 3. Adding flexibility to water allocation policies enhances resiliency to unexpected risks.

In mature water economies, the role of the state and the administration has very few options to generate additional water supplies. Most cost-effective opportunities to expand supply have already been developed, so resource management and flexible allocation mechanisms show greatest potential. Thus, the regulatory role of the public administration is enhanced and brought to sharper focus because systems are run very tight, and management mistakes could have relevant consequences.

It is no coincidence that the firmest, best defined and strongly enforced water rights are found in semi-arid countries (the Western U.S., Mexico, Chile, Israel, Spain, Australia). Managing scarcity and achieving use efficiency require that users obtain a water right and be given by the state some assurance that it will be protected. Saving some exceptions, rights are not granted at perpetuity, but have long-term duration. Combining adequate doses of firmness in rights definition and of flexibility to obtain the highest social value in all circumstances is a daunting task for water administrations and institutions.

Both the economy and the natural conditions of water systems evolve, and the need for changing the allocation of rights eventually appears. Adding flexibility and ensuring that it operates in the desirable fashion is a requirement for well-functioning water allocation mechanisms, as suggested by the seminal work of Howe et al. (1986). They defined six criteria: (1) flexibility, (2) security of tenure, (3) response to the opportunity cost, (4) predictability of outcome, (5) equity and (6) representation of public values.

Flexibility in the allocation of existing water supplies is of vital importance, so that water can be shifted from use to use and place to place as climate, demographic, and economic conditions change over time. There is a need for both short-term and long-term flexibility. It must also be realised that for flexibility to exist in an operational sense, it is not necessary that all water be subject to reallocation, only that there exist a tradable margin within each major water-using area that is subject to low-cost reallocation.

Security of tenure for established users ensures investment. Only if the water user can be assured of continued use will the user invest in and maintain water-using systems.

Whether or not the user is confronted with the real opportunity cost of the resources available is another important determinant to the outcome. For example, a perpetual contract for the supply of water at a fixed price will, over time, fail to reflect the changing opportunity costs involved in continued use. Allocation of a physical quota determined by a central authority so as to reflect other demands and the available supply may implicitly confront the user with the water's opportunity cost. A competitive market that sets a market-clearing price directly confronts the potential user with the real opportunity cost.

Perception of uncertainty plays also a major role, and this can be measured as the predictability of the outcome of the process. Older mechanisms may become outmoded but may be retained because they are familiar, because they cause no surprises. Change to a new allocative process, while promising some advantages, may increase uncertainty about the outcome.

A water allocation process should be perceived by the public as equitable or fair. For example, water users should not impose uncompensated costs on other parties. Parties giving up water should be compensated, as should those injured by changes in points of diversion or return flows.

Finally, a socially responsible water allocation process must be capable of reflecting public values that may not be adequately considered by individual water users.

In most cases where water is a limiting factor in the countries mentioned above, water has been found to follow non-stationary processes. The rules of Howe et al. are thus insufficient to address the complexity involved by tight water allocation systems and non-stationary environments.

11.4.1 The Australian approach

There are three types of tradable water rights in Australia (National Water Commission, 2010). First is water access right: it gives an access right to a share of the water resources plan area granted under or by law of the state. Second, water delivery right enables the holder to have water delivered by a water corporation and a right to have a share of the available flow in a delivery system. Third is irrigation right, which is issued by an irrigation entity and granted from the entity's bulk water access entitlement.

Also, there are basically three types of water exchanges: trade of water access entitlements (permanent markets), trade of seasonal water allocation (temporary markets, spot markets and lease markets), and environmental water buybacks by the Government (see Table 11.3 by Rey et al. 2011).

11.4.2 The Spanish approach

The 1999 reform of the Spanish Water Law introduced the legal possibility of voluntary exchanges of public water concessions, mainly through two types of exchanges: voluntary agreements between two right-holders, and water banks. Although the trading activity was allowed

Table 11.3 Principal water exchanges and method of operation

Exchange	*Ownership*	*Regions serviced*	*Products traded*	*Method of operation*
Water move	Victorian Government (operated by Goulbum-Murray Water)	Vlc. And southern NSM	Entitlements and allocations	Weekly pool
Water exchange	National Stock Exchange of Australia	Vlc., NSW and SA	Allocations, forward contracts	Posted sell and buy bids
Murrumbidgee Water Exchange	High Security Irrigators Murrumbidgee	NSW	Entitlements and allocations	Posted sell and buy bids
Murray Irrigation Exchange	Murray Irrigation Ltd	NSW	Entitlements and allocations	Posted sell and buy bids

Source: Rey et al. 2011, based on NWC, 2010

since then, the regulation imposed several barriers that limit and hamper water exchanges (Garrido et al., 2012): legal barriers (related to water rights definition); market barriers (to avoid monopolistic behaviour, limits to the spatial extent of trading, the priority system established by law that only allows water exchanges between right-holders of the same or higher level in this priority range); institutional barriers (area-of-origin barriers, opposition to sell water out of the sector) and environmental barriers (to avoid negative effects on the environment).

Exchanges of water rights involving two water users in the same basin require the approval of the River Basin Agency, which has a two-month period to study the case and assess if the water exchange is going to have negative impacts and if all the needed conditions are met. To avoid the transfer of "paper water rights", only the average of the actual volume used by the seller in the last five years would be allowed to be exchanged. If, after two months, the water authority has not respond to the request, it would be understood as an approval of the exchange.

The 1999 reform of the Water Law established that exchanges involving different river basins should have the approval of the Ministry of Environment. From 2005 to 2008, Spain suffered a severe drought period. In order to improve the situation of the most affected river basins, a Royal Decree was approved in 2005 to facilitate water exchanges between users in different river basins. Other Royal Decrees were approved during this drought period with the aim of solving some emergency situations in different areas.

Droughts in Spain are recurrent phenomena, so it is important for the country to find all the available means to reduce its negative impacts. Some reforms of the water law, both in the national and in the regional level, have been developed in order to improve water management in Spain (see Garrido et al., 2012). However, after the previous drought ended in 2008, there has not been any subsequent round of intensive trading that affords an assessment of the water law amendments, which in short are the following (Garrido et al., 2013). A recent reform of the Spanish Water Law in May 2012 highlights the need to simplify and accelerate the administrative procedures, and adds more flexibility and efficiency to the water management system. Focusing on groundwater resources, it proposes several measures to deal with water availability problems, including the encouragement of the transformation of private water rights into public water concessions. The regional government of Andalusia passed a more advanced legislative act in 2010. This new Andalusian Water Law presents some new innovations, all targeted to make markets more flexible. This approach could hopefully serve as a precedent for future amendments for the market regulation in the rest of Spain.

11.4.3 The case of Chile

Chile has broadly-defined water rights, although there is confusion about the priority of consumptive (predominantly irrigation) and non-consumptive (hydropower) rights, because they have different impacts on water management and availability (Grafton et al., 2010). The main principles of the 1981 Water Code include: disjunction of land property and water rights, free water allocation without justification of needs, no obligation of effective and beneficial use, free availability and purchase of water, registered ownership, recognition of existing rights, recognition of certain minor uses, and legal safeguard for the water rights.

This Water Code does not specifically address third-party effects or environmental impacts. Externalities as a result of water trading, however, may not be a major problem due to the low volume of trade. There is federal legislation mandating provision of water for the environment and for public good purposes (Grafton et al., 2010).

11.5 Conclusions

Adaptation to future uncertain climate conditions is a major challenge. This uncertainty is especially relevant for water for food production given both sectors link to ecosystems, cities and culture.

Climate change comes in conjunction with high development pressure and increasing populations, and water management is often mis-adapted to local conditions.

Climate change science has made the crucial observation that reducing vulnerability and appropriately planning for the future does not require accurate predictions of the future. This is because the present time already provides situations and instability which provide the experimental setting for learning about the future.

It is likely that the stress imposed by climate change on water intensifies the regional disparities in rural areas and the overall rural economies (Alcamo et al., 2007; EEA, 2012; Iglesias et al., 2011b; Iglesias, 2013).

We recognise major limitations in our assessment of irrigation and water policy. The study of water has major limitations. The recent past has demonstrated the high sensitivity of water availability to changes in climate and the resulting effects on the social system. Adaptation planning is inherently challenging and often restricted by a number of factors, including limitations in the participatory processes with the stakeholders that will have to adapt in the future; the exhaustive data requirements for evaluating adaptive capacity; the problems related to selecting adequate evaluation methods and criteria; and difficulties in forecasting crop response processes or challenges in predicting the future adaptive capacity of the water system (Alcamo et al., 2007). Uncertainties in climate change science and long planning horizons add to the complexity of adaptation decision-making.

Aggregated values of adaptive capacity often do not reflect local priorities for adaptation and therefore limit the use of result for policy development.

Given increasing demands, the prevalence and sensitivity of many simple water management systems to fluctuations in precipitation and runoff, and the considerable time and expense required to implement many adaptation measures, the agriculture and water resources sectors in many areas and countries will remain vulnerable to climate variability.

The following policy conclusions can be drawn from the findings reported in this chapter and the authors' own work. First, impacts of unfavourable climate and weather developments can be combated with proper planning, but the most vulnerable farmers have a wide set of choices that can enhance their adaptation potential. Agricultural and water policies should develop effective communication means, so that training programmes improve learning processes. Second, it is essential that scientists convey clearer messages to governments, NGOs and farmers about what findings are relevant and robust. Very often, politicians are unclear about what works best, in some cases because policy officials are exposed to a large body of unprocessed information with unclear links to policy implementation. The science-policy milieu fails to deliver arguments and storylines that farmers are able to process and translate into adequate business plans. Third, policies should focus on the immediate known risks and hazards, and assessments should be made as to how effective they are, because the more we learn about the present, the better we will be prepared to cope with very uncertain future developments. Finally, societies must prepare for 'black swans': the unpredicted and unlikely situations with catastrophic consequences.

Acknowledgements

We acknowledge the support of the CIRCE and BASE projects of the 7th Framework Research Programme of the European Commission.

References

Adger, W.N., Vincent, K. (2005). Uncertainty in adaptive capacity. C.R. Geoscience 337, 399–410.

Alcamo, J., Floerke, M., Maerker, M. (2007). Future long-term changes in global water resources driven by socio-economic and climatic changes. Hydrological Sciences 52(2), 247–275.

EEA (2012). Climate change, impacts and vulnerability in Europe 2012. An indicator-based report. EEA Report No 12/2012, ISSN 1725-9177.

Fankhauser, J., Smith, B., Tol, R.S.J. (1999). Weathering climate change: Some simple rules to guide adaptation decisions. Ecological Economics 30(1), 67–78.

Garrido, A., Rey, D., Calatrava, J. (2012). Water trading in Spain. In: Lucia de Stefano L. and Llamas M.R. (Eds.). Water, agriculture and the environment in Spain: can we square the circle? Botín Foundation and Francis and Taylor.

Garrido, A., Rey, D. Calatrava, J. (2013) La flexibilización del régimen de concesiones y el mercado de aguas en los usos de regadío. In Embid Irujo, A. (Ed.) Usos del Agua Concesiones, Autorizaciones y Mercados del Agua. Thomson Reuters Aranzadi, Cizur Menor (Navarra), pp.177–196

Garrote, L. (2013). A model to evaluate water policy. Water Resources Management (manuscript submitted for publication).

Grafton, R.Q., Landry, C., Libecap, G.D., McGlennon, S., O'Brien, J.R. (2010). An integrated assessment of water markets: Australia, Chile, China, South Africa and the USA. National Bureau of Economic Research. Working Paper 16203.

Howe, C.H., Schurmeier, D.R., Shaw Jr., W.D. (1986). Innovative approaches to water allocation: The potential for water markets. Water Resour. Res., 22(4), 439–445.

Iglesias, A., Garrote, L., Diz, A., Schlickenrieder, J., Martin-Carrasco, F. (2011a). Rethinking water policy priorities in the Mediterranean Region in view of climate change. Environmental Science and Policy, Vol. 14, 744–757.

Iglesias, A., Quiroga, S., Diz, A. (2011b). Looking into the future of agriculture in a changing climate. European Review of Agricultural Economics, 38(3), 427–447.
Iglesias, A., Raoudha, M., Moneo, M., Quiroga, S. (2011c). Towards adaptation of agriculture to climate change in the Mediterranean. Regional Environmental Change, 111, 159–166. doi:10.1007/S10113-010-0187-4
Iglesias, A., (2013). Water and people. In: Navarra, A. and Tubiana, L. (Eds.) Regional assessment of climate change in the Mediterranean, volume 2: Agriculture, forests and ecosystems services and people. Springer, The Netherlands (ISBN10: 9400757719 ISBN13: 9789400757714).
Ionescu, C., Klein, R.J.T., Hinkel, J., Kumar, K.S.K., Klein, R. (2009). Towards a formal framework of vulnerability to climate change. Environmental Modeling and Assessment, 14(1), 1–16.
Kearney, J. (2010). Food consumption trends and drivers. Phil. Trans. R. Soc. B. 365, 2793–2807.
National Water Commission (NWC) (2010). Australian water markets report 2009–2010. Australian Government, National Water Commission.
Nordhaus, W.D. (2002). Modeling induced innovation in climate-change policy. In Grubler, A., Nakicenovic N. and Nordhaus W.D., Technological change and the environment. RFF Press, Washington DC.
Rey, D., Calatrava, J., Garrido, A. (2011). Water markets in Australia, Chile and the U.S.A. Water Cap & Trade. IWRM-Net. Working Document.
Smit, B., Skinner, M.W. (2002). Adaptation options in agriculture to climate change: A typology. Mitigation and Adaptation Strategies for Global Change 7, 85–114.
Sunding, D., Zilberman, D. (2001). Ch. 4: The agricultural innovation process: Research and technology adoption in a changing agricultural sector. Handbook of Agricultural Economics. Volume 1, Part 1, 207–226.
Vorosmarty, C.J., McIntyre, P.B., Gessner, M.O., Dudgeon, D., Prusevich, A., Green, P., Glidden, S., Bunn, S.E., Sullivan, C.A., Liermann, C.R., Davies, P.M. (2010). Global threats to human water security and river biodiversity, Nature, 467(7315), 555–561.
Yohe, G., Malone, E., Brenkert, A., Schlesinger, M., Meij, H., Xing, X. (2006). Global distributions of vulnerability to climate change. Integrated Assessment Journal, 6(3), 35–44.

12
ADAPTATION IN AGRICULTURE

Doan Nainggolan,[1,2] *Mette Termansen and Marianne Zandersen*

[1]AFFILIATION FOR ALL AUTHORS: DEPARTMENT OF ENVIRONMENTAL SCIENCE (ENVIRONMENTAL SOCIAL SCIENCE GROUP), AARHUS UNIVERSITY, FREDERIKSBORGVEJ 399, 4000 ROSKILDE, DENMARK

[2]CORRESPONDING AUTHOR: DNA@DMU.DK

12.1 Introduction

The importance of agricultural production to supply food and fibers for humanity at different spatial and temporal scales cannot be emphasized enough with the view of expected population growth and increasing per-capita demand (e.g., Tubiello et al., 2007; Howden et al., 2007). Agricultural production has been shaped by a wide array of factors ranging from socio-economic to biophysical and climatic conditions. As a result, climate change unavoidably has important consequences on agricultural production (e.g., Schlenker and Roberts, 2009). Climate change presents both physical and socioeconomic impacts on agriculture (FAO, 2007). Numerous studies to model the impacts of climate change on agriculture have been undertaken covering various spatial extents from global to regional to specific countries (e.g., Adams et al., 1990; Rosenzweig and Parry, 1994; Reilly et al., 2003; Fischer et al., 2005; Piao et al., 2010). These studies have concluded large spatial disparities of impacts. In addition, the role of adaptation in determining the ultimate impacts of climate change on agriculture has been increasingly acknowledged (e.g., Rosenberg, 1992; Easterling, 1996; Mendelsohn and Dinar, 1999; Bryant et al., 2000; Burton and Lim, 2005; Howden et al., 2007).

This chapter reports on the present state of knowledge regarding agricultural adaptation to climate change with illustrations focusing largely on the European context. It is intended to serve as an overview of the subject matter. An exhaustive synthesis on adaptation in the agricultural sector at the global scale is however beyond the scope of this chapter. Instead, the discussions and illustrations presented in this chapter are focused on agricultural production particularly in relation to crops and livestock. Following the introduction, the chapter is structured into five main parts. The first section summarizes the likely impacts of climate change on agriculture, which subsequently provides a basis for identifying the different kinds of adaptations, discussed in the second section, that would be of particular importance for agriculture. The third section of the chapter focuses on the economics of agricultural adaptation to climate change and illustrates situations where and in which way public/government intervention is justified from a economic perspective. The fourth section of the chapter briefly discusses policy implications and the particular challenges in relation to adaptation to climate change in the agricultural sector. The final section concludes the chapter and highlights research areas that remain poorly understood yet critical for better informing adaptation action.

12.2 Adapting to what?

Gaining insights into what future climate may bring makes a useful starting point for exploring and discussing agricultural adaptation in response to changing climate. There appears to be a growing consensus in the scientific literature, that for Europe, agriculture will be affected by different climatic drivers: variability and changes in temperature and precipitation trends, increased atmospheric CO_2 fertilization, altered growing length, and extreme events or anomalies such as droughts, heat waves, floods and storms. This has been highlighted by various researchers based on reviews of literature, numerical modeling, or a combination of the two (e.g., Alcamo et al., 2007; Ciscar et al., 2011; Iglesias et al., 2012; Mechler et al., 2010a).

The literature suggests that future climate change will lead to non-uniform consequences across Europe, both due to variations in current sensitivities and due to variations in the magnitude of the changes. Available climate projections indicate temperature increases across Europe greater than the global mean with a little higher increase projected for northern Europe (Christensen et al., 2007; Alcamo et al., 2007; Schroter et al., 2005). IPCC AR4 reported projected temperature increase within a range of 2.2 to 5.3 °C (Christensen et al., 2007). An average temperature ranging from 2.5 °C to 5.4 °C has been considered for evaluating the impacts of future climate change on various economic sectors in Europe (Ciscar et al., 2011). While northern Europe is projected to experience the largest warming in winter, a contrast is likely to happen for Mediterranean region with the largest warming being in summer. Precipitation is projected to increase in the northern parts of Europe while the opposite is anticipated for the southern parts (Christensen et al., 2007; Alcamo et al., 2007; Schroter et al., 2005). Across most parts of Europe, variability in temperature is expected to increase in summer and decrease in winter (Christensen et al., 2007). Variability in precipitation from season to season and across regions has been projected (Alcamo et al., 2007). Despite the acknowledged uncertainty in the projection of extreme events, it is very likely that events such as extreme precipitation and floods, heat waves, droughts, and storms will become more frequent and intense and at increased duration in the future (Mechler et al., 2010b; Bindi and Olesen, 2011). While heat waves and droughts may become more of a concern in the south (especially over the Mediterranean region), more intense rainfalls and storms potentially pose a significant problem in the north. Of course, the ranges of projected values of the different climatic variables depend on the range of socioeconomic scenarios that drive future global anthropogenic GHG emissions as discussed in the fourth assessment report of IPCC (IPCC, 2007). In addition, although significant improvements have been made in climate change projection, the spatial resolution of existing projections is still too low to allow meaningful assessments of the consequences of climate change at specific locations. Nevertheless, it is clear that agricultural production in Europe will have to adapt to multiple manifestations of future climatic stimuli from changes in trends to climate variability as well as the extremes.

For agriculture in Europe, future climate change presents both opportunities and risk rendering a contrast between the north and the south (Alcamo et al., 2007; Iglesias et al., 2012; Olesen and Bindi, 2002; Schroter et al., 2005). Adaptation in agriculture in Europe, which will be further discussed in the next section, therefore means capitalizing on the opportunities while managing the risks. Based on a review of an extensive body of literature, Iglesias et al. (2012)

provided a comprehensive summary of the positive and negative implications of future climate change for European agriculture (Table 12.1). The northern parts of Europe are expected to experience net positive impacts as a combined consequences of enhanced atmospheric CO_2 fertilization, extended growing period of some crops, and expanded areas climatically suitable for crop production triggered by the projected warming. Some crops that are currently doing well only in the southern part are likely to become an attractive choice for farming in the north. There has been a warning, however, that beyond a certain threshold, increase in temperature may prove disadvantageous (see, for example, Ciscar et al., 2009). Besides, a warming climate could trigger significant threat from pest and disease infestations, a risk that is likely to have an effect all over Europe (Iglesias et al., 2012). Nonetheless, in southern Europe, particularly across the Mediterranean areas, the impacts will be largely negative. Areas suitable for crop cultivation will potentially shrink and crop yield may become lower with higher variability due to increased water shortage and extreme events (especially heat waves and droughts) (Olesen and Bindi, 2002). An assessment by Ciscar et al. (2011), linking biophysical and statistical models, provided quantitative estimates on the physical annual impacts of the 2080s climate and highlighted the regional disparities across Europe. To illustrate, their study showed that under the 2080s climate, crop yield in northern Europe can increase by just over 50%, while on the contrary southern Europe may experience a reduction of almost 30%. Results based on crop models by Iglesias et al. (2012) for the 2071 to 2100 climate indicate similar trends. For example, their findings showed that under the A2 scenario, crop yield in the Mediterranean region can suffer from a negative change of between 8 to 27%, as opposed to a positive change in yield in the Boreal region of between 41 to 54%.

Interestingly, quantitative assessments on the likely European-wide impacts of future climate change on livestock farming are not yet available. Existing scientific reviews on the subject matter only provide qualitative information, and this has not been the main emphasis of the assessments (see, for example, Olesen and Bindi, 2002; Bindi and Olesen, 2011; Iglesias et al., 2012). Nonetheless, as in the case of crop production, future climate change is likely to bring both beneficial as well as undesirable consequences for livestock farming directly on the animal physiological state and indirectly through the alteration of livestock farming conditions and feed supply (Maracchi et al., 2005; King et al., 2006; Moran et al., 2009; Hall and Wreford, 2012). For instance, a warmer climate would beneficially reduce energy consumption for winter heating, but on the other hand, extreme increase in temperature can cause too much stress for livestock, which may lead to higher mortality rates. As increased temperature is projected to induce dramatic spatial shifts in crop cultivation; this may trigger changes in the share of agricultural land devoted for grass production, which in turn affects grazing animal production (e.g., Bateman et al., 2011).

While many authors state that the impacts of extreme events are likely to be significant in agriculture, it is noted that there is a shortage of empirical analyses attempting to quantify these impacts. A study by Moriondo et al. (2010), however, is worth mentioning. It showed that, under a 2 °C scenario, heat and drought stresses could lead to reduction in summer crop yields in southern Europe although the magnitude of the impacts may vary between 5 to 13% reduction depending on crop types. In conclusion, the contrasting circumstances brought about by future climatic stimuli across different regions in Europe may reinforce the trends of agricultural extensification and land abandonment in the south versus intensification in the north (Olesen and Bindi, 2002; Bindi and Olesen, 2011).

Table 12.1 Regional disparities on the likely consequences of climate change for agriculture in Europe

Region	*Main risks*	*Key opportunities*	*Limiting factors or knock-on effects*
Boreal	Waterlogging, floods and decreased water quality due to increased precipitation	Expansion of areas suitable for crop cultivation, increase in crop yields and livestock productivity as a consequence of warmer climate and good supply of water	The potential for increase in agricultural production may be constrained by local environmental conditions such as soil types
Alpine	Loss of glaciers and alteration of permafrost, changes in precipitation pattern and increased frequency of extreme events	Increased crop and livestock production due to longer growing season and greater range of crops that can be grown	Secure supply of water is essential for realizing the opportunities, especially during critical period of crop growth; the realisation of the benefits may also be constrained by soil conditions
Atlantic North	Drier summers restrict growth of crops and forage	Increase in agricultural production particularly livestock farming	Soil types as a limiting factor for the realization of production increase in agriculture
Atlantic Central	Flooding triggered by sea level rising	Increase in agricultural production resulting from cereal yield increase and the region being climatically suitable for new crops	Level of crop yields will be contingent on water availability
Atlantic South	Crop yield decrease	Cultivation of heat and drought tolerant crops	Water use conflict; the extent to which opportunities can be exploited still needs thorough assessment
Continental North	Flooding	Increase in crop and livestock production with the former particularly supported by longer growing season and increase in range of crops that can be grown	The realization of opportunities may be limited by infertile soils and by water availability especially in summer
Continental South	Crop yield reduction as a consequence of hotter, drier summers	Encouraging the cultivation of non or less vulnerable crops, such as soya	Ensuring water availability is crucial for maintaining agricultural production
Mediterranean	Decreased total precipitation, drought and water scarcity, soil and water quality reduction, and desertification which altogether potentially cause significant shrink in the extent of areas suitable for agriculture and reduction in agricultural productivity	None	Agricultural land abandonment when farming becomes no longer viable

Adapted from Iglesias et al., 2012

12.3 What are the options?

Adaptation forms a critical response to what future climate will bring for agriculture – both the good and the bad. The importance of agricultural adaptation to climate change has been widely highlighted in the literature and accordingly several typologies of adaptation in general, which are also applicable in relation to agriculture in particular, have been proposed (e.g., Smit and Skinner, 2002; Olesen and Bindi, 2002; Füssel, 2007; Howden et al., 2007; Iglesias et al., 2012). Adaptation can be autonomous or planned, private or public, tactical or strategic, proactive (anticipatory) or reactive (responsive) or on-going (concurrent). It can encompass both short-term adjustments and long-term changes and investments. In practice, however, whether to classify a given measure as being an adaptation of one particular type or another is not always straight forward (Smit and Skinner, 2002). Furthermore, it can prove a challenging task to examine agricultural adaptation to climate change in isolation from the influence of other factors that drive on-going and future agricultural change, including environmental and socioeconomic factors and policy drivers.

In the case of agriculture, it is particularly important to distinguish adaptation options along two dimensions: individual farm level (private) versus public adaptation and reactive (on-going) versus anticipatory adaptation. To dismiss farmers' capacity to adapt is unwise, but to think on the other hand that farmers have perfect knowledge and all information at their disposal for their farm decision-making and to always act accordingly may also be unrealistic. It is acknowledged that when it comes to adaptation, farmers are neither completely ignorant nor perfectly knowledgeable (Tol et al., 1998). Moreover, different farm types respond differently to changes, including climate change, and the farm type heterogeneity highly determines the extent of the likely aggregate impacts at the regional level (Reidsma et al., 2010, Nainggolan et al., 2013). The significant role of farm-level adaptation within the overall picture of European agriculture responding to the likely consequences of climate change should not be underestimated (Reidsma et al., 2010). However, the real analytical challenge lies in selecting an approach that best models how farmers behave in adapting to an altered environment (Hanemann, 2000).

Available studies indicate that, in the simplest form, farm-level adaptation can autonomously take place in the form of short-term modifications through mostly tactical measures, such as adjusting sowing date and altering inputs (e.g., fertilizer use, water use, animal feeding mix) and through longer-term farm-scale reorganization by altering allocation of land across different types of use (e.g., crops versus forage production) (e.g., Leclère et al., 2013; Olesen and Bindi, 2002). However, in some cases, agricultural adaptation to climate change may demand measures or strategies beyond the short-term modification of farm management practices. Agricultural adaptation essentially centers on how farmers or land managers are able to respond effectively and efficiently given the knowledge and resources at their disposal and in interaction with external measures or supports, particularly at the policy level (extension and advice services, research, financial services and public/government incentives or supports). As climate change is expected to bring various beneficial and adverse consequences for European agriculture with noted regional disparities, a multitude of adaptation options need to be pursued accordingly.

A range of options for agricultural adaptation in Europe in response to risks and opportunities presented by climate change can be classified according to whether they require farm-level action versus public intervention (Table 12.2). Changes in future climate that will be favorable for agricultural expansion and productivity are expected for the northern region. In order to capitalize on these opportunities, farm-scale adaptation may combine tactical adjustments (such as on the timing of farm operations and the application of external inputs) and strategic decision-making (such as alteration of land allocation portfolio, crop diversification, and land use change). Beyond the farm level, important external actions and supports include particularly biotechnological breakthroughs for developing crops and varieties with potential higher productivity under future climate and dissemination of agricultural and climatic information and advice to farmers from relevant government agencies. The role of government regulatory bodies will be important in order to consistently evaluate and control the negative impacts of likely spatial expansion and increased intensity of agricultural production triggered by favorable climatic conditions. Beyond-farm adaptation is equally important when addressing the potential risks presented and augmented by climate change also in the northern region. This includes the development of a warning system, spatial zoning according to degree of vulnerability, and physical infrastructure construction in order to adapt to increased risks of sea water intrusion to and inundation of low lying agricultural lands and of natural hazards such as waterlogging and flood due to increased precipitation, which are likely to affect various parts of Europe, especially the northern region. Of course these efforts will need to work hand-in-hand with farmers or land managers' actions on site, for instance, through improvement of drainage systems and soil and floodplain management.

In many parts of Europe, particularly in the southern region such as in the Mediterranean, agricultural adaption to future climate may become more challenging as negative impacts of climate change are expected to be dominant. Accordingly, implementation of multiple adaptation options at the farm level and beyond will become a necessity (Table 12.2). One of the main concerns regarding the implications of future climate change in the southern part of Europe is the projection that more and more area will become unsuitable for agriculture, and that crop harvest will decline significantly. At the farm level, adapting to these risks will require not only tactical agronomic measures and strategic land allocation portfolio and land use change but also sound financial management, including purchasing crop insurance and making efforts to diversify household income sources by taking on board a wider range of livelihood activities. To this end, beyond-farm supports will be essential, particularly in the form of government intervention, to promote and facilitate diversification of farmers' income sources and to aid and encourage the launch of technological solutions, such as through the development of crops that are resilient to changing climate.

Another likely undesirable implication of climate change for agriculture in Europe manifests through escalated problems of soil degradation and water resources scarcity (both in terms of quantity and quality); the latter can be closely linked to anticipated increased drought under future climate. The challenge will become far more critical in parts of Europe where environmental issues such as land degradation and discrepancy between demand and supply of water for agricultural irrigation have been a long-term and on-going battle. To address these challenges, adoption of technical measures for better water and land resource use and management and conservation, cultivation of crops that need less water, and identification of new locations for agricultural production will form the key adaptation options at the farm level (Table 12.2). Beyond the farm level, in addition to facilitating research and the development of climate resilient crops and provision of information and advice, there is scope for government to influence and to encourage widespread implementation of technologies for more efficient use and conservation of water and land resources, such as the introduction of a water pricing mechanism within a particular catchment.

Table 12.2 Adaptation options for European agricultural production

Main risks (RS)/ opportunities (OP)	*(Farm-level) adaptation options*	*Adaptation options/measures/supports needed beyond farm scale*
Expansive spatial shifts in climatic suitability for crop choice and cultivation in the north (OP)	Altering portfolio of land allocation across different crops; changing land use; altering cultivation practices; diversifying crops; introducing new crops and varieties	Innovation – technological and biotechnological advancement – including development of new, more productive crop varieties; monitor and control unintended aggregate consequences of farm-scale change in production patterns; provision of information and advice (e.g., through extension services)
Climate regime that potentially favours increase in crop yields and livestock productivity (OP)	Adjusting sowing and planting dates; adjusting time of farm operations; altering the use of external inputs (e.g., fertilizer application in the case of crop production); expanding livestock farming to new areas; increasing stocking rate	Innovation – technological and biotechnological advancement – including development of new, more productive animal breeds; monitor and control unintended aggregate consequences of farm-scale change in production patterns; provision of information and advice (e.g., through extension services)
Increased hazards associated with increased precipitation (e.g., waterlogging, floods) (RS)	Improving drainage systems; improving soil physical properties management ; reducing grazing pressure; enhancing flood plain management; restoring/creating wetlands	Zoning system; integrated catchment management; development of early warning system; installation of hard defenses; encourage farmers to become 'custodian' of floodplains (e.g., through reward system)
Intrusion and inundation of agricultural lands due to sea level rising (RS)	Improving drainage systems; substituting crops; changing location of production from vulnerable areas	Zoning system, development of early warning system; installation of hard defences
Increased pest, disease, and weed problems (RS)	Integrated control of pest, disease and weed; livestock vaccination; introduction of pest-resistant species	Innovation – technological and biotechnological advancement – including development of pest-resistant varieties; provision of information and advice (e.g., through extension services)
Intensified drought and water scarcity problems due to decreased total precipitation (RS)	Implementing water conservation measures; improving irrigation efficiency; improving water allocation and distribution; changing location of production; introduction of drought-tolerant or less water intensive crops and varieties	Innovation – technological and biotechnological advancement – including development of climate-resilient varieties; provision of information and advice (e.g., through extension services); encourage efficient and water-saving practices and technologies through a mixture of approaches (e.g., technological investment support, regulation, water pricing, etc.)

(Continued)

Table 12.2 (Continued)

Main risks (RS)/ opportunities (OP)	*(Farm-level) adaptation options*	*Adaptation options/measures/supports needed beyond farm scale*
Crop yield decrease (RS)	Altering portfolio of land allocation across different crops; altering cultivation practices; diversifying crops; altering the use of external inputs (e.g., fertilizer application in the case of crop production); changing land use; changing farming system; introducing new crops and varieties; farm financial management especially through purchase of crop insurance and investment in crop shares and futures	Innovation – technological and biotechnological advancement – including development of more productive crop varieties; provision of information and advice (e.g., through extension services)
Deterioration of livestock conditions (RS)	Introducing more heat-tolerant species/breeds; adjusting time for different operations and breeding; altering grazing arrangement; altering pasture composition; complementing grazing with supplemental feeding; increasing shelter and heat protection	Innovation – technological and biotechnological advancement – especially development of climate-resilient varieties; provision of information and advice (e.g., through extension services)
Certain crops become unsuitable under the new climate regime (RS)	Substituting with different varieties or cultivars	Innovation – technological and biotechnological advancement – especially development of climate resilient varieties; provision of information and advice (e.g., through extension services)
Contraction of areas suitable for agriculture in the south (RS)	Changing land use; diversifying household income source	Programs to promote and facilitate livelihood diversification; provision of information and advice (e.g., through extension services)
Water quality deterioration (RS)	Minimizing nutrient leaching; increasing fertilization efficiency; aerating ploughing equipment	Innovation – technological and biotechnological advancement – including development of highly efficient fertilizers; provision of information and advice (e.g., through extension services)
Soil quality degradation and desertification (RS)	Implementing soil conservation and remediation measures	Zoning system; provision of information about potential and tested soil conservation measures; financial support to stimulate farm adoption of measures that demand high up-front investment

(Continued)

Table 12.2 (Continued)

Main risks (RS)/ opportunities (OP)	*(Farm-level) adaptation options*	*Adaptation options/measures/supports needed beyond farm scale*
Increased frequency, magnitude, and duration of extreme events with greater risk of production loss (RS)	Changing location of production from vulnerable areas; taking on board a wide range of financial management measures including crop insurance, investment in crop shares and futures and diversification of household income sources	Development of early warning system; provision of information and advice (e.g., through extension services); solidarity fund; appropriate compensation and assistance programs; promoting effective and efficient insurance scheme

Adapted from Iglesias et al., 2012; Smit and Skinner, 2002; Olesen and Bindi, 2002

Finally, climatic extreme events in the future across Europe are expected to occur more frequently at far greater magnitude and longer durations. In order to address the risk of agricultural production loss, there seems to be very limited technical and agronomic adaptation options that can be pursued (Table 12.2). Instead, adopting measures that can help reduce the negative financial and livelihood impacts on individual farmers, for example through crop insurance and investment in crop shares and futures, will be crucial at the farm level. To this end, government support to promote effective and efficient crop insurance schemes and to develop compensation and assistance programs that will not undermine insurance schemes will be important to help farmers deal with the likely adverse farm-level financial challenges triggered by climate change.

12.4 On the economics of adaptation to climate change in agriculture

The previous sections have given an overview of the likely climate impacts on agriculture and adaptation measures that could be considered to reduce the negative effects on the potential benefits for agricultural production. In order to understand the desirability of alternative adaptation choices, and the policies that might support or hinder their implementation, it is useful to analyze climate adaptation from an economic perspective.

Climate change is an international negative externality as a result of global economic development, fuelled by fossil fuels, primarily. As climate change impacts are global, all producing sectors are exposed to related environmental risks and subsequent damage costs along with welfare losses for consumers. Following conventional theory of externalities, environmental economics seek to identify the optimal level of externality to which the agricultural sector would need to adapt (Hanley et al., 2007). However, the interdependency between mitigation and adaptation where more emission reductions make less adaptation necessary, and vice versa, also needs to be incorporated into estimating and analyzing the optimal level of greenhouse gas emissions. On the other hand, adaptation might itself increase emissions, leading to additional pressure on future adaptation action. Understanding the economic arguments on climate change action rests on the key difference between mitigation and adaptation action, as mitigating climate changes can be understood as a public good whereas investments in adaptation has both private and public good characteristics.

The economics of adaptation in agriculture has mainly been studied through the attempts to quantify the costs of climate change aimed primarily at assessing the desirability of policies to reduce greenhouse gases. Early studies on the impacts of climate change on agriculture have followed the production-function approach (e.g., Rind et al. 1990; Cline, 1992), where impacts of climate change are estimated based on the effects of varying one or a few input variables, typically relying on calibrated crop-yield models. Often these studies have assumed that the existing pattern of land use would remain unchanged, maybe assuming some adjustments such as changes in fertilizer application or irrigation, but none assuming full adjustment to changing environmental conditions by the farmer (Mendelsohn et al., 1994). As a result, this type of study has predicted severe yield reductions due to climate change, representing an upper-bound estimate of the cost of damages from climate changes on agriculture.

Mendelsohn et al. (1994) were the first to suggest an alternative approach, called a 'Ricardian' or hedonic approach, based on comparative static estimates along a climate gradient on the change in equilibrium rents to land associated with a one-time change in mean climatic conditions. They estimate that for a 2.8 °C increase in mean temperature, changes in farmland rents in the United States would range from a 4.9% loss to a 1.2% gain. The production function approach for comparable crops found approximately three times higher damage costs compared to the Ricardian approach (Mendelsohn et al., 1994). This study was among the first to, controversially at the time, estimate positive impacts of climate change on agriculture.

Critiques find the Ricardian approach to be a lower-bound estimate of climate change damage costs on agriculture as the approach implicitly assumes zero adaptation costs, which is unlikely, as adjustments are unlikely to be instant and may involve investments (Quiggin and Horowitz, 1999). A further limitation of the Ricardian approach is that it is a deterministic approach where climate change has mainly been represented as changes in mean temperature and precipitation. This has reduced the emphasis on increase in extreme events such as droughts and extreme precipitation events in the assessment of the costs and benefits of climate change and the scope for adaptation to generate net benefits.

Despite critiques of both the production function and the Ricardian approach, the studies have highlighted the important role of autonomous adaptation in agriculture and the risk of biased estimates if it is not taken into account.

The optimal level of adaptation in agriculture is therefore critically dependent on (i) the risks posed by climate change determined by past and current mitigation efforts beyond the level of farm-level decision-making, (ii) realization of future climate and associated damage costs only partially known to the farmer and (iii) the costs and effect of adaptation (Figure 12.1). In simple terms, this illustrates that there are different sources of uncertainty in determining the optimal level of adaptation. One source is due to the link between mitigation and adaptation; the other source due to the uncertainty about the future climate, impacting the damage costs. Under a low-emission future, realizing a lower-bound damage on agricultural production due to climate change, adaptation at the level of $\boldsymbol{a}$ (Figure 12.1) would be optimal as the marginal reductions in damage costs (i.e., the marginal benefits of adaptation) is equal to the marginal costs of adaptation. At an upper-bound damage under the low-emission case, $\boldsymbol{a_2}$ would be the optimal level of adaptation, and the difference between $\boldsymbol{a}$ and $\boldsymbol{a_2}$ would indicate the range in adaptation that could be optimal given current knowledge on the agricultural impacts of climate change. In the same way, for the high-emission case, the range between $\boldsymbol{b}$ and $\boldsymbol{b_2}$ would represent candidates for an optimal adaptation level. The difference between $\boldsymbol{a}$ and $\boldsymbol{b}$ indicates the influence on adaptation choices of mitigation outcomes, given knowledge about the future

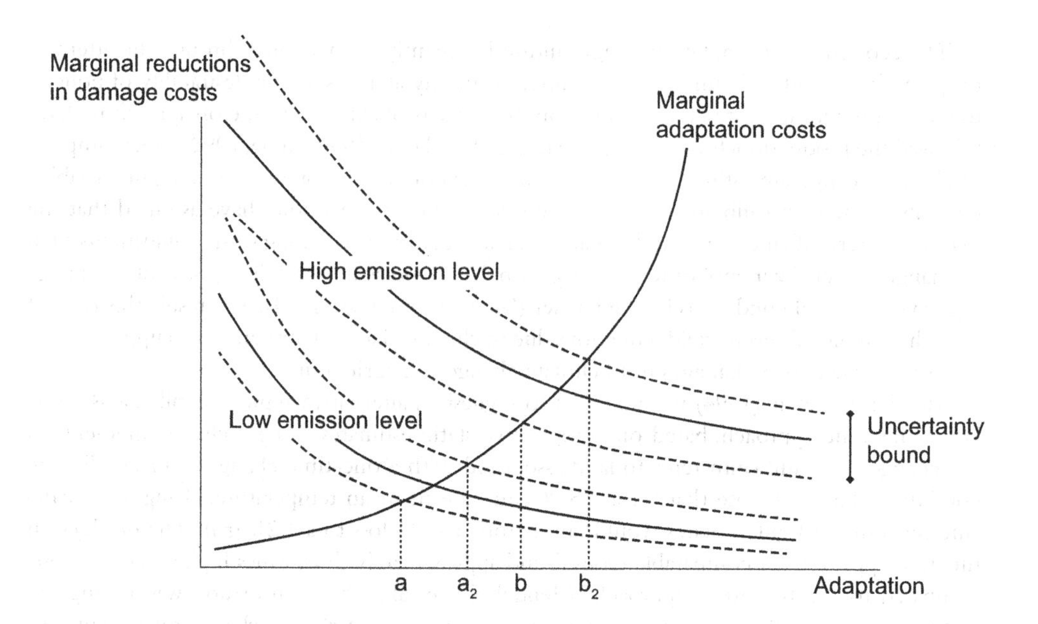

Figure 12.1 Optimal level of adaptation given future climatic outcome and level of mitigation effort

climate. It should be noted, however, that under a very severe or extreme situation, the optimal abatement might be to abandon agricultural production because the cost of adapting becomes too high or due to the limit to what can be done to adapt to such a situation. It should also be noted that measuring adaptation is not as simple as measuring mitigation. Mitigation can be measured as reductions in CO_2 equivalents, whereas adaptation cannot be measured using a single physical unit.

The preceding arguments highlight the importance of the uncertainty about future climate outcomes. In agriculture, however, inter-annual variability and the risk of droughts and water logging are of particular importance, even when the outcomes would not be classified as extreme events. In this sense, climate change presents an increase in risks to farmers. As farmers can be assumed to be risk-averse (Binswanger and Sillers, 1983), climate change poses a negative impact even when the mean yields are not affected. It should be noted that the arguments outlined above are illustrated for cases where climate change without adaptation results in loss of net benefits. This is not the case in all regions (see the section below on costs and benefits of adaptation). This does not, however, invalidate the arguments, but reference to damage costs would not be correct.

12.4.1 Estimates of costs and benefits of agricultural adaptation

Farm-level agricultural production impacts have been fairly extensively studied in the climate change adaptation literature. There are a large number of studies that assess the benefits of adaptation for different crops and regions. Easterling et al. (2007) provide a synthesis of 69 published studies on the impacts of climate change on crop yields for maize, wheat and rice. The adaptation benefit is measured as the difference between agricultural yield with and without adaptation from no climate change to 5 degrees warming. The studies show a large variation between regions and between studies within the same region, but adaptation does provide statistically significant benefits, although the benefits vary across regions, crops and level of climate change. On average, the temperate regions show net benefits of climate change allowing for adaptation, whereas the tropical regions on average are expected to see a net loss for climate change levels above 2–3 °C warming. The literature on the cost of adaptation in different regions is less extensive. This is suggested to be due to the fact that most studies have focused on farm-level adaptation where the costs are small (Agrawala et al., 2008). Public initiatives to support adaptation have not been assessed in terms of benefits and costs to the same extent. It has been suggested that adaptation to climate change for agriculture, forestry and fisheries will cost between $11.3 billion (with mitigation scenario) and $12.6 billion (without mitigation scenario) USD globally per year in 2030 (McCarl, 2007). The study estimates three types of costs: physical infrastructure such as irrigation, crop breeding of, for example, drought resistant varieties and agricultural advisory services. The costs are based on fairly crude assumptions of a percentage increase in current costs. This has been criticized due to the lack of analysis or thorough justification for setting the figures for the assumed additional funding requirement (Agrawala et al., 2008; Wheeler and Tiffin, 2009). Comparing McCarl's estimate above with those from existing bottom-up assessments proves problematic as the latter has focused exclusively on specific individual measures (Wheeler and Tiffin, 2009). Despite the aforementioned challenge, Wheeler and Tiffin (2009) suggest that the expected cost of climate change adaptation for agriculture globally is likely to be higher than the estimate provided by McCarl (2007). Benefit estimates on such public investment seems to be lacking.

In the context of European agriculture, Iglesias et al. (2012) provided qualitative evaluation of potential costs and benefits of a range of adaptation measures as low, medium, and high. By transforming these qualitative values into 1 (low), 2 (medium) and 3 (high), they carried out a simplistic quantification of effort/benefit ratio where the effort component takes into account average of time scale, technical difficulty and potential costs. It offers an important starting point for new studies to start quantifying benefits and costs of climate adaptation in European agriculture.

12.4.2 The economic rationale for government intervention in agricultural adaptation

The conventional economic rationale for government intervention in private decision-making is the case of market failure. This rationale applies to the challenge of adaptation to climate change. In particular, governmental adaptation measures can be economically justified if they take the form of public goods or social insurance. In the context of agriculture, the adaptation measures outlined in Table 12.2 fall under three forms of market failure. First, knowledge or information is a public good and governments have a role in providing investment in research and education to enable efficient adaptation. This could take the form of research into, for example, drought-resistant crops and investment in agricultural extension services to provide farmers with the best available information on the risks and the adaptation options available. The second form of market failure relates to government's role in developing insurance markets to allow individual risk-averse farmers to trade their risk with a risk neutral insurance provider.

This generates a welfare gain and can therefore be justified in economic terms (Dannenberg et al., 2009). The well-known problems associated with moral hazard (Arrow, 1963) and adverse selection (Rothschild and Stiglitz, 1976) are equally relevant in the context of insurance as an adaptation measure. The third form of market failure concerns the network externality associated with the provision of irrigation infrastructure and the security of the supply of water. A further public good aspect related to agriculture and climate change relates to the security of food supply. On conventional economic grounds, food is a private good and efficient operation of markets should ensure that there would be no role for government intervention. However, severe climate events may put basic human needs at risk, justifying arguments for intervention beyond welfare economic rationales (Dannenberg et al., 2009).

12.5 The broader policy context

Agricultural adaptation to climate change should not be viewed in isolation but rather in relation to other factors that are equally important in shaping the future of agriculture. Some policies can promote effective adaptation while others may undermine adaptation (Urwin and Jordan, 2008). The key challenge lies on the need to mainstream or integrate adaptation into policy design while making sure it does not trigger negative feedback to the environment in general and to the climate change mitigation goal in particular. Addressing such challenge necessitates a combination of evaluating existing policies and addressing climate impacts of the new ones so that they support rather than impede adaptation (Urwin and Jordan, 2008).

In the context of agriculture in Europe, some argue that there is no need to launch new, separate policies to facilitate adaptation. Instead, efforts should be geared towards strengthening policies that promote adaptation and modifying ones that undermine it (Iglesias et al., 2012). The mainstreaming or integration of adaptation strategies into agro-environmental schemes and relevant directives, especially the Common Agricultural Policy (CAP) and the Water Framework Directives, has been highlighted in the literature (Olesen and Bindi, 2002; Iglesias et al., 2012). Empirical analysis explicitly assessing the impacts of different policy initiatives in relation to agricultural adaptation in general and in the European context in particular is still lacking. Nonetheless, there has been a call for not only focusing on reforming existing relevant policies into less rigid structures but also for the need to develop policies that recognize the differential potential impacts of climate change across European regions and that promote high flexibility in agricultural production systems (e.g., in terms of land use) (Olesen and Bindi, 2002). Policy developments need to seriously consider the heterogeneity of farms in terms of diversity in spatial-biophysical and socioeconomic characteristics and the likely diverse farm-level responses to future climate and climate policies (Reidsma et al., 2007; Nainggolan et al., 2013). Finally, it is important to acknowledge that the ultimate challenge for policy implementation for European agriculture goes beyond the urgency to facilitate agricultural adaptation to climate change in terms of maintaining and/or increasing production. The challenge further includes the need to sustain the multifunctional role of agriculture encompassing a diverse suit of environmental, economic and social functions for the benefits of the society as a whole.

12.6 Conclusions

Existing studies have consistently concluded that for Europe, climate change will bring both desirable and undesirable consequences, which geographically vary – with the northern region expected to experience net benefits. To respond to the spatial variation in the implication of climate change, the adoption of a wide range of adaptation options will necessarily form the strategy to capitalize on the opportunities on the one hand and to deal with the negative consequences of changing climate on the other. At the farm level, some adaptation options are agronomic and technical in nature while others are focused more on managerial and financial aspects. While to a certain degree farm-level adaptation can be implemented autonomously, in many other situations, initiatives and measures at higher levels are crucial to support and incentivize farm-level action. Research and advisory services are needed to support and to help inform adaptation not only in tackling risks but also in realizing benefits. For example, biotechnological breakthroughs in developing crop and livestock varieties that will be compatible with and more productive under future climatic conditions are essential in aiding farmers to adapt. In addition, there is an important scope for government supports and interventions at different levels (regional, national and supranational) to influence farm-level decision-making on adaptation and to take appropriate measures to minimize detrimental ancillary effects, particularly the ones due to likely increased intensification of agricultural production in response to opportunities. On another level, it is important to highlight that farm responses to certain risks are very likely to demand cooperation with sectors beyond agriculture, for example in addressing the higher risk of adverse events such as flooding triggered by projected increased rainfall for some areas. Adaptation options aimed to financially safeguard farmers from the risk of agricultural income loss are likely becoming more and more important in view of future climate change, especially with the projected increase in the return frequency and magnitude of extreme events. The scope for technical adaptation options for agriculture in the face of extreme climates seem to be limited.

Research suggests that autonomous adaptation can provide significant benefits for adaptation in agriculture, although not evenly distributed across regions. Economic theory provides guidance on the types of policy instruments that should be considered, to complement autonomous adaptation, as climate change becomes an increasingly pressing issue in agriculture. In particular, investments in research and education and enabling the operation of efficient insurance schemes seem to be particularly relevant policy instruments in the context of agriculture.

Despite the growing body of literature on adaptation in the agricultural context in general and in Europe in particular, important knowledge gaps on various facets of the subject matter remain. Among the pressing agenda for future research in relation to climate change adaptation in agriculture are:

1. An adequate picture regarding future climate forms an important input to inform adaptation. However, the resolution of future climate projections from existing models is still quite coarse. In order for climate model outputs to become highly useful in developing adaptation strategies at the scale where agricultural production decision-making operates, there is certainly a scope for climate model improvement in terms of the resolution of the projections as well as the better treatment of uncertainty. In addition, providing such information in forms usable for different stakeholders should be part of the priority.
2. Clearly, current understanding about the economics of adaptation to climate change in agricultural context in general and in Europe in particular is still lacking. Economic assessment of the implications of climate change on agriculture needs to better factor in more strategic adaptation options, hence not merely focusing on short-term agronomic adjustments. There is clear need for more research into thorough cost and benefit analysis of agricultural adaptation and the effectiveness of alternative policy instruments to encourage this. Further, there is growing call for researching agricultural adaptation to climate

change through integrated lenses by explicitly taking into account the possible interacting influence of a wide array of factors, including different policy objectives. This leads to the need for empirical investigation on the synergies, trade-offs, and conflicts between diverse policy domains, including between adaptation and mitigation strategies at different spatial and temporal scales.

3. It is clear that present empirical studies on the impacts of climate change and on the role of climate change on livestock farming in Europe are limited. It would be interesting to investigate this matter in relation to the projected North-ward shifts in land-use patterns as a consequence of anticipated future trends on temperature and precipitation. More specifically, it would be of high relevance to study the implications of potential agricultural intensification (through a combination of expansion of suitable areas for agriculture and the greater range of cultivable crops) on grazing livestock productions in the northern regions of Europe.

Acknowledgement

The preparation of this chapter has been supported by the Norden Top-level Research Initiative sub-programme 'Effect Studies and Adaptation to Climate Change' through the Nordic Centre of Excellence for Strategic Adaptation Research (NORD-STAR) and by the EU Framework 7 Bottom-Up Climate Adaptation Strategies Towards a Sustainable Future (BASE) project (Grant No. 308337).

References

Adams, R.M., Rosenzwig, C., Peart, R.M., Ritchie, J.T., McCarl, B.A., Glyer, J.D., Curry, R.B., . . . Allen, J.L.H. 1990. Global Climate Change and US Agriculture. *Nature,* 345, 219–224.

Agrawala, S., Crick, F., Jetté-Nantel, S. & Tepes, A. 2008. Empirical Estimates of Adaptation Costs and Benefits: A Critical Assessment. *In:* Agrawala, S. & Fankhauser, S. (eds.) *Economic Aspects of Adaptation to Climate Change: Costs, Benefits, and Policy Instruments.* OECD.

Alcamo, J., Moreno, J.M., Nováky, B., Bindi, M., Corobov, R., Devoy, R.J.N., Giannakopoulos, C., . . . Shvidenko, A. 2007. Europe. *In:* Parry, M.L., Canziani, O.F., Palutikof, J.P., Van Der Linden, P.J. & Hanson, C.E. (eds.) *Climate Change 2007: Impacts, Adaptation and Vulnerability. Contribution of Working Group II to the Fourth Assessment Report of the Intergovernmental Panel on Climate Change.* Cambridge, UK: Cambridge University Press.

Arrow, K.J. 1963. Uncertainty and the Welfare Economics of Medical Care. *American Economic Review,* 53, 941–973.

Bateman, I.J., Abson, D., Andrews, B., Crowe, A., Darnell, A., Dugdale, S., Fezzi, C., . . . Termansen, M. 2011. Chapter 26: Valuing Changes in Ecosystem Services: Scenario Analysis. *In: The UK National Ecosystem Assessment Technical Report.* Cambridge: UK National Ecosystem Assessment, UNEP-WCMC.

Bindi, M. & Olesen, J.E. 2011. The Responses of agriculture in Europe to Climate Change. *Regional Environmental Change,* 11, S151–S158.

Binswanger, H.P. & Sillers, D.A. 1983. Risk Aversion And Credit Constraints In Farmers' Decision Making: A Reinterpretation. *The Journal of Development Studies,* 20, 5–21.

Bryant, C.R., Smit, B., Brklacich, M., Johnston, T.R., Smithers, J., Chiotti, Q. & Singh, B. 2000. Adaptation in Canadian Agriculture to Climatic Variability and Change. *Climatic Change,* 45, 181–201.

Burton, I. & Lim, B. 2005. Achieving Adequate Adaptation in Agriculture. *Climatic Change,* 70, 191–200.

Christensen, J.H., Hewitson, B., Busuioc, A., Chen, A., Gao, X., Held, I., Jones, R., Kolli, R.K., . . . Whetton, P. 2007. Regional Climate Projections. *In:* Solomon, S., Qin, D., Manning, M., Chen, Z., Marquis, M., Averyt, K.B., Tignor, M. & Miller, H.L. (eds.) *Climate Change 2007: The Physical Science Basis. Contribution of Working Group I to the Fourth Assessment Report of the Intergovernmental Panel on Climate Change.* Cambridge, United Kingdom and New York, NY, USA: Cambridge University Press.

Ciscar, J.C., Iglesias, A., Feyen, L., Goodess, C.M., Szabó, L., Christensen, O.B., Nicholls, R., . . . Soria, A. 2009. Climate Change Impacts in Europe. Final report of the PESETA research project. European Commission Joint Research Centre, Institute for Prospective Technological Studies.

Ciscar, J.C., Iglesias, A., Feyen, L., Szabó, L., Van Regemorter, D., Amelung, B., Nicholls, R., . . . Soria, A. 2011. Physical and Economic Consequences of Climate Change in Europe. *Proceedings of the National Academy of Sciences in the United States of America* 108, 2678–2683.

Cline, W. 1992. *The Economics of Global Warming*. Washington, DC: Institute of International Economics.

Dannenberg, A., Mennel, T., Osberghaus, D. & Sturm, B. 2009. The Economics of Adaptation to Climate Change: The Case of Germany. *ZEW Discussion Paper No.09-057*. Mannheim, Germany.

Easterling, W.E. 1996. Adapting North American Agriculture to Climate Change in Review. *Agricultural and Forest Meteorology,* 80, 1–53.

Easterling, W.E., Aggarwal, P.K., Batima, P., Brander, K.M., Erda, L., Howden, S.M., Kirilenko, A., . . . Tubiello, F.N. 2007. Food, Fibre and Forest Products. *In:* Parry, M.L., Canziani, O.F., Palutikof, J.P., Van Der Linden, P.J. & Hanson, C.E. (eds.) *Climate Change 2007: Impacts, Adaptation and Vulnerability. Contribution of Working Group II to the Fourth Assessment Report of the Intergovernmental Panel on Climate Change.* Cambridge, UK: Cambridge University Press.

FAO 2007. *Adaptation to Climate Change in Agriculture, Forestry and Fisheries: Perspective, Framework and Priorities,* Rome, Italy: FAO.

Fischer, G., Shah, M., Tubiello, F.N. & Van Velhuizen, H. 2005. Socio-economic and Climate Change Impacts on Agriculture: An Integrated Assessment, 1990–2080. *Philosophical Transactions of the Royal Society B-Biological Sciences,* 360, 2067–2083.

Füssel, H.M. 2007. Adaptation Planning for Climate Change: Concepts, Assessment Approaches, and Key Lessons. *Sustainability Science,* 2, 265–275.

Hall, C. & Wreford, A. 2012. Climate Change Adaptations: The Views of Stakeholders in the Livestock Industry. *Mitigation and Adaptation Strategies for Global Change,* 17, 207–222.

Hanemann, W.M. 2000. Adaptation and Its Measurement – An Editorial Comment. *Climatic Change,* 45, 571–581.

Hanley, N., Shogren, J. & White, B. 2007. *Environmental Economics: In Theory & Practice,* Palgrave Macmillan.

Howden, S.M., Soussana, J.F., Tubiello, F.N., Chhetri, N., Dunlop, M. & Meinke, H. 2007. Adapting Agriculture to Climate Change. *Proceedings of the National Academy of Sciences of the United States of America,* 104, 19691–19696.

Iglesias, A., Quiroga, S., Moneo, M. & Garrote, L. 2012. From Climate Change Impacts to the Development of Adaptation Strategies: Challenges for Agriculture in Europe. *Climatic Change,* 112, 143–168.

IPCC 2007. *Climate Change 2007: Fourth Assessment Report of the Intergovernmental Panel on Climate Change,* Cambridge, United Kingdom and New York, NY, USA: Cambridge University Press.

King, J.M., Parsons, D.J., Turnpenny, J.R., Nyangaga, J., Bakari, P. & Wathes, C.M. 2006. Modelling Energy Metabolism ff Friesians in Kenya Smallholdings Shows How Heat Stress and Energy Deficit Constrain Milk Yield and Cow Replacement Rate. *Animal Science,* 82, 705–716.

Leclère, D., Jayet, P.A. & De Noblet-Ducoudre, N. 2013. Farm-level Autonomous Adaptation of European Agricultural Supply to Climate Change. *Ecological Economics,* 87, 1–14.

Maracchi, G., Sirotenko, O. & Bindi, M. 2005. Impacts of Present and Future Climate Variability on Agriculture and Forestry in the Temperate Regions: Europe. *Climatic Change,* 70, 117–135.

McCarl, B.A. 2007. Adaptation Options for Agriculture, Forestry and Fisheries. *A Report to the UNFCCC Secretariat Financial and Technical Support Division.*

Mechler, R., Hochrainer, S., Aaheim, A., Salen, H. & Wreford, A. 2010a. Modelling Economic Impacts and Adaptation to Extreme Events: Insights from European Case Studies. *Mitigation and Adaptation Strategies for Global Change,* 15, 737–762.

Mechler, R. & Kundzewicz, Z.W. 2010b. Assessing Adaptation to Extreme Weather Events in Europe-Editorial. *Mitigation and Adaptation Strategies for Global Change,* 15, 611–620.

Mendelsohn, R. & Dinar, A. 1999. Climate Change, Agriculture, and Developing Countries: Does Adaptation Matter? *World Bank Research Observer,* 14, 277–293.

Mendelsohn, R., Nordhaus, W.D. & Shaw, D. 1994. The Impact of Global Warming on Agriculture: A Ricardian Analysis. *American Economic Review,* 84, 753–771.

Moran, D., Topp, K., Wall, E., Wreford, A., Chadwick, D., Hall, C., Hutchings, M., . . . Wu, L. 2009. Climate Change Impacts on the Livestock Sector. *Report to Defra No AC0307*. SAC Commercial Ltd.

Moriondo, M., Bindi, M., Kundzewicz, Z.W., Szwed, M., Chorynski, A., Matczak, P., Radziejewski, M., Mcevoy, D. & Wreford, A. 2010. Impact and Adaptation Opportunities for European Agriculture in Response to Climatic Change and Variability. *Mitigation and Adaptation Strategies for Global Change,* 15, 657–679.

Nainggolan, D., Termansen, M., Reed, M., Cebollero, E. & Hubacek, K. 2013. Farmer Typology, Future Scenarios and the Implications for Ecosystem Service Provision: A Case Study from South-Eastern Spain. *Regional Environmental Change,* 13, 601–614.

Olesen, J.E. & Bindi, M. 2002. Consequences of Climate Change for European Agricultural Productivity, Land Use and Policy. *European Journal of Agronomy,* 16, 239–262.

Piao, S.L., Ciais, P., Huang, Y., Shen, Z.H., Peng, S.S., Li, J.S., Zhou, L. P., . . . Fang, J.Y. 2010. The Impacts of Climate Change on Water Resources and Agriculture in China. *Nature,* 467, 43–51.

Quiggin, J. & Horowitz, J.K. 1999. The Impact of Global Warming on Agriculture: A Ricardian Analysis: Comment. *The American Economic Review,* 89, 1044–1045.

Reidsma, P., Ewert, F. & Lansink, A.O. 2007. Analysis of Farm Performance in Europe Under Different Climatic and Management Conditions to Improve Understanding of Adaptive Capacity. *Climatic Change,* 84, 403–422.

Reidsma, P., Ewert, F., Lansink, A.O. & Leemans, R. 2010. Adaptation to Climate Change and Climate Variability in European Agriculture: The Importance of Farm Level Responses. *European Journal of Agronomy,* 32, 91–102.

Reilly, J., Tubiello, F., Mccarl, B., Abler, D., Darwin, R., Fuglie, K., Hollinger, S., . . . Rosenzweig, C. 2003. US Agriculture and Climate Change: New Results. *Climatic Change,* 57, 43–69.

Rind, D., Goldberg, R., Hansen, J., Rosenzweig, C. & Ruedy, R. 1990. Potential Evapotranspiration and the Likelihood of Future Drought. *Journal of Geophysical Research,* 95, 9983–10004.

Rosenberg, N.J. 1992. Adaptation of Agriculture to Climate Change. *Climatic Change,* 21, 385–405.

Rosenzweig, C. & Parry, M.L. 1994. Potential Impacts of Climate Change on World Food Supply. *Nature,* 367, 133–138.

Rothschild, M. & Stiglitz, J. 1976. Equilibrium in Competitive Insurance Markets: An Essay on the Economics of Imperfect Information. *Quarterly Journal of Economics,* 90, 629–649.

Schlenker, W. & Roberts, M.J. 2009. Nonlinear Temperature Effects Indicate Severe Damages to US Crop Yields Under Climate Change. *Proceedings of the National Academy of Sciences of the United States of America,* 106, 15594–15598.

Schroter, D., Cramer, W., Leemans, R., Prentice, I.C., Araujo, M.B., Arnell, N.W., Bondeau, A., . . . Zierl, B. 2005. Ecosystem Service Supply and Vulnerability to Global Change in Europe. *Science,* 310, 1333–1337.

Smit, B. & Skinner, M.W. 2002. Adaptation Options in Agriculture to Climate Change: A Typology. *Mitigation and Adaptation Strategies for Global Change,* 7, 85–114.

Tol, R.S.J., Fankhauser, S. & Smith, J.B. 1998. The Scope for Adaptation to Climate Change: What Can We Learn From the Impact Literature? *Global Environmental Change,* 8, 109–123.

Tubiello, F.N., Soussana, J.F. & Howden, S.M. 2007. Crop and Pasture Response to Climate Change. *Proceedings of the National Academy of Sciences of the United States of America,* 104, 19686–19690.

Urwin, K. & Jordan, A. 2008. Does Public Policy Support or Undermine Climate Change Adaptation? Exploring Policy Interplay Across Different Scales of Governance. *Global Environmental Change-Human and Policy Dimensions,* 18, 180–191.

Wheeler, T. & Tiffin, R. 2009. Costs of Adaptation in Agriculture, Forestry and Fisheries. *In:* Parry, M., Arnell, N., Berry, P., Dodman, D., Fankhauser, S., Hope, C., Kovats, S., . . . Wheeler, T. (eds.) *Assessing the Costs of Adaptation to Climate Change – A Review of the UNFCCC and Other Recent Estimates.* London: International Institute for Environment and Development & Grantham Institute for Climate Change.

13
ADAPTATION IN COASTAL AREAS

Iñigo J. Losada and Pedro Díaz-Simal[1]

[1]BOTH AT ENVIRONMENTAL HYDRAULICS INSTITUTE "IH CANTABRIA", UNIVERSIDAD DE CANTABRIA (SPAIN)

13.1 Introduction

Coastal areas consist of both natural and human systems. Natural systems include beaches, barriers and sand dunes as well as rocky shores. Estuaries and deltas are also extremely relevant features of coastal morphology. One of the main characteristics of these features is that they host some of the most important ecosystems, including rocky shore animals; vegetated coastal habitats, such as sea grass beds like Posidonia, salt marshes, mangrove forests and macro algal beds and not less relevant, coral reefs. All of them provide important ecosystem services including carbon sequestration, regulation by promoting sedimentation or wave damping or providing shelter for nursing of different species, among others (Duarte et al. 2013).

The sea has attracted mankind for centuries. Consequently the natural system has been altered heavily by the development of the human system consisting of the built environment and all the human activities, mostly focused at taking advantage of the many coastal and ocean resources available.

As an interface between land and the ocean, the coexistence and interaction between the natural and human systems are completely framed by atmospheric and ocean conditions and their variability and long-term changes.

Consequently, there is an urgent need to assess, with limited uncertainty, the impacts of climate change that coastal areas have to face. As part of this assessment, the estimation of the costs of inaction or action, by the implementation of adaptation measures, is of outstanding importance for relevant stakeholders and decision makers.

The aim of this chapter is to provide an updated review of the different approaches existing in literature for assessing the costs of impacts and adaptation in coastal areas and discussing the different sources of uncertainty.

The main drivers of change in coastal areas are discussed first. It will be shown that human drivers may be of higher importance than climate drivers when assessing potential impacts.

The next section is devoted to a short summary of potential impacts in coastal areas. A section on the needs and options for adaptation follows. The rest of the chapter deals with the methods and tools that have been used so far for assessing the costs of inaction and the cost of adaptation. It will be shown that only a few different approaches have been used in practical applications considering sea level rise (SLR) as the major threat for coastal areas.

13.2 Coastal areas and drivers of change

Today, decision makers are mostly concerned about impacts originated by extreme climate events or by coastal degradation mainly induced by human action. Consequently, climate change is envisioned as one amongst many of the problems coastal areas have to deal with.

An understanding and quantification of the relevant climate-induced and non-climate-induced factors or stressors that may alter the coastal areas is key in order to understand the impacts coastal systems may have to face.

Regarding climate change, relative sea level rise (RLSR) is the most global and significant threat to be faced by coastal systems. There are several components contributing to sea level rise, including ocean thermal expansion and the melting of glaciers and the ice sheets from Greenland and Antarctica (Church et al. 2010).

There are three significant issues to be considered if costs and benefits of adaptation to rising sea levels are to be assessed. First, sea level rise exhibits a large regional variability with large deviations from the global mean sea level rise in many locations. Consequently, regional projections of sea level rise are to be used in the vulnerability assessments. Otherwise, large-scale assessments considering uniform sea level rise values may deal to erroneous potential damage costs estimates.

Second, what matters in terms of coastal impacts study is relative sea level rise (RLSR), meaning that the local subsidence of coastal land or the glacial isostatic rebound or abrupt tectonic movements, have to be accounted for. For example, in cities located in deltas, subsidence over the last 100 years may have reached several meters. Consequently, there are coastal spots where subsidence is even an order of magnitude higher than the currently observed sea level rise. Other parts of the world's coastline are experiencing a land-lift due to isostatic rebound or as a result of recent earthquakes. These land uplifts may partly or fully offset sea level rise. Failure to consider RLSR as the major driver for coastal inundation may end in misleading economic assessments or maladaptation.

Lastly, as it was already pointed out in AR4 (Solomon et al. 2007), recent results (Meehl et al. 2012) confirm that the sea level will continue to rise beyond the 21st century even if important mitigation measures are enforced. This is a point to be considered in any analysis tackling the economics of adaptation in coastal areas.

Other relevant climate-related drivers are extreme sea levels, mostly storm surges, which have already originated important disasters during this century, such as Katrina (Day et al. 2007) or Sandy (Vance 2012). A global analysis carried out by Menendez and Woodwarth (2010) shows that increasing frequency of extreme sea level events observed so far can be directly linked to rising mean sea level. Consequently, it can be inferred that an increase in global mean sea level rise will lead to increasing frequency of extreme events throughout the century. Past experiences in evaluating costs associated with extreme events can serve as a relevant baseline to estimate projected costs of damages and adaptation in the future.

There are some other climate-related drivers such as changes in wind and wave direction and intensity (Reguero et al. 2013; Semedo et al. 2011; Seneviratne et al. 2012) and changes in sea surface temperature, freshwater input and ocean acidification (Hoegh-Guldberg 2007).

However, as will be shown later, to date almost no studies consider the role of these drivers in economic assessments of impacts and adaptation.

Climate-related drivers are only additional stresses on systems that are already under intense and growing pressure due to human-induced drivers. In fact, today it can be said that most of the problems that coastal systems are facing are mostly due to non-climate-related drivers such as: interruption of sediment transport due to the construction of ports, marinas or land reclamation; reduction in sediment and freshwater delivery due to the construction of river dams or structures for water diversion for irrigation (Dai et al. 2009; Syvitski 2008); increasing coastal urbanisation (McGranahan et al. 2007); habitat loss due to coastal development; human induced-subsidence due to the extraction of water and subsurface resources; pollution; hypoxia due to increasing nutrient fluxes and others.

As with regards coastal development, according to McGranahan et al. (2007), coastal migration and urbanisation are the main reasons while coastal population is growing faster than the national average trends. SREX (IPCC 2012) has already pointed out that this socioeconomic development is one of the major drivers of coastal flood damage due to increasing exposure.

Finally, it has to be said that climate-induced changes may not always represent the major threat for coastal systems; however, when impacting on areas where human-induced stressors have reduced the capacity of coastal systems to cope with climate variability and climate change, they may contribute to increase the adaptation needs. Moreover, understanding the role human and climate-driven drivers play when they interact adds complexity to the attribution of impacts due to climate change.

13.3 Impacts

Due to the coexistence of natural and socioeconomic systems in the coastal area, climate change impacts can be found for both systems independently, and many times they have been treated as such in the literature. However, it has to be noted that in many places they have to be treated together considering the fact that impacts and consequences may be reinforced or weakened depending on systems interaction.

To date, most of the impacts detected in the natural coastal system can be attributed to human action. Eroded shorelines are mostly a consequence of human-induced trapping of sediment by construction of port infrastructure and waterfronts or by dams capturing sediment in rivers, one of the main sources of sediment for the coastlines.

Climate change can also contribute to erosion of beaches and coastal barriers due to rising sea levels, increasing intensity or frequency of coastal extreme events or changes in wave climate.

The EUROSION project, funded by the European Commission, concluded in 2004 that 20% of the total EU coastlines were facing serious erosion problems and that annually 15 km^2 of land was lost or seriously impacted by erosion.

Increasing ocean temperature and sea level rise are contributing to the decline or migration of important vegetated ecosystems such as kelp forests, mangroves or sea grass meadows. However, it has to be said that part of this decline is also due to human action, especially coastal development, extraction and pollution. Another example of the impacts on natural systems due to the combined effect of climate change and human action are coral reefs' decline and bleaching.

Table 13.1 Climate drivers and associated effects/impacts

Climate driver	*Effect or impact*
Relative sea level rise	Submergence, erosion, seawater intrusion, rising water tables, wetland permanent losses, dry land permanent losses
Winds	Storm surges, sand dunes transport, damage to infrastructures and housing in coastal areas, impacts on port operations and navigation
Waves	Flooding, erosion, overtopping, impacts on structures, navigation, port operations
Storms	Storm surges, extreme flooding, erosion, saltwater intrusion, infrastructures damage, overtopping
Sea surface temperature	Coral bleaching, species mortality or migration, water quality, stratification
Freshwater input	Water quality, sediment delivery, floods

Besides the natural importance of these ecosystems contributing, for example, to increase biodiversity, they provide important services to society. Coral reefs are one of the main locations for fishing, especially in developing countries. They provide a source of economic resources thanks to tourism in nations like Australia and contribute to protection against storm surges (Sheppard et al. 2011), consequently reducing damage costs or avoiding protection costs due to the construction of artificial structures.

However, to date most of the efforts found in the literature have been devoted to assessing the economic cost of impacts on human systems. Among the most important impacts considered are erosion and flooding due to extreme events and submergence and land loss due to relative sea level rise or the impact of salinity intrusion, the latter especially relevant for wave resources availability.

Most of the work available has tried to assess the number of people exposed under different scenarios (Sterr 2008; Nicholls et al. 2011, Hinkel et al. 2010) or the impacts on different relevant socioeconomic sectors such as critical coastal infrastructures (Horton et al. 2010; Zimmerman and Faris 2010); coastal tourism and recreation (Moreno and Amelung 2009) or fisheries, aquaculture and agriculture (De Silva and Soto 2009).

A detailed review of climate change impacts in coastal areas can be found in AR4-IPCC 2007, SREX and in the forthcoming IPCC-AR5.

In Table 13.1 a summary of the most relevant climate drivers and the associated potential impacts can be found.

13.4 Need and options for adapting to climate change in coastal areas

Based on the above, it is needless to say that failure to take actions to reduce the impacts of climate change in a timely manner will increase the likelihood of damage to the well-being of ecosystems, consequently undermine the services they provide, and what is more important, increase damage to the socioeconomic system, increase the risk of life losses and lead to economic losses.

There is a clear need for adaptation to climate change. Adaptation plays a vital role in coastal areas, especially in low-lying areas such as deltas, island states or in coastal areas highly vulnerable due to the already detected trend to increase exposure.

The selection of appropriate adaptation options can be a challenging exercise for decision makers. The risk of implementing options that may fail increasing the aimed adaptive capacity or may result in increasing vulnerability is relatively high if the relevant technical, financial, cultural and social aspects are not considered in the frame of an uncertain changing climate.

There is an ample range of adaptation options including both tangible and intangible actions that may be considered to address climate change impacts associated with inundation, flooding and erosion or other impacts in coastal areas, and none of them should be discarded a priori without a careful analysis of the short-term and long-term implications of its implementation.

In general, tangible adaptation options are identified with engineered or technological solutions like the construction of coastal protection structures, while intangible solutions are more institutional or socially oriented options linked with policies, gaining public awareness, education, insurance or land planning.

In a more general context, the different adaptation measures can be classified according to Hallegatte (2009) who defined five broad practical strategies able to cope with climate change, namely: 1) no-regret strategies, 2) reversible strategies, 3) safety margin strategies, 4) soft strategies and 4) strategies that reduce decision-making time horizons.

No-regret measures are those yielding benefits even in absence of climate change. For example, reducing coastal vulnerability by avoiding coastal natural or man-induced erosion or by limiting urbanisation in flooding prone areas will increase coastal adaptive capacity against extreme flooding for future climate conditions and avoid disaster losses in the present climate.

Reversible and flexible strategies should be favoured, if possible, in order to avoid maladaptation or to face current uncertainties in climate or vulnerability projections. Considering the previous example, limiting urban development in flooding-prone coastal areas may have short-term financial and social costs, but protection costs under unexpected sea level rise figures or extreme flooding may result in deaths and unavoidable high protection costs due to the fact that retreat may no longer be socially possible. Monitoring sea level rise and the occurrence of extreme flooding events and the reduction of uncertainties in climate projections may result in a more robust decision-making, reversing the urban development limitation.

Safety margins strategies are especially suited to add resilience to new infrastructures under an uncertain future. This strategy is based on the hypothesis that, in general, the marginal cost to increase the security margin of a new infrastructure during the design and constructions phases is small compared to the total cost of construction and will have a lower cost than refitting the infrastructure during the operation and management phases. This is especially relevant for new ports or extensions. The same can be said regarding coastal protection infrastructures against extreme flooding or erosion due to storm surge or severe wave storminess.

Among the classification proposed in Hallegatte (2009), soft strategies are also a relevant option. Those include institutional or financial tools such as the "institutionalisation" of long-term planning, relocation, introducing flexible insurance and taxes, early warning systems, building codes, public awareness initiatives, education programs and others. All of them can be applied in coastal areas.

One example of institutionalisation of long-term planning could be the application of such a strategy to the Spanish Maritime Terrestrial Public Domain (DPMT), which determines the boundaries of the coastal domain under public supervision and therefore, among others, limits the development of the built environment through different legal instruments. The DPMT is set by the evaluation of the historical maximum flooding levels. For example, housing or certain industrial infrastructure may occupy the DPMT thanks to a concession, currently lasting almost until the end of the century. Considering the fact that climate change is contributing to sea level rise and changes in storm surge and wave climate (the three main components of the total flooding levels), it would be highly recommendable to institutionalise the review of the Spanish DPMT on a regular basis, say every five years, with a 20-year perspective in order to analyse the potential impact of climate change on the DPMT and its consequences on several environmental, social and economic aspects, and to establish sufficient buffer areas and the legal tools to *adapt* the concept of DPMT to the consequences to be expected.

A final strategy proposed by Hallegatte (2009) is reducing the decision-making time horizons. For example, reducing the lifetime of investments and concessions in the DPMT and introducing an administrative procedure for reviewing and updating the boundaries is better than a total restriction of any development or drastic decisions to remove already existing activities or infrastructures. What seems to be completely inefficient is to extend concessions in the long-term without considering potential climate change-induced variations in the flooding prone areas, and providing legal support for those who in the future will have the right to demand public funding to protect their assets from extreme events and inundation.

An especially relevant strategy in coastal adaptation is ecosystem-based adaptation. Ecosystem-based adaptation relies on the fact that certain coastal ecosystems are able to provide certain services, mainly based on the so-called regulatory functions that may contribute to help society face some climate change impacts. It has been shown that wetlands, mangroves, Posidonia fields or coral reefs can provide an effective buffer for coastal protection against hurricanes, storm surges or extreme waves. Therefore, maintenance, restoration or re-engineering of these ecosystems can be an efficient option.

It has also to be pointed out that the conservation, restoration and use of vegetated coastal habitats in eco-engineering solutions for coastal protection provide a good strategy thanks to the vegetated coastal habitats to act as CO_2 sinks, therefore, delivering significant additional capacity for climate change mitigation (Duarte et al. 2013).

There are several publications collecting and discussing the pros and cons of different adaptation option and technologies (USAID 2009; UNEP 2010; U.S.EPA; 2009; Burkettand Davidson 2012 and others). Many of them are based on previously existing knowledge on coastal shore protection and consequently, mainly focused on flooding and erosion problems.

13.5 Costs of adaptation in coastal areas

As has been said, decision makers are exposed to a broad range of adaptation options when tackling climate change impacts in coastal areas. Besides social and environmental issues, often fundamental aspects to discard a given adaptation strategy, a key issue is the assessment of whether the cost of implementing and maintaining a given adaptation strategy is worth the benefits to be expected. In general, this is especially difficult considering the fact that adaptation strategies are expected to last long into the future where climate change and its impacts, together with the potential development of coastal areas is still uncertain.

An initial problem to be tackled is to define and find agreement, between the different stakeholders involved, on what are the objectives of the adaptation strategy to be implemented. This may become a real difficult issue to be solved, considering the large number of diverse stakeholders, uses and interests accumulated in the coastal areas. Some stakeholders may consider that the objective of cancelling all climate change projected impacts to maintaining or even improving the current status, taking advantage of climate change, may require the reversal of human-induced impacts, therefore increasing resilience. This may not be compatible with keeping the activity originating the stress. For example, restoring the long shore sediment transport interrupted by a coastal development, such as a marina, may reduce coastal erosion or even restore the sediment budget natural balance in a certain coastal area, increasing resilience against sea level rise or extreme flooding. Needless to say, this adaptation option may go against the interest of the marina's further development plans or even become a source of economic losses due to a reduction in operations capacity or simply due to an increase in annual costs associated to the operation of a sand by-pass to be implemented as part of the required adaptation strategy.

Consequently, adapting to climate change will imply conflict resolution and trade-offs between the different stakeholders and interests in the coastal area. This should be understood as advocating, first, a flexible, continuous, participatory adaptation process benefiting from an appropriate identification of every actor and role in adaptation and, second, the implementation of a good metric for monitoring and evaluation of the outcome of the adaptation strategy implemented. Due to the coexistence between the complex natural and extensively developed socioeconomic system in coastal areas, neither one nor the other have yet achieved the required development as to be able to provide strategies without risk of maladaptation.

When it comes to the costs of climate change in coastal areas, most studies have been focused on sea level rise, trying to determine the costs of inaction versus the costs of adaptation plus the residual damages. In general, EEA (2007) defines the full cost of climate change as the balance of the costs of mitigation plus the costs of adaptation plus the costs of inaction minus the benefits of mitigation and the benefits of adaptation. This definition, establishing a balance between costs and benefits, points in the direction of cost-benefit analysis (CBA) to help stakeholders and decision makers to select the best option to be taken. Consequently, this approach requires quantifying every cost and benefit in monetary terms, which opens new challenges such as the valuation of the services provided by coastal ecosystems. To date, there are still many assessments in which the damages of climate change are still given in physical units rather than in monetary values.

For the last decade, two different tools have been the most referred to for estimating the impacts of SLR as the most relevant driver of impacts in coastal areas. The Climate Framework for Uncertainty, Negotiation and Distribution (FUND) model (Tol 2007) and the Dynamic Interactive Vulnerability Assessment (DIVA) (Hinkel and Klein 2009) model have been used extensively in different countries and at the continental scale to give an initial assessment of the impacts of SLR with and without adaptation, including a monetary quantification of some of the most relevant impacts and benefits provided by adaptation.

The FUND model was initially devoted to the cost-benefit analysis of mitigation. Tol (2007) provides a version dedicated to sea level rise adaptation considering three different SLR scenarios (0.5 m, 1 m and 2 m). The model distinguishes 16 different world regions and runs in time steps of five years from 1995 to 2100. The model is able to provide for the different scenarios dry land lost, wetland lost, the cost of displaced people and the cost of protection considering fixed evolutions in GDP and population over time. So far limited practical applications of the tool seem to be available.

On the contrary, DIVA has been used at the global, national and subnational scales to analyse the major direct impacts of climate change in coastal areas in several practical applications.

DIVA analyses the impacts of a wide range of SLR scenarios thanks to a global database including over 10,000 coastal segments defined as a combination of natural, administrative and socio-economic characteristics. For each of the segments, a set of 30 indicators is defined, providing the user with information on the physical and economic impacts to be expected. DIVA provides information on increased flood-risk (extreme conditions) and inundation (permanent), erosion, coastal wetland changes or losses and saltwater intrusion. The model also includes the possibility to explore a limited set of adaptation measures including their costs.

A detailed review of a first extensive set of the work carried out on studies aiming at assessing the costs of sea level rise impacts and adaptation can be found in Agrawala and Fankhauser (2008).

Most of the cost assessment studies use the direct cost method, for example, by calculating inundated surfaces or assets and multiplying the inundated surface by exogenous prices. This is relatively simple to do if market prices and their projections are known. However, the problem arises when economic losses are to be valued when losses in ecosystem services or functions are the consequence of sea level rise or of temporary extreme flooding events. A recent review on valuing coastal ecosystem services can be found in Barbier (2012).

A few studies have applied general equilibrium analysis trying to evaluate how climate change impacts are transmitted across sectors and countries and how the impacts are affected by the macroeconomic environment. Some examples are Bosello et al. (2007), devoted to the estimation of the economy-wide implications of sea level rise in 2050 using a static computable general equilibrium model. Results show that general equilibrium effects increase the welfare costs of sea level rise, but not necessarily in every sector or region. In the absence of coastal protection, economies that rely most on agriculture are hit hardest. With full coastal protection, GDP increases, particularly in regions with substantial dike building. In a more recent study Ciscar et al. (2011) analysed the combined impact on agriculture, coasts, river floods and tourism in the European economy. They found an average welfare loss of 0.2–1.0% of income but with large regional differences, with losses in Southern Europe and gains in Northern Europe.

Independently of using one or another approach, during the last part of the last decade, a relevant number of studies have addressed the costs of sea level rise at different geographic scales, mostly evaluating the consequences on resources at risk. At a global scale, Nicholls and Toll (2006), using several economic criteria, estimated the direct costs, including coastal protection, residual land loss and wetland loss to both sea level rise and coastal protection, of dry land and wetland due to inundation under a global sea level rise of 0.35 m for all countries in the world for 4 SRES scenarios. One of the main conclusions was that it would be economically justified to protect most of the highly populated low-lying areas.

A new set of relevant studies on this topic have been published, considering different scales. Anthoff et al. (2010) evaluated the impacts of inundation following Nicholls and Toll's (2006) approach considering 0.5 to 2 m sea level rise by 2100 under four scenarios, using the net present value of total costs for the period 2005–2100 as an impact indicator. They found the total costs without adaptation to range between 800 and 3,300 billion US$ in 2100 and from 200 to 2,200 billion US$ with adaptation, confirming the benefits of introducing adaptation measures against no action. In a later work, Nicholls et al. (2011) evaluate the annual adaptation cost for inundation and erosion from 0.5–2.0 m sea level rise to in 25–270 billion US$/year.

Hanson et al. (2011) present a first estimate of the exposure of the world's large port cities to coastal flooding due to sea level rise and storm surge comparing the present (2005) and the 2070s projections under different scenarios of socioeconomic and climate changes. The analysis suggests that about 40 million people (0.6% of the global population or roughly 1 in 10 of the total port city population in the cities considered) are currently exposed to a 1 in 100-year coastal flood event. The total value of the assets exposed in 2005 across all port cities considered was estimated to be US$3,000 billion; approximately around 5% of global GDP in 2005. By the 2070s, the asset exposure is calculated to increase to more than 10 times the baseline levels or approximately 9% of projected global GDP in this period.

As has been previously said, DIVA has been extensively used during the last years to address global and regional studies. Hinkel et al. (2010) estimated that for the European Union, the total monetary damage caused by flooding, salinity intrusion, erosion and migration without adaptation would reach 17 billion US$ per year under global mean SLR scenarios ranging from 0.35–0.45 m. Introducing adaptation measures would reduce these costs to 2.6 to 3.5 billion US$ per year. Another continental assessment was carried out for Africa by Hinkel et al. (2011) who, for the same impacts as in the previous study, estimated expected annual damage costs without adaptation of 5–9 billion US$ under scenarios of 0.64–1.26 m of mean global SLR by 2100. The evaluation of adaptation strategies estimated that investing 2 to 6 billion US$ annually would reduce the damages costs in half by 2100.

In a recent paper, Bosello et al. (2012) used DIVA to estimate the physical impacts of SLR and the direct economic cost and the GTAP-EF model to evaluate the indirect economic implications of sea level rise for Europe for a range of SLR scenarios for the 2020s and 2080s. The GTAP (Global Trade Analysis Project) model is a standard multiregion, multisector, computable standard general equilibrium model (CGE), distributed with the GTAP database of the world economy. (GTAP-EF) is specifically developed to study the general equilibrium effects of climate change, as a refinement of GTAP-E, an energy-environmental oriented version of GTAP. Detailed information can be found atwww.gtap.org.

According to Bosello et al. (2012), by the end of the century, the largest economic losses are to be faced by Poland and Germany (483 million US$ and 391 million US$, respectively). Coastal protection can reach a protection level higher than 85% in the majority of European countries. Factor substitution, international trade and changes in investment patterns are found to contribute to positive or negative indirect economic implications, while direct economic impacts of SLR was found to be always negative. The overall effect on GDP was also assessed by countries.

Previous approaches are very useful to provide an integrated assessment of the economics of adaptation at very large scales, providing an excellent tool to compare different countries or sectors at continental or global scales. However, this is carried out introducing a series of simplifications such as coarse spatial resolution, limitations in the representation of the drivers such as assuming a homogeneous sea level rise globally, lack of resolution of the Digital Terrain Model, insufficient spatial resolution of some of the indices used to assess vulnerability and resources evaluation and others. Consequently, downscaling vulnerability and economic assessment of climate change to national, subnational or local scales requires introducing new methods and tools able to reduce uncertainties.

Besides, estimating the cost and benefits of introducing adaptation options is not an easy task either. In the assessments above, most of them carried out using DIVA, only a limited number of options have been considered, namely upgrading of defense structures and beach nourishment. Besides the fact that this is only a very small set of the different options available, the implementation cost of these two options may vary tremendously from one situation to another, resulting in global annual costs that may be way off the real values. Still, it is an excellent tool to provide inter-comparison between regions and countries and to be used for sensitivity analysis.

13.6 Conclusions

The economic assessment of the costs of climate change impacts and adaptation has been reviewed in this chapter. It has been found that only a limited number of approaches have been applied in practical cases and that important differences in cost estimates may arise depending on the approach and input data used in the assessments.

Even if it is not the only one, sea level rise has been the major source of impacts considered in the economic assessments so far. A major source of uncertainty and differences between the different economic valuations originates in the uncertainties associated with the projections of sea level rise. But for a few studies, most of the analyses do not consider regional relative sea level rise, which may lead to erroneous inundation damage estimates.

The studies available in the literature do cover different spatial scales, ranging from global to national and subnational scales. The results of the available work assessing costs at global and national scales can be considered as first relevant estimates. However, especially at global and continental scales, the number of simplifications introduced to specify the climate drivers, the simplification hypothesis in the impact models and the uncertainties associated with the vulnerability low resolution databases may result in high deviations from the actual cost estimates. Consequently, they should be taken with care. They can be especially useful to carry out sensitivity analysis or to provide inter-comparisons between countries or regions.

Impact evaluation is also a key part in cost estimates. Oversimplified impact modelling may result in under- or over-estimated damages and consequently in erroneous cost estimates.

Some of the existing models have implemented adaptation options, providing the opportunity to compare the costs of inaction vs. the cost of introducing different kinds of adaptation measures. There is still a long way to go until the ample range of adaptation options can be fully formulated in adaptation costs assessments. Reference costs are needed for the different options together with estimates of how these costs may evolve into the future. Also, the additional benefits they provide need to be accounted for.

The role of existing climate change countermeasures such as coastal protection structures, insurance policies, etc. and their future evolution should be considered when assessing actual costs by the end of the century.

Methods based on Cost Benefit Analysis require expressing all the costs and benefits in monetary terms. This is often a problematic issue when trying to determine the costs, for example, of ecosystems services and some tangible and intangible adaptation measures or their associated benefits.

It can be concluded that during the last 10 years, important efforts have been carried out in order to establish consistent methodologies to analyse the economics of climate change in coastal areas. A limited number of tools and estimates are already available. However, there is still a lot of work to be done in order to reduce the currently existing high uncertainties.

References

Agrawala, S. and Fankhauser, S. (Eds.) (2008). *Economic Aspects of Adaptation to Climate Change: Costs, Benefits and Policy Instruments*, Executive Summary. OECD, Paris.

Anthoff, D., Nicholls, R.J., and Tol, R.S.J. (2010). "The economic impact of substantial sea-level rise". *Mitigation and Adaptation Strategies for Global Change*, 15(4), 321–335

Barbier, E.B. (2012). "Progress and challenges in valuing coastal and marine ecosystem services". *Review of Environmental Economics and Policy*. 6(1), 1–19.

Bosello, F., Nicholls, R.J., Richards, J., Roson, R., and Tol, R.S.J. (2012). "Economic impacts of climate change in Europe: sea-level rise". *Climatic Change*, 112(1), 63–81.

Bosello, F., Roson, R., and Tol, R.S.J. (2007). "Economy-wide estimates of the implications of climate change: Sea level rise". *Environmental and Resource Economics*, 37(3), 549–571.

Burkett, V.R. and Davidson, M.A. (Eds.) (2012). *Coastal Impacts, Adaptation and Vulnerability: A Technical Input to the 2012 National Climate Assessment*. Cooperative Report to the 2013 National Climate Assessment, 150 pp.

Church, J., Woodworth, P.L., Aarup, T., and Wilson W.S. (Eds.) (2010). *Understanding Sea-level Rise and Variability*. Wiley-Blackwell Publishing Ltd. 428 pp.

Ciscar, J.-., Iglesias, A., Feyen, L., Szabó, L, Van Regemorter, D., Amelung, B., Nicholls, R., Watkiss, P., Christensen, O.B., Dankers, R., Garrote, L, Goodess, C.M., Hunt, A., Moreno, A., Richards, J., and Soria, A. (2011). "Physical and economic consequences of climate change in Europe". *Proceedings of the National Academy of Sciences of the United States of America*, 108(7), 2678–2683.

Dai A., Qian T., Trenberth, K.E., and Milliman J.D. (2009). "Changes in continental freshwater discharge from 1948 to 2004". *Journal of Climate*, 22, 2773–2792.

Day Jr., J.W., Boesch, D.F., Clairain, E.J., Kemp, G.P., Laska, S.D., Mitsch, W.J., Orth, K., Mashriqui, H., Reed, D.J., Shabman, L., Simenstad, C.A., Streever, B.J., Twilley, R.R., Watson, C.C., Wells, J.T., and Whigham, D.F. (2007). "Restoration of the Mississippi Delta: Lessons from Hurricanes Katrina and Rita". *Science*, 315 (5819), 1679–1684.

De Silva, S.S. and Soto, D. (2009). "Climate change and aquaculture: Potential impacts, adaptation and mitigation". In: *Climate Change Implications for Fisheries and Aquaculture: Overview of Current Scientific Knowledge*. FAO Fisheries and Aquaculture Technical Paper No. 530. Rome. pp. 151–212.

Duarte, C.M., Losada, I.J.; Hendriks, I., Mazarrasa, I., and Marbá, N. (in press). "The role of coastal plant communities for climate change mitigation and adaptation". *Nature Climate Change*.

European Environment Agency (EEA) (2007). *Climate Change. The Cost of Inaction and the Cost of Adaptation*.

Hallegatte, S. (2009). "Strategies to adapt to an uncertain climate change". *Global Environmental Change-Human and Policy Dimensions*, 19(2), 240–247.

Hanson, S., Nicholls, R., Ranger, N., Hallegatte, S., Corfee-Morlot, J., Herweijer, C., and Chateau, J. (2011). "A global ranking of port cities with high exposure to climate extremes". *Climate Change*, 109(1), 89–111.

Hinkel, J., Brown, S., Exner, L., Nicholls, R.J., Vafeides, A.T. and Kebede, A.S. (2011). "Sea level rise impacts on Africa and the effects of mitigation and adaptation: an application of DIVA". *Regional Environmental Change*, 12, 207–224.

Hinkel, J. and Klein, R.J.T. (2009). "The DINAS-COAST project: Developing a tool for the dynamic and interactive assessment of coastal vulnerability". *Global Environmental Change*, 19(3), 384–395.

Hinkel, J., Nicholls, R.J., Vafeidis, A.T., Tol, R.S.J., and Avagianou, T. (2010). "Assessing risk of an adaptation to sea level rise in the European Union: An application of DIVA". *Mitigation and Adaptation Strategies for Global Change*, 5(7), 1–17.

Hoegh-Guldberg, O., Mumby, P.J., Hooten, A.J., Steneck, R.S., Greenfield, P., Gomez, E., Harvell, C.D., Sale, P.F., Edwards, A.J., Caldeira, K., Knowlton, N., Eakin, C.M., Iglesias-Prieto, R., Muthiga, N., Bradbury, R.H., Dubi, A., and Hatziolos, M.E.(2007). "Coral reefs under rapid climate change and ocean acidification". *Science*, 318(5857), 1737–1742.

Horton, R., Rosenzweig, C., Gornitz, V., Bader, D., and O'Grady, M. (2010). "Climate risk information: Climate change scenarios and implications for NYC infrastructure. New York City Panel on Climate Change". *Annals of the New York Academy of Sciences*, 1196(1), 147–228.

IPCC (2012). *Managing the Risks of Extreme Events and Disasters to Advance Climate Change Adaptation. A Special Report of Working Groups I and II of the IPCC.* Field, C.B., Barros,V., Stocker,T.F., Qin, D., Dokken, D.J., Ebi, K.L., Mastrandrea, M.D., Mach, K.J., Plattner, G.-K., Allen, S.K., Tignor, M. and Midgley, P.M. (Eds.) Cambridge University Press, Cambridge, UK and New York, USA, 582 pp.

McGranahan, G., Balk, D., and Anderson, B. (2007). "The rising tide: Assessing the risks of climate change and human settlements in low elevation coastal zones". *Environment and Urbanization,* 19, 17–37.

Meehl. G.A., Hu, A., Tebaldi, C., Arblaster, J.M., Washington, W.M., Teng, H., Sanderson, B., Ault, T., Strand, and White III, J.B. (2012). "Relative outcomes of climate change mitigation related to global temperature versus sea level rise". *Nature Climate Change,* 2, 576–580.

Menendez, M. and Woodwarth, P.L. (2010). "Changes in extreme high water levels based on quasi-global tide-gauge dataset". *Journal of Geophysical Research,* 115, C10011.

Moreno, A. and Amelung, B. (2009). "Climate change and coastal marine tourism: Review and analysis". *Journal of Coastal Research.* Spec. Issue 56, 1140–1144.

Nicholls, R.J., Marinova, N., Lowe, J.A., Brown, S., Vellinga, P., De Gusmao, D., Hinkel, J., and Tol, R.S.J. (2011). "Sea level rise and its possible impacts given a 'beyond 4° C world' in the twenty-first century". *Philosophical Transactions of the Royal Society A: Mathematical, Physical and Engineering Sciences,* 369(1934), 161–181.

Nicholls, R.J. and Tol, R.S.J. (2006). "Impacts and responses to sea-level rise: A global analysis of the SRES scenarios over the twenty-first century". *Philosophical Transactions of the Royal Society A: Mathematical, Physical and Engineering Sciences,* 363(1841), 1073–1095.

Reguero, B. G, Mendez, F.J., and Losada, I.J. (2013). "Variability of multivariate wave climate in Latin America and the Caribbean". *Global and Planetary Change,* 100, 70–84.

Semedo, A., Sušelj, K., Rutgersson, A., and Sterl, A. (2011). "A global view on the wind sea and swell climate and variability from ERA-40". *Journal of Climate,* 24(5), 1461–1479

Seneviratne, S.I., Nicholls, N., Easterling, D., Goodess, C.M., Kanae, S., Kossin, J., Luo, Y., Marengo, J., McInnes, K., Rahimi, M., Reichstein, M., Sorteberg, A., Vera, C., and Zhang, X. (2012). "Changes in climate extremes and their impacts on the natural physical environment". In: Field, C. (Ed.) *IPCC (2012).* Cambridge University Press, pp. 109–230.

Sheppard, C.C., Crain, C.M., and Beck, M.W. (2011). "The protective role of coastal marshes: A systematic review and meta-analysis". *Plos One,* 6(11).

Solomon, S., Qin, D., Manning, M., Chen, Z., Marquis, M., Averyt, K.B., Tignor, M., and Miller, H.L. (Eds.) (2007). *Contribution of Working Group I to the Fourth Assessment Report of the Intergovernmental Panel on Climate Change.* Cambridge University Press, Cambridge, United Kingdom and New York, NY, USA.

Sterr, H. (2008). "Assessment of vulnerability and adaptation to sea-level rise for the coastal zone of Germany". *Journal of Coastal Research,* 24(2), 380–393.

Syvitski, J.P.M. (2008). "Deltas at risk". *Sustainability Science,* 3, 23–32.

Tol, R.S.J. (2007). "The double trade-off between adaptation and mitigation for sea-level rise. An application of FUND". *Mitigation and Adaptation Strategies for Global Change,* 12. 741–753.

UNEP (2010). *Technologies for Climate Change Adaptation. Coastal Erosion and Flooding.* Available from: http://www.unep.org/pdf/TNAhandbook_CoastalErosionFlooding.pdf . [Accessed: 07/01/2013]

USAID (2009). *Adapting to Coastal Climate Change: A Guidebook for Development Planners.* USAID, Rhode Island. Available from: http://www.crc.uri.edu/download/CoastalAdaptationGuide.pdf [Accessed: 07/01/2013]

U.S.EPA (2009). *Synthesis of Adaptation Options for Coastal Areas. Washington, DC, U.S. Environmental Protection Agency.* Climate Ready Estuaries Programs. EPA 430-F-08-024, January 2009.

Vance, J.H. (2012). "Hurricane Sandy: Flood defence for financial hubs". *Nature,* 491(7425), 527.

Zimmerman, R. and Faris, C., (2010). "Infrastructure impacts and adaptation challenges". In: Rosenzweig, C. and Solecki, W. (Eds.) New York City Panel on Climate Change, 2010: *Climate Change Adaptation in New York City: Building a Risk Management Response.* Prepared for use by the New York City Climate Change Adaptation Task Force. Annals of the New York Academy of Science, New York, NY, pp. 63–85.

14

CLIMATE CHANGE ADAPTATION AND HUMAN HEALTH

Aline Chiabai[1] *and Joseph V. Spadaro*[1]

[1] BASQUE CENTRE FOR CLIMATE CHANGE (BC3)

14.1 Introduction

The aim of this chapter is to present a synthesis of the current and projected health burdens of climate change and costing of adaptation measures to climate change with emphasis on health protection. In this section, we focus on current and projected health impacts, whereas adaptation measures, assessment methods and costs are discussed in the remaining sections of this chapter.

Good health is fundamental to sustainable development, just as much as concerns about economic prosperity and environmental protection. The severity of the potential consequences of climate change will depend to a large extent on the health status of the population affected. Globally, population health has undergone positive change over the second half of the 20th century; however, important regional differences exist within and between individual countries, especially between rich and low income nations. Children have higher disease risk factors than adults, and they bare a heavier burden of disease (Ezzati et al., 2003; McMichael et al., 2004).

Health in its most basic definition is more than just the absence of illness or infirmity; rather, it is a state of complete physical, mental and social well-being. Clinical medicine is concerned with the welfare of the individual, whereas the focus of public health is on protecting and improving the health of communities. Several indicators are used to track the status and development of public health:[1]

- Average life expectancy (lifespan),
- Population total and cause-specific mortality (cardiopulmonary and cardiovascular disease, for instance),
- Pregnancy outcome (maternal mortality, perinatal and infant mortality rates),
- New cases of infectious diseases and other disease risk factors,
- Prevalence of chronic diseases,
- Malnutrition, and
- Healthcare access.

As shown in Figure 14.1, there are various pathways through which climate change may directly or indirectly affect human health. Since the mid-19th century, the average global land and sea surface temperature has increased by approximately 0.6 (±0.2) degrees Celsius for the base period 1961–1990; most of this increase has been realized since the mid-1970s, with the greatest changes[2] occurring at mid to high latitudes. Nine of the warmest years on record have occurred since the year 2000, with 2012 ranking as the 10th warmest year and 2005 and 2010 as the hottest in the 132-year global weather record database. According to the Intergovernmental Panel on Climate Change (IPCC), if no specific actions to curtail GHG emissions are taken in future years, the global surface temperature could rise by between 1.5 and 6 degrees Celsius from 1990 levels by the end of the 21st century (McMichael et al., 2004). Extreme temperatures have a direct effect on human morbidity and mortality, increasing, for example, the baseline rate of allergic diseases, hyperthermia and cardio-pulmonary incidences caused by degradation of air quality (e.g., increases in particulate pollution from combustion of fossil fuels and higher ground-level ozone concentrations driven by distributional changes in precursor pollutants, such as volatile organic compounds and nitrogen oxides, and the amount of ultraviolet radiation, UVR, reaching the earth's surface).

Extreme temperature increases acute mortality by advancing the death of the most vulnerable individuals in society. Displaced mortality (*harvesting*) is a direct consequence of climate change on temperature. High temperatures also act as a contributing cause to excessive mortality in those individuals at high risk of thermal stress due to socioeconomic standing, physical activity or health status (Tables 14.1 and 14.3). A significant share of the total

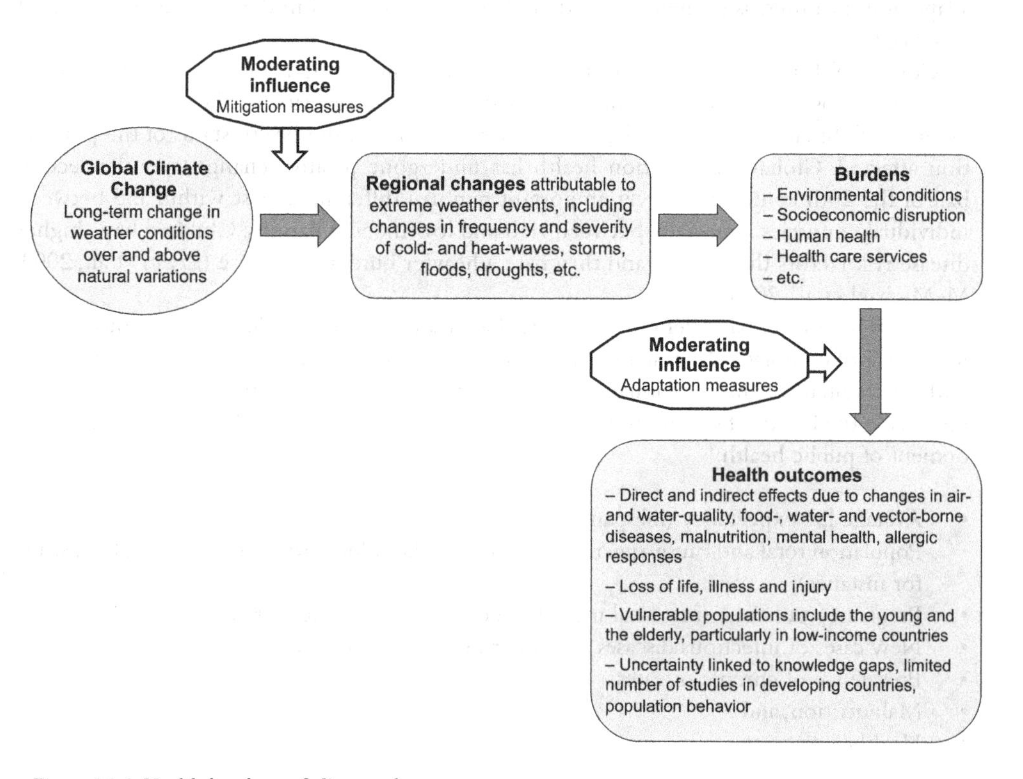

Figure 14.1 Health burdens of climate change

heat-wave mortality in Europe in 2003 was in fact the consequence of the indirect effects of high temperature on sensitive populations, in particular the aged over 75 (Toulemon et al., 2008). Follow-up studies have shown that inadequate monitoring of health risks and failure to act in a timely manner compounded the mortality burden. Awareness, greater access to air-conditioned shelters and improved social cohesion (providing support to those living alone, especially the elderly) would have offset a considerable share of the total heat-wave mortality.

The severity of climate-sensitive diseases related to local and regional changes in temperature, precipitation, floods, droughts, storms and fires depends to a large degree on mitigating and adaptive measures (*moderating influences*); population health status; access to medical care and clinical treatment (prevention and control of disease); preparedness of the healthcare system (monitoring, surveillance and response); community vulnerability (public awareness, access to housing, clean water and sanitation infrastructure, etc.) and institutional and political support. Based on present climate-health modeling forecasts, without proper design, implementation and monitoring of adaptation strategies, low-income countries and in particular children living in these areas, unfortunately, will face the greatest health burdens of a changing climate in the coming years (Tables 14.1–14.3). Sub-Saharan populations are especially at high risk of exposure to food- and water-borne diseases, are prone to epidemics of malaria and other infectious diseases, are at risk of increased cases of diarrhea and malnutrition (under nutrition and lack of protein and other micronutrients), will face significant shortages of plentiful and safe water and will have inadequate access to basic sanitation. The increased severity and frequency of extreme weather events will lead to prolonged seasonality or temporal distribution of disease transmission and increase the geographical range of diseases, impacting primarily areas where the diseases are already endemic or extending into adjacent locations.

Along with the increased potential of disease burden, famine, lack of access to local natural resources (especially water) and socioeconomic disruption also comes the increased risk of conflict and population displacement, primarily movement to urban areas, which can contribute to overcrowding and further exacerbate the impact of increasing ambient temperatures (*urban heat island effect*), and contribute to greater competition for already limited resources, to the spread of disease (including HIV) and even to mental distress.

Impacts are not limited exclusively to less developed areas across the world; health effects will also be a potential concern in the most developed countries (Haines et al., 2006), although the cost burden in rich countries will be on preserving local infrastructure protecting fresh water supplies (*climate proofing*) and preventing losses of personal property, especially along coastal zones, from storm devastation and sea-level rise (World Bank, 2010a). Vulnerability to climate change decreases with increasing per capita wealth, but will not protect the world's population from climate sensitive diseases and deaths, nor avoid human injuries and personal and economic losses caused by natural weather disasters. Adaptive capacity needs to be improved everywhere.

There is a growing body of evidence suggesting that climate change is already affecting human health adversely and is contributing to excessive morbidity and deaths, especially in the lower income regions around the world where the adaptive capacity is weakest. The areas most affected are countries in Sub-Saharan Africa, the Eastern Mediterranean and Southeast Asia (Tables 14.2 and 14.3). Present estimates of global climate-driven death burdens of diarrhea, malaria and malnutrition range between 2% and 3% of baseline totals for the year 2000

Table 14.1 Public health impacts of climate extremes

Impact category	*Severity increase, frequency and geographical range of climate sensitive events*				
	Temperature† *(heat-waves)*	*Floods*	*Drought*	*Storms*	*Fires*
Increase of existing chronic diseases	Extensive	Local to widespread	Extensive	Local to widespread	Local to widespread
Increased mortality	Moderate to high‡	Few to many (flash flood, landslides)	Few	Few to many (2005 Katrina hurricane)	Few to moderate
Increase of severe injuries	Moderate to many heat stroke cases	Few	Unlikely	Few	Few to moderate
Increased infectious diseases	Unlikely	Extensive	Possible	Extensive	Unlikely
Degradation of health care system	Unlikely	Local to widespread	Unlikely	Extensive	Local
Malnutrition	Unlikely	Unlikely	Common	Unlikely	Possible
Lack of safe water	Unlikely	Local to widespread	Extensive	Extensive	Local
Lack of sanitation	Unlikely	Local to widespread	Likely among displaced persons	Extensive	Likely among displaced persons
Lack of shelter	Local to widespread	Local to widespread	Local to widespread	Extensive	Local
Population displacement	Unlikely	Unlikely	Likely	Unlikely (low)	Unlikely

†In a warming world, there will be fewer health impacts attributed to cold effects.
‡The European heatwave of summer 2003 has been blamed for an excess mortality of 35,000 cases, of which 42% occurred in France. Most deaths involved the aged, persons 75 years and older. Many deaths could have been prevented through greater social engagement (looking after vulnerable individuals) and widespread availability of and access to air conditioned spaces.
Source: Adapted from Ebi (2011)

(McMichael et al., 2004). And, climate-attributable estimates are likely to double or even triple by 2030 (UNFCCC, 2007). However, uncertainties surrounding these climate-health relationships are considerable. These uncertainties stem from gaps in knowledge, lack of systematic and longitudinal empirical studies, modeling complexities of the spatial-temporal behavior of disease distribution and transmission rate, the role of biological adaptation, socioeconomic development and adaptation, health care and public health infrastructure.

Table 14.2 Projected estimates of additional cases of diarrhea, malaria and malnutrition by world region in 2030 for different atmospheric CO_2 stabilization scenarios (in thousands)

WHO region†	*Additional cases of*					
	Diarrhea		*Malaria*		*Malnutrition*	
	550 ppmv	*750 ppmv*	*550 ppmv*	*750 ppmv*	*550 ppmv*	*750 ppmv*
Africa	42,000	50,300	14,200	17,700	330	440
Americas	1,500	1,500	320	260	90	200
Eastern Mediterranean	5,800	5,800	2,500	3,200	340	530
Europe	790	790	–	–	–	–
Southeast Asia	63,100	73,600	–	70	2,300	3,300
Western Pacific	–	–	420	480	70	210
Total	**113,100**	**132,000**	**17,400**	**21,800**	**3,100**	**4,700**
Current burden	***4,514,000***		***408,000***		***46,400***	

†http://www.who.int/healthinfo/global_burden_disease/definition_regions/en/index.html
Source: Adapted from UNFCCC (2007)

14.2 Adaptation measures in the health sector

Health adaptation refers to any intervention meant to avoid or reduce risks related to climate change on human health. Adaptation is not a new concept as humans have always adapted to the variability of climatic conditions, and measures developed to address future climate change are therefore the same as in the current context.

Table 14.3 Climate change–attributable health impacts by region and by cause in the year 2000

WHO region	*Number of deaths and disease burden (DALYs) in thousands*							*All cause per million people*	
	Diarrhea		*Malaria*		*Malnutrition*		*Temperature*		
	Deaths	*DALYs*	*Deaths*	*DALYs*	*Deaths*	*DALYs*	*Deaths*[†]	*Deaths*	*DALYs*
Africa	13	414	23	860	17	616	2	90	3,067
Americas	1	17	<1	3	–	–	1	5	117
Eastern Mediterranean	8	291	3	112	9	313	1	42	1,586
Europe	< 1	9	–	–	–	–	<1	<1	18
Southeast Asia	23	640	–	–	52	1,918	8	56	1,704
Western Pacific	2	89	1	43	–	–	<1	2	104
Total	**47**	**1,459**	**27**	**1,018**	**77**	**2,846**	**12**	**28**	**925**

[†]Cardio-Vascular Disease (CVD) short-term temperature-related advanced (premature or displaced) deaths (acute mortality). These estimates do not include the mortality burden associated with indirect effects of extreme temperatures. Heat-related displaced deaths for the 2003 heatwave in France contributed only a modest share of the total heatwave mortality burden (Le Tertre et al., 2006; Toulemon et al., 2008).

Table notes:

- Disability adjusted life years (DALY) measures the burden of disease due to ill-health, disability and premature death (http://www.who.int/healthinfo/global_burden_disease/metrics_daly/en/).
- Diarrhea, malaria and malnutrition are important climate-sensitive health burdens.

Source: Adapted from McMichael et al. (2004)

Adaptation includes a wide variety of actions starting from physiological and physical acclimatization, behavioural changes, surveillance programs, medical treatments, health educational campaigns and structural and technological solutions. Actions are tailored to address specific diseases, usually at the local level. Tables 14.4–14.8 present a brief synthesis of the main types of interventions, assembled in broad categories, for a number of climate-related health outcomes, including heat stresses, extreme weather related impacts and vector-, food- and water-borne diseases. It must be noted that in some circumstances, there is a need for transnational coordination to set up appropriate programs, such as in the case of flooding, as decisions to reduce risks must often be taken at the catchment level with issues related to transboundary water.

Health interventions can be classified into different categories (as can be seen in Tables 14.4–14.8), some of which are relevant for the economic assessment of costs and benefits of adaptation. The first important distinction is between preventive and reactive measures. Preventive actions are directed to prevent health risks, while reactive actions are put in place to reduce the impact once occurred. Warning systems, flood protection and water and sanitation programs are examples of preventive actions, while curative care and administration of drugs are examples of reactive interventions. Surveillance and monitoring programs are essential for infectious diseases in order to predict and reduce eventual epidemics and to gather information about contributing factors. The assessment of costs and benefits is framed differently depending on the type of intervention. Preventive measures target vulnerable areas and

Table 14.4 Adaptation measures: Heat waves

Potential health risks	*Adaptation measure*	*Description*
Hypothermia, dehydration, heat exhaustion, heat stroke, respiratory and cardiovascular diseases in general and related mortality	Building design and technical solutions for indoor environment	Thermal buildings insulation, use of fans coolers and air conditioning help in reducing the indoor temperature exposure. In developed countries, people spent a lot of time in indoor environments (home, work, schools) (Kovats and Jendritzky, 2006).
	Urban planning	Effective measures to reduce urban heat island effects, such as green spaces, green roofs, planting trees in streets and other open places; increasing ventilation between buildings; open spaces, white roofs, fountains. Trees and green spaces provide shades and have a cooling effect while they also contribute to water management and flooding control.
	Heat health warning systems (HHWS)	It aims at warning and advising population and decision makers during heat waves (Koppe et al., 2003). The system includes both preventive and reactive measures, such as accurate heat wave forecasting, dissemination of warning, identification of vulnerable groups, effective response measures, etc.
	Emergency plans and reactive measures	They serve the purpose of assisting the affected population with medical and domiciliary services and transport to the neared hospital when needed. They include for example financial and domiciliary assistance services, accompaniment and transport to emergency medical services.
	Educational campaign	It aims at increasing people's awareness during heat waves suggesting behavioural changes, in order to reduce both mortality and morbidity. The most vulnerable groups of the population are specifically targeted, such as the elderly who have a poorer perception of the phenomenon.

populations, while reactive measures aim at minimizing the impact of exposed populations (curing illness and avoiding deaths). Prevention is playing an important role in adaptation and the related health benefits are expected to exceed the costs (Chiabai et al., 2010; Markandya and Chiabai, 2009).

A second important classification is between planned and autonomous adaptation. The first is defined as the result of a deliberative decision-making process promoted by the health sector or outside the health sector, while the second is the spontaneous adaptation in human behaviour which occurs naturally without any public intervention. The latter includes, for example, changing behaviour, such as in clothing, drinking, eating and hygiene, including healthy lifestyles, practices related to water storage, etc. The issue of autonomous or private adaptation is of great interest as it can be considered as a baseline for deciding upon planned adaptation. On the other hand, behavioural changes are triggered by planned actions such as health educational campaigns.

Table 14.5 Adaptation measures: Extreme weather events

Potential health risks	*Adaptation measure*	*Description*
Deaths, injuries, mental health effects (anxiety, depression), infectious disease outbreaks	Structural measures aiming at reducing intensity and occurrence of flooding	Engineering and technological solutions to reduce exposure of people as well as their material assets, such as flood protection structures (e.g., dams, dykes, walls and raised banks, pump stations), river channelization, bridges. Other measures related to the ecosystem such as reforestation, soil protection, restoration of riparian zones.
	Land-use and urban planning aiming at reducing impacts of flooding	These include flood-resistant buildings, building guidelines and urban planning regulation.
	Early warning systems and real-time forecasting	The system includes both preventive and reactive measures, such as weather forecasting, dissemination of warning (behaviour during and after the flood, general hygiene), identification of vulnerable groups, effective response measures, etc.
	Emergency plans and reactive measures	Pre-disaster recovery plans, first aid and emergency plans, temporary evacuation; post-flooding counselling; measures to treat mental health impacts (psychological or pharmacological measures, or financial assistance [Ebi, 2006]). These measures can be part of an early warning system.
	Disease surveillance	Surveillance of infectious diseases, including water-borne diseases, as well as mental health of affected population. Necessity of vaccination in case of outbreaks, control for drinking water, advice for boiling water and general hygiene, etc.

Lastly, we can distinguish between health interventions related to the environment or not. Environmental health interventions refer to all the measures meant to prevent and control disease, injury and disability associated with the environment. It is estimated that 25% of the global burden of disease and mortality is due to environmental factors, including exposure to hazardous substances, natural disasters, climate change, nutritional problems and the built environment (WHO, 2006). Environmental health interventions in the climate change context can refer to water, air quality, ecosystems, extreme weather events, oceans and coastal zones (Portier et al., 2010). Examples of these interventions include improved water and sanitation systems; improved food production, storage and packaging; food-borne surveillance; safe building materials; ventilation; reforestation; vector control (chemical, biological, environmental management and personal protection) and reduced air pollution (WHO, 2000). Non-environmental health intervention refers instead to curative medicines and treatments (e.g., oral rehydration therapy, antibiotics), case detection, breastfeeding, vaccination, impregnated bed nets and health education. Both types of interventions are relevant when assessing health adaptation.

Table 14.6 Adaptation measures: Vector-borne diseases

Potential health risks	*Adaptation measure*	*Description*
Malaria, leishmaniasis, dengue, lyme, tick-borne encephalitis, West-Nile virus	Vector control	It can include destruction of the vector habitat with establishment of new conditions, larval control, use of insecticides indoor, repellents and impregnated bednets. Planning of city parks and controlled burning of vegetation can be effective for reducing lyme disease (Lindgren and Jaenson, 2006).
	Vaccination	It can be targeted to humans or animals, but not always available.
	Diagnosis and treatment	Early detection of the disease and treatment with appropriate medicines and drugs (e.g., for malaria rapid diagnostic test RDT and artemisinin-based combination therapy ACT). Problems of drug resistance.
	Information and health education	Increased awareness in the exposed population about risks (symptoms, diagnosis) and suggested preventive behaviour (proper clothing, use of repellents, impregnated bednets, self-inspection, etc.)
	Surveillance and monitoring	Disease surveillance and monitoring includes activities meant to observe, predict and control the spread of disease and reduce the impacts on population in case of outbreaks. This can be done by increasing the knowledge about contributing factors.

14.3 Prioritizing goals in health protection

Adapting to the health impacts of climate change is a complex and multifaceted process where different institutions need to be involved. In order to develop a portfolio with an appropriate combination of measures to reduce health risks, decision makers need information on the possible impacts of climate change (current and future risks), available adaptation options, financial resources needed and benefits provided. When making decisions affecting a population's health, policy makers face a trade-off between preserving life and the costs associated with this decision. The costs of many programs, even if producing considerable health benefits, are often too high to be supported, especially in poor countries. As a matter of fact, in a world of scarce resources, rational decisions can only be based on the comparison between the costs associated with an intervention and its expected benefits to society.

In this respect, different approaches are available to set priorities and choose the most suitable mix of measures. The most popular approaches include the cost-effectiveness analysis (CEA), the cost-benefit analysis (CBA), and the multi-criteria analysis (MCA) (Markandya and Watkiss, 2009; UNFCCC, 2009). CEA compares monetary costs of an intervention with the health impacts estimated in physical terms. It is usually expressed as a ratio where the numerator includes the costs of the health measure and care provided, and the denominator is the health benefit in natural units, which can be represented by number of deaths or cases avoided or by other indicators such as quality-adjusted life-years (QALY) or

Table 14.7 Adaptation measures: Food-borne diseases

Potential health risks	*Adaptation measure*	*Description*
Salmonellosis, campylobacter, diarrhoea	Surveillance and monitoring	Observation of outbreaks and control of transmission, based on information about number of people affected, contributing factors, food contamination and places of purchasing and consumption, control of sources (WHO, 2008; WHO Surveillance Programme for Control of Food-borne Diseases in Europe, Kovats and Tirado, 2006). Zoonosis program to control disease in animals (e.g., controls for salmonella).
	Microbiological risk assessment	The objective is to assess a model for exposure and dose-response by identifying how disease incidence in humans is influenced by factors related to food preparation, processing, handling, storage, consumption, etc. (for salmonella and campylobacter in eggs and chicken) (Kovats and Tirado, 2006).
	Technological solutions as a prevention	Food sanitation and hygiene: refrigeration, chlorination of drinking water, pasteurization of milk, shellfish monitoring, sanitary slaughter and processing of meat, poultry and seafood, irradiation, microbial reduction for raw agricultural commodities. It is required that food industries can identify points in food production where contamination may occur (Altekruse et al., 1997).
	Diagnosis and treatment	Early detection of the disease and treatment with appropriate medicines and drugs (e.g., oral and intravenous rehydration, antibiotic, penicillin and antimicrobial therapy in severe cases, gastric lavage).
	Food safety education	Avoiding high risky food (such as runny eggs and raw shellfish), separating cooked and raw food, washing hands, cutting boards and contaminated surfaces.

disability-adjusted life-years (DALY).[3] The objective is to identify, among a set of options, the one providing the lowest cost for an expected outcome, or the highest benefit given the available resources. CBA, on the other hand, consists of placing a monetary value on both costs and benefits of a program, which makes them comparable on a common unit of measurement. An adaptation program can be considered attractive if total benefits exceed total costs.[4] The monetization of human health, required in the application of CBA, has triggered many controversies and disagreements in the scientific community due to the ethical issue that it is impossible to monetize loss of life (Hutton, 2000; Cookson, 2003). Different methodologies have been, however, developed by economists for this purpose, the most popular of which is based on non-market valuation through the well-known techniques of revealed and stated preferences (e.g., Mitchell and Carson, 1989).

Table 14.8 Adaptation measures: Water-borne diseases

Potential health risks	*Adaptation measure*	*Description*
Cholera, diarrhoea, typhoid, infectious hepatitis, amoebic and bacillary dysentery, cryptosporidiosis, skin and eye infections (trachoma, skin ulcers, scabies, conjunctivitis)	Structural measures: water and sanitation programs	Basic and low technology improvements: improved access to water sources and measures to protect water from contamination such as stand post, collected rain water; improved sanitation including sewer connection, septic tank, simple pit latrine, ventilated improved pit latrine. Further improvements: water disinfection at the point-of-use (chlorine), high technology improvements such as safe drinking water, household connection to regulated water supply and sewerage system, treatment of sewage (Hutton and Haler, 2004).
	Surveillance and monitoring	Disease surveillance and monitoring includes activities meant to observe, predict and control the spread of disease and reduce the impacts on population in case of outbreaks, including vaccination or water treatment. Sampling of surface water sources during and after rainfall extreme events to detect microbiological contamination (Kovats and Tirado, 2006).
	Diagnosis and treatment	Early detection of the disease and treatment with appropriate medicines and drugs (e.g., oral and intravenous rehydration, antibiotic and antimicrobial therapy in severe cases).
	Information and health education	Increased awareness in the exposed population about risks (symptoms, diagnosis) and suggested preventive behaviour (personal hygiene education, washing hands, bottled water etc.)

Non-market valuation developed within environmental economics in the 1970s, and was later adopted in health economics. However, its application in the two fields, environmental and health economics, has followed totally different patterns (Johnson and Adamowicz, 2010). Environmental goods and services are public goods and as such, their supply and safeguard is of the responsibility of the public or government sector. In this context, environmental economics is mainly concerned with the correction of market failures, and the use of non-market valuation (and CBA) has been urged by numerous governmental directives which have provided substantial resources for funding, as extensively explained by Johnson and Adamowicz (2010) (US Environmental Protection Agency, case of Exxon Valdez oil spill). Health economics, on the other hand, has undergone a different development, and assessments have focused more on private goods such as individual preferences for health treatments measured in physical units of avoided cases/deaths or risk reduction. The monetization of human health using non-market valuation methods is therefore much more incomplete and less advanced, with little use of CBA to inform decision-making in the health sector.[5] For this reason, public health

Table 14.9 Methodological approaches for setting priorities in health adaptation: Synthesis

Method	*CEA*	*CBA*	*MCA*
Objectives	Compare monetary costs with physical benefits	Compare monetary costs and benefits	Define a number of objectives to be attained for each option assigning weights
Output	Cost-effectiveness ratio ($/case or death avoided): identification of lowest cost for a certain outcome or highest benefit given available resources	Cost benefit ratio, net present value (C − B < 0, >0, =0; C/B <1, >1, =1) andinternal rate of return	Ranking of options evaluated against specific weights
Advantages	Physical benefits easier to quantify than monetary benefits	Comparability of benefits measured with the same unit of measurement, possible input to CEA and MCA	Assessment of distributional impacts, use of evaluation criteria different from the monetary one and when an impact cannot be quantitatively measured
Limitations	Benefits not comparable when expressed in different units (in measures with multiple benefits)	Problems in monetizing health benefits, ethical considerations, equity issues, no account for distributional impacts, not always possible to include all affected groups	Subjectivity of the attribution of weights and final ranking (depends on the stakeholders' views), complexity and timespan of the consultation process

makes a predominant use of CEA and cost per quality adjusted life years (QALY) (Mills and Bradley, 1986; Drummond and McGuire, 2001). As acknowledged by UNFCCC (2009), the use of CBA in the adaptation context is actually restrained due to lack of data, especially on the benefit side, and because the distributional impacts and related social inequality are usually disregarded in this approach. Indeed, even in the case that the benefits are not worth the costs, there might be nevertheless some positive distributional impacts to vulnerable groups which might justify the implementation of the measure. In such cases, a multicriteria analysis is able to assess the distributional impact of the intervention by assigning weights to winners and losers affected (Dasgupta, 1972). However, to date very few applications of MCA exist in the health sector (Baltussen and Niessen, 2006).

MCA can offer good guidance to policy makers on setting priorities as it takes into account different criteria simultaneously. It defines a number of objectives to be attained for each option, considering that some impacts cannot be translated into monetary values. These objectives (or decision criteria) are evaluated against specific weights attributed, according to which a score is assigned based on stakeholders' preferences, and the option with the highest score is chosen. The three methods, CEA, CBA and MCA, are briefly reported in Table 14.9, highlighting the main advantages and limitations.

While the main benefits of adaptation are a reduction of the expected residual damages of climate change, other important benefits may also arise, known as ancillary (or indirect) benefits, in terms of both health (e.g., reduced mortality and morbidity risks due to decrease of local pollutants, or arising from current climate variability) and non-health gains (e.g., impacts on economic productivity due to reduced illness, on health expenditures due to reduced medical costs and on environmental assets such as recreational or passive use values).

14.3.1 Going beyond CEA

When health interventions have the characteristics of public goods (e.g., vector control programs), the two fields, health and environmental economics, share some common methodological issues. This is the case of health impacts related to environmental quality, as observed in a climate change context (effects of temperature increase or extreme rainfall on human health). Questions related to the use of scarce resources, latency and discounting, altruism and future generations, uncertainty and irreversibility and ancillary benefits are common to both fields when it comes to economic assessment. In this setting, there are opportunities for "hybridization between environmental and health economics" by extending the existing knowledge on CBA and non-market valuation in the assessment of health adaptation, and applying a common methodological framework (Johnson and Adamowicz, 2010). Smith (2003) argues that some guidelines would be helpful to identify similarities and divergences between the two fields and the extent of application of the NOAA (National Oceanic and Atmospheric Administration) standards developed for environmental goods valuation (see, for example, Frew, 2010).

Many studies currently available in the literature[6] are not able to quantify the overall expected health benefits of many interventions, as they ignore important additional potential benefits at the individual and social level, such as reduction in suffering, psychological discomfort, loss of leisure time and productivity, avoided costs (travel for treatment, child care, health care costs, etc.) and altruism (avoiding illness to others). All these benefits might be estimated using non-market valuation techniques "borrowed" from environmental economics. Non-market values such as value of statistical life have been actually used in some cases in Integrated Assessment Models to quantify the health benefits (Bosello et al., 2006). We discuss two main approaches which can be used in this context and their applicability in the context of health adaptation. These are the conventional market prices, and the willingness to pay approach (WTP) based on revealed or stated preferences (Drummond et al., 1997).

Market prices are used within the cost of illness and human capital approaches to value mortality and morbidity risks. The human capital approach is based on the loss of human capital due to loss of productivity and foregone earnings. It estimates the work loss days or restricted activity days due to morbidity or the stream of foregone revenues attributable to premature deaths (Rice and Cooper, 1967; Landefeld and Seskin, 1982; Grossman, 2000). A proxy for the value of life is calculated as the discounted sum of individuals' future contribution to the social product.

A preferred method is the WTP approach, which is defined as the maximum amount of money an individual is willing to pay in order to obtain a risk reduction in mortality or morbidity. This measure is used to value goods which are not traded in regular markets and they are, for this reason, known as "non-marketed assets or goods". Health and wellbeing is one of them. The WTP approach is the only one that can capture values[7] such as pain and suffering, dread, psychological costs, disruption of daily life, loss of leisure time and wellbeing in general.

Deaths are valued through the use of the value of statistical life (VSL), estimated from the WTP and defined as the rate of substitution between income and a small reduction in mortality risk (Hammitt and Treich, 2007).

WTP and VSL can be estimated through revealed or stated preference methods. Revealed preferences use market-based data to infer the willingness to pay for mortality risk reductions. This can be done looking at the labor markets (compensating or hedonic wage studies) where risk preferences are estimated through wage differentials in risky jobs and industries. Other approaches look at consumers' behavior, for example in terms of house location in sites benefiting from good environmental quality (Maddison and Bigano, 2003). The advantage of revealed preference methods is that they rely on real market data rather than hypothetical scenarios. On the other hand, there are a number of problems which limit their application in the context of health adaptation. First, people having risky jobs are usually not representative of the population targeted in adaptation to climate change. The latter depends on the type of health outcome considered, but in general, children and older people are more vulnerable, as well as people having poor health status. Second, people tend to value differently voluntary and involuntary risks, as well as risks related to different contexts (risky jobs, smoking, climate change).

Approaches based on stated preferences are able to overcome these limitations. The estimation of the WTP is done following two main methodologies, contingent valuation where respondents state their WTP for changes in health risks (e.g., Viscusi, 1993; Hammitt and Graham, 1999; Klose, 1999; Smith, 2003; Frew, 2010), or discrete choice modeling where respondents are asked to choose between alternative scenarios with different health risks (e.g., Bennett and Blamey, 2001; Telser and Zweifel, 2002; Carson and Louviere, 2010). One important issue with stated preference techniques is that respondents are presented with hypothetical scenarios, which can undermine the validity of the study if not presented accurately (Mitchell and Carson, 1989; Bateman et al., 2002).[8]

In conclusion, we argue that further research is needed to incorporate CBA and non-market valuation in the assessment of health adaptation, while incorporating appropriate analysis of uncertainty.

14.4 Costing adaptation to climate change

14.4.1 Adaptation costs

Adaptation refers to measures intended to reduce environmental and infrastructural impacts, economic losses and community vulnerability to climate change in contrast to mitigation interventions, whose scope is to limit or control climate variability (GHG emissions). Adaptation costs are deficits over and above the costs of country development necessary to cope with future climate change. Interventions aim to: protect natural resources (ecosystems, forests, fisheries and water supply), maintain agricultural productivity (food security), preserve landscape and physical property (preservation of coastal zones and infrastructure maintenance and *climate proofing*), limit socioeconomic disruptions and population displacement and lessen risks and disease burdens from exposure to climate extremes. Indirect consequences of climate change are also important. Prices of products are likely to be substantially higher because costs for energy, raw materials, insurance and taxes will increase. Higher costs for doing business influence negatively industry competitiveness and economic growth, leading to loss of employment and overall economic security, and therefore impacting adversely human development. Perturbations to investment and financial flows to help maintain viable markets for goods and services (resource diversion) increase the risk of economic downturn and conflict, and may pose serious threats to community and national security.

Access to technology and financing, availability and access to medical services and public awareness and education are crucial to enhance community resilience to global climate change. Strategic planning, mitigation and/or adaptation responses can significantly reduce, but not eliminate completely, the adverse effects of climate change. The risks and burdens to future generations, the residual climate change–induced impacts and damage costs are still significant (Figure 14.2). Adaptation is coping with climate change (reactive measures); whereas mitigation addresses the causes that lead to long-term[9] trend changes in meteorological conditions (preventive actions). Mitigation activities focus on structural (technology) and behavioral (people) options to abate GHG emissions. While adaptation is more cost effective than mitigation in limiting residual climate costs in the short term, mitigation investments will result in lower marginal and aggregate residual costs over the long term future (De Bruin et al., 2009). The difference in lag times is attributable to atmospheric inertia. Along with other exogenous parameters, such as GDP and population growth rates, the discount rate applied to future costs and benefits is a key parameter in the macro-economic cost-benefit assessment. As illustrated in Table 14.10, the total cost can vary several fold between different discounting methods. Mitigation costs are especially sensitive to assumptions about the discount rate, with variations across methods easily exceeding an order of magnitude. This aspect is very relevant to policy and decision making, as financial budgets are finite and misallocation of resources severely impacts future economic growth and welfare.

The global economic impacts of climate change could be significant. Tol et al. (2004) estimated an annual burden between 300 and 350 billion USD (B$), or about 1% of world GDP. A study by the World Bank (2006) put the estimate at 0.5% to 2% of world GDP, assuming a 2.5°C mean global temperature increase in the absence of adaption measures. Ciscar et al. (2009) estimated that the health impacts of climate change in 2020 could fall in the range of 13 and 30 billion $€_{2005}$. By 2100, the estimate would escalate to between 50 and 180 billion $€_{2005}$ without acclimatization and between 8 and 80 billion $€_{2005}$ with acclimatization (under the A2 scenario). The IPCC (2007)[10] estimated a global economic value for the loss of life in the range of 6 to 88 billion USD_{1990}. The assessment by De Bruin et al. (2009) suggests there will be strong regional disparities in GDP losses, with India and Africa especially impacted. Health effects dominate damage costs in Africa (75% of total cost, or 4%–5% GDP loss), whereas damages in India arise from agricultural losses, health effects and from consequences attributed to extreme weather events (4%–5% GDP loss).

Current adaptation cost estimates range in the tens of billions of dollars per year (Figure 14.2). An assessment by the United Nations Framework Convention on Climate Change (UNFCCC, 2007) puts the global price tag between B$ 50 and 170 by 2030. Roughly, half of this investment would be needed in developing countries, whose populations will suffer the greatest burdens of climate change. The remaining half would be spent in rich countries to reduce disruption and loss of economic activity and to lessen damages to property, especially impacts to urban infrastructure and coastal areas. Hurricane Katrina in 2005, for example, has been blamed for 1,800 deaths in the US (people who died in the hurricane and subsequent floods), and as having contributed to an economic loss in excess of B$ 100 (NOAA, 2011), or about 1% of 2005 US GDP.

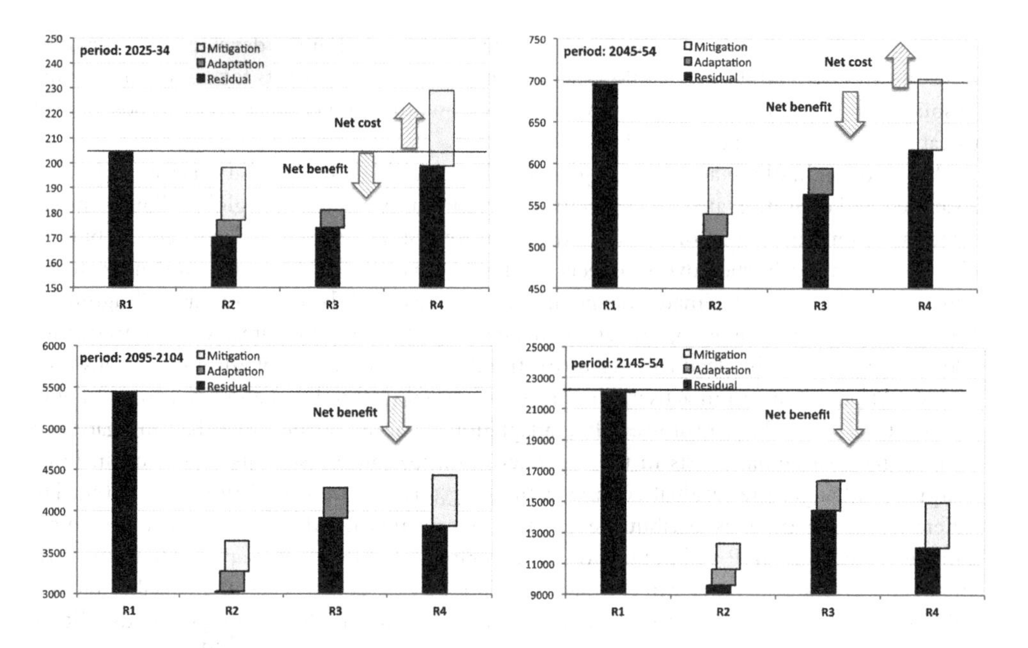

Figure 14.2 Global costs of climate change in billions USD per year
Source: Adapted from De Bruin et al., 2009

Reference Scenario:
R1 – No Mitigation or Adaptation measures (no control or climate policies)
R2 – Both Mitigation and Adaptation measures are applied (optimal control)
R3 – Only Adaptation measures are applied (no mitigation, optimal adaptation)
R4 – Only Mitigation measures are applied (optimal mitigation, no adaptation)

From a methodological point of view, the costs of adaptation can be estimated using a top-down (macro level) or a bottom-up (disaggregated) assessment (Markandya and Watkiss, 2009). The first approach provides aggregate results in terms of geographical areas and sectors, whereas the latter method delivers results disaggregated by geographical range and by sector. Within the macro-level studies, there are two main broad methodologies (UNFCCC, 2009): (i) economic integrated assessment modeling (IAM) and (ii) investment and financial flow analysis (I&FF). The former provides estimates of adaptation costs and benefits and residual climate-caused damages at a global and continental scale. The I&FF approach is a financial analysis and considers only the costs of adaptation at a continental scale, with future estimates for broad economic sectors, such as health, agriculture, infrastructure, coastal zones, etc. Most studies rely on a top-down analysis (Stern 2006; UNFCCC, 2007 and World Bank, 2010a). De Bruin et al. (2009) used a dynamic integrated model for climate and the economy (DICE) for projecting long-term mitigation, adaptation and residual damages under new climate conditions (Figure 14.2 and Table 14.10).

Several studies have attempted to quantify the costs of adaptive actions (e.g., World Bank, 2006; Stern Review, 2006; OXFAM, 2007; UNDP, 2007; UNFCCC, 2007 and World Bank,

Table 14.10 Climate cost composition for the optimal scenario (2005–2105, trillion $) under time-sensitive discount rate assumptions

Discount method (variable)	*Adaptation*	*Mitigation*	*Residual*	*Total*
Nordhaus (4.5 → 4%)	10.5	16.5	139	166
UK Treasury (3.5 → 2.8%)	9.9	48.1	137	195
Stern review (2.5 → 1.4%)	4.3	342	79.6	426

Source: Adapted from De Bruin et al. (2009)

2010a) and their effectiveness (Agrawala and Frankhauser, 2008; Hope 2009 and De Bruin et al., 2009). Figure 14.3 compares adaptation cost ranges for low and high income countries (numerical values are presented in Table 14.11). Estimates across studies vary because of different choices of input assumptions and accounting methodologies. Some studies cost adaptation by estimating what percentage of gross domestic and foreign investment budgets, plus foreign aid is climate sensitive (IF&F). A mark-up factor is then applied to account for costs of climate-proofing investments (UNFCCC 2007). However, there is little empirical evidence as to what are appropriate values for these input factors (Parry et al., 2009). The 2006 Stern Review assumed between 2% and 10% of gross domestic investments (GDI), 10% of foreign investments (FDI) and 20% of official development assistance (ODA) would be needed to cope with future climate change risks. A mark-up factor in the range of 5% to 20% was assumed. The annual cost of adaptation for developing countries in 2010 was therefore estimated to range between B$ 4 and 37 (Figure 14.3). The World Bank (2006) study assumed 2%–10% for GDI, 10% for FDI, 40% for ODA and a mark-up factor 10%–20%. The UNDP assessment included the additional costs to improve disaster response systems (B$ 2 per year) plus social protection programs designed to reduce population poverty (B$ 42 per year). The OXFAM study used a different approach that scaled up investments of community-specific projects and country-level National Adaptation Programme of Action (NAPA) programs to all other developing countries, in which limited or no information was available using scaling parameters (country GDP, for instance).

The UNFCCC (2007) and World Bank (2010a) (Economics of Adaptation to Climate Change, EACC) global studies implemented a more detailed approach (Tables 14.2 and 14.11), disaggregating the results by geographical region (Africa, Asia, Europe, etc.) and then by major economic sector, including health (malnutrition and a few infectious diseases). The World Bank has also carried out a limited number of country-specific case studies[11] (Bangladesh, Bolivia, Ethiopia, Ghana, Mozambique, Samoa and Viet Nam) using a sectoral-level, bottom-up approach that incorporated public, private and community-based adaptation actions and macroeconomic modeling to capture cross-sectoral interactions (Computable General Equilibrium, CGE, modeling).

The World Bank (2010a) analysis projected what the world might look like in 2050 considering alternative climate forecasts and sectoral development pathways (baselines), including a growth trajectory without climate change. Estimates of impacts to developing countries from climate change were assessed for various economic activities (agriculture, fisheries, and ecosystem services), human health (limited to prevention and treatment of cases

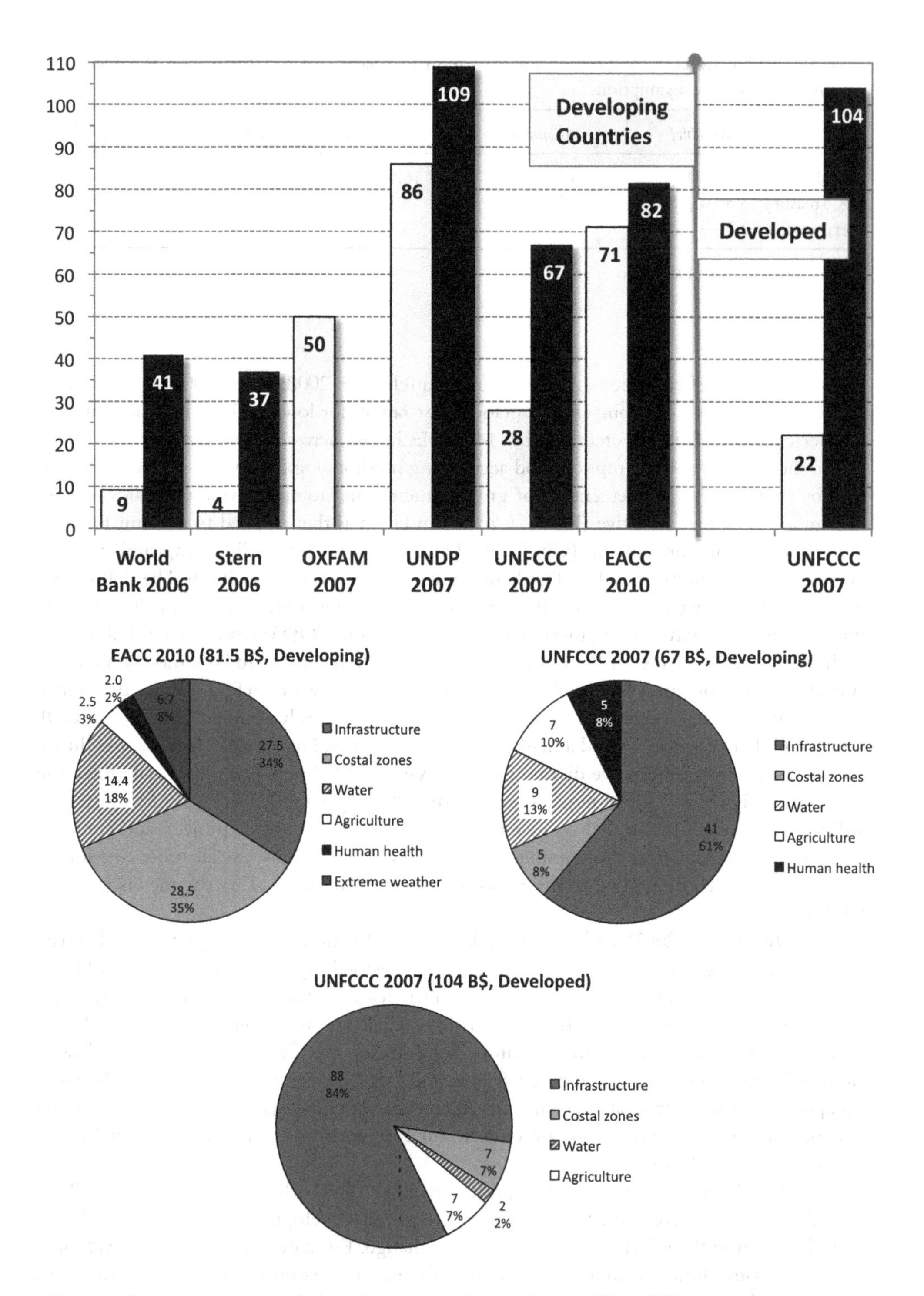

Figure 14.3 Annual estimates of adaptation costs to climate change, billion USD per year.
UNDP estimates include costs for disaster response systems (2 B$) and poverty reduction programs (40 B$).
Source: Authors' work

Table 14.11 Comparisons of adaptation cost estimates of climate change in billions of USD per year

	World Bank 2006	Stern Review 2006	OXFAM 2007	UNDP 2007	UNFCCC 2007 (low)	UNFCCC 2007 (high)	EACC 2010 (X-sum, low)	EACC 2010 (X-sum, high)	UNFCCC 2007 (low)	UNFCCC 2007 (high)
Method	IF&F	IF&F	NAPA	IF&F	Sectoral	Sectoral	Sectoral	Sectoral	Sectoral	Sectoral
Timeframe	2010	2010	2010	2015	2030	2030	2010-50	2010-50	2030	2030
Low	**9**	**4**	**50**	**86**	**28**		**71**		**22**	
High	**41**	**37**		**109**		**67**		**82**		**104**
PRS				40						
DRS				2						
Infrastructure					2	41	13.0	27.5	6	88
Costal zones					5	5	27.6	28.5	7	7
Water					9	9	19.7	14.4	2	2
Agriculture					7	7	3.0	2.5	7	7
Human health					5	5	1.5	2.0	–	–
Extreme weather					–	–	6.4	6.7	–	–
	DEVELOPING COUNTRIES								DEVELOPED	

Table notes:

- I&FF – Investment and Financial Flows for addressing climate sensitivity and climate proofing (UNDP 2007 incl., PRS and DRS actions); PRS – Poverty Reduction Strategies and DRS – Disaster Response System
- NAPA – upscaling of National Adaptation Programme and Action budgets; 8 to 33 B$ (total) needed to cover most urgent priorities
- EACC 2010 – Economics of Adaptation to Climate Change (World Bank 2010a); adapt up to a level at which future welfare is restored
- X-sum – potential gains offset costs at the country level and only net positive costs (> 0) are aggregated across countries

Source: Authors' own work

of diarrhea and malaria), environmental conditions (water supply and forests) and physical damage to natural and built-up infrastructure. Other than the cost implications of extreme weather events, no cross-sectoral feedback loops were analyzed. Costs of (optimal)[12] adaptation included measures adopted by countries to offset the adverse risks of climate change – that is to say, restore sectoral welfare in the future world to pre-climate change standards. Only public decisions or planned adaptive actions were assessed. The potential benefits of adaptation measures taken at the household- or community-level, by NGOs and the private sector, were not considered.

Details of the cost composition by sector, by geographical region (developing countries) and by time slice, cost aggregation method and assumptions about future climate variability are presented in Tables 14.12 and 14.13 and Figure 14.4 (World Bank, 2010a). Three aggregation methods are offered: gross, net and partial (X-sum) accounting. For the gross estimates, the benefits or gains realized from improved climate conditions are excluded; in the net assessment, deficits are offset by gains within and across countries, whereas in the partial accounting framework, the potential gains balance costs at the country level and only net positive costs (>0) are aggregated across countries (that is to say, benefits in one country are not offset by deficits in another). To capture as wide as possible impact range, two future climate projections were considered: an extreme wet and an extreme dry scenario.

Adapting to a warmer world, in which global ambient temperature will increase 2 °C above pre-industrial levels by 2050, will cost on average between 70 and 100 $B\$_{2005}$ per year over

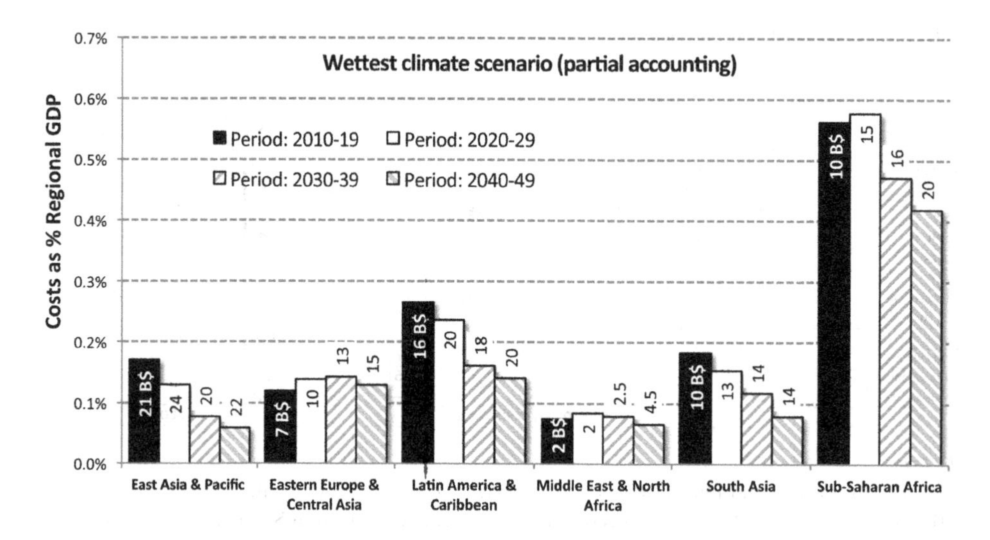

Figure 14.4 Projected adaptation costs by region and decade (billions $\$_{2005}$ per year, undiscounted)
Source: Adapted from World Bank, 2010a

the time horizon 2010–2050, which represents roughly 0.2% of GDP of developing countries. On a sectoral basis, infrastructure, costal zones and water supply and flood control categories dominate the sum, with a sub-total of around 85%. Cost estimates for coastal zones are higher than previous results (UNFCCC, 2007) due to additional investments related to sea level rise and storm surges. Wet-scenario estimates are higher than dry-scenario results primarily because of additional costs to protect and repair infrastructure. The level of adaptation cost is similar in magnitude to the current level of financial assistance received by developing nations.

East Asia and the Pacific region have the highest adaptation costs, whereas the Middle East and North Africa have the lowest. Not surprisingly, the absolute annual cost increases over time, and the change is uneven across regions. As a percent of regional GDP, the cost of adaptation is highest for Sub-Saharan countries, largely because of water supply investments to deal with future changes in precipitation rates. As nations become wealthier, their exposure (vulnerability) to climate change decreases.

Over the 40-year time period 2010–2050, the average adaptation cost in the health sector is roughly 2 B$$_{2005}$ annually (World Bank, 2010b). This cost estimate is for prevention and treatment of climate-sensitive excess cases of diarrhea and malaria. More than 90% of the adaptation cost is for managing the burden of diarrheal disease. The health burden is expected to decline over time, although unevenly across regions. South Asia and Sub-Saharan countries suffer the highest burden of new incidences, and consequently bear the highest costs of adaptation. Sub-Saharan Africa accounts for almost all new cases of malaria (6 to 8 million new cases by 2050) and, over time, for an increasing share of the total burden of diarrheal disease – reaching 80% of new cases (21,000 to 24,000) by 2050. In terms of disability adjusted life years (DALYs), Africa suffers the greatest loss of any other region, with nearly two-thirds of the diarrheal cases (670,000 to 714,000) by 2030 and upwards of 85% (752,000 to 863,000) by 2050.

Table 14.12 Regional adaptation costs for developing countries (billions $\$_{2005}$ per year, undiscounted)

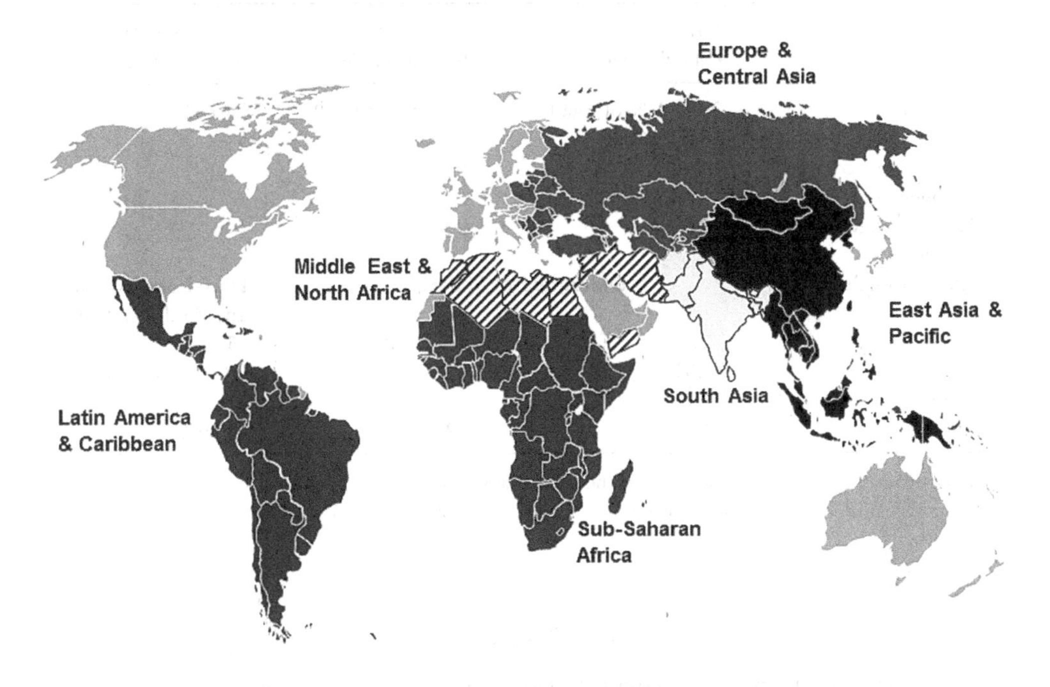

Region	*Gross aggregate cost (excluding benefits of climate change)*	*Net aggregate cost (including benefits of climate change)*	*Partial aggregate cost (sum of net positive costs within and across countries)*
	Wettest climate projection		
East Asia & Pacific (EAP)	25.7	21.7	21.7 (27% of total)
Europe & Central Asia (ECA)	12.6	11.1	11.2 (14%)
Latin Amer. & Caribbean (LAC)	21.3	18.7	18.7 (23%)
Middle East & N. Africa (MENA)	3.6	2.3	2.4 (3%)
South Asia (SAS)	17.1	12.3	12.4 (15%)
Sub-Saharan Africa (SSA)	17.1	14.9	15.1 (19%)
All regions	**97.5**	**81.1**	**81.5**
	Driest climate projection		
East Asia & Pacific (EAP)	20.1	17.7	17.9 (25% of total)
Europe & Central Asia (ECA)	8.1	6.5	6.9 (10%)
Latin Amer. & Caribbean (LAC)	17.9	14.5	14.8 (21%)
Middle East & N. Africa (MENA)	3.5	2.4	2.5 (4%)
South Asia (SAS)	18.7	14.6	15.0 (21%)
Sub-Saharan Africa (SSA)	16.4	13.8	14.1 (20%)
All regions	**84.8**	**69.6**	**71.2**

Source: Economics of Adaptation to Climate Change (EACC), World Bank (2010a)

Table 14.13 Climate adaptation costs for developing countries by sector (billion $ per year)

Sector	*Wettest climate (Partial accounting)*	*Driest climate (Partial accounting)*	*Comparison to (UNFCCC, 2007)*
Infrastructure	27.5 (34% of total)	13.0 (18% of total)	2 to 41
Coastal zones	28.5 (35%)	27.6 (39%)	5
Water supply & flood management	14.4 (18%)	19.7 (28%)	9
Agriculture, forestry & fisheries	2.5	3.0	7
Human health	**Diarrhoea & Malaria**		**Diarrhoea, Malaria and Malnutrition**
	2.0 (2.5% of total)	**1.5 (2% of total)**	**5[†] (7% to 18% of total)**
Extreme weather events	6.7	6.4	
All sectors	**81.5**	**71.2**	**28 to 67**

[†]Malnutrition contributes a mere 2% of the total adaptation cost attributable to human health.
Source: Economics of Adaptation to Climate Change (EACC), World Bank, (2010a)

14.4.2 Uncertainty

Uncertainties of current climate costs (adaptation, mitigation and residual damages) are large due to knowledge gaps in forecasting long-term weather-related effects on natural and social systems and in projecting long-term trends of major socioeconomic modeling parameters (GDP and population). Lack of longitudinal health data (empirical and disaggregated), especially from low-income countries that will be disproportionally affected by climate change, also contributes to uncertainty. Current modeling is still limited to assessing only a few of what are believed to be the most important climate-sensitive health burdens, namely, diarrhea, malaria, malnutrition, direct effects of thermal stress (displaced mortality) and the effects of air pollution on morbidity and mortality of cardio-pulmonary diseases. Developing nations have much higher local incidence rates of malaria, diarrhea and malnutrition than populations living in richer countries, and are, therefore, more vulnerable to the adverse consequences of climate change. The lack of economic resources and limited adaptive capacity of low-income countries to address the additional pressures of future climate change further aggravates the potential health burden.

A great number of studies have focused attention on malaria transmission in Sub-Saharan Africa (only a handful of studies have carried out assessments in other regions); yet, despite the known causal links established between climate and malaria transmission, expressed in the epidemiological literature as an increase in the relative risk above baseline (McMichael et al., 2004), there is much uncertainty about the dynamics of the disease and the role played by non-climatic factors, such as preventive and reactive actions undertaken by the population and the role and effectiveness of clinical care. Although environmental conditions may be favorable for malaria transmission, the number of incidences may be low if there are no infected vectors or there is no human contact (Casimiro et al., 2006). Higher economic prosperity will also limit the spread and occurrence of malaria infections.

14.5 Conclusions

The chapter provides an overview of economics of adaptation in the health sector, in terms of methodological approaches for assessment and estimates available. Health adaptation includes a wide variety of actions starting from physiological acclimatization, behavioural changes, surveillance programs, medical treatments, health educational campaigns, structural and technological solutions. Actions are usually designed at local level to address specific diseases.

Different approaches can be used to prioritize goals in health adaptation and identify the most appropriate set of measures, the most popular being the cost-effectiveness analysis (CEA), the cost-benefit analysis (CBA), and the multi-criteria analysis (MCA). A critical analysis of these methods is provided in the context of environmental and health economics, taking into account the development of these approaches in the related literature. The appropriateness of specific techniques to monetize human health is discussed in the context of adaptation, highlighting their main advantages and limitations.

The main conclusions from a methodological point of view can be summarized in the following points. The use of CBA for assessing adaptation is somehow limited, while CEA is the most used methodology. The reason is due to the limited availability of data on monetary benefits, and the impossibility of taking into account distributional impacts among the population and to vulnerable groups in particular (UNFCCC, 2009). However, we argue that further research is needed to integrate CBA and non-market valuation in the assessment of health adaptation, while incorporating appropriate analysis of uncertainty. Indeed, the studies currently available in the literature are not able to quantify the overall expected health benefits of many interventions, as they ignore important additional potential benefits at the individual and societal level, such as reduction in suffering, psychological discomfort, loss of leisure time and productivity, avoided costs (travel for treatment, child care, health care costs, etc.) and altruism (WTP to avoid illness for others). These benefits can be estimated using non-market valuation techniques "borrowed" from environmental economics, and guidelines are needed to standardize the use of these techniques in health economics.

Ancillary benefits and interventions with multiple benefits across sectors are important factors, but they are difficult to evaluate and to include in a CBA. CEA might be easier to apply in these contexts, but when multiple benefits arise from different areas and are evaluated with different metrics, comparison becomes difficult as well. Multicriteria analysis could play a role in this case, when it comes to assess adaptation options in local contexts where different stakeholders can be involved.

The studies reviewed on costs and benefits of adaptation highlight a limited coverage in general, and regarding the health sector, availability of data is mainly restrained to costs with less information on benefits (Watkiss, 2011 related to the EU project ClimateCost; Hutton, 2011; UNFCCC, 2009; Parry et al., 2009). For this reason, studies outside the climate change context and addressing climate-sensitive diseases are often used (Hutton, 2011; Markandya and Chiabai, 2009; Kiszewski et al., 2007; Hutton and Haler, 2004; Stenberg et al., 2007). The review reports results from different approaches to assess cost of adaptation, top-down and bottom-up, the first providing aggregated results and the second working at a disaggregated scale (Markandya and Watkiss, 2009). A sectoral analysis is available only in disaggregated approaches, while adaptation benefits are mainly assessed in integrated assessment models (top-down).

It is not easy to evaluate whether or not health adaptation is justified based on the existing studies, as comparison with benefits is usually lacking. In a more general context of adaptation (all sectors aggregated), the study of Hope (2009) provides an appraisal of both costs and benefits in different scenarios using a simulation model. His results suggest that benefits largely offset the costs of adaptation.

A number of critical research gaps can be highlighted in the studies assessing costs and benefits of adaptation. Some of the limitations of macro-level studies using I&FF analysis (UNFCCC, 2009) include, for example, the high level of aggregation with limited information at the sectoral scale, the lack of estimates on the benefit side, and the limited empirical evidence on which the climate change-attributed cost of adaptation is calculated. There is still the accounting issue as to whether the adaptation deficit should be assessed in the adaptation costs or as part of country development (Markandya and Watkiss, 2009). As regards the economic IAM, the main limitations concern their complexity and transparency, uncertainty analysis, non-linearity, thresholds and choice of discount rate. Finally, the disaggregated studies (bottom-up approaches) are mainly criticized as they do not include all sectors (health impacts are only partially considered) and all residual damages. Furthermore, costs of adaptation in the health sector are quite low if compared to other sectors as many health endpoints are still missing, which leads to an underestimation of the overall costs of adaptation.

Very few studies, thus far, have undertaken a comprehensive (bottom-up) and dynamic (cross-sectoral) macro-economic analysis Hope (2009), De Bruin et al. (2009) and a few country-level studies lead by the World Bank (http://climatechange.worldbank.org/content/country-case-studies-economics-adaptation-climate-change). Consequently, the health-related adaptation costs of climate change cited in the above text by the UNFCCC (2007) and the World Bank (2010a) are low-ball figures. At best, current literature results of climate costs (and impacts) should be interpreted as preliminary, *order of magnitude risks of climate sensitive impacts and costs*. Obviously, more research is needed to standardize the modeling framework for costing climate change adaptation and to carry out country-based assessments, especially in countries with significant risk of exposure to adverse consequences of climate change (i.e., developing countries). Proper estimates and strategic planning reduce the physical and economic impacts of climate change exposure and help build global and national resilience. Insights are relevant to long-term sustainable development and to set climate policy, particularly when it comes to addressing equity and fairness in the distribution of climate funds.

Acknowledgement

This review has been developed in the course of the BASE project (Bottom-up Climate Adaptation Strategies towards a Sustainable Europe), financed under FP7 program Grant agreement n. 308337.

Notes

1 World Health Organization, WHO, Global Burden of Disease, http://www.who.int/evidence/bod
2 National Aeronautics and Space Administration, NASA, http://www.nasa.gov/topics/earth/features/2012-temps.html
3 The quality-adjusted life year (QALY) is a measure of disease burden which refers to both quality and quantity of life used to guide health-care resource allocation (Weinstein, 2009), while the disability-adjusted life-years (DALY) is the number of years lost due to ill-health, disability or early death (Mathers et al., 2001).

4 Some authors show how to combine the two approaches, the WTP method traditionally used by environmental economists in CBA and quality-adjusted life years commonly used in CEA by health economists (Kenkel, 2006). In the same context, Klose (2003) analyses the links between QALYs and WTP. In contrast, Hammitt (2002) argues that the choice between the two approaches to value health will depend on which factor is given priority: application of egalitarian concepts or no restriction of individual preferences. QALY is favoring egalitarian principles in combining preferences across population, while the WTP approach allows for individual preferences to be determined by income, baseline risks and dread, for example.
5 The use of non-market valuation in the health context started some 10 years later than in the environmental field (1980s), but currently a number of studies exist in this milieu (Richardson, 1999; Ryan et al., 2001; Olsen and Smith, 2001; Hanley et al., 2003). For environmental health programs, see, for example, Krupnick and Portney (1993), Hanley and Spash (1993), US EPA (1999) and Alberini et al. (2007). For non-environmental health programs, see, for example, Masiye and Rehnberg (2005) and Schultz Hansen et al. (2012) focusing on malaria treatments and diagnostic.
6 See the next section for an extensive discussion of available estimates.
7 In the environmental economics context, these values are known as "non-use values".
8 Biases related to communication of risk to the respondents can be overcome using indifference-risk elicitation (Hammitt and Graham, 1999) or through the use of visual aids to communicate levels of risks and reductions (Corso et al., 2001).
9 Short-term meteorological variability is termed weather.
10 http://www.ipcc.ch/publications_and_data/ar4/wg2/en/contents.html
11 http://climatechange.worldbank.org/content/country-case-studies-economics-adaptation-climate-change
12 That is to say, countries should adapt up to point where marginal benefit exceeds marginal adaptation cost.

References

Agrawala, S., and Fankhauser, S., 2008. *Economic Aspects of Adaptation to Climate Change: Costs, Benefits and Policy Instruments*. Paris, France: OECD Publishing.

Alberini, A., Tonin, S., Turvani, M., and Chiabai, A., 2007. Paying for permanence: Public preferences for contaminated site cleanup, *Journal of Risk and Uncertainty* 34, 155–178.

Altekruse, S.F., Cohen, M.L., and Swerdlow, D.L., 1997. Emerging foodborne diseases, perspectives, *Emerging Infectious Diseases* 3(3), 285–293.

Baltussen, R., and Niessen, L., 2006. Priority setting of health interventions: The need for multi-criteria decision analysis, *Cost Effectiveness and Resource Allocation* 4(14).

Bateman, I.J., Carson, R.T., Day, B., Hanemann, M.W., Hanley, N., Hett, T., Jones-Lee, M., Loomes, G., Mourato, S., Ozdemiroglu, E., Pearce, D., Sugden, R., and Swanson, R., 2002. *Economic Valuation with Stated Preference Techniques: A Manual*. Cheltenham, UK: Edward Elgar.

Bennett, J., and Blamey, R., 2001. *The Choice Modelling Approach to Environmental Valuation*. Cheltenham UK: Edward Elgar.

Bosello, F., Roberto R., and Tol, R.S.J. 2006. Economy-wide estimates of the implications of climate change: Human health, *Ecological Economics*, 58(3), 579–591.

Carson, R.T., and Louviere, J.J., 2010. Experimental design and the estimation of willingness to pay in choice experiments for health policy evaluation. In: E. McIntosh, P.M. Clarke, E.J. Frew, J.J. Louvere, eds. 2010. *Applied Methods of Cost-Benefit Analysis in Health Care, Handbooks in Health Economic Evaluation*. Oxford University Press.

Casimiro, E., Calheiros J., Santos, D., and Kovats, S., 2006. National assessment of human health impacts of climate change in Portugal: Approach and key findings, *Environmental Health Perspectives* 114, 1950–56.

Chiabai, A., Balakrishnan, S., Sarangi, G., and Nischal, S., 2010. Human health. In: A. Markandya, A. Mishra, eds. 2010. *Costing Adaptation: Preparing for Climate Change in India*. New Delhi, India: TERI Press.

Ciscar, J.C., and Soria, A., 2009. *Climate Change Impacts in Europe: Final Report of the PESETA Research Project*, European Commission Joint Research Centre.

Cookson, R., 2003. Willingness to pay methods in health care: A sceptical view, *Health Economics* 12, 891–894.

Corso, P.S., Hammitt, J.K., and Graham, J.D., 2001. Valuing mortality-risk reduction: Using visual aids to improve the validity of contingent valuation, *Journal of Risk and Uncertainty* 23(2), 165–184.

Dasgupta, P., 1972. A comparative analysis of the UNIDO Guidelines and the OECD Manual, *Bulletin of the Oxford Institute of Economics and Statistics* 34(1), 33–51.

De Bruin, K., Dellink, R., and Agrawala, S., 2009. *Economic Aspects of Adaptation to Climate Change: Integrated Assessment Modelling of Adaptation Costs and Benefits.* Paris, France: OECD Environment Working Papers 6.

Drummond, M.F., and McGuire, A., 2001. *Economic Evaluation in Health Care: Merging Theory with Practice.* Oxford: Oxford University Press.

Drummond, M.F., O'Brien, B., Stoddart, G., and Torrance, G., 1997. *Methods for the Economic Evaluation of Health Care Programmes.* 2nd Edition. New York: Oxford University Press.

Ebi, K., 2011. Resilience to the health risks of extreme weather events in a changing climate in the United States, *International Journal of Environmental Research and Public Health* 8, 4582–4595.

Ebi, K.L., 2006. Floods and human health. In: B. Menne and K.L. Ebi, eds. 2006. *Climate Change and Adaptation Strategies for Human Health.* WHO, Springer.

Ezzati, M., Hoorn, S.V., Rodgers, A., Lopez, A.D., Mathers, C. D,. and Murray, C.J., 2003. Estimates of global and regional potential health gains from reducing multiple major risk factors, *Lancet* 362, 271–280.

Frew, E.J., 2010. Benefit assessment for cost-benefit analysis studies in health care: A guide to carrying out a stated preference willingness to pay survey in health care. In: E. McIntosh, P.M. Clarke, E.J. Frew, J.J. Louvere, eds. 2010. *Applied Methods of Cost-Benefit Analysis in Health Care, Handbooks in Health Economic Evaluation.* Oxford University Press.

Grossman, M., 2000. The human capital model. In A. Culyer and J. Newhouse, eds., 2000. *Handbook of Health Economics.* Vol. 1A. Amsterdam: Elsevier, 347–408.

Haines, A., Kovats, R.S., Campbell-Lendrum D., and Corvalan, C., 2006. Climate change and human health: Impacts, vulnerability and mitigation, *Lancet* 367, 2101–2109.

Hammitt, J., and Graham, J., 1999. Willingness to pay for health protection: Inadequate sensitivity to probability? *Journal of Risk and Uncertainty* 18, 33–62.

Hammitt, J., and Treich, N., 2007. Statistical vs. identified lives in benefit-cost analysis, *Journal of Risk and Uncertainty* 35, 45–66.

Hammitt, J.K., 2002. QALY versus WTP, *Risk Analysis* 22(5), 985–1001.

Hanley, N., Ryan, M., and Wright, R., 2003. Estimating the monetary value of health care: Lessons from environmental economics, *Health Econ* 12, 3–16.

Hanley, N. and Spash, C. L., 1993. *Cost-benefit Analysis and the Environment.* Cheltenham, UK: Edward Elgar.

Hope, C., 2009. The costs and benefits of adaptation. In: M.L. Parry, N. Arnell, P. Berry, D. Dodman, S. Fankhauser, C. Hope, S. Kovats, N. Nicholls, D. Satterthwaite, R. Tiffin, and T. Wheeler, eds., 2009. *Assessing the Costs of Adaptation to Climate Change: A Review of the UNFCCC and Other Recent Estimates.* London: International Institute for Environment and Development and Grantham Institute for Climate Change.

Hutton, G., 2000. *Consideration in Evaluating the Cost-Effectiveness of Environmental Health Interventions.* Sustainable Development and Healthy Environments Cluster. World Health Organization. WHO/SDE/WSH/00.10.

Hutton, G., 2011. The economics of health and climate change: Key evidence for decision making, *Globalization and Health* 7(18).

Hutton, G., and Haller, L., 2004. *Evaluation of the Costs and Benefits of Water and Sanitation Improvements at the Global Level.* Geneva, Switzerland: World Health Organization. WHO/SDE/WSH/04.04.

IPCC (Intergovernmental Panel on Climate Change), 2007. *Climate Change 2007: Impacts, Adaptation and Vulnerability. Contribution of Working Group II to the Fourth Assessment Report of the IPCC,* M.L. Parry, O.F. Canziani, J.P. Palutikof, P.J. van der Linden and C.E. Hanson, eds., Cambridge, UK: Cambridge University Press, 976.

Johnson, F.R., and Adamowicz, W.L., 2010. Valuation and cost-benefit analysis in health and environmental economics. In: E. McIntosh, P.M. Clarke, E.J. Frew, J.J. Louvere, eds., 2010. *Applied Methods of Cost-Benefit Analysis in Health Care, Handbooks in Health Economic Evaluation.* Oxford University Press.

Kenkel, D., 2006. WTP- and QALY-based approaches to valuing health for policy: Common ground and disputed territory, *Environmental & Resource Economics* 34, 419–437.

Kiszewski, A., Johns, B., Schapria, A., Delacollette, C., Crowell, V., Tan-Torres, T., Ameneshewa, B., Teklehaimanot, A., and Nafo-Traore, F., 2007. Estimated global resources needed to attain international malaria control goals, *Bulletin WHO 2007* 85, 623–630.

Klose, T., 1999. The contingent valuation method in health care, *Health Policy* 47, 97–123.

Klose, T., 2003. A utility-theoretic model for QALYs and willingness to pay, *Health Econ* 12(1), 17–31.

Koppe, C., Jendritzky, G., Kovats, R.S., Menne, B., 2003. *Heatwaves : Impacts and Responses*. Copenhagen: World Health Organization.

Kovats, R.S., and Jendritzky, G., 2006. Heat-waves and human health. In: B. Menne and K.L. Ebi, eds., 2006. *Climate Change and Adaptation Strategies for Human Health*. WHO, Springer.

Kovats, R.S., and Tirado, C., 2006. Climate, weather and enteric disease. In: B. Menne and K.L. Ebi, eds., 2006. *Climate Change and Adaptation Strategies for Human Health*. WHO, Springer.

Krupnick, A.J., and Portney, P.R., 1993. Controlling urban air pollution: A benefit cost assessment. In: R. Dorfman and N. Dorfman, eds., 1993. *Economics of the Environment: Selected Readings*. 3rd edition. New York: W.W. Norton & Company, Inc., pp. 421–437.

Landefeld, J.S., and Seskin, E.P., 1982. The economic value of life: Linking theory to practice, *Am J Public Health* 72(6), 555–566.

Le Tertre, A.A., Lefranc, D., Eilstein, C., Declerq, S., Medina, M., Blanchard, B., Chardon, P., Fabre, L., Filleul, J.F., Jusot, L., Pascal, H,. Prouvost, S., Cassadou, and Ledrans, M., 2006. Impact of the 2003 heatwave on all-cause mortality in 9 French cities, *Epidemiology* 17, 75–79.

Lindgren, E., and Jaenson, T.G.T., 2006. Lyme Borreliosis in Europe: influences of climate and climate change, epidemiology, ecology and adaptation measures. In: B. Menne and K.L. Ebi, eds., 2006. *Climate Change and Adaptation Strategies for Human Health*. WHO, Springer.

Maddison, D., and Bigano, A., 2003. The amenity value of the Italian climate, *Journal of Environmental Economics and Management*, 45, 319–332.

Markandya, A., and Chiabai, A., 2009. Valuing climate change impacts on human health: Empirical evidence from the literature, *International Journal of Environmental Research and Public Health*, 6(2), 759–786.

Markandya, A., and Watkiss, P., 2009. *Potential costs and benefits of adaptation options: A review of existing literature*. UNFCCC Technical Paper. Available at: http://unfccc.int/resource/docs/2009/tp/02.pdf (Accessed 24 January 2013).

Masiye, F., and Rehnberg, C., 2005. The economic value of an improved malaria treatment programme in Zambia: Results from a contingent valuation survey, *Malaria Journal* 4(60).

Mathers, C.D., Vos, T., Lopez, A.D., Salomon, J., and Ezzati, M., eds., 2001. *National Burden of Disease Studies: A Practical Guide*. Edition 2.0. Geneva: World Health Organization.

McMichael, A.J., Campbell-Lendrum, D., Kovats, S., Edwards, S., Wilkinson, P., Wilson, T., Nicholls, R., Hales, S., Tanser, F., Le Sueur, D., Schlesinger, M., Andronova, N., Ezzati, M., Lopez, A.D., Rodgers, A., and Murray, C.J.,L., 2004. Global climate change. In: M. Ezzati, A.D. Lopez, A. Rogers, and C.J.L. Murray, eds. *Comparative Quantification of Health Risks*, vol. 2. Geneva: World Health Organization, 1543–1649.

Mills, A., and Bradley, D., 1986. *Methods to Assess and Evaluate Cost-Effectiveness in Vector Control Programs*. Geneva: World Health Organization.

Mitchell, R.C., and Carson, R.T., 1989. *Using Surveys to Value Public Goods: The Contingent Valuation Method*. Washington, DC: Resource for the Future.

NOAA (National Oceanic and Atmospheric Administration), 2011. *The Deadliest, Costliest and Most Intense United States Tropical Cyclones from 1851 to 2010 (and Other Frequently Requested Hurricane Facts)*. National Climatic Data Center, National Hurricane.

Olsen, J.A., and Smith, R,. 2001. Who have been asked to value what? A review of 54 WTP-based surveys on health and health care. *Health Econ* 10, 39–50.

Oxfam, 2007. *Adapting to Climate Change: What is Needed in Poor Countries and Who Should Pay*? Oxfam Briefing Paper 104, Oxfam, Oxford, UK.

Parry, M., Arnell, N., Berry, P., Dodman, D., Frankhauser, S., Hope, C., Kovats, S., Nicholls, R., Satterthwaite, D., Riffin, R., and Wheeler, T., 2009. *Assessing the Costs of Adaptation to Climate Change: A Review of the UNFCCC and Other Recent Estimates*. Imperial College London, UK: International Institute for Environment and Development and Grantham Institute for Climate Change.

Portier, C.J., Thigpen Tart, K., Carter, S.R., Dilworth, C.H., Grambsch, A.E., Gohlke, J., Hess, J., Howard, S.N., Luber, G., Lutz, J.T., Maslak, T., Prudent, N., Radtke, M., Rosenthal, J.P., Rowles, T., Sandifer, P.A., Scheraga, J., Schramm, P.J., Strickman, D., Trtanj, J.M., and Whung, P.Y., 2010. *A Human Health Perspective on Climate Change: A Report Outlining the Research Needs on the Human Health Effects of Climate Change.* Research Triangle Park, NC: Environmental Health Perspectives/National Institute of Environmental Health Sciences. Available at: www.niehs.nih.gov/climatereport (Accessed 24 January 2013).

Rice, D.P., and Cooper, B.S., 1967. The economic value of human life, *American Journal of Public Health* 57, 1954–1966.

Richardson, J., 1999. *The role of willingness-to-pay in resource allocation in a national health scheme*, Working Paper 80, Centre for Health Program Evaluation.

Ryan, M., Scott, D.A., Reeves, C., Bate, A., van Teijlingen, E.R., Russell, E.M., Napper, M., Robb, C.M., 2001. Eliciting public preferences for health care: A systematic review of techniques. *Health Technol Assess* 5(5).

Schultz Hansen, K., Pedrazzoli, D., Mbonye, A., Clarke, S., Cundill, B., Magnussen, P., and Yeung, S., 2012. Willingness-to-pay for a rapid malaria diagnostic test and Artemisinin-based combination therapy from private drug shops in Mukono district, Uganda, *Health Policy Plan*. First published online May 15, 2012. Available at: http://heapol.oxfordjournals.org/content/early/2012/05/14/heapol.czs048.full.pdf+html.

Smith, R.D., 2003. Construction of the contingent valuation market in health care: A critical assessment, *Health Economics* 12, 609–628.

Stenberg, J., Johns, B., Scherpbier, R.W., and Edeger, T.T., 2007. A financial road map to scaling up essential child health interventions in 75 countries. *Bulletin WHO 2007* 5, 305–314.

Stern, N., 2006. *Stern Review: Economics of Climate Change*. Cambridge, UK: Cambridge University Press.

Telser, H., and Zweifel, P., 2002. Measuring willingness-to-pay for risk reduction: An application of conjoint analysis, *Health Economics* 11, 129–139.

Tol, R.S.J., Downing, T.E., Kuik, O.J., and Smith, J.B., 2004. Distributional aspects of climate change impacts, *Global Environmental Change* 14, 2–30.

Toulemon, L., and Barbieri, M., 2008. The mortality impact of the august 2003 heat wave in France: Investigating the "harvesting" effect and other long-term consequences, *Population Studies* 62(1), 39–53.

UNDP (United Nations Development Programme), 2007. *Human Development Report 2007/08*. New York, USA: Palgrave Macmillan.

UNFCCC (United Nations Framework Convention on Climate Change), 2007. *Investment and Financial Flows to Address Climate Change*. Climate Change Secretariat, Bonn, Germany.

UNFCCC (United Nations Framework Convention on Climate Change), 2009. *Potential Costs and Benefits of Adaptation Options: A Review of Existing Literature*. Technical paper. FCCC/TP/2009/2.

US Environmental Protection Agency, 1999. *The Benefits and Costs of the Clean Air Act Amendments of 1990-2010*, Washington, DC: Report to the U.S. Congress.

Viscusi, W., 1993. The value of risks to life and health, *Journal of Economic Literature* 31, 1912–1946.

Watkiss, P., ed., 2011. *The ClimateCost Project*. Final Report. Volume 1: Europe. Published by the Stockholm Environment Institute, Sweden.

Weinstein, M.C., Torrance, G., and McGuire, A., 2009. QALYs: The basics, *Value in Health*, 12(1).

World Bank, 2006. *Investment Framework for Clean Energy and Development*. Washington, DC: World Bank.

World Bank, 2010a. *Economics of Adaptation to Climate Change (EACC): Synthesis Report*. Washington, DC: World Bank.

World Bank, 2010b. *Costs of Adapting to Climate Change for Human Health in Developing Countries*. Discussion paper No.11 as part of the Economics of Adaptation to Climate Change study.

World Health Organization (WHO), 2000. *Considerations in Evaluating the Cost-Effectiveness of Environmental Health Interventions*. Geneva, Switzerland: WHO.

World Health Organization (WHO), 2006. *Preventing Disease Through Healthy Environments*. Geneva, Switzerland: WHO.

World Health Organization (WHO), 2008. *Foodborne Disease Outbreaks. Guidelines for Investigation and Control*. Geneva, Switzerland: WHO.

15

MULTI-SECTORAL PERSPECTIVE IN MODELLING OF CLIMATE IMPACTS AND ADAPTATION

Miles Perry[1,2] *and Juan Carlos Ciscar*[1,2]

[1] INSTITUTE FOR PROSPECTIVE TECHNOLOGICAL STUDIES, JOINT RESEARCH CENTRE, SPAIN

[2] THE VIEWS EXPRESSED ARE PURELY THOSE OF THE AUTHORS AND MAY NOT IN ANY CIRCUMSTANCES BE REGARDED AS STATING AN OFFICIAL POSITION OF THE EUROPEAN COMMISSION.

15.1 Introduction

By 2020, developed countries will provide US$ 100 billion a year to address the needs of developing countries related to climate policy, both mitigation and adaptation policies (Copenhagen Accord, UNFCCC 2009). Among the developed countries, the European Union adopted a pan-European adaptation strategy in 2013 (European Commission 2013). Planning and funding adaptation on such a scale requires knowledge of the impacts of climate change and potential for adaptation that is both detailed and broad-based. It is necessary to identify the most important areas for action (geographical and sectoral) in order to design appropriate cost-effective policies to reduce the negative consequences of climate change, and maximise potential benefits.

With this in mind, the purpose of this chapter is to review the state of the art in model-based assessments that estimate the economic damages of climate change and the scope for adaptation from a multi-sectoral, bottom-up perspective. The literature on climate change adaptation at a whole-economy level features two main types of integrated assessment model (IAMs). These are top-down and bottom-up. Up to now, top-down models have been used in most of the literature that provides integrated assessments of the costs of climate change and costs and benefits of adaptation. These assessments typically feature an aggregated characterisation of the economy and concentrate on analysing the optimal mix of damages, mitigation and adaptation from the perspective of intertemporal optimisation. Examples of the models used include PAGE, FUND, WITCH and DICE, while recent assessments include Agrawala *et al.* (2010) and Bosello *et al.* (2012a).

The bottom-up approach typically models the details of specific cause-effect mechanisms from climate change. For instance, river floods are simulated using hydrological models, taking into account the time and spatial pattern of rainfall and the topography of the terrain around the river basin. In this way, specific links between physical climate change and economic damages are captured in greater detail than would be expected from a top-down perspective. While this approach has the advantage of sectoral detail, comprehensive modelling of climate impacts and adaptation requires a multi-sectoral perspective, since climate change can have interrelated effects across a wide range of activities, from agriculture to human health (see, e.g., Stern 2007).

Given that climate change will affect sectors and regions of the global economy differently, it is arguable that there is a need for more bottom-up assessment capable of assessing how the different impacts of climate change and costs and benefits of adaptation interact within a linked global economy. This chapter therefore takes a step in this direction by reviewing three studies that take a multi-sectoral and multi-regional approach to assessing climate damages and/or adaptation.

The chapter pays special attention to the methods employed to ensure the different impacts are brought together consistently within an overarching framework. In the identified studies, this consists primarily of incorporating sectoral impacts in a common computable general equilibrium (CGE) modelling framework.

The chapter is divided into four sections, including this introduction. Section 15.2 identifies and reviews the methodology of three main studies that take a bottom-up, multi-sectoral approach to climate change impacts and/or adaptation. Section 15.3 compares the results of these studies. Section 15.4 discusses the issue of planned vs. autonomous adaptation to climate change, an issue of particular relevance to multi-sectoral, multi-regional assessments. The chapter concludes with a discussion of issues for further research.

15.2 Review of multi-sectoral, bottom-up assessments

This section reviews the main features of the modelling of climate impacts and the treatment of adaptation in multi-impact assessments. It gives an overview of three studies that have used CGE modelling as a tool for examining the costs of climate change and/or adaptation across different individual sectors of the economy. These are the PESETA project coordinated by the European Commission Joint Research Centre (JRC), the EACC study of the World Bank and the use of the ICES model by researchers at the Fondazione Eni Ennio Mattei (FEEM) in Italy. An overview of the studies' main features appears in Table 15.1. While none of the studies rely on CGE exclusively, they all employ the technique as a way of combining analysis from several different sectors in a single, quantitative framework.

A CGE is typically a multi-country, multi-sector model linking economies through endogenous bilateral trade. The model uses a system of simultaneous equations to replicate the behaviour of agents within a competitive market economy where consumers seek to maximise welfare, firms seek to maximise profit and the pattern of production and consumption in each sector is determined accordingly.

CGE models have two main characteristics which are advantageous for the analysis of climate change at a multi-sectoral level. First, they are typically based on an economy's national statistics (input-output tables), which means they are able to cover a number of sectors in detail while including all the interactions between industries, households and government across the economy (albeit at an aggregated level). Second, CGE models use a system of simultaneous equations to replicate the interactions of supply, demand, scarcity and prices across the economy in a way that is consistent with economic theory.[1] Together these characteristics allow the models to estimate the overall impact of climate change on the economy as a whole, as well as the differentiated impacts on sectors that are directly affected.

Table 15.1 Main features of the reviewed studies

	PESETA	*ICES*	*World Bank (Global Track)*
Sectors	4 Agriculture Coast River floods Tourism	7 Agriculture Coast River floods Tourism Energy Forest Human health	8 Agriculture Coast Water supply and flood protection Forestry and ecosystem services Human health Infrastructure Fisheries Extreme weather events
Geographical coverage	24 EU member states	Global (14 regions)	Developing world
Time horizon	2080s climate in 2010 economy	2010–2050	2010–2050

Table 15.2 Implementation of sectoral climate impacts in GEM-E3

	Impact	*Model implementation*
Agriculture	Yield change	Productivity loss for crops
Coastal areas	Migration cost Sea floods	Additional obliged consumption Capital loss
River floods	Residential buildings damages Production activities affected	Additional obliged consumption Production loss and capital loss
Tourism	Change in tourism expenditure	Change in export flows

Source: based on Ciscar et al. 2011

15.2.1 PESETA

The methodology applied in PESETA (Ciscar *et al.* 2011) has two main steps. In the first step, a series of biophysical impact models were run to compute the biophysical impacts generated by a set of climate change scenarios. In the second step, those impacts were valued in economic terms with a CGE model, GEM-E3.[2]

The various impact categories are integrated into GEM-E3 by changing components of the production structure and supply-side (capital and labour) of the economic sectors, and of the consumption structure of households. Table 15.2 summarises this process.

Regarding agriculture, the yield changes due to climate change, computed from crop models (Iglesias *et al,* 2012), are implemented in the GEM-E3 model as a change in total factor productivity (TFP) in the agriculture sector. The main economic damages in coastal areas, due to sea level rise, are related to migration and sea floods. Migration costs are interpreted as additional 'obliged' consumption (reducing welfare). These costs are related to the income level of the affected population and can have objective components (e.g., costs associated with moving to a new house) and subjective components (the disutility due to forced migration). Sea flood costs are implemented as a capital loss.

River flood damages have two main components. The first component, accounting for around 2/3 of the overall damage, is damage to residential buildings. It is assumed that they are repaired, leading to additional consumption expenditure (obliged consumption), interpreted as a welfare loss. The rest of the damage affects productive activities, both production and capital losses.

Finally, regarding tourism, climate change induces changes in tourism expenditure across Europe. When one country, for instance, receives fewer tourists, this means the economy has lower exports. Thus, the changes in tourism expenditure are interpreted in the CGE model as changes in exports, allocated to the market services sector.

The PESETA analysis was made in comparative static terms, i.e., shocking the current economy (as of the year 2010) by the future climate (as of the 2080s).

In the PESETA project, an effort was made to have a consistent approach to adaptation (Christensen *et al.* 2012). Private adaptation actions are considered in all studies. Regarding agriculture, farmers were assumed to respond to climate change by implementing optimal management practices, such as crop changes, fertilizers use and water irrigation. Migration to safer areas is modelled in the coastal systems assessment. Tourism flows are adjusted within the year in the various seasons in the tourism assessment.

For the coastal systems assessment, the PESETA project goes further and explicitly considers public adaptation measures, using a simplified cost-benefit framework. The optimal protection level is determined by the equalisation of marginal costs and benefits. Two hard engineering adaptation measures are made available. First of all, dikes are built to protect the coast. The costs of dikes are compared to the benefits in terms of lower sea flood damages, salinisation costs and migration costs. The second measure is beach nourishment, which is decided by comparing the nourishment costs (basically a function of cubic metre of sand) with its benefits. The benefits depend on agricultural land value if there are not tourists, and where there are tourists, the benefits depend on the number of tourists and their expenditure. It is important to note that while adaptation measures requiring additional expenditure (fertilizer, irrigation, dikes and beach nourishment) are considered in the wider PESETA project, they are not included in the CGE assessment discussed in this chapter (Ciscar et al. 2011).

15.2.2 ICES model (FEEM)

Researchers at the Fondazione Eni Enrico Mattei have produced a number of studies using CGE analysis to calculate the costs of climate change across different regions and sectors. Earlier studies focused on one or two sectors or direct climate impacts in isolation (though the inclusion within a CGE framework takes account of the wider economy by definition). These include Bigano *et al.* (2008) on sea-level rise, Bosello *et al.* (2006) on health impacts, Ronneberger *et al.* (2009) on agriculture and Calzadilla *et al.* (2011) on the efficiency of water use.

More recently, studies by Bosello *et al.* (2012b) (building on Eboli *et al.* [2010]) have used the ICES model to estimate the costs of climate change across multiple sectors. Both studies follow a similar procedure to PESETA, using physical impact data to create shocks whose overall economic impact is analysed through a CGE model. There are two major differences between the PESETA and ICES models. First, ICES is a dynamic model used to estimate the gradual build-up of economic impacts as climate change occurs over the period 2010–2050. This contrasts with the PESETA approach of imposing the climate conditions of the 2080s on the economy of today. Second, while both studies feature a set of economies linked by international trade, Bosello *et al.* (2012b) (henceforth referred to as ICES) has 14 aggregated regions (see Table 15.3) and is global in scope, whereas PESETA models EU Member States individually (with constant rest-of-world prices).

The sectoral climate impacts of ICES are given in Table 15.4. This shows that the methods chosen to convert physical impacts into CGE shocks are similar (but not identical) to those employed in PESETA. Both studies employ supply-side shocks (to stocks or productivity of primary factors of production) and demand-side shocks. In ICES, the demand-side

Table 15.3 Regional disaggregation of the ICES model

Region	*Description*
USA	United States
MEUR	Mediterranean Europe
NEUR	Northern Europe
EEUR	Eastern Europe
FSU	Former Soviet Union
KOSAU	Korea, S Africa, Australia
CAJANZ	Canada, Japan, New Zealand
NAF	North Africa
MDE	Middle East
SSA	Sub Saharan Africa
SASIA	India and South Asia
CHINA	China
EASIA	East Asia
LACA	Latin and Central America

Source: based on Bosello *et al.* 2012b

Table 15.4 Implementation of sectoral climate impacts in ICES

Impact	*Implementation in ICES model*
Coast	• Reduction in stock of productive land and capital
Tourism	• Change in demand for market services sector • Correction of regional incomes to reflect change in international monetary flows
Agriculture	• Change in land productivity
Energy	• Change in demand for oil, gas and electricity
Forest	• Change in productivity of natural resource input to the forestry sector
River floods	• Reduction in stock of productive land • Reduction in capital productivity • Reduction in labour productivity (average loss of one working day per week among affected population)
Human health	• Change in labour productivity related to heat and humidity

Source: based on Bosello *et al.* 2012b

shocks involve making exogenous adjustments to consumers' demand for energy and tourism, together with compensating changes in demand for other goods and services (so that the consumers' budget constraint is still met). These demand changes are then entered into the CGE model as an input, with the final demand levels being computed by the model. This is different from the PESETA approach of charging migration cost and damages to residential buildings as shocks to the fraction of consumption that is considered compulsory.

15.2.3 Global economic analysis by the World Bank

The study *Economics of Adaptation to Climate Change* (EACC) by the World Bank (2010a) provides an estimate of the total cost of adaptation to climate change for the developing world. The aim of the study was to provide an estimate of adaptation costs to inform international climate negotiations and inform developing countries' strategies for dealing with the risks posed by climate change. Its approach is different from PESETA and ICES since it pays more attention to adaptation and employs a scenario-based approach instead of a multi-region CGE.

The study follows a "twin track" approach: the Global Track uses global datasets to estimate the adaptation cost for the entire developing world, while the Country Track features seven country case studies (for Bangladesh, Bolivia, Ethiopia, Ghana, Mozambique, Samoa and Vietnam) and includes CGE analysis.

To ensure consistency between countries, sectors and analytical tracks, a common methodology is established. The study defines adaptation cost as the expenditure necessary to restore welfare to the level that would be enjoyed in the absence of climate change.[3] This begins with construction of baseline trajectories for regional GDP per capita for the period 2010–2050, using a common set of regional GDP projections[4] and population forecasts.[5] These GDP trajectories are similar to those of the UNFCCC A2 scenario and are therefore thought to be plausible in the event of climate change. Two A2 climate change scenarios are then considered from a choice of 26 global circulation models (GCMs) in order to express the widest possible range of precipitation outcomes to 2050.[6] The wettest scenario is obtained from the National Centre for Atmospheric Research (NCAR) model and the driest from the Commonwealth Scientific and Industrial Research Organization (CSIRO) model.

In the Global Track assessment (World Bank 2010b), sectoral welfare outcomes are measured in physical terms, with the cost of adaptation being defined as the additional cost of achieving the baseline level of welfare over the 2010–2050 period, but in the presence of climate change. Table 15.5 shows the welfare metrics and adaptation measures considered, alongside the estimated adaptation costs for each climate scenario.

The measures analysed are predominantly public adaptation measures (as opposed to private, autonomous adaptation) and employ 'hard' as opposed to 'soft' techniques (capital-intensive investments as opposed to institutions and policies). In each case, this is not because hard solutions are necessarily the most effective but because the cost and effectiveness of these types of initiative are easier to quantify on a multi-country basis. Sectors where the treatment of adaptation is notably different are coastal zones and extreme weather events. For coast, the cost of adaptation (based on the DIVA model, which is also used in ICES and PESETA) is calculated as the optimal level of abatement plus the cost of residual damages (rather than the cost of reducing damages to zero). For extreme weather events, the cost of adaptation is based on a soft measure – improvements in female education. This is because a number of studies have shown a significant link between risk of death from floods and droughts and levels of female education.

Table 15.5 Sector estimates of developing country adaptation cost, World Bank EACC study

Sector	*Welfare metrics*	*Adaptation measures*	*Adaptation cost estimates* ($bn/year)*	
			NCAR (wet)	*CSIRO (dry)*
Infrastructure	Level of services	Design standards; Climate-proofing maintenance	27.5	13
Coastal zones	Optimal level of protection plus residual damage	River and sea dikes; Beach nourishment; Port upgrade	28.5	27.6
Water supply & flood protection	Level of industrial and municipal water availability; availability of flood protection	Reservoir storage; Recycling; Rainwater harvesting; Desalination; Flood protection dikes and polders	14.4	19.7
Agriculture	Number of malnourished children and per capita calorie consumption	Agricultural research; Rural roads; Irrigation infrastructure expansion and efficiency improvements	2.5	3
Fisheries	Level of revenue	Fisheries buybacks; Individual transferable quotas; Fish farming; Livelihood diversification measures; Marine protected areas		
Forestry and ecosystem services	Stock of forests; level of services	No planned adaptation necessary since climate studies show broadly positive impact on forest productivity to 2050		
Human health	Health standard defined by burden of disease	Prevention and treatment of disease	2	1.5
Extreme weather events	Number of deaths and people affected	Investment in human resources	6.7	6.4
Total			**81.6**	**71.2**

*Cost estimates are taken from the Synthesis Report (World Bank 2010a) since this is more up-to-date than the Global Track report (World Bank 2010b).
Source: World Bank (2010a).

For the Country Track assessment, sectoral analysis is undertaken that is similar in nature to the Global Track but tailored to each country. This is augmented by a social component and by single-country CGE assessments (see the next section). The social component (World Bank 2010c) involved undertaking fieldwork assignments focussed on vulnerability hotspots within each country. This process culminated in Participatory Scenario Development workshops which identified interactions between climate change and development, including how impacts of climate change and potential benefits from different adaptation strategies are differentiated between individuals, households and subnational regions. An important feature of the social analysis is that it goes beyond the conceptual boundaries of the modelling assessments by considering autonomous adaptation (in addition to planned adaptation) employing both 'hard' and 'soft' approaches. It also examines factors that are difficult to model quantitatively, such as differences in access to resources between social groups, and the nature of governance and institutions.

15.2.4 CGE analysis

The CGE modelling employed in the EACC Country Track analysis is similar to PESETA and ICES in the sense that the studies use CGE to combine the results of individual sectoral impacts and to estimate their importance at a macro-economic level, taking into account interactions with the rest of the economy. However, there are a number of key differences between the three assessments:

- First, the EACC CGE assessment uses a series of separate single-country models while PESETA and ICES link together groups of affected economies (the countries of the EU and larger world regions, respectively) thereby capturing the effects of climate change in neighbouring countries through trade interactions. The single country models of EACC typically have greater sectoral detail, especially in the agricultural sector where a large number of crop types are included. This is particularly relevant in a developing country context where a greater proportion of the population is dependent on agriculture for their livelihood. In PESETA and EACC, the rest of the world is not modelled explicitly, whereas the ICES model implements most climate shocks on a worldwide basis.
- Second, EACC and ICES use dynamic CGE models built around a baseline to 2050, whereas PESETA employs a comparative static assumption.
- Third, EACC is more explicit in using CGE modelling to explore adaptation options (as opposed to quantifying impacts). This is partly due to the single, developing country context, in which foreign development assistance can be considered as an exogenous source of adaptation finance.[7] For example, the CGE element of the Ghana study considers three scenarios where an amount of development assistance equal to the climate-induced welfare loss[8] is received and invested in different adaptation strategies.

Table 15.6 shows the results from the CGE analysis of adaptation for the case of Ghana. Each row represents a different climate scenario derived from the four combinations of GCM and SRES emissions scenario which show the wettest and driest outcomes both for the world (global) and for Ghana. Column 1 shows that in the absence of adaptation, total welfare losses to 2050 of \$2.7bn–\$13.1bn are expected, depending on the climate scenario. Column 2 shows that some of these losses can be reduced by reallocating resources within the country's road budget. Instead of spending money on new roads, those resources are allocated to the maintenance of the road network, making it more resilient to climate change. This involves making investments in improved road quality in the short term (and therefore spending less on road expansion) in order to reduce maintenance costs in the long term (and therefore enable greater road expansion). These investments are expected to reduce welfare loss in all scenarios except Ghana Dry (where the extra damage costs are too small to justify the reallocation of funds away from network expansion).

Columns 3 through 5 show three adaptation cases where it is assumed that Ghana undertakes the road design measures but also receives development assistance equal to the loss from

Table 15.6 Welfare change 2010–2050 for different Ghana adaptation options simulated through CGE modelling ($bn, present value)

	No adaptation (1)	*Road design only (2)*	*Adaptation scenario (includes road design improvement plus one of these 3 options)*		
			Agriculture (3)	*Hydro & agriculture (4)*	*Education (5)*
Global dry	-13.118	-10.308	-0.121	-0.941	-2.09
Global wet	-10.095	-5.854	-2.973	2.116	0.584
Ghana dry	-2.709	-3.009	-1.193	-1.782	-1.308
Ghana wet	-4.050	-0.766	1.936	1.358	1.795

Source: World Bank 2010a

Note: each adaptation scenario features road design and one of the other investment types (agriculture, hydro and agriculture, education).

Column 1. This assistance is then spent on one of the following: expansion of irrigation (which increases agricultural productivity in the CGE model) (3); a combination of investment in new hydro power projects and expansion of irrigation (4); or increased spending in broad-based education (which increases labour productivity in the CGE model) (5). The results imply that some of the adaptation packages, for example in the Ghana Wet scenario, are capable of restoring welfare to the baseline level or higher (which means the climate damages are more than compensated). They also show that the relative effectiveness of agriculture, energy and education investments varies depending on the climate scenario.

However, Table 15.6 also shows in most cases (particularly Ghana Wet), a substantial share of the adaptation is achieved 'for free' due to the budget-neutral road improvements (the difference between 1 and 2). This implies that it may be more effective to award lump-sum compensation payments than spend resources on the additional adaptation options shown. Despite these results, the EACC study does not state explicitly whether or not these programmes would be an effective use of funds. This is understandable given (i) the inherent uncertainty faced by policy makers (who do not know which, if any, of the climate scenarios shown is most likely); and (ii) the assumptions needed to characterise the effectiveness of each adaptation action (especially soft actions such as broad-based education) within a CGE model. Instead, the study uses the CGE results to support more general recommendations. For example, the results from Table 15.6 imply that in a dry climate-change scenario, a combination of some compensation and targeted high-return investments in agriculture may be the most effective use of adaptation funds, whereas targeted hydro investment may be more effective in wet conditions.

15.3 Comparison of results

This section summarises the results of the three studies and discusses the different types of insight that they provide. Table 15.7 gives an overview of the results, though it is important to note that these figures are not directly comparable since each study has a different scope (in terms of time, place and climate scenario) and different objectives.

Table 15.7 Costs of climate damage and/or adaptation

Type of cost	*PESETA*	*ICES*	*EACC (Global track)*
Climate scenarios	2 × A2. 2 × B2 2.3°C–3.1°C global temperature increase	A1B 1.9°C global temperature increase	A2 (NCAR & CSIRO)
Timescale	Climate: 2080s Economy: 2010	2010–2050	2010–2050
Monetary value	€20bn–€65bn Annual GDP loss vs. baseline for EU	–	€55–76bn* Estimated annual cost of climate change adaptation for developing countries 2010–2050
% of regional GDP/Welfare –	-0.2%– -1% (welfare) –	-0.5% –	0.1%–0.7% in 2010s 0.1%–0.5% in 2040s –
Details	Annual reduction in welfare vs. baseline for EU • Fall of up to 1.4% in Southern Europe • Gain of 0.5%–0.8% in Northern Europe	Change in global GDP in 2050 vs. Baseline. • Fall of 1.5%–3% in NAF, SSA, SASIA & EASIA • Fall of around 0.2% in MEUR and EEUR. Gain of around 0.2% in NEUR, USA and CHINA	Estimated total cost of climate change adaptation for developing countries as % of GDP

*Original EACC results are reported as $70–98bn. Conversion made at spot rate of 07 November 2012.

In PESETA, imposing a range of climate estimates for the 2080s on today's economy is estimated to reduce EU GDP by up to €65bn, with most damages occurring due to river floods, sea level rise and agriculture. In terms of welfare, gains in Northern Europe are outweighed by losses in the South. Ciscar *et al.* (2012a) point out that the GDP results understate climate damages since expenditure such as repair of residential buildings adds to GDP but subtracts from welfare (because compulsory consumption related to climate change detracts from consumers' discretionary spending while still contributing to GDP). However, PESETA gives a similar value for EU-level losses (0.2%–0.6%) under both metrics.

ICES estimates that under a dynamic A1B climate scenario, GDP would be 0.5% lower than in the baseline. Compared to PESETA, damages for Europe are lower, which is logical since the study considers a climate of 2050 rather than the 2080s. Elsewhere, the United States benefits from gains in tourism and mild damages in other sectors, while gains in China are due to agriculture (despite losses in tourism). Losses in Africa, South Asia and East Asia are overwhelmingly due to agriculture. At a global level, ICES finds agriculture, tourism and sea level rise to be the most significant sectors, with tourism alone accounting for a 0.1% reduction in global GDP. This contrasts with the PESETA results where the importance of tourism is minor.

Instead of estimating the cost of climate damages, the EACC Global Track analysis estimates non-monetary damages and calculates the cost of restoring welfare to baseline standards (adaptation) through expenditure on individual projects (see Table 15.5). At under 0.7% of GDP in the affected regions, the adaptation cost estimates are considerably lower than the ICES estimate of GDP loss for most of Africa and Asia (1.5%–3%). However, it is important to note that in the EACC Global Track, GDP is an exogenous estimate that is not directly affected by climate damages or adaptation spending, whereas in PESETA and ICES, GDP is an output of the CGE model and is therefore directly influenced by climate change.

The damages reported in Table 15.7 should be considered an underestimate of the "true" cost of climate change, since it is always possible that significant damages exist in areas whose impacts have not yet been considered (for example, future versions of PESETA will incorporate damages to roads resulting from warmer and more extreme weather conditions).

15.4 Planned versus autonomous adaptation

In the literature, a distinction is made between planned and autonomous adaptation (see Smit & Pilifosova 2001 for a detailed discussion of the concepts). These concepts are also referred to as "public" (planned) vs. "private" (autonomous) in EACC and "policy driven" (planned) vs. "market driven" (autonomous) in Bosello *et al.* (2009). In this section, we discuss the importance of these concepts in a modelling context before commenting on the relevance of this distinction in the three assessments reviewed in this chapter.

15.4.1 Planned vs. autonomous adaptation in principle

Planned adaptation is often interpreted as the result of a deliberate policy decision on the part of a public agency (Smit & Pilifosova 2001). In the context of modelling studies such as those reviewed in this chapter, planned adaptation measures can be thought of as 'game changing' initiatives that alter the parameters through which climatic changes cause damages, or through which people are able to react to them. For example, a public programme to build flood defences or investment in education and warning procedures should both reduce the damage caused by sea level rise.

Autonomous adaptation consists of measures taken by private actors in order to improve or maintain their welfare in the presence of climate change. Aaheim *et al.* (2012) and Bosello *et al.* (2009) argue that substitution effects allowed in CGE analysis (changes in producer and consumer behaviour) therefore count as autonomous adaptation, since they describe a process through which private actors take steps to reduce climate damages relative to what they might otherwise have been.

Substitution effects occur as sectors and regions suffer different direct climate impacts, leading to changes in relative prices, which allow producers and consumers to re-allocate their production and consumption. This type of general equilibrium adjustment would typically be absent in modelling exercises that apply aggregate damage functions. For example, a CGE model may predict that a fall in agricultural productivity equivalent to 0.5% of (baseline) GDP leads to a final GDP loss of only 0.2%, once consumers and producers respond to the change in relative prices by substituting to other products. The remaining 0.3% could be thought of as damage avoided due to general equilibrium adjustment.

Other types of methodology make different assumptions about the extent of autonomous adaptation. These range from the *Ricardian approach* (Mendelsohn *et al.* 1994), in which no types of adaptation are specified but all possible measures are considered implicitly,[9] to the hypothetical 'dumb farmer' assumption (a term attributed to Rosenberg [1992]) where one assumes private agents take no autonomous adaptation measures whatsoever.

Studies such as those reviewed in this chapter, which specify modelled relationships for production technologies and/or consumer behaviour, lie between these extremes since they allow some substitution in response to climate change but remain constrained by the technologies or preferences specified in the models employed. For example, in a process-based agricultural model, a farmer may be free to adapt to climate change by changing patterns of water and fertiliser use, but she will still be constrained by the specified technologies. From this point of view, CGE models should offer a wider range of adaptation options than a sector-specific model since the farmer is free not only to change input combinations in agriculture but also to re-deploy her resources in a different sector (such as tourism). At the same time, intermediate and final consumers are able to switch consumption into or away from agricultural goods and trade internationally in order to maximise their welfare and profits in the circumstances.

In principle, the CGE models' greater ability to consider substitution effects can therefore be considered an advantage. However, there are two main caveats to bear in mind before deciding whether these substitution effects should really be considered "adaptation".

- First, it is not guaranteed that general equilibrium losses will be smaller than direct costs. For example, if the rest of the economy depends heavily on the output of the affected sector (because there are low substitution possibilities), the overall loss in the example above could be greater than the direct cost. In the previous example, this would be a general equilibrium loss greater than the original 0.5%.
- Second, if one is confident that the CGE representation of the economy is reasonable (i.e., if the CGE model represents the best estimate of how economic agents react to changes, whether or not they are due to climate change), then the general equilibrium loss of 0.2% should be considered the "true" cost of the climate shock, since the 0.5% loss would never actually occur. In this way, estimates of damage costs or adaptation requirements inevitably rest on some implicit assumption regarding the degree to which 'business-as-usual' actors will respond to a change in their environment. In an agricultural context, Schneider *et al.* (2000) refer to this as the difference between the "dumb farmer", the "realistic farmer" and the "clairvoyant farmer".

15.4.2 Planned vs. autonomous adaptation in the studies reviewed

Of the studies reviewed in this chapter, the EACC Global Track methodology considers predominantly planned adaptation. This consists of assumed dose-response relationships between specific investments (e.g., coastal defences, female education) and restoration of baseline welfare levels. However, autonomous adaptation is considered elsewhere in EACC through the social analysis and country-level CGE studies.

As discussed in the previous subsection, the CGE analysis of PESETA, ICES and EACC considers the possibility of autonomous adaptation since substitution decisions by consumers and producers are an inherent part of these models' optimisation procedures. However, in order to assess the value of this adaptation (and compare it to planned adaptation measures), it is necessary to examine in detail the production technologies and consumer preferences used in such models and decide to what extent they exceed the "reasonable farmer" assumption of bare minimum action.

Furthermore, in order to observe the extent to which autonomous adaptation contributes to damage reduction, it would be necessary to compare the expected CGE results with a counterfactual situation in which less autonomous adaptation takes place. The closest exercise to this is undertaken in ICES, where the authors compare the direct cost of the climate change (shock as a percentage of GDP) to the CGE outcomes. The study finds that in most cases, final costs are lower than direct costs, when both are expressed as a percentage of GDP, implying that the possibility of substitution by consumers and firms should lessen the impact of climate change. However, in some sectors, general equilibrium interactions are found to amplify the economic cost of climate shocks. This is the case for tourism in the Middle East and Sub-Saharan Africa and for agriculture in East Asia. A recent study by Ciscar *et al.* (2012b) also found general equilibrium losses in excess of direct costs when using a global CGE to model to examine the effects of agricultural yield change, which may be due to the limited possibility to substitute away from agriculture at a global level.

15.5 Conclusion

This chapter has reviewed several studies implementing a multi-sectoral, bottom-up approach to understanding the channels through which climate change can affect the economy and how adaptation can reduce these impacts. The literature reviewed implements an integrated approach (whether CGE or the more scenario-based approach of the World Bank) to compare costs across sectors and impact types.

Of the studies reviewed in this chapter, ICES and PESETA concentrate mainly on estimating impacts in a coherent way while EACC concentrates on estimating the costs of adaptation. However, none of these studies (which to our knowledge represent the state of the art in multi-sectoral, bottom-up assessments) are able to make a comprehensive assessment of the costs, benefits and trade-offs involved in climate change mitigation, damages and adaptation. This shows that the bottom-up methodology requires further development in order to match the scope of some top-down assessments.

On the other hand, the ability of the CGE methodology to model economic interactions is a distinct advantage. In reality, climate change (like any change in the economic environment) will cause economic agents to make decisions, which in turn affect the costs of damages and adaptation. Top-down models that do not allow for these substitution effects could therefore significantly overstate (or understate) the costs of climate changes. The findings of Bosello *et al.* (2012b) suggest the former is more likely, but more research is needed in this area.

Another fruitful research line should involve improving existing top-down models by deriving climate impact assessments that are more consistent globally (e.g., the recent assessment of Arnell *et al.* 2013) and better connected to bottom-up empirical evidence (Fisher-Vanden *et al.* 2012).

Furthermore, greater insight is needed into the relative contributions of autonomous and planned adaptation. This is a more important consideration for multi-sectoral bottom-up assessments than for aggregated top-down studies. This is because by taking a less aggregated approach, bottom-up models consider the details of economic agents' autonomous decisions. The parameters of such models will capture some decisions that are straightforward (only a "dumb farmer" would fail to act) and some that reflect a more fundamental change in practice (and therefore capture genuinely additional adaptation). Where a model fails to parameterise a certain action explicitly (for example, the farmers' decision to choose the optimal planting date), it is important to know whether or not the action is *Ricardian* (in which case one can assume implicitly that the action is taken).

Further research is also needed into the contrast between damage costs and the costs and benefits of adaptation. Of the reviews studied, EACC goes furthest in this area since it estimates the cost of the adaptation measures suggested. This is greatly assisted by the fact the adaptation consists of external development assistance (rather than resources raised from within the economy itself).

In conclusion, it appears that multi-sectoral, bottom-up studies are complicated by their detailed consideration of several economic and technological interactions that are aggregated together in top-down studies. Since bottom-up assessments consider a number of specific physical and economic relationships in detail, a great deal of further research will therefore be required in order for bottom-up assessments to match the top-down approach in scope. However, the bottom-up approach adds value by considering the interactions between physical and economic damages, and between sectors and regions of the global economy. Understanding such interactions will become more important as resources for climate change adaptation begin to be deployed and damages from climate change are experienced.

Notes

1 See Böhringer & Löschel (2006) for a general discussion of the use of CGE models for energy, economy and environmental analysis.
2 See www.gem-e3.net
3 In EACC, welfare is measured in physical terms. This differs from PESETA, where the concept of welfare refers to change in consumers' economic wellbeing, which is measured using the concept of equivalent variation.
4 The average of projections from three integrated assessment models: FUND (Anthoff & Tol 2008), PAGE2002 (Hope 2006), and RICE (Nordhaus 1996), and the world energy projections of the International Energy Agency and Energy Information Administration.
5 The 2006 middle-fertility projections of the United Nations Population Division.
6 For the period 2010–2050, variation in precipitation between GCMs is greater than variation in temperature.
7 In PESETA or ICES, the sources of international financial flows for adaptation would have to be modelled explicitly, whereas in a single-country model, they can come from an unspecified foreign source.
8 In this case, welfare is defined as loss of absorption (the sum of all spending within the economy), whereas in PESETA, equivalent variation is used to quantify welfare change.
9 Mendelsohn *et al.*'s approach imagines that as expected temperature changes, farmers switch input combinations and outputs (from arable to grazing to retirement home) in order to perform the highest value economic activity on a given parcel of land. Empirical Ricardian models will typically observe the econometric relationships between climate-related variables and performance-related measures, such as revenue.

References

Aaheim, H.A., Amundsen, H., Dokken, T. and Wei, T. (2012). "Impacts and adaptation to climate change in European economies", *Global Environmental Change* 22(4): 959–968.
Agrawala, S., Bosello, F., Carraro, C., de Bruin, K., De Cian, E., Dellink, R. and Lanzi, E. (2010). "Plan or react? Analysis of adaptation costs and benefits using integrated assessment models". OECD Environment Directorate, Environment Working Paper No. 23.

Anthoff, D. and Tol, R.S.J. (2008). The Climate Framework for Uncertainty, Negotiation and Distribution (FUND), Technical description, version 3.3. Available at: http://www.fnu.zmaw.de/fileadmin/fnu-files/staff/tol/FundTechnicalDescription.pdf.

Arnell, N.W., Lowe, J.A., Brown, S., Gosling, S.N., Gottschalk, P., Hinkel, J., Lloyd-Hughes, B., Nicholls, R.J., Osborn, T.J., Osborne1, T.M., Rose1, G.A., Smith, P. and Warren, R.F. (2013). "The global assessment of the effects of climate policy on the impacts of climate change". doi:10.1038/NCLIMATE1793

Bigano, A., Bosello, F., Roson, R. and Tol, R. (2008). "Economy-wide impacts of climate change: a joint analysis for sea level rise and tourism", *Mitigation and Adaptation Strategies for Global Change* 13(8):765–791.

Böhringer, C. and Löschel, A. (2006). "Computable general equilibrium models for sustainability impact assessment: Status quo and prospects", *Ecological Economics* 60(1):49–64.

Bosello, F., Carraro, C. and De Cian, E. (2009). "An analysis of adaptation as a response to climate change", Working Paper, Department of Economics, Ca'Foscari University of Venice No. 25/WP/2009.

Bosello, F., De Cian, E. and Ferranna, L. (2012a). "Choosing the optimal climate change policy in the presence of catastrophic risk", European Investment Bank, EIB Working Papers.

Bosello, F., Eboli, F. and Pierfederici, R. (2012b). "Assessing the economic impacts of climate change". Review of Environment, Energy and Economics. FEEM.

Bosello, F., Roson, R. and Tol, R. (2006). "Economy-wide estimates of the implications of climate change: Human health", *Ecological Economics* 58(3):579–591.

Calzadilla, A., Rehdanz, K. and Tol, R. (2011). "Water scarcity and the impact of improved irrigation management: A computable general equilibrium analysis", *Agricultural Economics* 42(3) 305–323.

Christensen, O.B., Goodess, C.M. and Ciscar, J.C. (2012). "Methodological framework of the PESETA project on the impacts of climate change in Europe", *Climatic Change* 112(1): 7–28.

Ciscar, J.C., Iglesias, A., Feyen, L., Szabó, L., Van Regemorter, D., Amelung, B., Nicholl, R., Watkiss, P., Christensen, O.B., Dankers, R., Garrote, L., Goodess, C.M., Hunt, A., Moreno, A., Richards, J. and Soria, A. (2011). "Physical and economic consequences of climate change in Europe", *Proceedings of the National Academy of Science*, January 31, 2011. doi:10.1073/pnas.1011612108

Ciscar, J.C., Iglesias, A., Perry, M. and van Regemorter, D. (2012b). "Agriculture, climate change and the global economy". Paper presented at Conference on Macroeconomics and Climate Change, Cowles Foundation for Research in Economics, Yale SOM, June 11–12, 2012.

Ciscar, J.C., Szabó L., van Regemorter, D. and Soria, A. (2012a). "The integration of PESETA sectoral economic impacts into GEM-E3 Europe: Methodology and results", *Climatic Change* 112(1): 127–142.

Eboli, F., Parrado, R. and Roson, R. (2010). "Climate change feedback on economic growth: Explorations with a dynamic general equilibrium model", *Environment and Development Economics* 15(5): 515–533.

European Commission (2013). "An EU Strategy on adaptation to climate change", COM (2013) 216 final. Available at: http://eur-lex.europa.eu/LexUriServ/LexUriServ.do?uri=COM:2013:0216:FIN:EN:PDF

Fisher-Vanden, K., Wing, I.S., Lanzi, E. and Popp, D. (2012). "Modeling climate change feedbacks and adaptation responses: recent approaches and shortcomings", *Climatic Change*, doi:10.1007/s10584-012-0644-9.

Hope, C. (2006). "The marginal impact of CO2 from PAGE2002: An integrated assessment model incorporating the IPCC's five reasons for concern", *Integrated Assessment* 6(1): 19–56.

Iglesias, A., Garrote, L., Quiroga, M. and Moneo, M. (2012). "A regional comparison of the effects of climate change on agriculture in Europe", *Climatic Change* 112(1): 29–46.

Mendelsohn, R., Nordhaus, W. and Shaw, D. (1994). "The impact of global warming on agriculture: A Ricardian analysis", *American Economic Review* 84(4): 753–771.

Nordhaus, W. and Yang, Z. (1996). "a regional dynamic general-equilibrium model of alternative climate-change strategies", *American Economic Review* 86(4): 741–765.

Ronneberger, K., Berrittella, M., Bosello, F. and Tol, R. (2009). "KLUM@GTAP: Introducing biophysical aspects of land-use decisions into a computable general equilibrium model a coupling experiment", *Environmental Modeling and Assessment* 14(2): 149–168.

Rosenberg, N (1992). "Adaptation of agriculture to climate change", *Climatic Change* 21: 385–405.

Smit, B. and Pilifosova, O. (2001). "Adaptation to climate change in the context of sustainable development and equity". Chapter 18 of *Climate Change 2001: Impacts, Adaptation and Vulnerability.* Cambridge University Press, Cambridge, on behalf of IPCC.

Stern, N. (2007). *The Stern Review of the Economics of Climate Change.* Cambridge Univ. Press.

UNFCCC (2009). *Copenhagen Accord.* Available at: http://unfccc.int/resource/docs/2009/cop15/eng/l07.pdf

World Bank (2010a). *Economics of Adaptation to Climate Change: Synthesis Report.* World Bank, Washington, DC.

World Bank (2010b). *The Cost to Developing Countries of Adapting to Climate Change: New Methods and Estimates.* World Bank, Washington, DC.

World Bank (2010c). *Economics of Adaptation to Climate Change: Social Synthesis Report.* World Bank, Washington, DC.

16
FLOOD RISK MANAGEMENT
Assessment for prevention with hydro-economic approaches

Sébastien Foudi[1] *and Nuria Osés-Eraso*[2]

[1]BASQUE CENTRE FOR CLIMATE CHANGE (BC3)
[2]DEPARTMENT OF ECONOMICS, UNIVERSIDAD PÚBLICA DE NAVARRA, SPAIN

16.1 Introduction

A warmer climate, with its increased climate variability, will increase the risk of floods (Parry et al. 2007). Indeed, a warmer climate will increase evaporation and the intensity of water cycling, and a warmer atmosphere can hold more water vapour and has a higher energy potential. As a consequence, rainfall should be more intense, and extreme events should be more frequent (Kundzewick and Schellnhuber 2004). Such changes are not far future projections. Climate modification has already been observed: the frequency of great floods has increased during the twentieth century (Milly et al. 2002).

The role of anthropogenic activities on climate perturbation and flood risk modification cannot be ruled out. Pall et al. (2011) revealed that in nine out of ten cases of their climatic simulations, the 20th-century anthropogenic greenhouse gas emissions increased the risk of floods occurring in England and Wales in autumn 2000 by more than 20%, and in two out of three cases by more than 90%. However, the precise magnitude of the anthropogenic contribution to the climatic risk remains uncertain.

The role of anthropogenic activities on the economic or socioeconomic risk is much less subject to uncertainties than its role on the climatic risk. Future flood damages depend heavily on the exposure to the flood, i.e. on land-use decisions. Other drivers of the socioeconomic flood risk have non negligible impacts on the damages. Among them are the flood warning and forecast systems, the emergency measures and the longer term precautionary measures and the experience of flood (Thieken et al. 2005). The size of the population located in the floodplain, the value capital accumulated and the value of the services exchanged as well as the sensitivity of these elements are others drivers of the future damages and reveal how vulnerable an economy and a society are to flood.

The nature of the flood is also a driver of the damages and of the risk. The intensity of precipitation, the volume and the timing of rainfall, flow velocity and flood duration (Penning-Rowsell et al. 2005) define the nature of the hazard the exposed elements will suffer. For example, during the first decade of the 21st century, the Ebro river basin in Spain experienced two major flood events whose intensity has been accentuated by the combination of meteorological events. In winter 2003 and spring 2007, the Atlantic and Mediterranean precipitation episodes delivered a large volume and flow of water into the basin. A rise in temperatures causes a thaw of the snow cover in the Pyrenees mountains and some plains of the basin that rose an already high flow of river. As a consequence of this combination of meteorological events, the Ebro River burst its banks in several places and in the city of Zaragoza (Ebro RBA 2011).

Climate change therefore poses challenges to water managers for flood risk prevention. Hydrological conditions under climate change are associated with a higher level of uncertainty than they were under a stable climate. Long-term planning measures of adaptation and prevention have therefore to take into account this uncertainty. In the face of an uncertain environment, water managers then should be able to identify the possible outcomes and assign them probabilities in order to make decisions. Building scenarios of hydrological and meteorological conditions could therefore help to make decisions (Hallegatte 2009). Cost-benefit analysis, cost-efficient analysis or multi-criteria analyses are tools that can be implemented in this context to support the decision. In this chapter, we present how a cost-benefit analysis can be coupled to a hydro-economic model to support the decision-making of water managers to prevent flood risk.

Hydro-economic approaches are widely used in the literature to assess flood damages and flood risk. Flood damage assessment requires the identification of the elements at risk and the valuation of them. Flood risk assessment relates the value of the damage to a flood probability (i.e. to a degree of risk). Therefore, different degrees of protection can be achieved in accordance with the degree of protection a society is willing to pay for. The willingness to pay for protection will depend upon various factors such as the risk itself, the risk aversion and the wealth of the society (Gollier 2001). The level of acceptable risk a society is willing to support and its willingness to live with floods influence the set of measures of prevention. Shall it be protected against large floods with a low probability of occurrence or against small floods with large probability when budget is limited? The traditional civil infrastructure measures of protection have a capacity of flood prevention for a given hazard, and these measures tend to increase the damage when their capacity is exceeded. Future prevention measures would have to be taken into different hazards' context and uncertainty. More flexible measures should complement the existing infrastructure measures. These measures translate the capacity and willingness of a society to live with floods. They have for objectives to recover the natural dynamics of rivers and to improve the flowing of water.

The chapter is organized as follows. In Section 16.2 we describe the initial steps of the flood risk assessment: the typology of damages and the spatial and temporal scales of the studied area. A flood risk assessment model is presented in Section 16.3. Examples from case studies complement and illustrate the model. Section 16.4 shows how a cost-benefit analysis can be applied to decide on flood prevention measures. Examples from cases study areas of the literature illustrate the section.

16.2 Flood damage assessment

16.2.1 Classification of flood damage

The impacts of flood are classified into direct and indirect impacts. The assessment of the damage also categorizes the impacts into tangible and intangible impacts. This typology is the most common in the literature (Rose 2004, Cochrane 2004, Merz et al. 2010) but interpretations and delineations of what is considered as direct or indirect may differ.

The direct impacts are those that result from the physical contact of water with the elements exposed. The indirect impacts are consequences of the direct effects and occur therefore at different scales – spatial and temporal – than the direct impacts. The tangible impacts are impacts that are easily measured in monetary terms. The intangible impacts refer to the elements that are not traded in the market or for which a value is difficult to assign.

Examples of the different types of socioeconomic damages are as follows:

- Direct and tangible: the damage to property, to stocks, to capital for production. One can find in this category the damage to buildings and their inventory, to infrastructure such as roads and irrigation systems and to agricultural production and cattle – and also the cleaning costs.
- Direct and intangible: loss of life, injuries, and damage to cultural heritage among others.
- Indirect and tangible: deals with the disruption of the consumption of flows of goods and services. One can find in this category the disruption of transport, supply of intermediate goods for production, supply of public services and electricity or water, among others.
- Indirect intangible: the post-traumatic stress, the trust in public interventions, modification of preferences or risk perception and acquisition of experience in flood prevention.

The ecosystem functions (de Groot et al. 2002) might also be affected by natural hazards, and as a consequence, the services delivered to societies might be affected, too. However, the impacts are not necessarily negative impacts on well-being. The way the functions will be affected depends heavily upon the type of functions, the characteristics of the flood and to what extent the services are valued by societies. For example, the impact of a flood on a wetland might be positive as long as the flow velocity is not too strong to take away vegetal and animal species. On the contrary, chemical pollution carried by the river flow to safe places has negative impacts and high flow velocity can erode agricultural and natural soils.

One has to notice that other types of typology can be found in the literature. Hallegatte and Przuski (2010) distinguish market and non-market impacts instead of tangible and intangible. Direct market losses are losses to goods and services traded on the market, for which a price exists. Non-market direct losses include all damages that cannot be repaired or replaced through purchases on the market, such as life.

Another possible distinction is between direct and higher-order effects (Rose 2004). The higher-order effects correspond to (i) indirect effects and (ii) induced effects. Indirect effects arise when the reduction of the supply of one firm generates a reduction of the demand to another firm. The induced effects refer to the interactions between the consumer and the firm.

Rose (2004) recalls a basic economic principle in the assessment of natural disasters impacts: to avoid the double counting of stocks and flows of goods and services because the value of an asset (stocks) is the discounted flow of its net returns. The double counting of goods and services would consist in counting an effect to both parties supporting the impacts: a producer and a consumer for example.

Moreover, flow measures are more representative of the cost of the disaster than stock measures as stocks are a single point in time value of production or asset and do not represent the disruption effects of the hazards like flow values. Flow values therefore enable the capturing of the time of disruption and recovery of the activity.

16.2.2 Spatial and temporal scales

The definition of the temporal and spatial scales of the hazard under study is a fundamental principle to set in order to classify the elements affected by the flood (Merz et al. 2010).

At the spatial scale, one can distinguish between the micro-, meso- and macro-scales. The micro-scale refers to an assessment based on single elements (buildings, element of infrastructure, a natural area). To measure the losses due to a given flood for a group, damages are estimated for each element at risk. Meso-scale studies are based on a spatial aggregation of units like land use units or residential and economic units. Macro-scale studies are large-scale studies: country, regional, municipal or river basin scales.

Floods generate damages to the economy and to the capital accumulated (houses, goods of consumption, cars, etc.) by a community. But for an outside community, the flood has positive impacts as it will receive orders of goods and services (repairs, machineries, pumps, etc.). The spatial delimitation of the study limits the extent to which the indirect effects should be estimated.

Floods also generate disruptions of economic activity; numerous flows of goods or services cannot be supplied even outside the flooded area (indirect effects). The size of these longer-term losses will depend upon the time necessary to recover the continuity of the transactions.

16.3 Flood risk assessment

16.3.1 Methodology for flood risk assessment

Generally, risk is defined as a product of a hazard and its consequences. In particular, flood risk is a product of flood hazard and the negative consequences of flooding. Therefore, flood risk assessment must consider an estimate of flood hazard and an estimate of its negative consequences. Hydraulic and hydrologic studies analyse flood hazard. The study of the consequences of these events needs to combine hydraulic and hydrological issues with elements of economic analysis. The consequences of flooding depend on the elements that are present at the location involved (exposure) and on the lack of resistance of those elements to the flooding (vulnerability). Therefore, the methodology (adapted from Kron 2002 and Penning-Rowsell et al. 2005) for flood risk assessment consists of four steps: (i) hazard assessment, (ii) exposure assessment, (iii) vulnerability assessment and (iv) risk assessment.

Flood hazard is the probability of occurrence of potentially damaging flood events (Schanze 2006). Floods events are characterized by volume of water, depth of water, flow velocity, spatial dynamics and temporal dynamics. In hydrology, the probability of occurrence of an avenue with certain characteristics is determined using probabilistic and statistical analysis based on past hydrological records. The average time that elapses between the occurrence of a certain flood event and the next occurrence of an event of the same magnitude is called the *return period*, t_r. That is, the return period is the number of years within which a given flood is expected to recur.

The return period is a measure of frequency, and it can be interpreted as the inverse of the probability that a certain flood will be exceeded in any one year, $t_r = \frac{1}{p_r}$, where p_r denotes the annual probability of overcoming the flood event (*exceedance probability*). Therefore, the return period can be used for flood hazard assessment. Hydraulic studies typically collect information on flood events of different return periods and describe their characteristics – mainly, volume of water, water depth, flow velocity and spatial distribution of flood spot. These variables play a leading role in estimating exposure and vulnerability to assess the consequences of flooding.

Exposure to flooding depends on the spatial distribution of flood spot. Flood exposure can have negative consequences in environments where human influence is remarkable through land use or river training. However, flood exposure cannot be classified as a threat in natural floodplains (Schanze 2006) as nature is prepared to endure flooding events.

To assess exposure to flooding events, it is necessary to identify the elements on the corresponding flood spot. First, we should distinguish between rural and urban areas. In rural areas, it is necessary to distinguish between natural areas for which floods pose no threat and areas where man has intervened (mainly through agriculture) and are susceptible to damage by flooding. In urban areas, the elements at risk are mainly residential properties, non-residential properties, cultural heritage and public infrastructures. In addition, humans living in the floodplain can also suffer the consequences of flooding.

The identification of these elements can be done using *Geographic Information Systems* (GIS) (Zenger and Wealands 2004). However, the existence of elements at risk does not mean that there will be negative consequences from flooding; these elements can be prepared to bear the stakes of some flooding.

Vulnerability refers to inherent characteristics of these elements which determine their potential to be harmed (Schanze 2006; Messner and Meyer 2005). An element at risk of being harmed is more vulnerable the more it is exposed to a hazard and the more it is susceptible to its forces and impacts (Messner and Meyer 2005). The degree of exposure to the hazard depends on flood characteristics such as velocity, depth or duration. The susceptibility measures how sensitively an element at risk behaves when it is confronted with a flood hazard and includes preparedness (before the flood), coping (during the flood) and recovery (after the flood) capabilities.

A vulnerability analysis for each of the elements at risk allows estimating flood damages (direct and indirect, tangible and intangible) for different return periods. For example, in residential properties, the vulnerability analysis shows that water depth, the duration of the flood and the type of building determine the direct damages of flooding (Penning-Rowsell 2005). We find a concave relationship between water depth and damage per square meter in residential buildings. This damage includes the effect of flooding in building fabric, in household inventory and the unavoidable cleaning costs. In a similar way, the vulnerability analysis for non-residential properties shows that water depth is again a leading variable for the estimation of flood damages. In this case, the vulnerability analysis must considered for each type of building not only the building structure but also the equipment (whether it is moveable or not), the fixtures and fittings and the stock. Summing up all damages, tangible and intangible, direct and indirect, the negative consequences of flooding are estimated.

Flood risk is the result of linking the negative consequences of each potential flood (return period) with the exceedance probability of such flood.

16.3.2 Flood risk measures

The above methodology can be resumed in a four-graph diagram (Figure 16.1) (adapted from Penning-Rowsell et al. 2005). The graph in the upper left corner (A) represents the flood hazard and shows the relationship between a flood return period and the annual exceedance probability. The graph in the upper right corner (B) represents a simplified

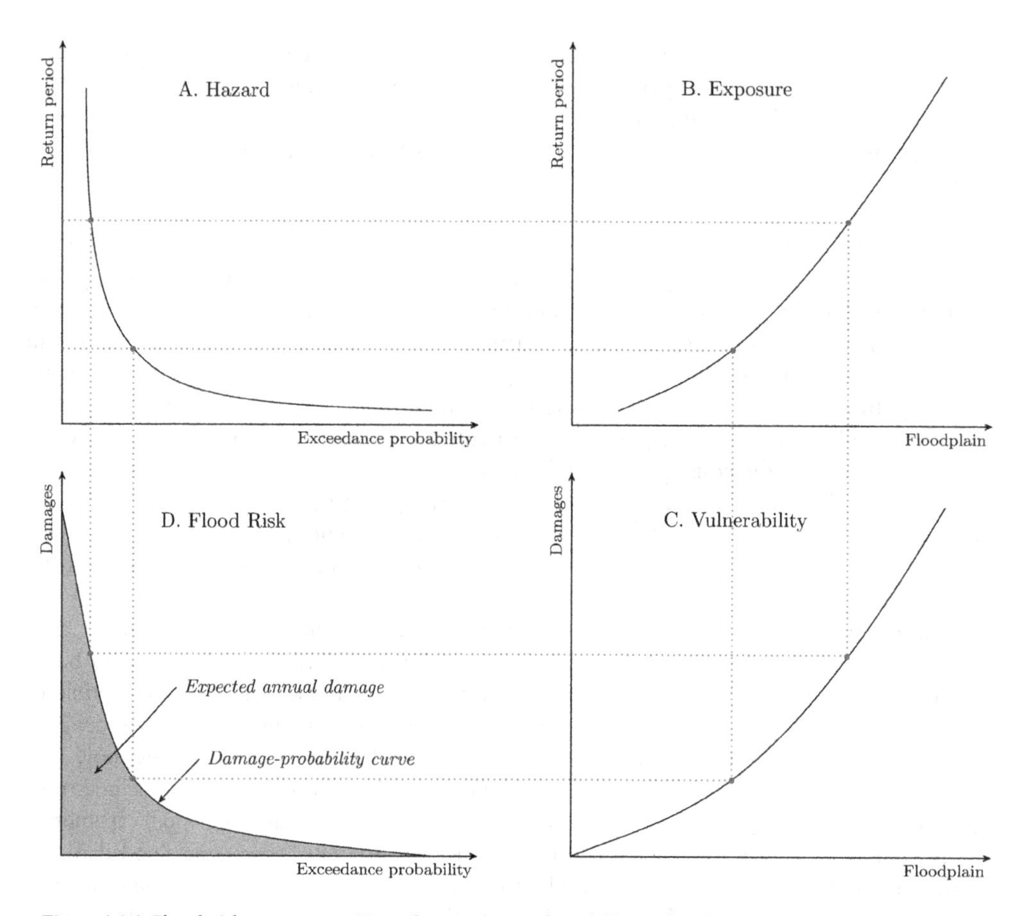

Figure 16.1 Flood risk assessment: Hazard, exposure, vulnerability and risk

version of exposure. It shows an increasing relationship between flooded area (elements at risk) and flood return period. The graph in the lower right corner (C) represents a simplified version of vulnerability showing a positive relationship between flooded area (elements at risk) and damage caused by flooding. Graph D in the lower left corner relates exceedance probability and negative consequences of flooding (damage). The relationship in graph D is call *damage-probability curve*.

Damage-probability curves, also called loss-probability curves, relate exceedance probability and flood damage. Let $x(p_r)$ be the estimate damage for a t_r-year flood. Estimating x for different return periods, the corresponding damage-probability depicts a continuous relationship between exceedance probability (return period or flood severity) and the flood damage. It represents a combination of flood hazard and the negative consequences of flooding – that is, it captures the flood risk.

A sufficient number of potential floods should be apprised so that an accurate picture can be developed of the shape of the damage-probability curves. Usually this means that at least five floods need to be appraised (Penning-Rowsell et al. 2005). The information in this curve can be used to calculate the *expected annual damage* that is the most frequently used measure of flood risk.

Expected annual damage is the average of flood damages computed over many years, although the scarcity of accurate long time series makes necessary the use of damage-probability curves (Arnell 1989, Lekuthai and Vongvisessomjai 2001) for its estimation. Assuming $x(p_r)$ represents the damage-probability curve, the expected annual damage is[1]

$$\mathbf{E(x)} = \int_0^1 \mathbf{x(p_r)dp_r} \tag{1}$$

The area under the damage-probability curve is equivalent to the expected annual flood damage (Lekuthai and Vongvisessomjai 2001).

Damage-probability curves and expected annual damages can be calculated for the total estimated flood damage or can be calculated for one of the damages, for example, for damages to residential properties, for damages in agriculture or for damages in health.

16.3.3 Examples from case studies

Flood risk in Amurrio (River Nervión, Basque Country, Spain)

According to the information of the Basque Water Hydrological Network, the Nervión River Basin covers an area of approximately 535 km^2. The river has a longitude of 55 km. The low-medium part of the river is in the province of Bizkaia while the headwaters belong to the province of Álava. From its headwaters, the basin supports a major population and industrial occupation, which increases the vulnerability of the area to potential flooding. Being a narrow valley, populated and industrialized, much of the margin is occupied by buildings and roads, so the vegetation associated with the river is very poorly represented. Although the service sector is of great importance in the municipalities of lower part such as Basauri, in the rest of the watershed, industrial activity appears as the first sector, Llodio and Amurrio being the most relevant industrial villages.

Amurrio has an extension of 9,640 ha, of which 190 are residential, 163 are for economic activities, 140 for general systems and the rest are not developable land. Population is slightly over 10,000 inhabitants according to recent data from the Basque Statistics Institute (Eustat). Several hydraulic studies have been developed to analyse potential flood events in the urban area of this village. Three potential flood events have been analysed: a 50-year flood event, a 100-year flood event and a 500-year flood event. Following the above methodology, we get the data in Figure 16.2.

The available data from the hydraulic studies allow only the estimation of three potential flooding events. Based on this sample, the estimated damage-probability curve (depicted in graph D, Figure 16.2) is $x = 21.378e^{-19.51p_r}$ $(R^2 = 0.9282)$. The corresponding estimated annual damage is $E(x)$ = 1.10 million euros (area under the damage-probability curve).[2]

Flood risk in the catchment of the River Urola (Basque Country, Spain)

The River Urola has a longitude of 55 km, and its catchment covers an area of 337 km^2. It has several tributaries, but the two main ones are the River Ibai-Eder in the middle of the route and the River Narrondo in the estuary. We restrict the present analysis to the River Urola and

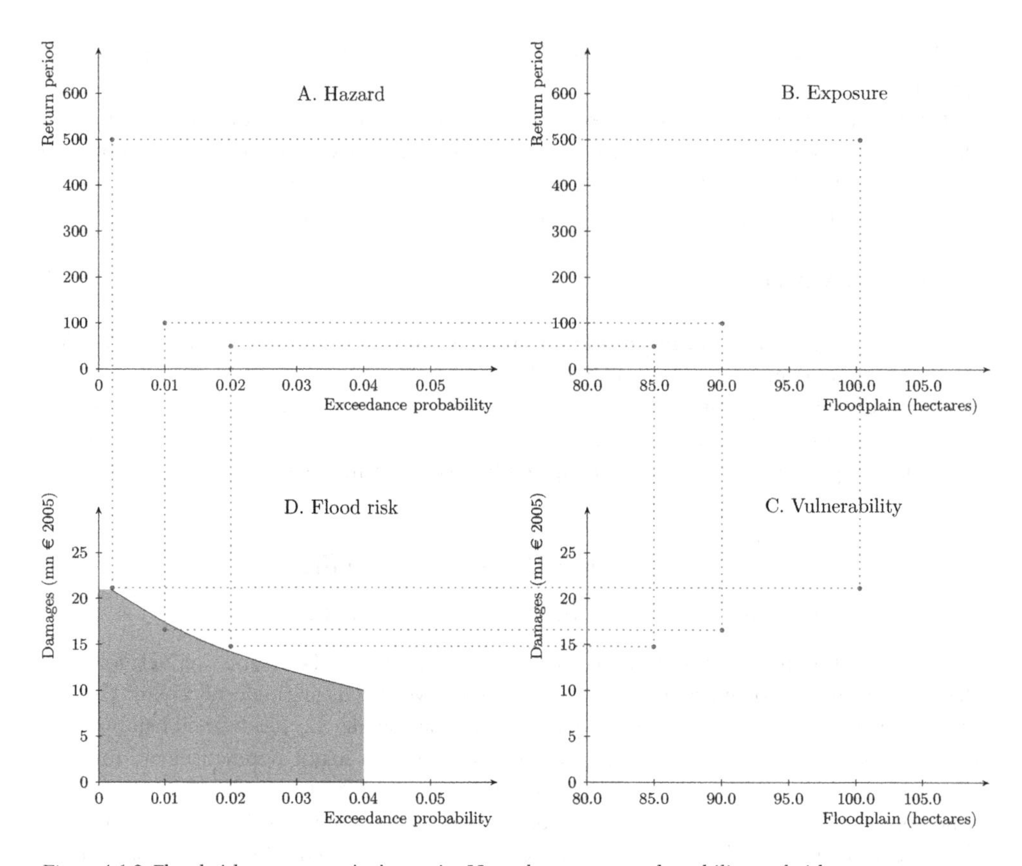

Figure 16.2 Flood risk assessment in Amurrio: Hazard, exposure, vulnerability and risk

these two tributaries. The River Urola has been the backbone of the industrial and urban development of its catchment area. There are seven villages (Legazpi, Urretxu, Zumárraga, Azkoitia, Azpeitia, Zestoa and Zumaia) along this river that are at risk of flooding; these are small villages, and only two of them surpass 10,000 inhabitants. The hydraulic studies in this catchment report contain data on three different potential floods: 10-year, 100-year and 500-year. The estimated damage-probability curves are represented in Figure 16.3. The numbers in parentheses in the caption are the expected annual damage.

Differences in risk of flooding along the river are clear if we look at the damage-probability curves and at the annual expected damage. For example, in villages like Zumarraga, damage barely increases the higher the return period (the smaller the exceedance probability). In others, like Zestoa, damage increases sharply the higher the return period. Other villages, like Azkoitia and Azpeitia, have similar expected annual damages (0.54 and 0.57 million euros) but different damage-probability curves. These differences show that the flood prevention measures need not to be the same everywhere, and the benefits these villages get from the implementation of the same type prevention measures will clearly differ.

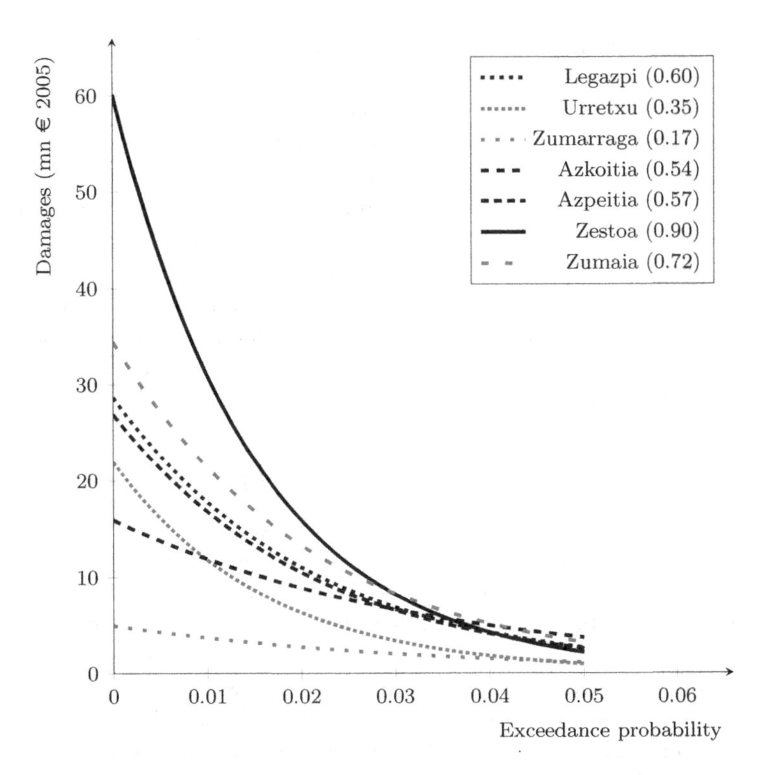

Figure 16.3 Damage-probability curves and expected annual damage in the Urola river basin

16.4 Risk prevention and cost-benefit analysis

According to the United Nations Office for Disaster Risk Reduction (UNISDR), **prevention** is defined as the outright avoidance of adverse impacts of hazards and related disasters. Prevention expresses the concept and intention to completely avoid potential adverse impacts through action taken in advance. But, the total elimination of the risk is very often not feasible, and mitigation measures have to take over. **Mitigation** consists in the lessening or limitation of the adverse impacts of hazards and related disasters. Mitigation measures encompass engineering techniques and hazard-resistant construction as well as improved environmental policies and public awareness. More structural measures are adaptation measures. **Adaptation** is defined by the UNISDR as the adjustment in natural or human systems in response to actual or expected climatic stimuli or their effects, which moderates harm or exploits beneficial opportunities.

The policies of prevention have evolved over time with experience and lessons learned from past disasters. Traditionally, the management of flood risk had been done with infrastructure protection against floods. Nowadays, the efficiency of these strong infrastructures to contain waters has been questioned (Werrity 2006, Brouwer and van Ek 2004). In a climate change context, in which it can be expected that floods were more frequent and flow larger volumes of water at a given point in time, the non-flexible infrastructures might not be adequate. In the

European Union, a shift in policy orientation occurred in the beginning of the 21st century with the Water Framework directive (2000/60/EC) and the Floods directive (2007/60/EC), together with the CAP reforms and Rural Development plans, in order to give more space to the environmental measures and favour good ecological status of rivers. Moreover, integrated flood management strategies are becoming more frequent and more relevant in the assessment of flood risk as these strategies recognize the interrelationship between risk management measures; their analysis; their cost and their effectiveness within changing social, economic and environmental contexts (Hall et al. 2003, Brouwer and van Ek 2004, Jonkman et al. 2008).

In this section, we present a classification of measures according to their dedicated objectives. The planned measures of adaptation and prevention to flood risks are costly and have to be taken within budget boundaries. We therefore illustrate and explain how a cost-benefit analysis through avoided damages can drive the decision-making of a social planner.

16.4.1 Types of measures or policies

Prevention policies are designed to reduce the damages to people and goods and therefore to lower the loss-probability curves. Reducing the risk can be achieved with various measures: measures dedicated to reduce the hazard, the exposure or the vulnerability. These measures are not mutually excludable but rather are complementary in prevention policy.

One can classify the prevention measures into four classes: the institutional, the infrastructural, the environmental and the socioeconomic measures. It is important to notice that prevention measures are not only in the hand of the social planner but are also taken by private agents or groups as own initiative or following the emergency plans (Thieken et al. 2005). Such measures can target the different components of the risk: the hazard, the exposure and the vulnerability. As illustrated in Section 16.3, these three components of the risk are related, therefore a modification of the hazard only will have an impact on the exposure, then on the vulnerability and finally on the risk. For this classification, only the first order impact of the measure is considered. Table 16.1 presents this classification with some non-exhaustive examples of measures.

The institutional measures are a basis for the coordination and implementation of public prevention measures. Knowledge about the river dynamics, the monitoring of the rivers, the implantation of flood risk maps and flood warning systems, appropriate land use planning, the existence of emergency plans from Civil Protection or municipalities and dams security norms are examples of this type of measures. They mostly target exposure and vulnerability reduction.

In 1985, a program of river monitoring and information systems was started in Spain and the Ebro river basin. The so-called Automatic Hydrologic Information System[3] has been operational since 1997 and has been the main alert system since 2011. Moreover, it was complemented with a system of decision support[4] in 2002. These two systems revealed their importance for crisis management during the flood of winter 2003 (January 28th to February 6th, 2003) in the Ebro river basin, Spain.

The question of the security of dams and reservoirs arises with the risk of artificially retaining such a volume of water upstream of cities. The risk of a dam's failure is an extreme risk – low probability and huge impacts – that could be prevented with specific and drastic technical norms. The issue is of importance in Spain with its more than 1,200 dams of large capacity of storage (>56 hm3). Four hundred and fifty of them were built before 1960 and more than 100 were already built by 1915. In the Ebro river basin, 293 dams have been built, with different capacities of storage, of which 72 are owned by the Spanish state and 221 by others owners.

Table 16.1 A classification of flood risk prevention policies with their objectives

Measures	*Elements of Risk*		
	Hazard	*Exposure*	*Vulnerability*
Institutional			
Flood warning		✓	
Emergency plan protection		✓	✓
Security norms		✓	
Land use planning		✓	
Infrastructural			
Permanent defenses	✓		✓
Temporary defenses			✓
Dam regulation	✓		✓
Environmental			
River restorations	✓	✓	
Facilitation of river flow	✓	✓	
Controlled flooded areas		✓	
Socioeconomic			
Insurance			✓
Self-insurance			✓
Self-protection		✓	✓

Therefore, the regulation of both non-state and state owners of dams is of interest for public security.

In the past, various accidents of dam collapses occurred all over Spain. On the 30th of April, 1802, at its first fill, the dam of Puentes in the Segura river basin (Spain) broke through. The 30 hm^3 of water escaped in one hour, and one hour after, a wave formed and reached the city of Lorca, killing 608 people. On the 10th of January, 1959, the dam of Vega de Tera (Guadiana river basin, Spain) broke through in its first fill. 144 people died in the village of Ribadelago. On the 20th of October, 1982, the dam of Tous (Júcar river basin, Spain) broke and caused the loss of 30 human lives (Caballero 2003, Rodriguez-Trelles and Mazaira 2003).

In reaction to these catastrophic events, a series of decisions had been taken to improve the security of dams and reservoirs management. After the first recorded catastrophe of the dams of Puentes in 1802, the School of Civil Engineering was created, followed a hundred years later by the first norms for dam projects. The Vega de Tera catastrophe of 1959 motivated the creation of a commission for dams monitoring, and in 1967 the governmental order relative to the elaboration of project, construction and exploitation of dams was adopted. The Tous dam collapse in 1982 gave impulsion to an audit program of dam security monitoring followed by a governmental order relative to the technical rules for security in dams and reservoirs in 1996 (Caballero 2003). The Tous collapse gave rise to a general awareness for improving the flood alert system and is a turning point in flood risk management in Spain (Serra-Llobet et al. 2013).

The Real Decree 9/2008, modifying the Real Decree 846/1986 relative to the Hydraulic Public Domain, adds a chapter relative to the security of dams and reservoirs. The objective of this chapter is to unify in a single norm the security standards. The finality is the protection of people, environment and property goods. To that end, it is the government who regulates the norms of security of the dams and reservoirs, defines commitments and liabilities of incumbents and monitors security procedures and the role of public administration.

The infrastructure measures are the strong engineering measures of defense towards water flows and stocks. One can find the permanent construction of defense, the temporary elements of protection (sand bags, for example), but also the reservoir regulation system. As depicted in Table 16.1, these measures are dedicated to hazard and vulnerability reduction. In the Ebro river basin (Spain), 293 dams and reservoirs have been built and some expansions are being implemented. The reservoirs system has a significant role to contain water flows and stock. For example, during the flood event of spring 2007 (March 19 to April 6, 2007), the Yesa reservoir in the river Aragon outflowed 150 m3/s and then 100 m3/s when the inflow was 250 m3/s. The reservoir of Yesa together with the reservoir of Itoiz contributed to absorbing the flood wave significantly in the sub-basins of Irati and Aragon (Ebro River Basin Authority, 2007).

The environmental measures are soft measures aiming at reducing the hazard and the exposure. In the case of the Ebro river basin in Spain, a national strategy of rivers restoration was launched in 2005 in order to minimize the risk of floods. The program consists of actions and measures of improvement of the ecological state of the river that facilitate water discharges and reduce the volume of water entering in the floodplain. The measures consist mostly of recovering the natural dynamic of rivers, improving the flowing of water with operation of residual elimination, river bed cleaning and forestation of river banks.

As an example, downstream the Cinca River (Ebro river basin, Spain) it was decided that a secondary dyke of protection be eliminated and that the natural vegetation of the river bank be restored. The population will still be protected for a 500-year return period flood by a principal dyke, but the elimination of the secondary dyke enabled the river to recover its natural dynamic and have more space. The elimination of the secondary dyke reduces the risk of dyke break of the principal dyke. Moreover, the soil recovers its capacity of absorption over time. And natural vegetation has replaced the productive species more vulnerable to flood than natural river bank species.

The socioeconomic measures aim at reducing the exposure and vulnerability of people and goods. This can be achieved with financial insurance systems, self-insurance measures that consists in lowering the size of the loss due to the event (land uses decisions in agriculture for example) and self-protection measures that consist of lowering the probability of being hit (Ehrlich and Becker 1972), for example following the recommendation of civil protection or elevating the goods when possible. The EU commission is amenable to the development of the insurance system as it is a most helpful tool for good risk and crisis management, in many circumstances. In Spain, the private insurance system has a long history in the agricultural sector since the Agricultural Insurance Act of 1978. The agricultural insurance system is a public/private system where the risk premium paid by farmers is partly publicly subsidized. In the case of the flood of 2007 in the autonomous region of Aragon (Spain), the insurance companies compensated 621,746.59€ to 1,350 farms. However, such an insurance system needs regulation in order to fulfil the established market competition rules in place. The EU directive 2006/C319/01 states the guidelines for 2007–2013 in order to not transcend the limits of the competition conditions of the World Trade Organization. It fixes a minimum damage to production of 30% of the normal production to be admissible for receiving public aids for losses. Moreover, it has fixed the share of public subsidy in the risk premium paid by farmers: up to 80% of the cost of the risk premium in the case of policies covering specific natural disasters such as floods. Furthermore, in the case of flood risk, public compensation payments are distributed to compensate the damages to the factors of production in order to help the production process to recover.

16.4.2 Cost-benefit analysis and the expected annual damage avoided

To assess the economic impacts of projects, including flood prevention projects, the cost-benefit analysis is the more suitable (Pearce 1998). In the case of the European Floods directive (2007/60/EC), this method is required to assess measures of transnational effects (Annex A.5 of the directive); otherwise, it is recommended that flood risk management plans "shall take into account relevant aspects such as costs and benefits [. . .]"(Article 7[3] of the directive). Some member states, such as Spain, in their transcription of the EU Floods directive, require the use of cost-benefit analysis to justify the investment in structural measures (Annex AH7 of the Real Decree 903/2010).

Other methods such as multi-criteria can be also a tool for decision-making (Janssen 1992). In this chapter, only the cost-benefit analysis is presented as it is required by the Real Decree 903/2010 in Spain for investment in structural measures of flood prevention.

The cost-benefit analysis consists of balancing the costs with the benefits a project can generate. If the discounted net benefit (also called the net present value) is positive, the implementation of the project is beneficial to the manager. The costs and benefits should account for economic, social but also ecological considerations to present an integrated balance of the project or policy measures (Brouwer and van Ek 2004).

The expected annual damage avoided

The advantages and disadvantages of the measures or projects are evaluated in monetary terms. The monetary evaluation can be based on various techniques (Shabman and Stephenson 1996): market prices, hedonic prices and contingent valuation methods. The hedonic price method consists of estimating the benefits through the changes in the real estate market (Donnelly 1989, Bin and Polasky 2004, Bin et al. 2008). The contingent valuation method consists of asking the willingness of citizens to pay for a flood prevention project (Brouwer et al. 2009, Botzen and van den Bergh 2012).

The most important benefit of the measure is the expected annual damage avoided. It measures how much damage will be avoided by the measure or project. It is derived as the difference between the annual average damages before a project and the annual average damages after project. Let's consider the example adapted from Erdlenbruch et al. (2008). Consider a project of flood prevention that would potentially reduce the damages of flood with low return periods but would increase the damages for floods with high return periods (i.e. with low probability of occurrence). The expected annual flood damages in the initial situation (without the project) are represented by the light grey area in the left hand graph of Figure 16.4. The expected benefits of the project are therefore represented by the dark grey area in the right-hand graph of Figure 16.4.

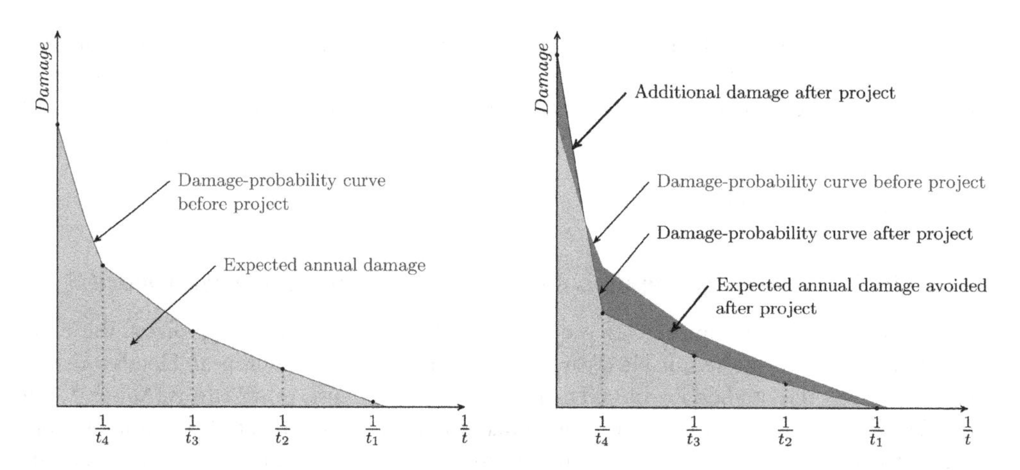

Figure 16.4 Derivation of the annual average damages avoided
Source: Adapted from Erdlenbruch et al. 2008

The net present value

The net present value (NPV) of the project is the indicator on which the social planner can base his/her decision of investment. It balances the investment costs with the benefits generated by the project. For a project of duration T, it is computed as

$$\mathbf{NPV} = -\mathbf{C}_0 + \sum_{t=1}^{T} \frac{\mathbf{1}}{(\mathbf{1}+\mathbf{r}_t)^t}(\mathbf{B}_t - \mathbf{C}_t) \tag{2}$$

where C_0 is the initial cost of the project, B_t the expected benefit at date t, C_t is the maintenance cost of period t and r_t is the discount rate of the period .

Two parameters of the NPV require a harmonization at the national level to make projects comparable: the discount rate and the time horizon of projects. The discount rate is an interest rate used to convert a future income stream to its present value. It is not necessarily constant but can vary with the period of time. The time horizon is the time during which the flows of the project – the costs and benefits – have to be accounted for.

Discounting recognizes that both individuals and societies prefer to get benefits in the present and to postpone any costs to the future. Discounting gives lesser weight to benefits and costs that occur in future years. A critical decision relies on the choice of the discount rate. The higher the discount rate, the lower the present value of either costs or benefits occurring in the future. Thus, a 100-million-euro benefit in 100 years would be valued today at 45,460€ if the annual discount rate is 8 percent, but at two million euros if the discount rate is 4 percent. Therefore, in order to compensate for the decrease of the future benefit in a long time horizon in present value, the discount rate should decline over time.

In France, the discount rate was fixed in 2005 at 4 percent for public investments with a time horizon of up to 30 years and then decreases to reach nearly 2 percent for a time horizon of 500 years (Commissariat Général du Plan 2005). In the UK, the interest rate at 30 years is 3.5 percent and then decreases to reach 1 percent after 300 years (HM Treasury 2003). The

European Union – DG Regional Policy (2008) has fixed a discount rate of 5 percent for the project funded in the context of the European Regional Development Fund.

16.4.3 Examples of cost-benefit analysis

Erdlenbruch et al. (2008) assessed the economic impacts of a dike project in the Orb delta in France designed for the protection of floods up to a 100-year return period. The elements at risk considered are the buildings, the agricultural sector, road infrastructure economic activity and camping. The risk analysis was conducted for the 10-year, 30-year, 50-year and 100-year return periods. Water depth is the damage driver used for the vulnerability assessment. The cost of the project amounts to 31.9 million euros, with a 3 percent maintenance cost. The expected annual damage avoided by the dike has been estimated at 4 million euros. Therefore, with a time horizon of 50 years and the discount rate structure for the French public investments, the net present value of the project amounts to 36 million euros and is an economic argument for the implementation of such a project of flood risk prevention.

Brouwer and van Ek (2004) evaluated the expected consequences of alternative flood protection measures in the Rhine and Meuse delta in the Netherlands. They studied the potential of flood plain restoration and land use changes for risk prevention in a country where dike enclosures had been the predominant protection measure. They assessed the benefits and costs of these two measures in an integrated way accounting for their ecological, economic and social impacts. The cost-benefit analysis was used for the economic impact assessment. The discount rate used was a 4 percent discount rate, and the time horizon varies between 50 and 100 years with the elements at risk. The economic costs in these projects refers to the costs of expropriation and compensation payments, operation and maintenance costs of infrastructure and measures and the damages to crops. They were evaluated for the agricultural, industrial, residential and floodplain elements at risk. The present value of the economic costs amounts to 5,540 million euros. The benefits account for the damages avoided by the measures and the recreational benefits. They amount to a present value of 3,400 million euros. The net present value is therefore a loss of 2,140 million euros. An extension of the cost-benefit analysis to the intangible impacts such as the social and environmental impacts (public safety, biodiversity conservation, landscape amenities) revealed the importance of assessing the damage in an integrated approach. The economic valuation of these intangible benefits is based on willingness to pay values per household per year for having these services or benefits. The total discounted intangible benefits reached 3,100 million euros. Therefore, the project of floodplain restoration and land use changes result in a net gain of 860 million euros.

16.5 Conclusion

Along with earthquakes, one of the most damaging natural phenomena is flooding. The wide occupation of river banks by man exposes many items to a likely flooding event. This chapter has shown some aspects that may be relevant and may help in properly managing these natural phenomena and their possible consequences.

The multiple measures that can be implemented in order to prevent the negative consequences of successive floods (whether they are institutional measures, infrastructure measures, environmental measures or socioeconomic measures), would benefit from a cost-benefit analysis to assess their impacts on the risk prevention. This cost-benefit analysis is not possible without holding a proper assessment of flood risk and a proper analysis of how this risk will be affected by the implementation of corresponding preventive measures. At this point, we propose a methodology for flood risk assessment to obtain a monetary estimate of flood risk that may be incorporated in the relevant cost-benefit analysis.

This methodology consists, on the one hand, in the estimation of flood hazard using information from different hydrological studies and, on the other hand, in the identification of the elements (natural and/or artificial) that are exposed to each potential flood hazard. This identification is done mainly crossing the information of the hydraulic studies with the geographic data of the flooded area. Having identified the elements exposed, a further analysis (micro-scale analysis) allows us to disentangle their vulnerability to being flooded. An economic analysis of vulnerability leads to monetary calculation of the flood damage associated with an x-year flood event.

Based on this methodology, in this chapter we have proposed two measures of flood risk, the damage-probability curves and the expected annual damage, which can be incorporated into cost-benefit analysis. We have shown how a flood prevention project modifies the damage-probability curve and consequently modifies the expected annual damage. In fact, the most important benefit of a certain prevention measure is the expected annual damage avoided.

The quantitative assessment of flood risk is relevant to public decision-making in risk management. This analysis becomes even more important in a context of climate variability where rainfall may be altered in both amount and timing, modifying (raising in most cases) the risk of flooding. A proper assessment of the potential risks and how they will be affected by preventive measures is necessary to implement a measure that has not only a hydrological basis but also a strong socioeconomic justification to support the investment.

Notes

1 If x is flood damage and $f(x)$ is the corresponding probability density function, the expected annual damage is $E(x) = \int xf(x)dx$.

2 The figure shows only part of the damage-probability curve, particularly the range of probabilities from 0 to 0.04, in order to observe in more detail the estimates of the initial sample. Consequently, the shaded area under the curve is only part of the expected annual damage. The value of 1.1 million euros corresponds to the expected annual damage calculated for the entire range of probabilities.

3 Sistema Automático de Información Hidrológica, SAIH.

4 Sistema de Ayuda a la Decision, SAD.

References

Arnell, N.W. 1989. Expected annual damages and uncertainties in flood frequency estimation. Journal of Water Resources, Planning and Management 115, 94–107.

Bin, O., Kruse J. B and Landry C.E. 2008. Flood hazards, insurance rates and amenities: Evidence from the coastal housing market. The Journal of Risk and Insurance 75(1), 63–82.

Bin, O. and Polasky S. 2004. Effects of flood hazards on property values: Evidence before and after hurricane Floyd. Land Economics 80(4), 490–500.

Botzen, W.J.W. and van den Bergh, J.C.J.M. 2012. Risk attitudes to low-probability climate change risks: WTP for flood insurance. Journal of Economic Behavior and Organization 82(1), 151–166.

Brouwer, R., Akter, S., Brander, L. and Haque, E. 2009. Economic valuation of flood risk exposure and reduction in a severely flood prone developing country. Environment and Development Economics 14(3), 397–417.

Brouwer R. and van Ek, R. 2004. Integrated ecological, economic and social impacts assessment of alternative flood control policies in the Netherlands. Ecological Economics 50(1), 1–21.

Caballero, C. 2003 Seguridad de presas: pasado, presente y futuro. IT n°62.

Cochrane, H. 2004. Economic loss: Myth and measurement. Disaster Prevention and Management 13, 290–296.

Commissariat General du Plan, 2005. Révision du taux d'actualisation des investissements publics. Rapport du groupe d'experts présidé par Daniel Lebègue.

de Groot, R.S., Wilson M.A. and Boumans R.M.J. 2002. A typology for the classification, description and evaluation of ecosystem functions, goods and services. Ecological Economics 41(3), 393–408.

Donnelly, W. 1989. Hedonic price analysis of the effects of a floodplain on property values. Water Resources Bulletin 24, 581–586.

Ebro River Basin Authority, 2007. Memoria 2007.

Ebro River Basin Authority, 2011. Descripción de la situación meteorológica y respuesta hidrológica de la cuenca del Ebro durante los episodios de avenida del Ebro en 2003 y 2007.

Erdlenbruch, K. Gilbert, E. Grelot, F. and Lescoulier, C. 2008. Une analyse coût-bénéfice spatialisée de la protection contre les inondations- Application de la méthode des dommages évités à la basse vallée de l'Orb. Ingénieries 53, 3–20.

Ehrlich, I. and Becker, G.S. 1972. Market insurance, self-insurance, and self-protection. Journal of Political Economy 80(4), 623–648.

European Union – DG Regional Policy, 2008. Guide to cost-benefit analysis of investment projects. Structural Funds, Cohesion Fund and Instrument for Pre-Accession.

Gollier, C. 2001. The economics of risk and time. The MIT press.

Hall J.W., Meadowcroft, I.C., Sayers, P.B. and Bramley, M.E. 2003. Intergated flood risk management in England and Wales. Natural Hazards Review 4(3), 126–135.

Hallegatte, S. 2009. Strategies to adapt to an uncertain climate change. Global Environmental Change 19, 240–247.

Hallegatte, S. and Przuski, V. 2010. The economics of natural disasters. Concepts and methods. The World Bank. Policy Research Working Paper WPS5507.

HM Treasury, 2003. The Green Book, Appraisal and Evaluation in Central Government, Treasury Guidance, TSO, London.

Janssen, R. 1992. Multiobjective decision support for environmental management. Kluwer Academic Publishing, Dordrecht, The Netherlands.

Jonkman S.N., Bockajova, M., Kok, M. and Bernardini, P. 2008. Integrated hydroeconomic and economic modeling of flood damage in the Netherlands. Ecological Economics 66(1), 77–90.

Kron, W. 2002. Flood risk = hazard x exposure x vulnerability. Flood defence, Wu et al. (eds), Science Press, New York Ltd.

Kundzewick, Z.W. and Schellnhuber, H.J. 2004. Floods in the IPCC TAR perspective. Natural Hazards 31, 111–128.

Lekuthai, A. and Vongvisessomjai, S. 2001. Intangible flood damage quantification. Water Resources Management 15, 343–362.

Merz, B. Kreibich H., Schwarze, R. and Thieken, A. 2010. Assessment of economic flood damage. Natural Hazards and Hearth System Sciences 10(8), 1697–1724, doi:10.5194/nhess-10-1697-2010

Messner, F. and Meyer, V. 2005. Flood damage, vulnerability and risk perception – challenges for flood damage research, UFZ Discussion papers 13/2005.

Milly, P.C.D., Wetherald R.T., Dunne, K.A and Delworth, T.L. 2002. Increasing risk of great floods in a changing climate. Nature 415, 514–517.

Pall, P., Aina, T., Stone, D.A., Stott, P.A., Nozawa, T., Hilberts, A.G.J., Lohmann, D. and Allen, M.R. 2011. Anthropogenic greenhouse gas contribution to flood risk in England and Wales in autumn 2000. Nature 470, 382–385.

Parry, M.L., Canziani, O.F., Palutikof, J.P., van der Linden, P.J and. Hanson, C.E. (eds). 2007. Contribution of Working Group II to the Fourth Assessment Report of the Intergovernmental Panel on Climate Change. Cambridge University Press, Cambridge, United Kingdom and New York, NY, USA.

Pearce, D.W. 1998. Cost-benefit analysis and environmental policy. Oxford Review of Economic Policy 14(4), 84–100.

Penning-Rowsell, E., Johnson, C., Tunstall, S., Tapsell, S., Morris, J., Chatterton, J. and Green, C. (2005): The benefits of flood and coastal risk management: A handbook of assessment techniques. Middlesex University Press.
Rodriguez-Trelles, A.M. and Mazaira J.P. 2003. Presas. Seguridad y percepción del riesgo. IT n° 62.
Rose, A. 2004. Economic principles, issues and research priorities in hazard loss estimation. In: Okuyama, Y. and S. Chang (eds.), Modeling spatial and economic impacts of disasters, pp. 14–36. Springer, Berlin.
Schanze J 2006. Flood risk management – a basic framework. In: Schanze, J.; Zeman, E.; Marsalek, J. (Eds.), Flood risk management: Hazards, vulnerability and mitigation measures, pp. 1–20. Springer.
Serra-Llobet, A., Tàbara, J.D. and Sauri D. 2013. The Tous dam disaster of 1982 and the origins of integrated flood risk management in Spain. Natural Hazards 65(3), 1981–1998.
Shabman, L. and Stephenson, K. 1996. Searching for the correct benefit estimate. Empirical evidence for an alternative perspective. Land Economics 72(4), 433–449.
Thieken, A.H., Muller, M., Kreibich, H. and Merz, B. 2005. Flood damage and influencing factors: New insights from the August 2002 flood in Germany. Water Resources. Research 41 (12), W12430, doi:10.1029/2005WR004177
Werrity, A. 2006. Sustainable flood management: Oxymoron or new paradigm? Area 38(1), 16–23.
Zenger, A. and Wealands, S. 2004. Beyond modelling: Linking models with GIS for flood risk management. Natural Hazards 33, 191–208.

PART IV

Other dimensions of adaptation

17
FAST-GROWING COUNTRIES AND ADAPTATION

Yan Zheng[1] *and Jiahua Pan*[1,2]

[1] INSTITUTE OF URBAN & ENVIRONMENTAL STUDIES, CHINESE ACADEMY OF SOCIAL SCIENCES

[2] RESEARCH CENTRE FOR SUSTAINABLE DEVELOPMENT, CHINESE ACADEMY OF SOCIAL SCIENCES

17.1 Introduction

The term "fast-growing countries" (FGCs) refers to countries with significant importance and good performance in economic growth during a given period. The BASIC and BRIC countries are typical FGC groupings taking a lead in the world economy, while East Asia and the Pacific, South Asia, Latin America and the Caribbean are top fast-growing regions. Normally, FGCs share some common features and challenges in development and adaptation. For example, large populations and the growth of aggregate wealth have been increasing their exposure to climate change. At the same time, the FGCs must take socioeconomic development, poverty reduction and greenhouse gas emissions into consideration. In fact, adaptation has become an extra burden for FGCs in covering their development deficit.

This chapter discusses the subject of "fast-growing countries and adaptation" and consists of five parts. The first part is a brief introduction. The second part deals with "Political economics of adaptation". This topic defines the main characteristics of FGCs, some representative regions (Asian Pacific regions, South Asian regions and Caribbean regions for example) and some major country groupings such as BASIC and BRIC. In this part, the authors pay special attention to analysing the role that FGCs are likely to play in world political and economic situations and in the world climate management regime as well as the challenges that FGCs are now facing. The third part approaches the determinants of the risks that FGCs are facing in the context of global climate change, namely increasing exposure and diverse drivers of vulnerability. This part devotes special attention to disaster damage costs and the cost of adaptation in FGC regions. Based on literature review, the fourth part undertakes case analyses on agriculture, migration due to climate change, adaptation to climate change on the part of cities in such FGCs as China, India, Vietnam, etc. This part also outlines the characteristics of FGCs and the challenges that they face as well as the importance of strengthening strategic adaptation planning, coordinated management and reducing greenhouse gas emissions on the part of FGCs. The last part concludes with highlights and implications.

17.2 Political economy of adaptation

17.2.1 Description of fast-growing countries (FGCs)

Fast-growing countries (FGCs)

Although there is no universally accepted definition, the term "fast-growing countries" (FGCs) is frequently used in the academic and international community to indicate those countries with good performance in economic growth. Macro-economists have paid considerable attention to analysing why and how some economies can kick off and sustain fast growth (Berg et al., 2012). FGCs can be measured by percent per annum in their aggregate GDP and in per capita terms. They can also be classified from a temporal perspective into, for instance, historical, present and potential fast-growing countries. Most studies of fast-growing economies use a 5 percent average growth rate in per capita GDP over a successive period (such as 10 years or more) as a benchmark for identifying fast-growing economies (Virmani, 2012).

Based on neo-classical economics, very few countries would be able to embark on a process of sustained fast growth for more than a quarter of a century during modernisation. During the late 20th and early 21st centuries, the world economy experienced rapid economic growth of large emerging markets, starting from Asia and now extending through other parts of the developing world (Eichengreen et al., 2011). From 1960 to 2006, seven fast-growing developing countries – Korea, Singapore, Malaysia, Thailand, Indonesia, Botswana, and Mauritius – were among the top performers in the world in terms of growth in GDP per capita. Subsequently, China ranks as the only country in history that has grown at an average per capita rate of over 7 percent per annum (7.4 percent) for three decades (Virmani, 2012). According to *Global Economic Prospects 2012*, 29 countries in 6 sub-regions are likely to account for the highest growth over the next two years based on the World Bank's growth estimates for 2013 and 2014: during that time East Asia and the Pacific, South Asia, Latin America and the Caribbean have ranked among the top three highest growth regions at the present rate and that of coming years (WB, 2012, see Table 17.1).

BRIC

Owing to rapid growth and increasing significance in the global economy, the BRIC group, made up of the four major developing countries Brazil, Russia, India and China, has become well-known to the international community as a symbol of the shift in global economic power away from the developed G7 economies towards the developing world (O'Neill, 2001).[1] Furthermore, the "E7" (seven Emerging Countries) also stand out, comprising BRIC plus Mexico, Turkey and Indonesia. By 2050, with the benefits of economic catch-up, low labour costs, technology transfer and population growth, the E7 will be around 50% larger than the current G7 (US, UK, Germany, France, Japan, Italy and Canada). China and India are expected to catch up with the US around 2025 and 2050, respectively. The list of the other economies expected to grow the fastest up to 2050 includes Vietnam, Nigeria, the Philippines, Egypt and Bangladesh (Hawksworth & Cookson, 2008: 3–4).

BASIC

The BASIC countries (Brazil, South Africa, India and China) were driven and led by China in UNFCCC COP 15 during the negotiation process of consolidating a common position on emission reductions and climate funding mechanisms for developing countries and trying

Table 17.1 Fast-growing developing countries by region (percent change from previous year, except interest rates and oil price)

	2010	*2011*	*2012e*	*2013f*	*2014f*
Developing countries	7.4	6.1	5.3	5.9	6.0
East Asia and Pacific	9.7	8.3	7.6	8.1	7.9
China	10.4	9.2	8.2	8.6	8.4
Indonesia	6.2	6.5	6.0	6.5	6.3
Thailand	7.8	0.1	4.3	5.2	5.6
Europe and Central Asia	5.4	5.6	3.3	4.1	4.4
Russia	4.3	4.3	3.8	4.2	4.0
Turkey	9.2	8.5	2.9	4.0	5.0
Romania	-1.6	2.5	1.2	2.8	3.4
Latin America and Caribbean	6.1	4.3	3.5	4.1	4.0
Brazil	7.5	2.7	2.9	4.2	3.9
Mexico	5.5	3.9	3.5	4.0	3.9
Argentina	9.2	8.9	2.2	3.7	4.1
Middle East and N. Africa	3.8	1.0	0.6	2.2	3.4
Egypt	5.0	1.8	2.1	3.1	4.2
Iran	2.9	2.0	-1.0	-0.7	1.5
Algeria	3.3	2.5	2.6	3.2	3.6
South Asia	8.6	7.1	6.4	6.5	6.7
India	9.6	6.9	6.6	6.9	7.1
Pakistan	4.1	2.4	3.6	3.8	4.1
Bangladesh	6.1	6.7	6.3	6.4	6.5
Sub-Saharan Africa	5.0	4.7	5.0	5.3	5.2
South Africa	2.9	3.1	2.7	3.4	3.5
Nigeria	7.9	7.4	7.0	7.2	6.6
Angola	3.4	3.4	8.1	7.4	6.8

Source: WB, 2012: 8
Notes: e = estimate; f = forecast

to convince other parties to sign up to the Copenhagen Accord. As a new geopolitical alliance, the BASIC countries are working together and playing an important role in building a global climate regime. While there are different views on the goal to keep global temperature increase well below 2 °C (as a result of various scientific uncertainties such as selection in climate modelling, assumptions in socioeconomic scenarios, attribution issues and so on, for a given amount of emissions the consequent maximum temperature rise can be predicted only in terms of probability), there is agreement among all developing countries on the priorities of socioeconomic development and fighting poverty. Based on this common position, the BASIC countries have set out to consider a practical way of getting 'equitable access to sustainable development' for developing countries. The concept has three essential parts: 1) access to carbon space, 2) sustainability, and 3) schedule for development. A BASIC Expert group was mandated in February 2011 to collaborate with this commission. The "carbon budget proposal"[2] is a representative collaboration of the BASIC teams, which clearly concludes that the critical issue for developing nations is the gap between their equitable shares of the global carbon space versus the physical shares that will be accessible to them. BASIC experts argue that failure to deal with mitigation in a way that keeps temperature increases below 2 °C will increase the adverse effects of climate change and then lead to a multiplier effect on the costs of adaptation; therefore, an equity-based framework for a global regime should link mitigation, adaptation and their respective costs (BASIC, 2012). To provide a better understanding of the situation of BASIC countries, some indicators relevant to development and climate change are listed in Table 17.2.

Challenges of FGCs in a changing climate

While fast-growing countries have an exceptional economic growth, they have to face many challenges in their sustainable development. Action for climate change is likely to conflict with or undermine development goals and economic growth (Cosbey, 2009). In comparison to the developed countries, a noteworthy feature is that most fast-growing countries (except, for example, Russia) are not rich in natural resources. Some Asian countries (China, S. Korea, Singapore, Japan, Hong Kong SAR and Thailand) have been continually suffering from scarcity in land and water resources, which is likely to set a threshold for their attaining sustainable development in the long run. On the other hand, increasing population and accumulated assets in FGCs will promote income and quality of life; however, they will also entail increasing risks due to greater exposure to climate disasters (IPCC, 2012; Nicholls et al., 2009; WB, 2010a).

Table 17.2 Major developmental indicators of BASIC countries

Country	*China*	*India*	*Brazil*	*South Africa*
Population billion (2011)	1.344	1.241	196.7	50.59
GDP (constant 2000 billion US$) (2011)	3246.01	963.40	916.13	187.25
Per capita GDP (constant 2000 US$) (2010)	2425	787	4699	3746
Land areas (sq. km) (2010)	9327480	2973190	8459420	1214470
Per capita CO_2 Emissions (2008)	5.3	1.5	2.1	8.9
HDI (2011)[1]	0.687	0.547	0.718	0.619
Total loss of climate related natural disasters (1983–2012) (000 US$)[2]	264,961,277	35,409,751	11,132,170	3,154,070
Affected population of climate related natural disasters (1983–2012)[3] (Millions of people)	2933	1266	51	19

Source: WB, WDI Database, 2012 http://data.worldbank.org/country

[1] *Source:* UNDP, Human Development Report, 2012. Table 1 Human Development Index and its components, p. 127–130.

[2] *Source:* EM-DAT: The OFDA/CRED International Disaster Database, www.em-dat.net – Université Catholique de Louvain – Brussels – Belgium. The "Climate-related natural disasters" include drought, extreme temperature, flood, mass movement wet, mass movement dry, and storm.

[3] *Source:* EM-DAT: The OFDA/CRED International Disaster Database, www.em-dat.net – Université Catholique de Louvain – Brussels – Belgium. The "Climate-related natural disasters" include drought, extreme temperature, flood, mass movement wet, mass movement dry, and storm.

17.2.2 Political economy of the adaptation of fast-growing countries

Similarities between FGCs

Rapid economic growth has not only promoted human welfare but also contributed to the development of the social economy. At the same time, it has also contributed to the deterioration of environment in the developing world (Bourguignon 2005; Brock &Taylor 2005). On the one hand, fast growth in major developing countries has raised incomes and living standards. FGCs such as China and India have made a prominent contribution to global poverty alleviation and human development. On the other hand, FGCs have undergone increasing population growth and the rise of middle classes in their process of urbanisation and industrialisation, resulting in a profound transformation in both economic and social structure (Kharas, 2010). Using an index of car owners in developing countries, the middle classes have been estimated to number about 4 billion in 70 developing countries (Dadush & Ali, 2012). Increasingly large populations and emerging middle classes will require FGCs to make further efforts if they are to sustain rapid growth.

However, most FGCs are not resource-rich countries and have not reached as high a level of human development as developed countries. Industrialisation and urbanisation must be substantially promoted and completed: this will inevitably entail some degree of material and energy consumption to ensure that most people can afford a decent life, including housing, utilities, infrastructure, and improving institutions and governance to deal with the changing climate (Pan & Zheng, 2011).

Developing countries suffer from a "double exposure" to both economic globalisation and climate change (O'Brien & Leichenko, 2000). The fast growth of emerging countries means rapid shifts in the relative weights of different regions, such as South versus North, East versus West, Asia versus Europe and the United States. This has geopolitical implications that extend far beyond the confines of economic issues (Eichengreen et al., 2011). This has occurred during the process of climate negotiation and global regime building.

Dual-challenges of FGCs: Development and dealing with climate change

There are major discrepancies between parties on the principle and responsibility of sharing emission reductions, so a legally binding climate agreement of post-Kyoto Protocol had still not been reached since the Copenhagen Accord at the 15th Conference of the Parties to the United Nations Framework Convention on Climate Change (UNFCCC) in December 2009. By addressing Carbon Emissions Kuznets Curves for 16 major economies, Figure 17.1 shows the three major country groupings with common economic and political positions in the negotiation of UNFCCC. As clearly indicated, FGCs are an emerging grouping in the corner of the map of global emissions and economic growth, compared with the other two groupings of developed countries. The upper group comprises resource-rich UMBRELLA[3] countries which are inactive and hampering the taking of action under the Kyoto Protocol (KP) framework; by contrast, the other group (EU and Japan) have been active in taking a lead in guiding a global low-carbon future.

To meet human basic needs including food, shelter, transport and other infrastructures, FGCs will obviously need to consume more energy to overcome their development deficit and their adaptation deficit,[4] such as fundamental and climate-proofing infrastructures, which will inevitably lead to correspondingly high carbon emissions in future developments. On the other hand,

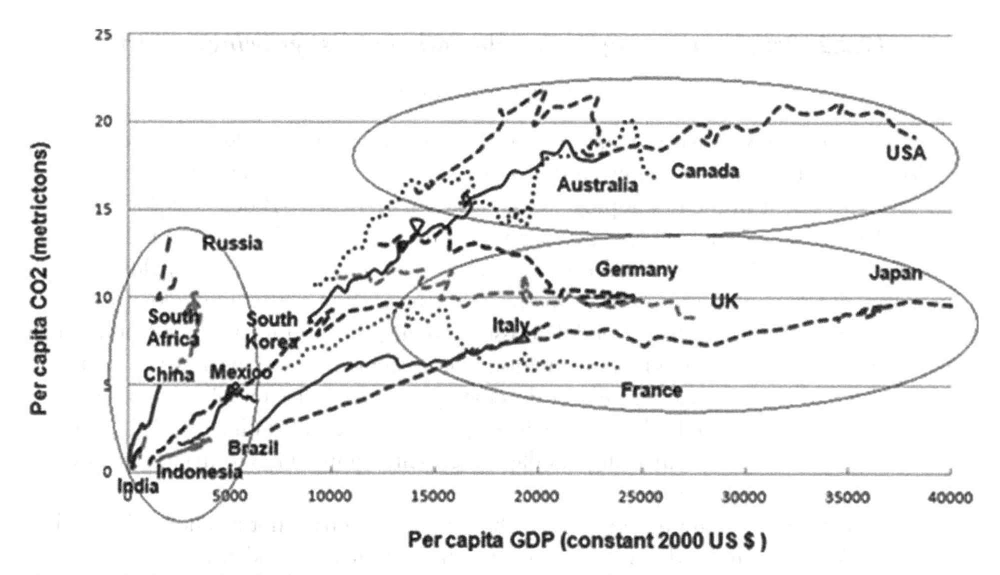

Figure 17.1 The relationship between economic growth and emissions in 16 major economies (1960–2006)
Source: Pan & Zheng, 2009, revised with WB indicators, http://data.worldbank.org/country

the economy of developing countries is unstable and not robust enough to withstand risks such as global economic recessions, so there will be wide fluctuations in their rapid expansion. The situation of FGCs and the priority allocated to promoting domestic development and contributing to the meeting of global Millennium Development Goal (MDG) targets mean that a commitment on global emission mitigation by FGCs would be not only unequal but harmful to their sustainable development (Pan & Zheng, 2009; Pan, 2011). However, so as to set an example for other developing countries and promote global regime building, the BASIC countries promised to declare their national voluntary emission reduction targets before COP 17 (2011).

Emerging role of FGCs in building a global climate regime

While there is a conceptual consensus on the principle of Common but Differentiated Responsibility, there still is a big discrepancy between South and North on the rationale and responsibility of adapting to climate change. The global climate regime has been making great efforts to mitigate GHG emissions since the ratification of the KP in 1998. However, adaptation did not become the fourth pillar of the global climate regime building until the Bali Action Plan had been ratified in 2007 at UNFCCC COP 15. Before the Fourth Assessment Report of the Intergovernmental Panel on Climate Change (IPCC), clearly indicating that "adaptation will be necessary to address impacts resulting from the warming which is already unavoidable due to past emissions" (IPCC, 2007: 18), adaptation was ignored for a long time by the international community, though it is considered especially relevant for developing countries in taking domestic actions for disaster prevention, reduction and relief (Schipper, 2007).

Adaptation to unavoidable damage from climate change is now considered as an additional financial burden for developing countries, with costs in the tens of billions of US dollars per year (Ackerman, 2009; Horstmann, 2008). With a strong appeal from the South, it was agreed to set up the Green Fund in the framework of the UNFCCC, and to mobilise long-term finance to support action in developing nations, with a level of USD 100 billion annually by 2020 after the Doha conference. However, the amount earmarked is still limited and not comparable to the estimated adaptation costs for developing countries. An "additional, adequate and sustainable" financing mechanism on adaptation has not yet been successfully formed.

It is not easy for FGCs to consolidate a common political position and take the lead in negotiation although their voices and powers have been heard and seen. In fact, with different situations as regards economic growth and priorities in dealing with climate change, FGCs can be classified into different sub-groupings: e.g. Russia and other East-European partners are OECD states and are regarded as transit economies, and will eventually leave the non-legally-binding-commitment parties group; and developing countries have potentially begun to diverge, for instance, the most vulnerable ones such as those in AOSIS[5] and the big emerging countries represented by BASIC. The BASIC countries have been involved in arguments about taking on compulsory emission reduction commitments in the coming years (for the post 2012 global regime), and at the same time they have to deal with and suffer increasing climate risks and economic losses due to climatic disasters by themselves and give up their limited adaptation funding so that the most vulnerable countries can be given priority.

17.2.3 Development-oriented adaptation and incremental adaptation

Links between adaptation and development

Economic development enables an economy to diversify and become less reliant on sectors such as agriculture, which are more vulnerable to the effects of climate change; it also makes more resources available for abating risk (WB, 2010a). There are many no-regrets/low-regrets measures that can provide joint-benefits to promote both development and adaptation.

There is a close link between adaptation and development; however, it is imperative to explicitly separate the concept of adaptation from that of development (Markandya & Watkiss, 2009). This problem can be addressed by defining "adaptation deficit" and "development deficit". The notion of "adaptation deficit" means that countries are underprepared for current climate conditions, much less for future climate change, i.e. it describes the gap between current and optimal levels of adaptation (WB, 2010a: 39). The notion of "development deficit" covers both costs for development and adaptation to climate change: it means that developing countries have deficits in their capacity not only to adapt to climate events but also to provide education, housing, health, and other services classed as basic necessities (WB, 2010a: 20). In this regard, a development deficit is a typical characteristic of many developing countries/regions.

Development-oriented and incremental adaptation

From a socioeconomic perspective, it was suggested that adaptation be categorised in two aspects – namely development-oriented and incremental adaptation – based on the development stage (Pan, Zheng & Markandya, 2011). Climate change may increase the frequency and intensity of climate risks, making the existing facilities inadequate and necessitating additional investment. A classic definition of adaptation applies only to incremental adaptation and adaptation deficit; however, these narrowly defined notions do not reflect the real situation of the fast-growing developing countries considered here.

Table 17.3 Explanation of incremental adaptation and development-oriented adaptation

Adaptation type	*Status quo (BAU scenario)*	*Scenarios of future change*
Incremental adaptation (developed regions)	Regular risk: 60 Climate change risk: 0 Investment in adaptation: 60 Climate-related net loss: 0 Incremental investment required: 0	Regular risk: 60 Climate change risk: 30 Investment in adaptation: 90 Climate-related net loss: 0 Incremental investment required: 30
Development-oriented adaptation(less developed regions)	Regular risk: 60 Climate change risk: 0 Investment in adaptation: 30 Climate-related net loss: 30 Incremental investment required: 30	Regular risk: 60 Climate change risk: 30 Investment in adaptation: 30 Climate-related net loss: 60 Incremental investment required: 60

Source: Pan, Zheng, & Markandya, 2011

Note: (1) Adaptation inputs are assumed to be proportional to efficiency, i.e. one unit of adaptive input can reduce the corresponding risk level by one unit; (2) Inevitable risks are ignored.

As Table 17.3 shows, incremental adaptation refers to the additional resources and measures required to address newly emerging risks, since regular risks have been considered in the process of satisfying basic development requirements. By contrast, adaptation in most developing countries should place equal emphasis on both normal existing and newly emerging risks, because the infrastructures there are either non-existent or not robust enough to withstand most impacts of current climate hazards, let alone future climate change. Incremental adaptation is closely related to 'climate change justified' actions, which have a robust defence in terms of benefits under a wide range of possible scenarios. Comparatively, development-oriented adaptation contains many 'no' or 'low' regrets measures (e.g. eliminate poverty, reduce air pollution, conserve bio-diversity, protect water resources and enhance the public health system) which aim to enhance adaptive capability. Even though these adaptive actions might over-evaluate climate risks, they are still indispensable in the process of socioeconomic development (Pan, Zheng & Markandya, 2011).

17.3 Climate risk of FGCs and adaptation costs

17.3.1 Exposure, vulnerability and risk in FGCs

Exposure and vulnerability are key determinants of disaster risk and of impacts (IPCC, 2012: 8). FGCs have become increasingly exposed and vulnerable to climate risks due to fast growth and changes in socioeconomic conditions.

Increasing exposure to climate risk in FGCs

There is no impact if without exposure to climate risk. "Exposure can be conceptualized as the presence of people; livelihoods; environmental services and resource; infrastructures; or economic, social, or cultural assets in places that could be adversely affected (by weather and

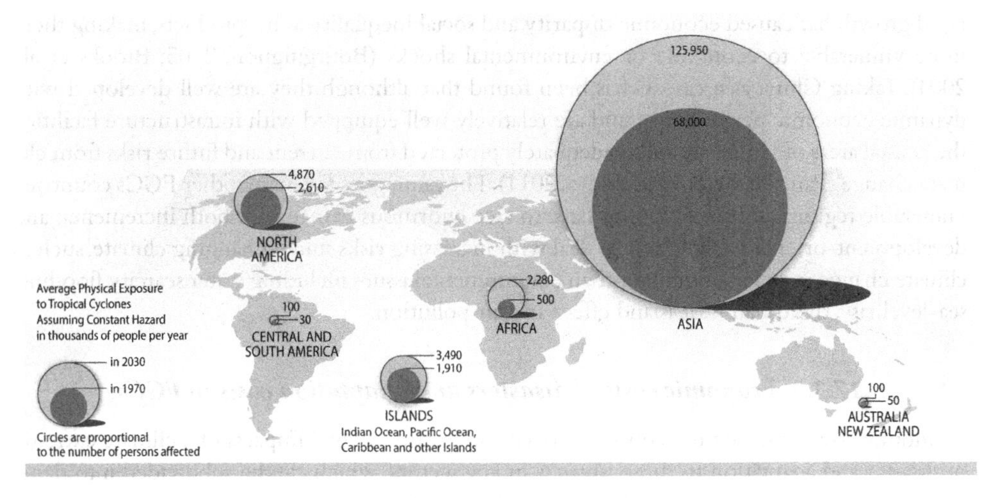

Figure 17.2 Average physical exposure to tropical cyclones assuming constant hazard (in thousands of people per year)
Source: Handmer et al., 2012: 240

climate extremes)" (IPCC, 2012: 5). Population growth, settlement patterns, urbanisation, land use and so on are important driving factors of changing exposure (IPCC, 2012).

"Increasing exposure of people and economic assets has been the major cause of long-term increases in economic losses from weather- and climate-related disasters" (IPCC, 2012: 234). Compared with other regions, Asia's mega-deltas are highly exposed to extreme events such as flooding and tropical cyclones since these areas have been experiencing fast economic growth and rapid urbanisation. In the period from 2000 to 2008, Asia experienced the highest number of weather- and climate-related disasters. As Figure 17.2 shows, physical exposure (yearly average number of people exposed) to tropical cyclones is estimated to have increased from approximately 73 million in 1970 to approximately 123 million in 2010, with Asia ranking as number one on the exposure list.

Driven by rapid urbanisation and economic growth, some emerging developing countries are expected to face increased future exposure to extremes, especially in highly urbanised areas (IPCC, 2012). Taking China as an example, the coastal areas have been the fastest-growing and most densely populated areas in China in the past decades, which means that they are also the country's top-ranking zone as regards the risk of natural disasters, suffering from typhoons, flooding, heat waves, drought, sea-level rise and so on (Shi, 2011).

Multi-driving factors of vulnerability and potential risks

Vulnerability is defined as "the propensity or predisposition to be adversely affected". It is composed of (i) sensitivity of human beings, infrastructure, environment, society and ecosystem to a hazard event and (ii) lack of coping and adaptive capability in climate change adaptation (IPCC, 2012: 72).

Vulnerability may be increased by shortcomings in development processes such as environmental degradation, rapid and unplanned urbanisation in hazardous areas, failures of governance and the scarcity of livelihood options for the poor (IPCC, 2012: 10). In many FGCs, rapid growth has caused economic disparity and social inequality as by-products, making them more vulnerable to economic or environmental shocks (Bourguignon, 2005; Brooks et al., 2004). Taking China as a case, it has been found that although they are well developed with dynamic economic performance and are relatively well equipped with infrastructure facilities, the coastal areas of China are still inadequately protected from current and future risks from climate change (Pan, Zheng & Markandya, 2011). The same goes for many other FGCs countries: vulnerable regions and social groups have to face enormous demand for both incremental and development-oriented adaptation to deal with increasing risks under changing climate, such as climate change induced migrants, urban environmental issues including water scarcity, flooding, sea-level rise, the urban heat island effect and air pollution.

17.3.2 Economic costs of disasters and adaptation costs in FGCs

Economic costs arise due to economic, social and environmental impacts of a climate extreme or disaster and adaptation to those impacts in key sectors, which can be subdivided into damage costs and adaptation costs (IPCC, 2012: 264). In this section, these two categories of costs in FGCs are discussed separately.

Economic impacts of climate disasters

"The severity of the impacts of climate extremes depends strongly on the level of the exposure and vulnerability to those extremes" (IPCC, 2012: 10). While rich countries account for most of the total economic and insured losses from disasters, fatality rates and economic losses expressed as a proportion of gross domestic product (GDP) are much higher in developing countries. During the period from 2001 to 2006, losses amounted to about 1% of GDP for middle-income countries, about 0.3% of GDP for low-income countries and less than 0.1% of GDP for high-income countries (IPCC, 2012: 234).

The occurrence of weather/climate related disasters and impacts is often geographically distributed (see Figure 17.3 and IPCC, 2012: 270). In fast-growing countries, increasing population and assets at risk are the main drivers for increasing impacts and disaster damage. East Asia and the Pacific, South Asia, and Latin America and the Caribbean rank as the top three fast-growing regions and also suffer greater climate risk than many other regions owing to greater exposure and vulnerability.

In Asia, the geographical distribution of flood risk is heavily concentrated in India, Bangladesh and China, where high human and material losses are caused (IPCC, 2012: 254). For example, Bangladesh is one of the most vulnerable countries to climate risks: about two-thirds of the nation is less than five meters above sea level and is susceptible to river and rainwater flooding. It is estimated that the damage from a single typical severe cyclone with a return period of 10 years may well rise nearly fivefold to over $9 billion by 2050, accounting for 0.6 percent of GDP (WB, 2010a: 27). Taking China as a case, more than 70% of the economic loss from natural disasters is accounted for by weather and climate-related disasters, mainly from flooding, drought, typhoons and so on (see Figure 17.4). By making great efforts in the financing of infrastructures and disaster risk reduction, China has seen a decline in mortality from weather-related disasters from 5000 to 2000 from the 1980s to 2010, and a substantial decrease in economic losses as a percentage of total GDP from 3–6% in 2000 to 1–3% in the past decade (see Figure 17.5).

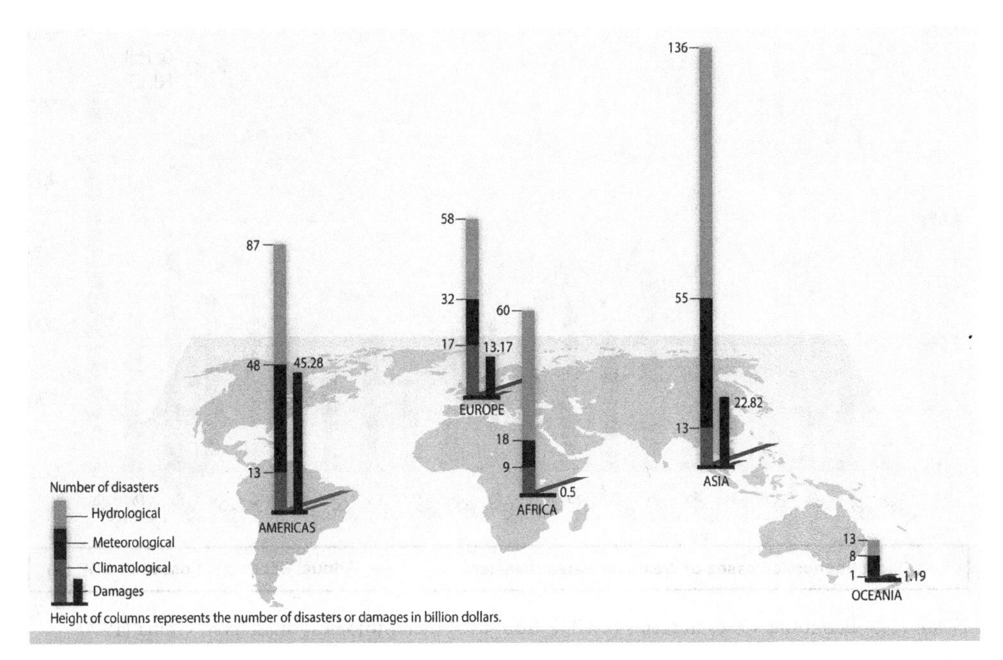

Figure 17.3 Weather- and climate-related disaster occurrence and regional average impacts from 2000 to 2008. The figure shows the number of climatological (e.g. extreme temperature, drought, wildfires), meteorological (e.g. storms) and hydrological (e.g. flooding, landslides) disasters recorded for each region, along with an evaluation of damage (2009 US$ in billions).
Source: Handmer et al., 2012: 270

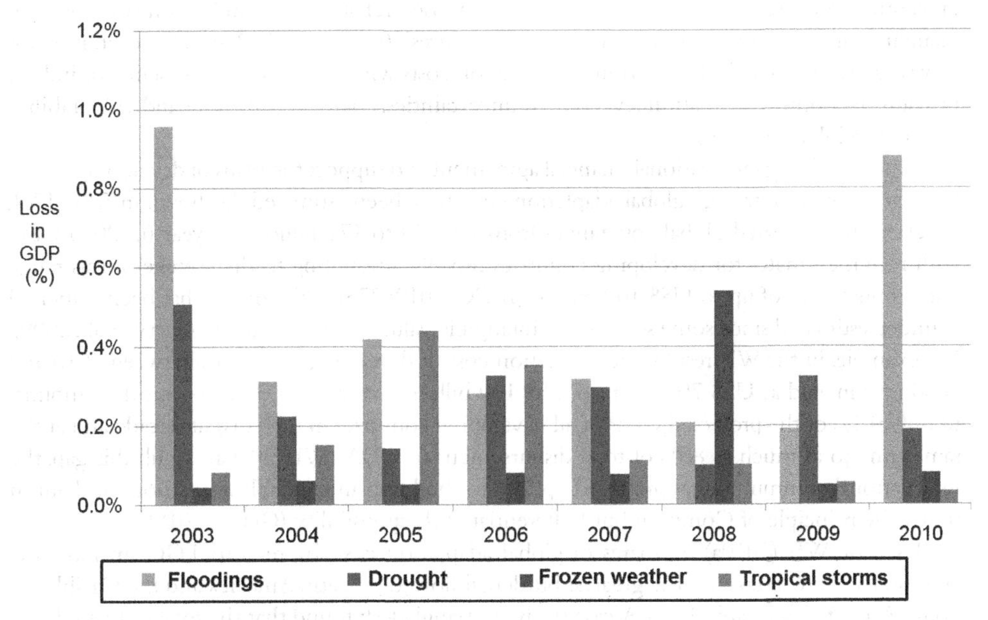

Figure 17.4 Economic losses in terms of GDP from different meteorological disasters in China (2003–2010)
Source: CMA, 2011

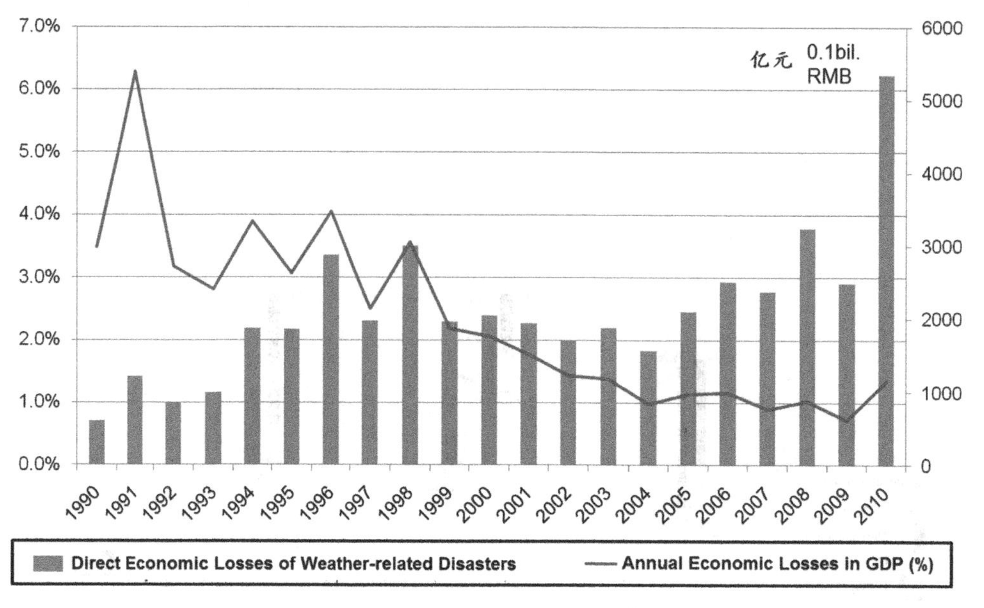

Figure 17.5 Economic losses in terms of GDP from total meteorological disasters in China (1990–2010)
Source: CMA, 2011

Adaptation costs in FGCs

"Adaptation costs are associated with adaptation and facilitation in terms of planning, risk prevention, preparedness, emergency disaster responses, rehabilitation and reconstruction, risk financing and implementation of adaptation measures" (IPCC, 2012: 264). However, it is not easy to draw up a practical definition of adaptation costs with a development baseline or indeed to assess the optimal adaptation level due to uncertainties, changing exposure and vulnerability and methodological issues.

To help building international financial agreement and support the plans of developing countries for adaptation strategy, global adaptation costs have been estimated. As shown in Table 17.4, for 2030 the estimated global cost ranges from US$ 48 to 171 billion per year (in 2005 US$) with recent estimates for developing countries broadly amounting to the average of this range with annual costs of up to US$ 100 billion (IPCC, 2012: 273). This amount has been criticised as under-estimated since some sectors and intangible values are not included (Parry et al., 2009). For example, in the WB results, the adaptation cost of developing countries between 2010 and 2050 is estimated at US$ 70 billion to US$ 100 billion a year at 2005 prices, but this amounts to only 0.2% of the projected GDP of all developing countries in the current decade and at the same time to as much as 80% of total disbursement of ODA (WB, 2010a). To fill this gap, the international community has proposed options for building an equitable adaptation mechanism under the principle of Common but Differentiated Responsibility (Grasso, 2010).

Based on WB (2010a) estimates of global adaptation costs by regions, FGCs in East Asia and the Pacific will take on the greatest burden, followed by Latin America and the Caribbean, South Asia, etc. (See Table 17.5). A case study on Bangladesh found that the total estimated cost for the country would be $2.4 billion in initial investment and $50 million in annual recurrent

Table 17.4 Estimates of global costs of adaptation to climate change

Study	*Results*	*Time frame and coverage*	*Sectors*	*Methodology and comment*
World Bank (2006)	9–41[1]	Present, developing countries	Unspecified	Cost of climate-proofing foreign direct investments, gross domestic investments, and Official Development Assistance.
Stern (2007)	4–37[1]	Present, developing countries	Unspecified	Update of World Bank (2006)
Oxfam (2007)	>50[1]	Present, developing countries	Unspecified	World Bank (2006) plus extrapolation of cost estimates from National Adaptation Programmes of Action and NGO projects
UNDP (2007)	86–109[2]	In 2015, developing countries	Unspecified	World Bank (2006) plus costing of targets for adapting poverty reduction programs and strengthening disaster response systems
UNFCCC (2007)	48–171 (28–67 for developing countries)[2]	In 2030, developed and developing countries	Agriculture, forestry, and fisheries; water supply; human health; coastal zones; infrastructure; ecosystems (but no estimate for 2030 for ecosystem adaptation)	Additional investment and financial flows needed for adaptation in 2030
World Bank (2010)	70–100[2]	Annual from 2010 to 2050, developing countries	Agriculture, forestry, and fisheries; water supply and flood protection; human health; coastal zones; infrastructure; extreme weather events	Impact costs linked to adaptation costs, improvement upon UNFCCC (2007): climate-proofing existing and new infrastructure, more precise unit cost, inclusion of cost of maintenance and port upgrading, risks from sea level rise and storm surges, riverine flood protection, education investment to neutralise impacts of extreme weather events.

Source: Handmer et al., 2012: 273

[1]In 2000 US$

[2]In 2005 US$

Table 17.5 Global costs of adaptation by region

East Asia and Pacific will shoulder the biggest burden (Global costs of adaptation by region)							
Aggregation type/Scenario	*East Asia & Pacific*	*Europe & Central Asia*	*Latin America & Caribbean*	*Middle East/North Africa*	*South Asia*	*Sub-Saharan Africa*	*Total*
Gross-sum/ Wet scenario	25.7	12.6	21.3	3.6	17.1	17.1	97.5
X-sum/ Dry scenario	17.9	6.9	14.8	2.5	15	14.1	71.2

Source: WB, 2010a: 21

Note: Gross-sum is the sum of the positive costs of adaptation in all sectors and all countries, excluding gains (negative costs) from climate change. X-sum deducts these eventual gains in countries with overall positive costs of adaptation.

costs, including adaptation investment on embankments, afforestation, cyclone shelters and early warning systems (WB, 2010b). Compared with these large adaptation costs, the climate funding currently available within UNFCCC and from international assistance focuses more on the most vulnerable developing countries such as OAIS and LDCs (Least Developed Countries), and still does not satisfy the needs for climate-proof investments in most of the FGCs.

17.4 Planning adaptation for FGCs: Cases and challenges

While development is critical to improve resilience to disaster risk, maladaptation[6] may pose a new problem in dealing with development and adaptation deficits. It is imperative for FGC countries to mainstream adaptation into development to plan a resilient and sustainable future.

17.4.1 Climate change and migration planning

Migration induced by environmental and climate change has been the object of more and more attention from researchers and policy makers since the 1990s (IPCC, 2007). There is complex interaction between demographic change, environmental change and migration, and global demographic change could be a driver of migration within the context of anticipated climate change (Hugo, 2011). Climate migrants may be driven by different climate risks, such as flooding, typhoons, drought, sea-level rise and so on. Climate change-induced migrants could be temporary or permanent, depending on the type of factors that drive and push them and on policy intervention. With many climate risks and high exposure to climate-related disasters, the Asia and Pacific region has been evaluated as a global hot spot for climate change induced migration (IOM, 2009; ADB, 2012).

Urbanisation is often a big pulling factor for climate migrants, and drought or flooding takes an important role in pushing residents away from suffering livelihood loss. In fast-growing countries, the "push" factors are often greater than the "pull" factors for potential migrants who make the decision to resettle permanently. For example, in Bangladesh the number of people living in cities is set to triple while the rural population will fall by 30 percent by 2050, which poses a long-term challenge to move people and economic activity into less climate sensitive areas (WB, 2010b).

Box 17.1 Ecological migration in Ningxia, China

In China, poverty-stricken areas are in essence ecologically fragile and prone to climate change risks. Situated in the arid north of China, Ningxia Autonomous Region is most vulnerable to climate change owing to desertification, low human development and a scarcity of water resources. Among the poorest provinces in China, Ningxia had per capita GDP of USD 3,800 and rural income at USD 535 per capita in 2010. In the past decades, Ningxia has experienced a prominent warming trend and a decline in rainfall has been recorded. As a result, the livelihood of more than one million rural people in severe poverty has been highly impacted. From 1983 to 2010, the Ningxia government facilitated the resettlement of 786,000 rural people from severe drought areas to places where water resources are accessible (Zhang, 2012: 259). As a successful policy of planned adaptation at the local level, another 150,000 migrants will be resettled with the implementation of a Five-Year Migration Plan.

17.4.2 Vulnerable cities and urban planning in FGCs

Rapid urbanisation with growing populations and economic assets has increased exposure to disaster risk, especially for cities in developing countries (WB, 2010a; IPCC, 2012). According to the United Nations World Urbanisation Prospects Report 2011, half of the world's 7 billion people live in cities, and that number is set to increase by 2.3 billion by the year 2050. In Asia, the urban population will increase from 1.9 billion to 3.3 billion during the same period. This rapid urbanisation process has not only exacerbated the pressure on the urban living environment and infrastructure, but also left more and more people and physical assets exposed to the threat of climate risks. Many coastal cities suffer frequent and intense disasters due typhoons/cyclones, floods, landslides, droughts etc. For example, Asia is home to more than one third of the largest port cities, and increasing population and assets will put these cities at risk in future decades (Nicholls et al., 2008).

Adaptation planning has been taken as an effective approach to address the impact of climate change, which can be lessened by risk awareness, governance capacity and scientific support such as climate modelling, vulnerability assessment and policy identification. Policies and activities have been enforced at national and local levels, for instance National Adaptation Strategies in European countries, the Climate Change Adaptation Strategy for London, adaptation planning in New York (in the PlaNYC, NYS 2011 Commission), and Australian cities etc. International initiatives such as ICLEI's Cities for Climate Protection Programme have also been developed to support "Resilient Cities" (ICLEI, 2011). However, compared with mitigation action, adaptation is still not well developed or implemented in practice. It is found that policy experiments on adaptation only account for 12% in a sample of 100 global cities with more than 600 cases (Broto & Bulkeley, 2013).

Box 17.2 Vulnerable mega-cities in China

Since the 1980s, China has seen fast urban sprawl during development and urbanisation. Between 1978 and 2010, the total number of cities in China increased three-fold from 193 to 657; GDP increased more than 100-fold during the same period. At present, China possesses one-fourth of all the world's cities with populations of more than 500,000. More than 70% of these cities and half their total inhabitants are located on the coasts. However, many coastal mega-cities such as Shanghai, Tianjin, Guangzhou, Ningbo and Hong Kong rank among the top 20 coastal cities in terms of high climate risk by 2050 (Nicholls et al., 2008).

Using a five-dimensional vulnerability assessment framework, the vulnerabilities to climate change of the top four biggest mega-cities have been compared (Pan et al., 2011). The result indicates that the four mega-cities have similar social development levels, but Shanghai and Chongqing are more vulnerable in general than Beijing and Tianjin because they are more climate-sensitive (measured in terms of economic losses from weather-related disasters, climate-sensitive sectors, etc); have more vulnerable populations (the elderly, life expectancy, illiteracy rates, etc); have higher ecological vulnerability (such as forest coverage rate, water resources, etc) and have less environmental governance capability (see Figure 17.6). In another case, Shanghai has been ranked as the most vulnerable city to flooding out of nine global coastal cities in both the 2009 and 2100 scenarios, which account for higher exposure to storm surge and sea-level rise, large populations close to the coastline, lower resilience and so on (Balica, Wright & Meulen, 2012).

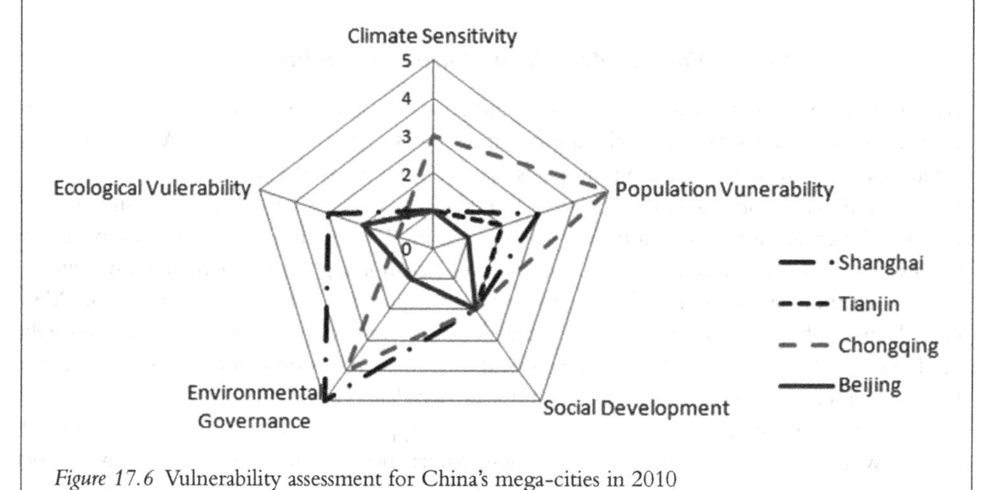

Figure 17.6 Vulnerability assessment for China's mega-cities in 2010
Source: Pan et al., 2011

17.4.3 Joint-benefits of mitigation and adaptation

For most developing countries, mitigating global climate change is a long-term, arduous challenge, while adapting to climate change is a pragmatic, imperative task. Synergising mitigation and adaptation means seeking a "win-win" scheme, and developing policies capable of controlling greenhouse gas emissions and benefiting climate change adaptation at the same time. Case studies of synergy in FGCs may well be good cases for research into adaptation economics in optimising social welfare, but successful experiences will be replicable in other developing countries where circumstances are similar.

Integrating adaptation and emission reduction policies has great practical significance for the development of climate change policies and planning: some research issues have been addressed in this field (Klein et al., 2005; Wilbanks & Sathaye, 2007; IPCC, 2007). Some preliminary collaboration has also been considered in urban planning in North American cities (Zimmerman, 2011), and pilot projects in some fast-growing developing countries have also been encouraged (for example, see Box 17.3).

Box 17.3 Synergy of mitigation and adaptation in Vietnam

Given the limited financial and human resources in developing countries, climate policies with joint-benefits from mitigation and adaptation would be socially and economically efficient. To support this idea in the process of policy making, research was conducted in Vietnam with a multi-criteria analysis (MCA) technique in regard to prioritising mitigation and adaptation options.

With principles of "commercial availability", "sustainable development" and "adaptation benefits of mitigation", a list of options for mitigation and adaptation emerged. For adaptation options in selected natural and socioeconomic sectors, some policies have both a prominent emission reduction potential and positive benefits for adaptation, including developing small-scale hydropower, water management for reducing methane emissions from rice fields, protection of forests and forest generation etc. The potential mitigation options selected in Vietnam were the promoting of hydropower development in water management, the use of more robust seeds for forests and the development of flexible crop patterns in agriculture. These have been addressed as priority options with joint-benefits (Dang et al., 2003).

Cities are regarded as a policy laboratory for mainstreaming climate change issues into development strategy (Pan, Zhuang & Zhu, 2012: 17). As the subject of governance at the local level, cities are both the executors of national level emission reduction targets and activities and the victims of climate change risks and impacts. Therefore, they are equipped with favorable conditions for carrying out the collaborative governance of mitigation and adaptation. Great potential for building low carbon resilient cities in China and other FGCs is expected to be seen.

Box 17.4 China's potential for building low-carbon resilient cities

According to China's National Climate Change Programme,[7] the key areas for greenhouse gas emission mitigation are energy, industry, transport, building, agriculture, forestry, urban waste and so on; while those for climate change adaptation are agriculture, forestry, water resource management and coastal areas. Therefore, forestry and urban greening have been taken as a top priority for policy synergy in achieving both mitigation and adaptation. For instance, Some integrated objectives have been incorporated into Shanghai's 12th Five-Year Plan, including the achieving of a forest coverage rate of 15.5% in 2015, with multiple objectives of coastal protection, water conservation and carbon sinking; developing vertical urban greening and roof greening; setting the proportion of green area of new-build public constructions with the total appropriate public roof greening area at more than 95% etc.

Under pressure to meet the national emission reduction commitment promised by the Chinese premier at the Copenhagen Conference in 2009, governments at different levels have paid much more attention to mitigation targets than adaptation. In 2010, a low carbon demonstration programme was launched by the National Development Reform Committee in five selected provinces and eight cities. However, as climate risk and economic impacts increase, China's municipal governments are becoming aware of the challenge and opportunities that they face in building "low carbon resilient cities" in the long run (Zheng, Wang & Pan, 2013).

Table 17.6 Checklist for addressing win-win options of mitigation and adaptation

Key areas/adaptation measures	*To mitigate the heat island effect*	*To reduce urban flooding risk*
Urban ecosystem	Coastal shelterbelts, urban wetlands, urban forest, water conservation forest, carbon sink forest, etc.	Coastal shelterbelts, urban wetlands, urban forest, water conservation forest, green belts along roads, etc.
Water resources and watershed management	Water diversion projects, urban waterway regeneration, water reuse and rainwater and storm water utilisation techniques, scalar water pricing system, etc.	Urban waterway regeneration, urban underground drainage network reformation, reservoir storage and readjustment, irrigation facilities, small watershed governance, flood discharge and flood storage projects, etc.
Energy and power system	Renewable energy generation techniques such as community rooftop solar power utilisation, high energy-efficiency and energy-saving techniques, wind power, tidal power, geothermal power, garbagepower, power demand side management, smart grid techniques, etc.	Renewable energy generation techniques, smart power grid and green power electric appliances
Public transport	Vertical virescence along roads, public transport (urban BRT, solar powered vehicles, free bikes), etc.	To enhance the total urban transport management capability (transit share)
Human settlement	Energy conservation buildings, renewable energy building techniques, roof greening, low-carbon green communities	Green roofs, water-permeable bricks in pedestrian areas and community, rain/storm water management, water storage techniques, etc.
Urban planning	Land use, population policy, industrial layout, low-carbon city planning, regulation on urban green coverage rate, etc.	Regulations on urban land planning and climate-proof infrastructures, such as urban green coverage rate, building codes, for flood protection and storm drainage, etc.

Source: Adapted from Zheng, Wang, & Pan, 2013

Checklist for win-win options

Various options for collaborative management in adaptation and emission reduction are available with regards to different cities with diverse climate risks and adaptation goals, respectively. For instance, the urban heat island effect can make heat waves worse, and typhoons and rainstorms often cause urban floods. The checklist in Table 17.6 provides a practical way of allocating priority win-win options in different fields when adaptation objectives are targeted, such as mitigating the urban heat island effect and coping with urban flooding.

17.4.4 Adaptation planning and future challenges

High exposure and vulnerability are generally the outcome of skewed development processes (IPCC, 2012). By contrast, the most effective adaptation can offer development benefits in the relatively near term, as well as reductions in vulnerability over the longer term (WB, 2010a). This point is significant for fast-growing countries aiming for sustainable development in a changing climate. Some policy implications for FGCs taking pathways towards a sustainable, resilient future are suggested below.

First, smart development and planning for adaptation are critical for FGCs. Economic development is the most fundamental and cost effective approach for adapting to climate change, but it should not be business as usual. Adaptation planning and climate-smart development can make countries more resilient to the effects of climate change and reduce maladaptation (WB, 2010a). In fact, not all risks and losses can be resolved through development. Development itself can aggravate exposure and vulnerability, while a failure to adapt may lead to very large losses, in terms of both the destruction of infrastructure and lost opportunities for future growth. There can be no "one case fits all" strategy for FGCs in taking actions, since there are significant differences between the geographical, cultural and institutional backgrounds and stages of development of FGCs.

Second, low-regret strategies are more practical than addressing "optimal" adaptation in decision-making. Cost-benefit analysis is helpful in assessing optimal options; however, "low-regret" actions should be taken as a priority principle in decision-making under different climate and socioeconomic scenarios. FGCs can evaluate and deploy potential low-regret options for priority development even without climate change, such as early warning systems, risk communication between decision makers and local citizens, land use planning, ecosystem management and restoration, health care, water supply, sanitation, irrigation and drainage systems, climate-proofing of infrastructures, development and enforcement of building codes, better education and awareness and so on (IPCC, 2012).

Third, improving governance and leadership is important at both national and international levels. Interactions between the goals of climate change mitigation, adaptation, and disaster risk management are sure to be a challenge in governance between levels and sectors for both developing and developed countries. For global regime building, FGCs need greater economic performance to consolidate their bargaining power in negotiations on adaptation and mitigation. At the national level, FGCs will have to make transformational changes from traditional policies to adaptive governance in dealing with the changing environment and with climate change. In fact, institutional and social dimensions are critical barriers to developing and implementing adaptation strategies (Biesbroek et al., 2013). Learning by doing, partnership building, climate risk assessment and information sharing in decision-making have been highlighted in the practice of adaptive planning for adaptation (Mirfenderesk & Corkill, 2009; Williams, 2011); however, it is not easy to put these golden rules into practice while striking the right balance between development and adaptation.

17.5 Conclusions

FGCs are countries with rapid economic growth over a particular period of time. These countries share some common characteristics: for example, higher increases in GDP growth rate and income, consumption of natural resources and greenhouse emissions. The BRIC and BASIC countries can be taken as examples: their population, GDP and emissions have risen by a significant percentage, and they are playing an important role in the world political economy. In short, FGCs are still in the process of rapid industrialisation and urbanisation, but in the meantime they are also faced with multiple challenges such as development, poverty reduction, adaptation to climate change etc.

FGCs are faced with increasing vulnerability to climate change. Overpopulation and the increasing aggregation of wealth have increased their exposure to climate change. Specifically, the large proportion of the population living in poverty exacerbates pressure on their governments' poverty relief efforts. Shortcomings in urban planning, the degradation of the natural environment and uneven regional development seem bound to increase the vulnerability of FGCs to climate change. It should be noted that the real nature of vulnerability for FGCs is different from that of developed countries. In FGCs, vulnerability is characterised by a development deficit, which means development-oriented adaptation.

There are two implications of adapting to climate change from a political economy perspective: on the one hand, the FGCs represented by BASIC have been exhibiting increasing vulnerability to climate change, thus generating a strong need for adaptation. On the other hand, these countries are also faced with mounting pressure to reduce the greenhouse gas emissions caused by their rapid economic development. Although FGCs are playing an important role in developing the world climate mechanism, they have different interests and appeals. As a matter of fact, all FGCs have reached a consensus on the view that "adapting to climate change may lower the costs of economic development"; however, they are starved of funds from the international community. Instead, they have to increase their adaptation capabilities by relying on their sustainable economic growth.

FGCs are faced with multiple challenges and demands. As a result, they must take socioeconomic development, poverty reduction, greenhouse gas emissions and sustainable development into consideration. To achieve this goal, these countries can improve collaboration and communication on both the national and international levels. For instance, special attention can be paid to climate change-induced migration in arid areas, and to urban adaptation planning and climate risk management in coastal cities. Furthermore, FGCs need to factor transformation into their decision-making processes to strengthen leadership and effectiveness in the implementation of their national adaptation strategies, planning and policies.

Acknowledgements

This paper is based on research within the "Global climate governance and its political economy" project funded by Chinese Academy of Social Science, "Climate change impacts on coastal cities in Yangtze Delta Region in China and adaptation" project (No.70933005), funded by China's National Science Foundation Commission, and the "Adapting to Climate Change in China"(ACCC) project funded by SWISS-DEC and UK-DFID.

Notes

1 The BRIC countries were responsible for about 18% of the world's gross domestic product (GDP), occupy over a quarter of the world's land area and more than 40% of the world's population. In March 2012, South Africa appeared to join BRIC, which is now called BRICS.

2 This approach was first developed by researchers from the Chinese Academy of Social Sciences to ensure sustainability, equity and efficiency through building a global carbon budget compatible with a

2 °C goal, allocating emission space base on equal accumulated emissions per capita and emission trading. This approach was included in the BASIC (2011) report.

3 The Umbrella Group is a loose coalition of non-EU developed countries including Australia, Canada, Japan, New Zealand, Norway, the Russian Federation, Ukraine and the US. See "Party groupings" (http://unfccc.int/parties_and_observers/parties/negotiating_groups/items/2714.php).

4 An adaptation deficit arises when the current infrastructure is inadequate to cope with present climate variations. Action to correct this situation can possibly be justified even without reference to future climate change (although it may still not be the top priority) (Parry et al., 2009).

5 The Alliance of Small Island States (AOSIS) is a coalition of some 43 low-lying and small island countries, most of which are members of the G-77. Particularly vulnerable to sea-level rise, AOSIS countries are united by the threat that climate change poses to their survival and frequently adopt a common stance in negotiations. See "Party groupings" (http://unfccc.int/parties_and_observers/parties/negotiating_groups/items/2714.php).

6 Maladaptation arises when an action taken ostensibly to avoid or reduce vulnerability to climate change impacts adversely on, or increases the vulnerability of, other systems, sectors or social groups (Barnett & O'Neill, 2010).

7 China's National Climate Change Programme, http://www.china.org.cn/english/environment/213624.htm

References

Ackerman, F., (2009). Financing the Climate Mitigation and Adaptation Measures in Developing Countries, G-24 Discussion Paper Series, No. 57, December 2009. © United Nations.

Asia Development Bank (ADB), (2012). Addressing Climate Change and Migration in Asia and the Pacific, Mandaluyong City, Philippines: Asian Development Bank.

Balica S.F., N.G. Wright, F. van der Meulen, (2012). A flood vulnerability index for coastal cities and its use in assessing climate change impacts, *Natural Hazards*, 64, 1, 73–105.

Barnett, J., S. O'Neill, (2010). Editorial: Maladaptation. *Global Environmental Change,* 20, 2, 211–213.

BASIC, (2012). Equitable Access to Sustainable Development: Contribution to the Body of Scientific Knowledge, BASIC expert group: Beijing, Brasilia, Cape Town and Mumbai, Beijing: Property Intellectual Press, ISBN 978-7-5130-1420-5, p. 242.

Berg, A., J.D. Ostry, J. Zettelmeyer, (2012). What makes growth sustained?, *Journal of Development Economics*, 98, 149–166.

Biesbroek, G.R., J.E.M. Klostermann, C.J.A.M. Termeer, P. Kabat, (2013). On the nature of barriers to climate change adaptation, *Regional Environmental Change*, 3, 1–7.

Bourguignon, F., (2005). The effect of economic growth on social structures. In: Philippe Aghion & Steven Durlauf (eds.), Handbook of Economic Growth, edition 1, volume 1, chapter 27, pp. 1701–1747. Elsevier.

Brock, W.A., T.M. Scott, (2005). Economic growth and the environment: A review of theory and empirics. In: Philippe Aghion & Steven Durlauf (eds.), Handbook of Economic Growth, edition 1, volume 1, chapter 28, pp. 1749–1821. Elsevier.

Brooks, N., W. N. Adger, P. M. Kelly, (2004). The determinants of vulnerability and adaptive capacity at the national level and the implications for adaptation, *Global Environmental Change*, 15, 151–163.

Broto V. C., Bulkeley, H., (2013). A survey of urban climate change experiments in 100 cities, Global Environmental Change, 23, 92–102.

CMA (Chinese Meteorological Agency), (2011). Annex VIII: China's climate disasters. In: W.G. Wang, G.G. Zheng, Y. Luo, J.H. Pan, Q.H. Chao, Y. Zheng, G. Q. Hu (eds.) Green Book of Climate Change, Annual Report on Actions to Address Climate Change: Durban Dilemma and China's Strategic Options, Beijing, China: Social Sciences Academic Press.

Cosbey, A., (2009). Developing Country Interests in Climate Change Action and the Implications for a Post-2012 Climate Change Regime. United Nations Conference on Trade and Development, International Institute for Sustainable Development (IISD), United Nations, New York and Geneva.

Dadush, U., Ali, S., (2012). In search of the global middle class: A new index, *Inernational Economics*, Carnegie Endowment for International Peace, http://www.carnegieendowment.org/files/middle_class-edited.pdf.

Dang H.H., A. Michaelowaa, D. D. Tuan, (2003). Synergy of adaptation and mitigation strategies in the context of sustainable development: The case of Vietnam, *Climate Policy*, 3, 1, (Supplement), S81–S96.

Eichengreen, B., D.Park, K. Shin, (2011). When Fast Growing Economies Slow Down: International Evidence and Implications for China. *NBER* Working Paper 16919, http://www.nber.org/papers/w16919.

Grasso, M. (2010). An ethical approach to climate adaptation finance, *Global Environmental Change*, 20, 1, 74–81.

Handmer, J., Y. Honda, Z.W. Kundzewicz, N. Arnell, G. Benito, J. Hatfield, I.F. Mohamed, P. Peduzzi, S. Wu, B. Sherstyukov, K. Takahashi, Z. Yan, (2012). Changes in impacts of climate extremes: human systems and ecosystems. In: C.B. Field, V. Barros, T.F. Stocker, D. Qin, D.J. Dokken, K.L. Ebi, M.D. Mastrandrea, K.J. Mach, G.-K. Plattner, S.K. Allen, M. Tignor, & P.M. Midgley (eds.), Managing the Risks of Extreme Events and Disasters to Advance Climate Change Adaptation. A Special Report of Working Groups I and II of the Intergovernmental Panel on Climate Change (IPCC). Cambridge University Press, Cambridge, UK, and New York, NY, USA, pp. 231–290.

Hawksworth, J., G. Cookson, (2008). The World in 2050, Beyond the Brics: A Broader Look at Emerging Market Growth Prospects. PricewaterhouseCoopers. p. 34. https://www.pwc.ch/user_content/editor/files/publ_tls/pwc_the_world_in_2050_e.pdf

Horstmann, B., (2008). Framing Adaptation to Climate Change: A Challenge for Building Institutions. Bonn: DIE, Discussion Paper/Deutsches Institut für Entwicklungspolitik, ISBN 978-3-88985-414-8.

Hugo, G., (2011). Future demographic change and its interactions with migration and climate change – original research article, *Global Environmental Change*, 21, Supplement 1, S21–S33.

ICLEI, (2011). Financing the Resilient City: A Demand Driven Approach to Development, Disaster Risk Reduction and Climate Adaptation – An ICLEI White Paper, ICLEI Global Report.

IOM, (2009). F.Laczko, C. Aghazarm (eds.), Migration, environment and climate change: assessing the evidence, *IOM Report*, ISNB 978-92-9068-454-1.

IPCC, (2007). Climate Change 2007: Impacts, Adaptation and Vulnerability. Contribution of Working Group II to the Fourth Assessment Report of the IPCC. Parry M.L., O.F. Canziani, J.P. Palutikof, P.J. van der Linden and C.E. Hanson (eds.) Cambridge University Press, Cambridge, UK, 982pp.

IPCC, (2012). Managing the Risks of Extreme Events and Disasters to Advance Climate Change Adaptation. A Special Report of Working Groups I and II of the Intergovernmental Panel on Climate Change. C.B. Field, V. Barros, T.F. Stocker, D. Qin, D.J. Dokken, K.L. Ebi, M.D. Mastrandrea, K.J. Mach, G.-K. Plattner, S.K. Allen, M. Tignor, & P.M. Midgley (eds.) Cambridge University Press, Cambridge, UK, and New York, NY, USA, 582 pp.

Kharas, H., (2010). The Emerging Middle Class in Developing Countries, Global Development Outlook. Working Paper No. 285. http://www.oecd.org/dev/44457738.pdf.

Klein R.J.T., E.L.F. Schipper, S. Dessai, (2005). Integrating mitigation and adaptation into climate and development policy: Three research questions, *Environmental Science & Policy*, 8, 579–588.

Markandya, A., P. Watkiss, (2009). Potential Costs and Benefits of Adaptation Options: A Review of Existing Literature. UNFCCC Technical Paper. F CDCeCce/mTPb/e2r0 20090/29. http://unfccc.int/resource/docs/2009/tp/02.pdf.

Mirfenderesk H., D.O. Corkill, (2009). The need for adaptive strategic planning: Sustainable management of risks associated with climate change. *International Journal of Climate Change Strategies and Management*, 1, 2, 146–159.

Nicholls, R.J., S. Hanson, C. Herweijer, N. Patmore, S. Hallegatte, J. Corfee-Morlot, J. Chateau, R. Muir-Wood, (2008). Ranking Port Cities with High Exposure and Vulnerability to Climate Extremes. OECD Environment Working Papers. Organisation for Economic Co-operation and Development, Paris.

O'Brien, K.L., R.M. Leichenko, (2000). Double exposure: Assessing the impacts of climate change within the context of economic globalization, *Global Environmental Change*, 10, 221–32.

O'Neill, J., (2001). Building Better Global Economic BRICs. Global Economics, Goldman Sachs Economic Research Group, Paper No: 66, November 30, 2001. http://www.goldmansachs.com/our-thinking/topics/brics/building-better.html.

Oxfam, (2007). Adapting to Climate Change. What Is Needed in Poor Countries and Who Should Pay? Oxfam Briefing Paper 104, Oxfam International, Oxford, UK, 47 pp.

Pan, J.H., (2011). Carbon budget for basic needs: Implications of international equity and sustainability, *Chinese Journal of Population, Resources and Environment*, 9, 2.

Pan, J.H., Y. Zheng, (2009). Carbon emissions and development rights, *Journal of Global Environment*, 5, 58–63.

Pan, J.H., Y. Zheng, (2011). The concept and theoretical implications of carbon emissions rights based on individual equity, *Chinese Journal of Population, Resources and Environment*, 9, 2.

Pan, J.H., Y. Zheng, A. Markandya, (2011). The concept and theoretical implications of carbon emissions rights based on individual equity, *Economía Agraria y Recursos Naturales*, 11, 1, 99–112.

Pan, J.H., Y. Zheng, X.L. Xie, H.Z. Mu, Z. Tian, H.X. Meng, . . . W. Wei, et al., (2011). Climate Capacity, Vulnerability and Adaptation: Evidence & Cases in China. RCSD/CASS Policy Brief for UNFCCC, Durban, South Africa.

Pan, J.H., G.Y. Zhuang, S.X. Zhu, et al., (2012). Low Carbon Cities: Economic Analysis and Development Indicators with Case Studies. Beijing, China: Social Sciences Academic Press, p. 371.

Parry, M. L., N. Arnell, P. Berry, D. Dodman, S. Fankhauser, C. Hope, S. Kovats, R. Nicholls, D. Satterthwaite, R. Tiffin, T. Wheeler, (2009). Assessing the Costs of Adaptation to Climate Change: A Review of the UNFCCC and Other Recent Estimates, International Institute for Environment and Development and Grantham Institute for Climate Change, London.

Schipper, E. L. F., (2007). Climate Change Adaptation and Development: Exploring the Linkages. Tyndall Centre for Climate Change Research Working Paper, 107.

Shi, P.J., (2011). Atlas of Natural Disaster Risk in China. Beijing: China Science Press, ISBN: 9787030311696, 244 pp.

Stern, N., (2007). The Economics of Climate Change: The Stern Review. Cambridge University Press, Cambridge, UK.

Virmani, A., (2012). Accelerating and Sustaining Growth: Economic and Political Lessons, 2012. International Monetary Fund, IMF Working Paper. http://www.imf.org/external/pubs/ft/wp/2012/wp12185.pdf

Wilbanks T.J., J. Sathaye, (2007). Integrating mitigation and adaptation as responses to climate change: a synthesis, *Mitigation and Adaptation Strategies for Global Change*, 12, 957–962.

Williams B.K., (2011). Adaptive management of natural resources: Framework and issues, *Journal of Environmental Management*, 92, 1346–1353.

World Bank (WB), (2010a). Economics of Adaptation to Climate Change (EACC): Synthesis Report. © World Bank. www.worldbank.org License: Creative Commons Attribution CC BY 3.0.

World Bank (WB), (2010b). Country Report Case Studies: Bangladesh. In: Economics of Adaptation to Climate Change. © World Bank, International Bank for Reconstruction and Development, www.worldbank.org. License: Creative Commons Attribution CC BY 3.0.

World Bank (WB), (2012). Global Economic Prospects: Managing Growth in a Volatile World. Volume 5. © World Bank, International Bank for Reconstruction and Development, www.worldbank.org. License: Creative Commons Attribution CC BY 3.0 .

Zhang, J.S., X.Y. Huang, Y. Tan, T. N. Wang, H. X. Ji, . . . L. Q. Zhang, et al., (2012). Research on ecological migration in Ningxia. In: Z.Y. Ma (ed.), A Study of Strategical Solutions to Global Climate Change in Ningxia. Yinchuan: Sunshine Press, 2012: pp. 230–370.

Zheng, Y., W.J. Wang, J.H. Pan, (2013). Low carbon resilient city: Concept, approach and policy options, *Urban Studies*, 20, 3, 11–15.

Zimmerman, R., C. Faris, (2011). Climate change mitigation and adaptation in North American Cities, *Current Opinion in Environmental Sustainability*, 3, 1–7.

18

ECONOMICS OF ADAPTATION IN LOW-INCOME COUNTRIES

Rodney Lunduka[1], *Saleemul Huq*[1], *Muyeye Chambwera*[2], *Mintewab Bezabih*[3] *and Corinne Baulcomb*[4]

[1] INTERNATIONAL INSTITUTE FOR ENVIRONMENT AND DEVELOPMENT (IIED), LONDON, ENGLAND

[2] UNITED NATIONS DEVELOPMENT PROGRAMME, BOTSWANA

[3] UNIVERSITY OF PORTSMOUTH, UNITED KINGDOM

[4] SCOTLAND'S RURAL COLLEGE (SRUC), EDINBURGH CAMPUS, SCOTLAND

18.1 Introduction

Even if current greenhouse gas emissions are reduced drastically, broad scientific consensus claims that climate change will continue over the next several decades (IPCC, 2007). Currently, the impact of climate change has been reducing agricultural output, industrial output, economic growth, and aggregate investment (Dell et al., 2008) in addition to loss of lives, property, culture and social norms. These have been and are anticipated to be greatest in low-income countries where most of the world's poorest people live (Parry et al., 2009; World Bank, 2010a), particularly in sub-Saharan Africa and in parts of South Asia. Clements (2009) estimated that the potential economic costs of climate change for Africa for example, will be 1.5–3 percent of its GDP by 2030. This presents a huge negative impact, first, due to the low-income countries' initial poverty and their relatively high dependence on environmental capital for their livelihoods (Dercon, 2011). Secondly, negative climate change impacts in the poor world are aggravated by lack of adaptive capacity, lack of political will and limited access to climate finances (Atta-Krah, 2012).

However, despite this limited adaptive capacity, limited political will and limited access to finances, several adaptation strategies are currently being practiced at both the national and community/household levels to cope with present climate variability in the low-income countries. For example, within the agricultural sector, these strategies range from the development and deployment of early warning systems, better agricultural management systems, improved crop cultivars, better and more efficient irrigation systems and good grain storage systems (Nkomo et al., 2006). These strategies are putting a tighter squeeze on the limited resources of the low-income countries now and will continue to demand even more in the future as climate change impacts upsurge. It is estimated that the potential cost of adapting to climate change in Africa alone will reach at least US$ 10–30 billion every year by 2030 (Clements, 2009).

Given the limited resources that the low-income countries have, adaptation to climate change implies trade-offs between alternative policy and livelihood strategy goals. To recommend and prioritise which strategies to use or implement, economics offers insight into these trade-offs and helps to explain the differences between the potential of adaptation and its achievement as a function of cost and benefits and their distribution to different social groups. One of the main challenges of cost-benefit analysis is accommodating the wide-ranging impacts of climate change on diverse individuals and groups. While some adaptations provide public benefits, such as protecting coastal areas from rising sea levels, many others generate more private gains for individuals, firms or consortia of these actors. In addition, many natural resources that are affected by climate change (for example water, land, forest, etc.) extend across geographical and political boundaries. Therefore, adaptation often involves decisions and actions by multiple stakeholders with differing shares in the costs and benefits of these initiatives.

This chapter presents the economics of climate change adaption in low-income countries, focusing on the stakeholders' cost-benefit estimation of climate change adaptation strategies. Using case studies from low-income countries, this chapter highlights important issues that need to be considered when conducting a cost-benefit analysis of climate change adaptation in low income countries. The rest of the chapter is organised as follows. The next section presents the linkage between economics and climate change and the following section briefly presents the importance of stakeholder cost-benefit analysis in low-income countries. The last section presents a stakeholder cost-benefit analysis in low-income country using case studies.

18.2 Climate change and economics

From the economics perspective, what does adaptation mean to these low-income countries? In general, climate change presents itself as an economic problem because it affects negatively (or positively) capital resources that are used in economic development, and hence can be viewed as a constraint of allocating scarce resources to attain development (Chambwera and Stage, 2010). Given that view, in an economic analysis, there is nothing unique about climate change because developing countries are subject to many internal and external constraints and stresses that affect their choice of development strategies, positively or negatively. Changes in terms of trade, natural disasters, wars and other external factors, as well as internal factors such as weak institutions, corruption and domestic strife, all affect the portfolio of choices available, and all need to be considered when a country determines its development strategy. Therefore, just like any other constraints in development and production, climate change can enter into a production function as a production shifter. Just like private development actors presumably seek to maximise their profits (or utility) regardless of the weather, climate change can be one of the many factors affecting their production and investment decisions. The same is presumably true of low-income countries. It is therefore, needless to say that countries have to consider climate change in their development strategies, because climate change will be an important constraint on the choices that are likely to be available for economic development.

However, unlike other production constraints like cash, land or labour, climate change has an added complication in that it is unpredictable; hence, it poses additional risks. The unpredictability is in geographical space, time and magnitude. Such a characteristic makes climate change very difficult to factor into production function. Climate change exacerbates problems of uncertainty for farmers, particularly smallholders, by extending beyond traditional risk assessment, because probability distributions for climate change outcomes are unknown. For example, for most countries, there is no consensus whether future climate will be wetter or drier, or how the frequency and severity of major storms will change.

The major concern that mostly low-income countries face is that climate change generates damages that alter country growth trajectories relative to the case without climate change. To sustain growth trajectory, deliberate coping strategies have to be put in place from policy to project level. When these coping behaviours are factored into development programs ('reactive adaptation' or 'anticipatory adaptation'), they impose a toll on economic growth in the form of additional cost (Lecocq and Shalizi, 2007). These costs are in the form of the actual infrastructure and institutional investment in addition to the foregone benefits because of the change in growth trajectory. This does not only affect development at the national level, but also the smallest decision maker, the individual. Failure to include everyone in a cost-benefit analysis of climate change may result in erroneous estimations, or such adaptation strategies may not have support from the very people that they aim to protect from climate change impacts. For example, a number of irrigation projects that were thought to be very good climate change adaptation strategies have failed or created more challenges in the form of conflicts in the pastoralist societies of Kenya,[1] Ethiopia and Tanzania due to lack of including the local community in initial analysis of possible strategies (Nangulu 2001; Fox, 2004).

18.3 Defining adaptation

Adaptation is now recognised as an essential part of the global response to climate change. Understanding what adaptation means is the first critical step in any action or analysis. Although there are formal definitions of adaptation to climate change in the literature on this issue, specific practical actions to address climate change and other existing needs are generally defined by the context concerned. The IPCC Fourth Assessment Report provides a useful synthesis of the term, defining adaptation as 'initiatives and measures to reduce vulnerability of natural and human systems against actual or expected climate change effects.' Various types of adaptation exist, e.g. anticipatory and reactive, private and public, and autonomous and planned (IPCC, 2007). The different definitions and characterisations of formal and non-formal adaptation and varying approaches to its implementation testify that this is an emerging concept. Levina and Tirpark (2006) cite four separate definitions of the term taken from IPCC TAR (2001), the website of the UNFCCC Secretariat, UNDP (2005). However, in this chapter, we use the IPCC definition.

18.4 Theoretical framework

Numerous analytical frameworks have been proposed to deal with adaptation projects. However, the most comprehensive analytical framework is by the Stern Review (REF). The framework is based on inter-temporal benefit/cost analysis in a welfare economics framework (Chambwera and Stage, 2010). It evaluates the benefits of adaptation projects/policies relative to a benchmark/baseline, the latter consisting of the costs of climate change for specific communities/regions/countries or broader geographical areas. This model evaluates the benefits of 'taking action' through adaptation not only to the costs of undertaking a project per se but also to the costs of 'inaction'. Arguably, such is the only manner to capture 'avoided losses' (i.e. encompassed as benefits) in an economic analysis (Nicholls, 2007). However, Chambwera and Stage (2010) argued that adaptation to climate change needs to be seen as an integral part of a country's development planning, rather than as a separate issue, and hence such developmental investments may produce benefits that go beyond just avoided losses. Good adaptation measures may lead to increased livelihood and hence adaptation measures that lead to better overall development outcomes are preferable to ones that focus exclusively on adapting to climate change impacts while ignoring other stresses.

With this in mind, we expand on the Stern Review framework (Figure 18.1) and add the benefits as a declining function as climate change impacts increase. These could be from any economic activity done by the community, e.g. farming or livestock production. From the Stern Review and presented by Chambwera and Stage (2010), an increase in the global mean temperature leads to climate change, whose impacts represent a cost to society (the costs of climate change), as depicted in Figure 18.1. Adaptation reduces these costs, but not completely, such that there will always be residual damage costs. This could be, for example, an eroded piece of land that is completely lost in flood-prone areas. The difference between the cost of climate change without adaptation and the residual cost of climate change after adaptation is the gross benefit of adaptation. Including the cost of adaptation reduces the benefit to the net benefit. However, given an effective and efficient adaptation strategy, benefits can be higher than the previous production systems, e.g. a drought-tolerant crop that is higher yielding or raised homesteads in flood prone areas that allow for continuous production during flooding season. Therefore, instead of the community being idle during flood season, they can still harvest additional output from the raised homestead and fields. This then shifts the benefits function to the right.

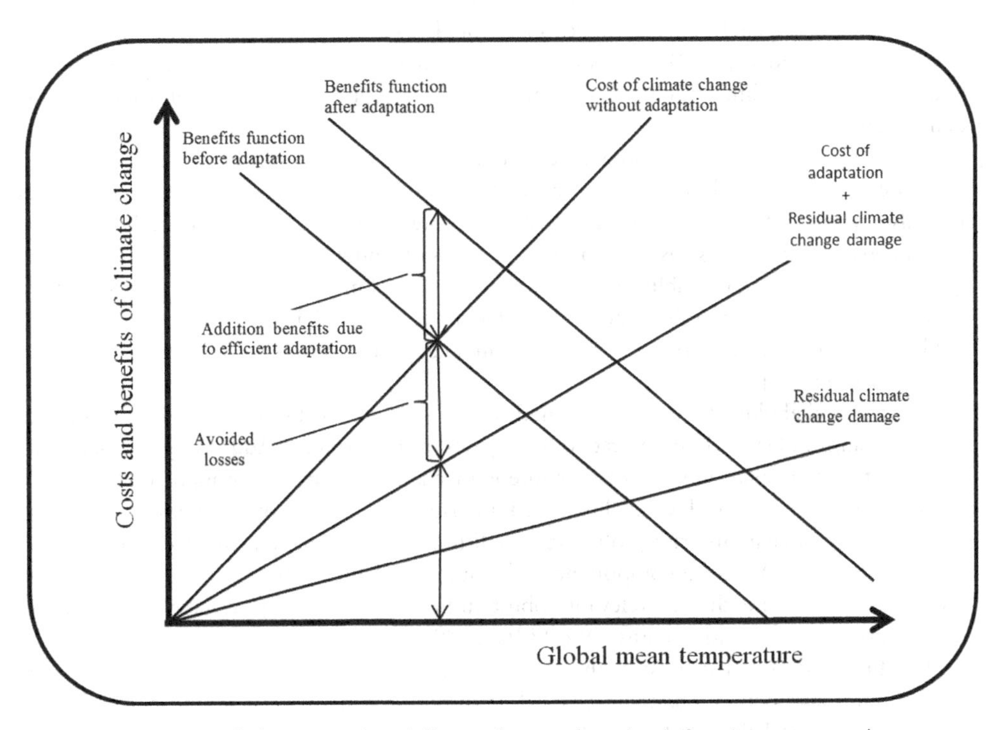

Figure 18.1 Framework for economics of climate change adaptation in low-income countries
Source: Adapted from Chambwera and Stage, 2010

Therefore, the new production system, which is a result of the adaptation strategies, shifts the benefits function to a higher level and gives additional output. This is an important factor as climate change adaptation can be taken as an opportunity on which to springboard development that can bring out benefits that surpass previous outputs. It also allows adaptation to be combined with usual developmental programmes that have been used to improve livelihood and build resilience (e.g. credit facilities, marketing, improved hybrid seeds, etc.). Therefore, the benefits of adaptation/developmental strategies for all stakeholders should be captured in assessing the effectiveness of such strategies. This involves engaging all stakeholders in the assessment.

18.5 Why a stakeholder cost-benefit analysis?

The literature on this topic identifies several reasons why stakeholder-based adaptation is needed in low- and middle-income countries. The first is the need for equity in stakeholder adaptation. Parry et al. (2009) note that climate change has a disproportionate impact on vulnerable populations, many of whom are poor. Therefore, it is important for adaptation planners to consider both the net benefits of adaptation and the way that the costs and benefits of adaptation options are distributed. In addition, steps must be taken to identify and help the poor and most vulnerable – including soliciting their views on adaptation priorities and ensuring an enabling environment (World Bank, 2010a).

The second compelling reason for adopting a stakeholder-focused approach is the importance of qualitative cost-benefit analysis (CBA) in climate adaptation, and the fact that stakeholder-based cost-benefit analysis is flexible enough to accommodate the qualitative aspects of this tool. Quantitative CBA tends to dominate other types of economic analysis, but it is widely criticised in climate policy as too simplistic for complex climate-economy analysis (van den Bergh, 2004).

The third reason why stakeholder-based CBA is suitable for climate change adaptation studies relates to a particularly important aspect of climate adaptation as it protects public goods. It is also worth noting that information about future climate patterns and the benefits and costs of adaptation options has some of the attributes of a public good (Leary, 1999). However, quantitative measuring of public good and attributing to climate change is not easy. Therefore, studies tend to leave it out of conventional cost-benefit analysis. Adaptive responses that protect public goods from the adverse impacts of climate change and variability will generate benefits that are external to private agents.

Fourthly, stakeholder-focused cost-benefit analysis is justified by the need to look beyond climate change policy and incorporate other aspects of the economy. Addressing the many barriers to effective adaptation requires a comprehensive and dynamic policy approach that covers a range of issues at various levels. Climate risk management, for example, ranges from farmers' understanding of changing risk profiles to establishing efficient markets that facilitate effective response strategies. A crucial component of this approach is the implementation of adaptation assessment frameworks that are relevant, robust and easily operated by all stakeholders, practitioners, policy makers and scientists (World Bank, 2010a).

The final reason is that most of the real costs of adaptation will be borne by the affected communities and households themselves, despite the best efforts of government and other external donors. Stakeholder-focused CBA helps people understand what those costs are likely to be, and to seek potential sources of external funding.

18.6 Steps in stakeholder cost-benefit analysis

This section presents the framework for stakeholder-based cost-benefit analysis of climate adaptation options. The first step in this approach is to specify the impacts of climate change that need to be addressed and identify a set of suitable adaptation strategies for so adapting. The second step involves identifying the groups, individuals and institutions that are believed to be mainly affected by the climate change scenario and likely to bear the costs and benefits of the adaptation strategies. The third step is the prioritisation by stakeholders of a range of adaptation strategies according to criteria that they set themselves. In the fourth step, monetary and non-monetary costs and benefits are ascribed to different aspects of the priority adaptation strategy, while the final step assesses how much each stakeholder is willing to pay for particular strategies and how the costs and benefits of adaptation are distributed between stakeholders.

18.6.1 Step 1: Identify the impacts of climate change

By its very nature, climate change is global and has wide-ranging implications for every imaginable aspect of life. Therefore, the rational first step in the cost-benefit analysis of a climate change adaptation project is to define the impacts of climate change on a certain location and particular economic sector (agriculture, water, etc.). It is essential to understand the scope of the climate change scenario in order to frame an adaptation strategy and consider possible alternative options. This involves defining the geographical area and main economic sector concerned, identifying baselines and looking at various climatic predictions.

One of the most important aspects of estimating the costs and benefits of adaptation options is setting the baseline, which defines what would happen to the main variables if the climate did not change. This is particularly difficult because adaptation assessments have to look into the future and predict levels of development and social change in a given period. Some researchers advocate the use of multiple baselines to accommodate the uncertainties involved in estimating the costs and benefits of adaptation and evaluating adaptation options (Parry et al., 2009; World Bank, 2009b). The next challenge is to identify climatic projections. This normally involves choosing two or more climate scenarios to capture the widest possible range of climate model predictions (World Bank, 2009b), a step that entails looking at past climate data in close collaboration with climate change modelers.

The purpose of this step is to ensure that all the adaptation strategies covered by the analysis are designed to address the impacts of climate change. Where adaptation strategies have yet to be identified, this step provides information that can be used to generate adaptation actions with stakeholders. The possible impacts of climate change are identified on the basis of likely future climate patterns. Such information is usually general, relating to a whole country or region, and is rarely specific to the area covered by a case study.

18.6.2 Step 2: Identifying stakeholders

It is important to identify the stakeholders concerned to ensure that everyone's interests are represented in the collaborative process, data collection and analyses. The extent to which all stakeholders are identified and their interests taken on board will largely determine the acceptability and sustainability of any adaptation actions.

The term 'stakeholder' refers to all significant groups or entities that are likely to be directly or indirectly affected by climate change, or who take part in, or are affected by, adaptation actions

Table 18.1 The four main categories of stakeholders

Sector	*Stakeholders*
Private	Households (e.g. small-scale farmers); Private firms and investors
Social	Community groups (Community-based organisations – CBOs)
Public	Government departments, Non-governmental organisations, Donor partners
Environment	Natural resources (forests, water, land, etc.), ecosystems

(Bryson, 2003). The extent to which stakeholder groups are disaggregated into upstream and downstream groups, livestock farmers and crop producers, male- and female-headed households, national and local governments, etc., depends on the extent to which their interests differ, and the scale of analysis. Greater disaggregation obviously entails additional engagement and analysis.

For example, the stakeholders in each case study site within the IDRC/IIED Economics of Climate Change Adaptation research project were grouped into four main categories: households, the private sector, the public sector, and the environment, as shown in Table 18.1.

The private sector was defined to include all actors that make decisions as individuals, such as households, private firms and for-profit organisations. The social sector was defined to include groups of individuals that make decisions based on general consensus, such as community-based organisations, natural resource committees, associations and people from certain areas or localities. Communities may contain particularly disadvantaged sub-groups (women, children, the disabled, the elderly and so on), who may need to be treated as separate groups. The public was defined to include governments, NGOs and donor partners who generally invest in public goods and services. The last category, the environment, is the most interesting of all, and was defined such that it spanned the natural environment, ecosystems and biodiversity. This category can be assessed based on the inputs and preferences of concerned interest groups (such as environmental NGOs), or based on scientific evidence related to ecosystem health and dynamics. These groups were selected within the IDRC/IIED Economics of Climate Change Adaptation research project based on the idea that the analysis was going to maintain a fairly local, rather than national or regional, focus. Further disaggregation was not pursued given the overall size of the project, but could benefit local case studies such as these.

18.6.3 Step 3: Identifying adaptation strategies

Stakeholders can choose from a number of climate change adaptation options, which will be more or less appropriate to their needs. In order to identify suitable adaptation strategies, stakeholders need to get together to evaluate the available options, following a number of steps set out by the World Bank (2009b):

- Identify current climate changes and their implications for local natural systems, rural livelihoods and local communities.
- Identify possible options to support local strategies for adaptation to climate change.
- Prioritise these possible response options as activities and initiatives that will form the basis of local action plans.

There are three components to the prioritisation exercise:

a) The first involves identifying and weighting a number of criteria. Stakeholders are asked to allocate 100 points between eight impact criteria, and another 100 points between six viability criteria.
b) The next step is to determine the characteristics of each response option, in order to develop a profile of each option identified by stakeholders that includes information on:
   (i) the underlying need for the response option,
   (ii) technical characteristics, and
   (iii) a rough indication of the associated costs and benefits.
c) The third and final step is to apply the impact and viability criteria to each response option and assign values to them, to generate a final prioritised ranking of the response options. Participants are given a matrix and asked to value each response option on a scale from 1 to 10, according to the extent to which they think it addresses each criterion.

These ratings are then weighted according to a previously agreed upon system, with 50 percent of the overall score proportionately assigned to the impact criteria and 50 percent to the viability criteria, providing a final ranking of the possible adaptation options.

Another popular approach to identifying adaptation options is rapid rural appraisal (RRA). This methodology is used in a wide range of project evaluation exercises, and was originally devised for the Farming Systems Research and Extension promoted by the Consultative Group on International Agricultural Research Centres (CGIAR) (FAO, 1997). It was developed to counteract the disadvantages of more traditional research methods, such as the time taken to produce results, the high cost of formal surveys, and the unreliability of data due to non-sampling errors. Rapid rural appraisal provides a bridge between formal surveys and unstructured research methods such as in-depth interviews, focus groups and observation studies. In low- and middle-income countries, it is sometimes difficult to apply the standard market research techniques employed elsewhere, due to a lack of baseline data, poor facilities for market research (no sampling frames, relatively low literacy among many target populations, lack of trained enumerators) and low awareness of the need for market research. RRA can help overcome these and other limitations by mapping the strategy options identified by stakeholders and identifying the most feasible and plausible strategies for detailed and quantitative evaluation.

Within the IDRC/IIED Economics of Climate Change Adaptation research project, the methods used to identify adaptation alternatives in the five case studies can be seen as a combination of RRA and the priority ranking advocated by the World Bank (2009b). It is worth noting that most of the adaptation actions that are identified will necessarily entail short-term adaptation to climate variability rather than longer-term adaptation.

18.6.4 Step 4: Measuring costs and benefits

Having identified the optimal adaptation strategy (as perceived by stakeholders), the next step is to evaluate its economic merit. This is done by analysts using qualitative and quantitative methods to establish a rough estimate of the costs and benefits for all stakeholders, both as individuals and as a group. It is important to understand various aspects of the costs and benefits before embarking on this evaluation.

a) *Evaluating the monetary costs of adaptation*

The concept of costing adaptation is based on comparisons between a future world without climate change and a future world with climate change. Each of the actions required to adapt to the new conditions caused by climate change will have certain costs attached (World Bank, 2009b), which are generally estimated by major economic sectors (Parry et al., 2009). The methods used for stakeholder-focused cost-benefit analysis need to be tailored to assess the qualitative and quantitative costs borne by different stakeholders, which may differ within and between groups. In the specific case studies discussed here, it was important to assess the range of costs that households have already incurred in adapting to climate-related hazards, as well as how much money (and what resources) institutions need to help households adapt to particular hazards in the future. These estimates are then used to judge how much investment or aid governments or donors will need to provide to promote particular adaptation interventions in rural areas. The information collected from stakeholder interviews is cross-checked with information from focus group discussions and expert interviews.

b) *Evaluating the monetary benefits of adaptation*

A standard way of estimating the benefits of adaptation to climate change is to calculate the expected impacts of climate change without adaptation and the expected damage avoided through adaptation. The gross benefit of adaptation is the difference between the expected damage caused by climate change with and without adaptation. The uncertainty surrounding climate change and its impacts makes estimating the costs and benefits of adaptation a very complex and somewhat arbitrary process, in addition to the challenges associated with evaluating physical and ecological changes in monetary terms (World Bank, 2009b).

c) *Estimating non-monetary costs and benefits*

Because projects inevitably generate costs and benefits that extend beyond their direct beneficiaries, it is important to examine their non-monetary costs and benefits as well as their economic value. While the negative impacts of certain climate-related events on human lives, livelihoods and ecosystems cannot be monetised, they have financial implications that may amplify or reduce the positive or negative effects of a project. This suggests that non-monetary costs and benefits may determine whether a project is actually worthwhile in the eyes of all stakeholders (for instance, whether a climate change adaptation project will have the expected significant positive effects on surrounding ecosystems).

d) *Methods used to evaluate an adaptation strategy*

The most popular method for quantitative evaluation of a project's net worth is to calculate its Net Present Value (NPV). Other non-market aspects such as social and environmental benefits and costs are evaluated qualitatively. These could be dealt with quantitatively, but in developing countries, due to lack of markets, it may not be possible to do this, and values are not based on market forces. As such, the only way to include them is to do qualitative valuation based on subjective assessments of local stakeholders. NPV is the difference between the discounted value of the future benefits and costs associated with a project, calculated at a required discount rate. The higher the NPV, the more economically viable the project, as it means that the benefits of the project exceed the costs through time. If the NPV is negative, the project is not economically viable. Another metric that can be used to assess the net worth of a project is the cost-benefit ratio (CBR). The net present value is calculated using the following formula:

$$NPV = \sum_{i=1}^{n} \frac{net\ benefit}{(1 + discounter\ rate)^{i}}$$

where n = the lifetime of the project and i = any given year.

The CBR is the ratio between the present value of a project's benefits and the present value of its costs. The higher the CBR, the more economically viable the project, as it means that it is earning more than the required rate of return. For example, a CBR of 1.04 means that for every dollar spent on a project, the benefits generated are valued at $1.04. A third possible metric that can be used to evaluate the performance of an adaptation strategy is the internal rate of return (IRR). The IRR is the break-even discount rate, at which the present value of a project's benefits equals the present value of its costs. The higher the internal rate of return, the more economically attractive the project. The internal rate of return is iteratively computed by setting the NPV value equal to zero. In order to calculate the internal rate of return, the equation above is modified to:

$$NPV = \sum_{i=1}^{n} \frac{net\ benefit}{(1 + IRR)^{i}} = 0$$

An iterative solver is used to compute the internal rate of return, using arbitrary starting values in Excel or other solvers.

A project, or in this case adaptation strategy, needs to generate positive net benefits to be economically worthwhile; in other words, the discounted value of its benefits needs to exceed the discounted value of its costs. This is equivalent to an NPV greater than zero, and an IRR higher than the capital costs.

e) *Analytical methods*

For the purposes of financial evaluation, a project is viewed as a commercial entity whose objective is to maximise private profits. Its success in doing so is judged by analysing its annual income and expenditure. The major benefit component in NPV analysis is the annual financial flows, or annual revenue earned. The capital cost of an adaptation project represents the time stream of investment over its lifetime. Investment expenditure in any year may include the purchase of capital goods, cost of acquiring land, and payments for skilled and unskilled labour and material inputs for project construction. Operating and maintenance costs include annual expenditure on energy, material inputs for maintenance and payments for skilled and unskilled labour. Investment goods and material inputs used by the project are evaluated at market prices, with the commodity market price taken as the producer price plus commodity tax minus commodity subsidy.

Cost-benefit analysis of a planned climate change adaptation project could perform discounted cash flow analysis using discount rates of 6% (market interest rate) and 50% (average time preference derived from experimental field surveys in numerous low- and middle-income countries [Yesuf and Bluffstone, 2009]). Although project lifetimes vary, a 30-year period can be used as a baseline.

f) *Uncertainty, discount rates and time horizons of adaptation projects*

Before assessing the costs and benefits of an adaptation project, it is important to identify three critical dimensions of the initiative:

(i) First, the degree to which uncertainty can be incorporated into the assessment. There will inevitably be considerable uncertainty about each phase in the chain of climate cause and effect: GHG emissions, effects on climate, ecological and hydrological consequences, social and economic responses, impacts on human health and world-wide welfare distribution (van den Bergh, 2004). Because uncertainty about the future impacts of climate change makes it difficult to identify the best adaptation options, adaptation measures need to be designed in a flexible manner so that their respective costs and benefits can be reported with a given margin for uncertainty (Parry et al., 2009).

(ii) The second critical parameter is the discount rate that will be used to convert benefit and cost streams into their equivalent present values. Present values are very sensitive to discount rates and assumptions about their consistency over time. Applying a range of discount rates allows planners to test the validity of results and ensure that the chosen discount rate is not close to a tipping point that reverses the decision, in which case further analysis is required (Parry et al., 2009). Discount rates for projects with short time horizons (20 to 30 years) should not be controversial, as the costs and benefits of adaptation measures are usually felt within a reasonably short time, and the ancillary benefits of investments make projects similar to other public investments (World Bank, 2009b).

(iii) Lastly, the time horizon of the evaluation is directly linked to the discount rate. This horizon depends on the lifespan of the options under consideration. The lifespan of infrastructure projects like dams and roads ranges from 50 to 70 years; therefore, all costs, including investment and maintenance costs and the benefits and expected impacts of climate change over the entire period, should be taken into account when assessing these options. By contrast, plans for adaptation to the health impacts of climate change can take a short- to medium-term view (5 to 20 years), which can subsequently be extended to cover longer periods if necessary (Parry et al., 2009).

g) *Data requirements*

(i) *Opportunity costs of land acquisition and preparation*

Most pre-production costs are associated with acquiring and clearing land for the planned operation and setting up the facilities. It may be assumed that land for this kind of investment is acquired through long-term leases, and that leases and licenses account for most of the acquisition costs.

(ii) *Cost of capital/funding*

It is assumed that the funds required for investment are obtained by borrowing a lump sum (to cover start-up costs) from a commercial bank at a market interest rate, which is given at 6 percent. It is assumed that operational expenses are covered by internal finances.

h) *Valuating community-based adaptation*

Most estimates of the welfare impacts of climate change made so far have used willingness to pay (WTP) measures. However, the fact that developing countries have not caused the problem implies that the appropriate welfare measure is willingness to accept (WTA) rather than WTP (Rajamani, 2002; Stage, 2010). Given the huge changes involved for many of the people who will be affected, the welfare losses may be considerably higher than the WTP measures indicated (Chambwera and Stage, 2010). However, both WTP and WTA presume that the society in question has a good understanding of the goods and services affected by climate change, and that they can easily estimate the values of these according to their needs. However, there is great uncertainty related to climate change and its impacts. It is also the case that most of the societies in developing countries are not fully integrated into economic systems. Hence, values estimated in developing country contexts may not conform to the expected economic rationale, and in turn, this may lead to gross under valuation. For example, most societies in developing countries have huge social and cultural values. These demand responsibility from the community to act together. In as much as an activity may not have any 'economic' value, the cultural and social responsibilities may require a community to invest time and resources in their action. Such values cannot be directly monetised as there is neither any financial cost nor benefits that are associated with them nor are there any cash transfers. However, it is in the community's rational decision to act and participate in activities. Therefore, valuation of cost and benefits in Community-Based Adaptation demands approaches beyond conventional economic methods. Stage (2010) also commented that in traditional cost-benefit assessments of adaptation plans, the benefits of avoided climate change impacts must be assigned monetary values. But these benefits come in various forms – such as steadier agricultural yields, reduced loss of lives, access to markets, gains in local knowledge and community empowerment – not all of which can be realistically quantified.

18.7 Empirical evidence of stakeholder cost-benefit analysis

Discussions with stakeholders about adaptation options should be informed by scientific climate analysis and local experience. That said, there was a gap between the theoretical considerations and the practical realities on the ground, and no clear separation between adaptation to climate change and coping with current climate variability. It seems that hard adaptation options generally deal with climate change, while autonomous actions cope with existing variability and development deficits. The fact that both are needed makes it difficult for stakeholders to separate them out in the analyses.

18.7.1 Significance of non-monetary benefits

Non-monetary benefits are highly significant, particularly for local households and the environment. Including different stakeholder groups in cost-benefit analysis helps identify benefits that are not directly captured in conventional cost-benefit analysis. For instance, in a study in Morocco on the impacts of switching from surface irrigation to drip irrigation, it was found that more than 50% of the water saved by the switching will increase long-term groundwater levels (Mohamed, 2013). By conducting a traditional cost-benefit analysis of the irrigation technologies alone without including the 'environment' as a stakeholder does not capture this huge benefit. In Malawi, estimating the cost and benefits of different adaptation strategies at a catchment level shows assessing the strategies separately led to maladaptation (wrong adaptation that ends up with more negative effects). Some adaption strategies (irrigation) have high benefits for a group of stakeholder (e.g. crop farmers) while having negative impacts on other stakeholders (e.g. the fishing community) (Lunduka, 2013a). However, incorporating soil and water conservation technologies (an additional cost) into irrigation benefits fisheries, as there is less siltation in the lake, and better long-term yields reduce the need to hunt birds. Box 18.1 presents the Malawi case study in detail.

Box 18.1 Analysis of effectiveness of autonomous adaptation strategies

Case study 1: Lake Chilwa Catchment, Malawi

Climate change and its impact

Lake Chilwa basin is a very important catchment that supports the livelihoods of more than 117,000 farming families. The resources in the basin, which include water, fish, birds and grass (used for thatching and constructing houses, boats, mats, fish traps, bird traps and baskets), are used and managed by diverse stakeholders with different, often conflicting, objectives. The increased incidence of drought and erratic rainfall caused by climate change have increased reliance on irrigation and led to more land being cleared for rice and irrigated maize. This has increased soil erosion, which causes siltation, reduces water flow into the lake and adversely affects fish stocks.

Available adaptation strategies

A multi-stakeholder approach was used to analyse the four main adaptation strategies in the Lake Chilwa catchment. The first autonomous adaptation to drought was irrigation, which has enabled a number of smallholder farmers to remain productive but has caused massive soil erosion into the river and lake and adversely affected the fish population. The government and various NGOs are promoting soil and water conservation technologies to address this problem, which are assessed as the second adaptation strategy. The third strategy, which is more specific to the fisheries sector, is pond construction for aquaculture and monitoring of the lake during the closed season (January to June) when fishing is banned to enable fish to breed. Because some people ignore the ban, the waters are patrolled to ensure that no one fishes illegally during the closed season. The fourth adaptation strategy is protecting bird sanctuaries.

Results of the stakeholder-focused cost-benefit analysis

The soil and water conservation (SWC) technologies deployed in the upper catchment area are very important because they have a direct bearing on the fishing and bird sectors, and irrigation only works as an adaptation measure if such technologies are implemented. Stakeholders took part in a qualitative ranking exercise that asked which of the four sectors bears most of the cost and enjoys most of the benefits of each adaptation strategy. A score of zero represents no cost or benefit, and five indicates the greatest cost or benefit.

These web diagrams show that stakeholders see irrigation as more costly for the environment than SWC technologies, and that the public sector pays more for irrigation than it pays for SWC, whose costs are mainly borne by the community and private sector (households). The environment and public sector are perceived as benefiting the most from SWC technologies, while the community and private sector benefit more from irrigation. Interestingly, stakeholders thought that the public sector benefits more from SWC technologies but puts less money into them, and that the environment pays for the high benefits ascribed to the private sector and community.

The qualitative results were used to inform the quantitative analysis. The quantitative analysis was therefore done on the adaptation strategies as a package rather than assessing them separately. This also included combining the effects of the strategies for both communities in the lower and

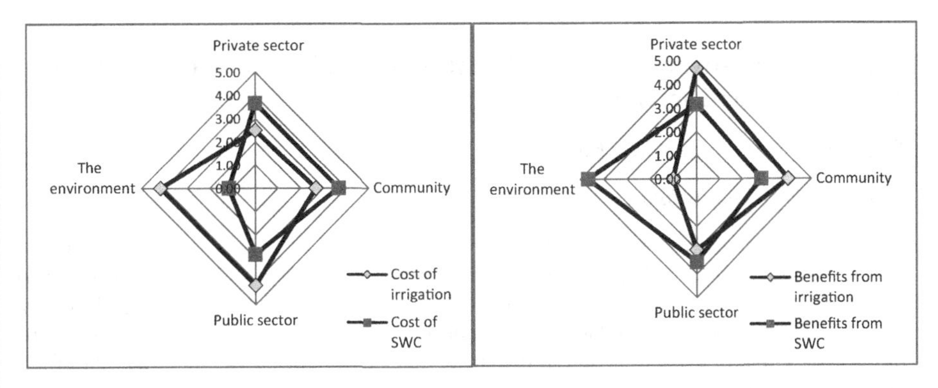

Figure 18.2 Stakeholder rankings of (i) the cost of irrigation and SWC technologies and (ii) the benefits of irrigation and SWC technologies

upper catchments. In addition, the two communities were also able to understand the implications of their activities and hence able to adjust the strategies, i.e. including SWC technologies in irrigation schemes in the upper catchment to reduce negative impact in the lower catchment.

Source: Lunduka, 2013b

18.7.2 Different stakeholder are faced with different constraints to adapt to climate change

Stakeholder cost-benefit analysis also reveals the constraints that different stakeholders face. These constraints may mean that effective and efficient technologies may not be adopted due to the constraints being binding. Therefore, the benefits of adaptation for specific stakeholders could be maximised by giving them access to appropriate technologies through others strategies, e.g. subsidies. For example, the Net Present Value (NPV) of drip irrigation for small-scale farmers in Morocco could be improved if they had the technology to extend drip irrigation to food crops, rather than limiting it to cash crops as is currently the case. Smallholder farmers are faced with a liquidity constraint to access such a technology. Reducing such a constraint through subsidies does only benefit the small-scale farmers themselves, but the wider stakeholders in the area, including the environment. This also gives them greater parity with large-scale farmers who allocate more land to cash crops, and fare better in the economic analyses (Mohamed, 2013).

Most capital costs for infrastructure, for example, are borne by government and external funding agencies, while local communities and the private sector tend to carry operational costs and the cost of private capital assets such as appropriate fishing gear, irrigation equipment, adaptive tourist facilities, etc. This tends to mean that total costs (at least at start-up) are weighted towards public stakeholders and local communities and the private sector expects public stakeholders to invest in infrastructures. This brings out the different weights from different stakeholders. Hence, simple aggregation of cost and benefits from the different stakeholders may be erroneous. Box 18.2 presents a case study from Bangladesh where weights from different stakeholders were used to assess the different adaptation options available.

Box 18.2 Prioritising different adaptation strategies and selecting the most cost-effective and acceptable

Case Study 2: Khulna City, Bangladesh

Climate change and its impacts

Khulna is the third-largest city in Bangladesh, covering about 45 km^2 of land in the southwestern coastal region. It is bounded by three rivers: Bhairab-Rupsha to the east, Khudi Khal-Mayur to the west, and Harintana Khal to the south, all of which are affected by tidal waves. In 2001, Khulna had a population of around 2 million and an annual growth rate of 2 percent (last available census report). About 46 percent of land in the city is now built up, with around 15 percent of this put to industrial use, 5 percent under commercial use and the rest used for residential and other purposes. Non-built-up areas are used for agriculture and fisheries. The whole metropolitan area is approximately 2.5 metres above the mean sea level. Its drainage system is linked to the western Mayur River because of local topography, and low-lying areas are often waterlogged due to natural flooding and mismanagement of the drainage infrastructure.

Predicted climate events for Khulna city under future IPCC scenarios A and B show that average monthly temperatures will rise by 1.7 °C by 2050 and annual rainfall will increase by 0.5 percent, while rainfall intensity (4.3mm/hr) will increase by 4.2 to 5.9 times per year. In addition to rising sea levels, floods are expected to increase and cover almost the whole Khulna area. Nearly 26 of the city's 31 wards are likely to be inundated if the model's predictions are correct.

Adaptation strategies

These climate change predictions were used to forecast future drainage blocking in Khulna city, and prepare an adaptation plan to mitigate the effects of climate change on the lives and livelihoods of its millions of inhabitants. Predicted climate events indicate that the most severe impacts will be felt after 2030, and that adaptation plans should reconsider the design and management of urban infrastructures. The predictions also suggest that the city should realign its institutions in order to develop an effective strategy to combat the effects of climate change. For the purposes of this analysis, adaptation activities were classified into four groups: i) Structural – construction of infrastructures, ii) Maintenance – maintaining existing infrastructures in order to deal with climate risks, iii) Managerial – requiring changes in the overall management of the city and the roles played by individuals, communities and government institutions and iv) Awareness – informing communities and households how to deal with climate risks.

Results of the stakeholder-focused cost-benefit analysis

There was quite a lot of divergence among the stakeholders in terms of which adaptation measures they preferred. For example, 75% of the structural measures were preferred by households, while the government, preferred 100%. The community preferred 63% and environmental stakeholders preferred 25% of the structural measures. The same differences were observed for the repair and maintenance, managerial, and awareness adaptation measures.

Project benefits based on stakeholder preferences

Using the cost of adaptations provided in the Asia Development Bank (ADB) study, the government prefers the most costly options while environmental stakeholders prefer the least costly adaptation options. It is also evident that while structural adaptation measures are preferred by the

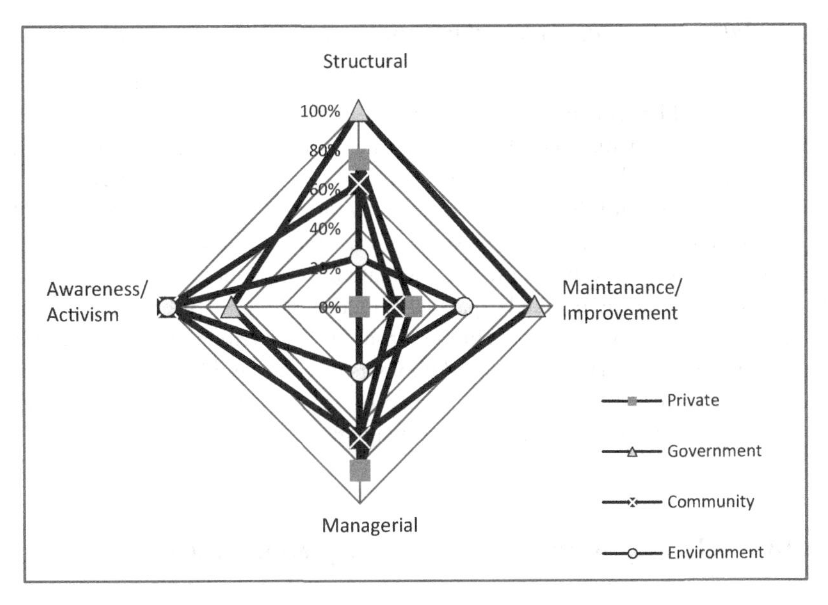

Figure 18.3 Qualitative rating of the strategies by stakeholders

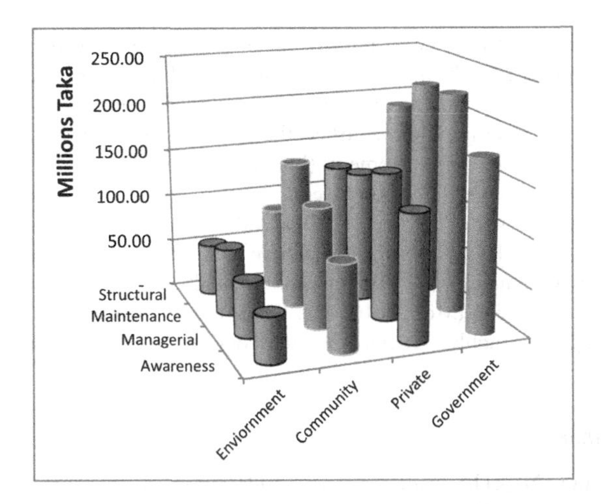

Figure 18.4 Cost of the different strategies to the stakeholders

government or public agencies, community and environmental groups preferred the awareness type of adaptation measures (ADB, 2011). Such understanding of preference and cost can ensure acceptability and hence sustainability of adaptation projects. The study further used these preferences to weigh the cost and benefits and came up with well agreed upon strategies that were cost effective, sustainable and acceptable by all stakeholders.

Source: Haque, 2013

18.7.3 A platform for stakeholders' negotiations on different adaptation strategies

A key challenge in cost-benefit analysis of climate change adaptation strategies is the integration of the costs and benefits from different stakeholders into a common framework in a way that can inform decision-making. This issue can be addressed at various stages of the project, starting with a compound cost-benefit assessment that covers all stakeholders, and then separating the costs and benefits for different stakeholders in order to inform the process in which they can debate which options to pursue. The value of this approach lies in enabling stakeholders to reach an informed consensus based on analysis that takes account of both monetary and non-monetary benefits. In Nepal in Lake Rupa, such a methodology was used where estimated costs and benefits were presented to the stakeholder groups to negotiate with each other. Given their social economic and financial positions, stakeholders in upper and lower catchments were able to agree on a very effective and efficient adaptation strategy and also how much they would each contribute to the implementation of the adaptation strategy.

Box 18.3 Selecting the most acceptable adaptation strategy

Case study 3: Rupa Watershed, Kaski, Nepal

Climate change and its impacts

The Rupa Tal watershed is home to aquatic, wetland and terrestrial ecosystems, and also to Lake Rupa Tal, the third largest lake in the Pokhara valley, which has an average depth of 1.95 metres and a surface area of 1.07 km^2. The region's ecosystem has deteriorated over the past few decades due human encroachment into the watershed, the extraction of forest resources and conversion of forest into agricultural land and settlements (Pradhan and Providoli, 2010).

Local communities across the region have observed higher temperatures and erratic and intense rainfall over the past 10 years (Thapa et al., 2011), although climate change varies in different parts of the region. Some pockets of western Nepal have seen a decline in pre-monsoon rainfall and an increase in monsoon rainfall, while post-monsoon rainfall on the southern slopes of hills in western Nepal is also increasing (MoE, 2010). The Kaski district is classified as moderately vulnerable to climate change, and highly vulnerable to landslides because of the erratic and intense rainfall (MoE, 2010).

Adaptation strategies

The government of Nepal has prioritised various adaptation options for the country (MOE, 2010) whose implementation will cost an estimated US$ 350 million. National Adaptation Programmes of Actions (NAPA) identify and prioritise adaptation measures to meet urgent and immediate needs in districts that are vulnerable to drought, flooding, landslides and overflow from flooded glacial lakes. In this study, different groups of stakeholders were asked to list the type of adaptation measures they currently employ and those that they plan to use.

Stakeholder workshop (negotiation meeting)

At the end of the study, a workshop was held to share the outcomes of the process and facilitate stakeholder negotiations on beneficiaries and their willingness to pay for adaptation measures. Various project options were presented, along with their estimated cost and benefits, internal rate of return (IRR) and cost-benefit analysis (CBA). After discussing the comparative analyses of the options and trying to find trade-off points for the various costs and benefits, participants decided which option was most socially acceptable, economically viable and environmentally sustainable.

Results of the stakeholder-focused cost-benefit analysis

The conditions around Lake Rupa and proposed construction of a dam to protect the lake feature in the official response to climate change set out in the National Adaption Plan of Action (NAPA). The main justification for building the dam was the generation of hydropower, which meant that its capacity had to be increased.

Local stakeholders around Lake Rupa were completely unaware of these proposals. During the focus group discussions, upstream and downstream stakeholders identified potential options according to their particular needs. Downstream communities are more interested in controlling the river by building a check dam in the Talbesi River to reduce siltation, control river cutting and protect agricultural land. The government project consists of two major components: constructing an earth-fill dam by the lake and a check dam in the Talbessi River. Cost-benefit analysis shows that it is more economically feasible to construct a check dam than an earth dam or a check dam and earth dam, although all three options (check dam, earth dam and both) are acceptable.

Although it requires substantial investment, the stakeholder workshop and negotiation meeting agreed to accept this option because it addresses issues that affect both upstream and downstream communities. It will protect the lake, increase economic opportunities and resolve local conflict if it is implemented openly. Therefore, it seems socially viable, economically acceptable and environmentally friendly.

Source: Dhakal and Dixit, 2013

18.8 Conclusion

Climate change threatens to stay with us into the unforeseeable future. The current climate change impacts of flood, drought and seawater intrusion are likely to continue affecting the livelihoods of many, in particular the poor. Climate change adaptation is the most logical strategy in the short term for the poor as mitigation may not be efficient given the low greenhouse gas emissions from the poor. In general, climate change presents itself as an economic problem because it affects negatively (or positively) capital resources that are used in economic development and hence can be viewed as a constraint of allocating scarce resources to attain development (Chambwera and Stage, 2010). Making economic estimates of losses from climate change and then the benefits of adaptation investments are inherently difficult for a number of reasons, including the uncertainty of projecting future climate change impacts and losses (Markandya and Watkiss, 2009). However, one of the important elements in making economic estimates is that the most vulnerable populations be given significant weightage in arriving at final decisions. This is why participatory CBA is so important in the case of adaptation, especially in low-income countries where a top-down tendency by government is the default position in most cases. The case study examples on Morocco, Malawi, Bangladesh and Nepal testify to the importance of including everyone involved in the analysis.

Note

1 In the mid-1980s, the Kenyan government introduced a number of irrigation schemes including the 9,000-hectare Turkwell Gorge project, Katilu (the dam costing over £270 million). The Pokot and Turkana were forcibly removed from the project area and though they were promised immediate resettlement, many were left landless. In the 1990s, the government pulled out from some of these schemes as they had been a dismal failure and contributed to food insecurity in the region (Nangulu, 2001).

References

Asian Development Bank (ADB) (2011). Adapting to Climate Change: Strengthening the Climate Resilience of the Water Sector Infrastructure in Khulna, Bangladesh. Mandaluyong City, Philippines.

Atta-Krah, A. (April 2012). Challenges in Building Resilience to Respond to Climate Change: A Focus on Africa and Its Least Developed Countries. United Nations Economic Commission for Africa.

Bryson, J.M. (2003). What to Do When Stakeholders Matter: A Guide to Stakeholder Identification and Analysis Techniques. A paper presented at the National Public Management Research Conference, Georgetown University Public Policy Institute, Washington, D.C.

Chambwera, M. and Stage, J. (2010). Climate Change Adaptation in Developing Countries: Issues and Perspectives for Economic Analysis. International Institute for Environment and Development (IIED), London, UK.

Clements, R. (2009). The Economic Cost of Climate Change in Africa. Pan African Climate Justice Alliance and Practical Action Consulting.

Dell, M., Jones, B.F., Olken, A. (2008). Climate Change and Economic Growth: Evidence From the Last Half Century. NBER Working Paper Series. Cambridge, Massachusetts, National Bureau of Economic Research.

Dercon, S. (2011). Is Green Growth Good for the Poor? Paper prepared for the World Bank project on Green Growth.

Dhakal, Arjun and Ajaya Dixit (2013). Economics of Climate Change in the Water Sector in Nepal: A Stakeholder-Focused Approach. A case study of the Rupa Watershed, Kaski, Nepal. International Institute for Environment and Development, London, UK.

Food and Agriculture Organisation (FAO) (1997). Marketing Research and Information Systems. Food and Agriculture Organization of the United Nations, Rome, ISBN 92-851-1005-3.

Fox, Bruce (2004). An Overview of the Usangu & Great Ruaha River Ecosystem Environmental Disaster. Unpublished paper.

Haque, E.A.K. (2013). Reducing Adaptation Costs to Climate Change Through Stakeholder-Focused Project Design – The Case Of Khulna City in Bangladesh. International Institute for Environment and Development (IIED), London, UK.

IPCC (2001). Climate Change 2001: The Scientific Basis. Contribution of Working Group I to the Third Assessment Report of the Intergovernmental Panel on Climate Change. J. T. Houghton, Y. Ding, D. J. Griggs, M. Noguer, P. J. van der Linden, X. Dai, K. Maskell, and C. A. Johnson (eds.). Cambridge University Press, Cambridge, United Kingdom and New York, NY, USA.

IPCC (Intergovernmental Panel on Climate Change) (2007). Climate Change 2007: Synthesis Report. Contribution of Working Groups I, II and III to the Fourth Assessment Report of the Intergovernmental Panel on Climate Change. IPCC, Geneva, Switzerland.

Leary, A. (1999). A Framework for Cost-Benefit Analysis of Adaptation to Climate Change and Climate Variability. IPCC Working Group II Technical Support Unit, Washington D.C.

Lecocq, Franck and Shalizi, Zmarak (2007). Balancing Expenditures on Mitigation of and Adaptation to Climate Change – An Exploration of Issues Relevant to Developing Countries. Policy research working paper number 4299. The World Bank, Washington D.C.

Levina, E. and Tirpak, D. (2006). Adaptation to Climate Change: Key Terms. Organization for Economic Cooperation and Development, COM/ENV/EPOC/IEA/SL T, Paris.

Lunduka, R.W. (2013a). Multiple Stakeholders' Economic Analysis in Climate Change Adaptation – Case Study of Lake Chilwa Catchment in Malawi. International Institute for Environment and Development (IIED), London, UK.

Lunduka, R.W., Bezabih, M. and Chaudhury, A. (2013b). Stakeholder-focused Cost Benefit Analysis in the Water Sector: A Synthesis Report. International Institute for Environment and Development (IIED), London, UK.
Markandya, A. and Watkiss, P. (2009). Potential Costs and Benefits of Adaptation Options: A Review of Existing Literature. UNFCCC Technical Paper. FCCC/TP/2009/2 80.
MoE (Ministry of Environment) (2010). National Adaptation Programme of Action (NAPA) to Climate Change. Ministry of Environment, Government of Nepal, Kathmandu, Nepal.
Mohamed, B. (2013). Better economics: Supporting Climate Change Adaptation with Stakeholder Analysis: A Case Study of Morocco. International Institute for Environment and Development (IIED), London, UK.
Nangulu, A.N. (2001). Food Security and Coping Mechanisms in Kenya's Marginal Areas: The Case of West Pokot. A PhD dissertation, Eberly College of Arts and Sciences at West Virginia University.
Nicholls, R. (2007). Adaptation Options for Coastal Zones and Infrastructure. A report to the UNFCCC Financial and Technical Support Division (http://unfccc.int/cooperation_and_support/financial_mechanism/financial_mechanism_gef/items/4054.php).
Nkomo J.C., Nyong, A.O. and Kulindwa, K. (2006). The Impacts of Climate Change in Africa. Final Draft Submitted to The Stern Review on the Economics of Climate Change.
Parry, M., Arnell, N., Berry, P., Dodman, D., Fankhauser, S., Hope, C., Kovats, S., Nicholls, R., Satterthwaite, D., Tiffin, R. and Wheeler, T. (2009). Assessing the Costs of Adaptation to Climate Change. A Review of the UNFCCC and Other Recent Estimates. International Institute for Environment and Development (UK) and the Grantham Institute for Climate Change, Imperial College London.
Pradhan, N. and Providoli, I. (2010). Valuing Water and Its Ecological Services in Rural Landscapes: A Case Study from Nepal. Mountain Forum Bulletin, January 2010.
Rajamani, L. (2002). The Principle of Common but Differentiated Responsibility and the Balance of Commitments under the Climate Regime. *Review of European Community & International Environmental Law*, 9(2): 120–131.
Stage, J. (2010). Economic Valuation of Climate Change Adaptation in Developing Countries. Annals of the New York Academy of Sciences, 1185: 150–63, New York Academy of Sciences.
Thapa, Keshav and the LIBIRD Team (2011). Exploring Climate Adaptation Mechanisms for Watershed Management. Climate Adaptation Design and Piloting Project, Nepal. 2011.
van den Bergh, C.J.M. (2004). Optimal Climate Policy is a Utopia: From Quantitative to Qualitative Cost-Benefit Analysis, *Ecological Economics*, 48: 385–93.
United Nations Development Programme (UNDP) (2005). Human Development Report, New York, NY.
World Bank (2009b). Vulnerability to Climate Change in Agricultural Systems in Latin America: Building Response Strategies. Latin America and Caribbean Region, The World Bank, Washington D.C.
World Bank (2010a). The Cost to Developing Countries of Adapting to Climate Change: New Methods and Estimates. Washington, D.C: World Bank.
Yesuf, M. and Bluffstone, R.A. (2009). Poverty, Risk Aversion and Path Dependence in Low-Income Countries: Experimental Evidence from Ethiopia. *American Journal of Agricultural Economics*, 91: 1022–1037.

19

REGIONAL AND LOCAL CLIMATE CHANGE ADAPTATION POLICIES IN DEVELOPED COUNTRIES

Valentine van Gameren[1]

[1] INSTITUTE FOR ENVIRONMENTAL MANAGEMENT AND LAND-USE PLANNING, UNIVERSITÉ LIBRE DE BRUXELLES, BELGIUM

19.1 Introduction

Adaptation to climate change requires multi-level actions at different scales – temporal (from short to long-term) and spatial (international, national, regional and local) – to be implemented by both public and private actors (i.e. decision makers, businesses, NGOs and individuals). In other words, adaptation constitutes a multi-level governance issue (Adger et al., 2005).

This chapter deals with the specific role and leadership of subnational authorities and policies in climate change adaptation, such as regions, provinces, cities and municipalities. By looking at the reasons and means of action at these scales of governance, the related challenges and success factors, as well as some concrete examples of implemented initiatives, we aim to give an overview of this issue in the context of developed countries. While adaptation has been primarily considered a problem specific to developing countries, we know today that developed states are also vulnerable to climate change. Indeed, their often presumed high adaptive capacity has been questioned following recent events, like the 2003 European heat wave or the 2005 Hurricane Katrina. Moreover, vulnerability and adaptive capacity differ widely across regions and populations (Wolf, 2011), leading to specific adaptation needs and contexts. As a consequence, a growing interest in adaptation in industrialized countries has progressively emerged from research and policy, including the regional and local levels.

19.2 Why is public action towards adaptation to climate change necessary at the regional and local levels?

Before looking at the specific need for action at the regional and local policy levels, we first have to explain why public action in general is necessary to adapt to the impacts of climate change. In this respect, the roles and functions of public actors in adaptation are multiple and well described in the literature (Aaheim et al., 2008; Agrawala et al., 2008; Tompkins et al.,

2010; Biesbroek et al., 2010; Hallegatte et al., 2010; Tubiana et al., 2010). We can summarise them as follows:

i. Providing information, awareness and tools on climate change impacts and adaptation in order to stimulate adaptation actions by all concerned stakeholders;
ii. Creating incentives for adaptation by non-state actors through regulations or market instruments;
iii. Protecting the least able populations and individuals to cope by addressing the causes of vulnerability and/or by compensating for the unequal distribution of climate impacts;
iv. Protecting and provisioning public goods such as nature conservation, flood protection infrastructures and early warnings of extreme events;
v. Coordinating action and regulating adaptation spillovers in order to avoid negative externalities on vulnerability, i.e. maladaptation;[1]
vi. Mainstreaming adaptation into public policies on climate-vulnerable sectors.

In other words, adaptation policies are needed to anticipate climate change impacts and to compensate the limits of spontaneous (or autonomous) private adaptation, related to various factors such as: uncertainties and the lack of knowledge and of perception about the need to adapt, the absence of direct and/or short-term benefits of adaptation, the lack of adaptive capacity and the lack of consensus and coordination towards action (Stern, 2006; Biesbroek et al., 2010). Such anticipatory and proactive public initiatives constitute the so-called *'planned adaptation'* which results from a *"deliberate policy decision, based on an awareness that conditions have changed or are about to change and that action is required to return to, maintain, or achieve a desired state"* (IPCC, 2007), in contrast to *'autonomous'* or *'spontaneous adaptation'* which refers to *"adaptation that does not constitute a conscious response to climatic stimuli but is triggered by ecological changes in natural systems and by market or welfare changes in human systems"* (IPCC, 2007).

In this context, subnational governments have specific roles in the building and implementation of climate change adaptation policies for various reasons. First, in contrast with the governance of climate change mitigation that aims at global effects, most adaptation measures will need to address local needs and will provide local effects (Füssel, 2007; Isoard, 2011). Climate change impacts are deeply territorialised and will not threaten regions in a similar manner, hence adaptation has to be specifically fitted to local geographic and socioeconomic conditions and vulnerability: a coastal zone will not face exactly the same risks than an agricultural rural area, a mountainous region or a big city will, even if situated in the same country. Second, subnational authorities, especially the most local ones, are generally close to local stakeholders – e.g. businesses, community organizations, households – which can facilitate the understanding of the local contextual factors to take into account in adaptation decisions as well as the implementation of a participatory decision-making process (Corfee-Morlot et al., 2010). Third, as a large part of adaptation measures will occur at a decentralised level, the presence of a local adaptation framework is crucial for the efficiency of individual actions (Isoard, 2011). Finally, regional and local levels can innovate and experiment in elaborating policies and concrete measures, thus becoming in some cases 'laboratories for creative policies' (Puppim de Oliveira, 2009; Corfee-Morlot et al., 2010). In view of these features, subnational governments at the region, municipality and city levels are thus key-players to grasp regional and local specificities of climate change effects and to plan adaptation measures.

Beyond these general considerations, the role of subnational authorities, and particularly of regional governments, depends on the institutional structures of each country. In federal or decentralised states, such as the United States, Germany, Spain or Belgium, autonomous entities generally have a large part of the legislative power, including over climate-vulnerable sectors, and usually have a more significant role in developing regional adaptation. Westerhoff et al. (2011: 1079) claim that abilities and functions of the regional and local levels to plan adaptation depend on (i) the legal and political capacity to act on adaptation; (ii) the existence of financial and human resources for adaptation; and (iii) the ability of regional authorities to engage in internal and external networks for resources and information sharing (cf. Section 19.4.1: Context of Emergence of Subnational Adaptation Policies).

Box 19.1 A particular local adaptation challenge: Adaptation in cities

As centres of population and economic activity, cities may be particularly vulnerable to climate change and consequently constitute a key decision level for adaptation action. In Europe for example, three quarters of the population live in urban areas (EEA, 2012).

Because of their concentration of people and goods and the specific characteristics of cities, climate change impacts may be exacerbated in urban areas. Higher temperatures and heat waves for example have major expressions in cities because of the so-called 'urban heat island', i.e. a phenomenon of microclimate, notably created by the envelopes and arrangements of buildings, and leading to an increased temperature of the urban air compared to rural surroundings, that might reach up to 10 °C or more (EEA, 2012). Flood risk, already important in case of urbanisation in risk areas, may also be increased because of the high proportion of non-porous surfaces in cities (Carter, 2011). In addition, climate change may accentuate the problem of water scarcity with the occurrence of more frequent and severe droughts, as cities already compete with other users for water (EEA, 2012).

Moreover, cities are exposed to a number of indirect impacts of climate change, such as loss of jobs, income sources and quality of life in case of damages to infrastructures and interruption of services.

The city level is appropriate to adapt to climate change because this scale of governance is neither too small, as cities have policy means of action, nor too big, as they constitute integrated systems, including infrastructure networks and economic and social fabric (ONERC, 2010). In this respect, cities are generally competent in major policy areas for adaptation in the framework of urban planning and development, such as land use and transportation planning, building codes, water infrastructures, disaster prevention and response, public housing and social welfare (Corfee-Morlot et al., 2010).

19.2.1 Success factors of regional and local adaptation strategies and policies

A key element and general principle for a successful adaptation strategy is to tailor policies and measures to the specific regional or local climate conditions and political and socioeconomic contexts.

Besides this prerequisite, the scientific and grey literature identifies several factors that may play a role in facilitating the planning and implementation of adaptation policies, including at the regional and local levels (Ribeiro et al., 2009, Isoard, 2011, EEA, 2012). These different factors are explained as follows:

i. Raising awareness of the different non-state stakeholders (such as local communities, private sector organisations, labour unions and non-governmental organisations) and ensuring dialogue with them on various aspects of climate change adaptation is crucial in all vulnerable sectors, especially those with long lead times. Knowledge transfers are important for supporting the development of adaptation strategies, for communicating about these strategies and for facilitating the integration of adaptation at the regional and local levels;
ii. Integrating new available scientific information about current extreme weather events, projections of climate change and assessments of vulnerability, impacts and risks into decision-making is a key factor for the formulation of an adaptation strategy, as well as translating this knowledge into relevant information at the regional and local scales;
iii. Horizontal collaboration between different sectors or policy departments and vertical coordination between different administrative levels are necessary to a coherent adaptation policy framework (cf. Section 19.2.3. Multi-Level Governance and the Need for Coordination);
iv. Developing and securing a sufficient resource base in terms of financial, human and institutional resources is essential for all steps of the adaptation process;
v. Measures that promote goals other than adaptation (e.g. nature conservation, regional or urban development, climate change mitigation) and deliver additional benefits (i.e. 'win-win' measures) are generally better accepted and possibly more easily financed;
vi. Combining technological measures ('grey' or 'hard' measures) with soft (behavioural adaptation, participation of stakeholders, etc.) and green or 'ecosystem-based' measures (increase of ecosystem resilience) constitutes a solution to insure diversification and flexibility of adaptation responses;
vii. Monitoring and evaluating the adaptation strategy or policy against a set of targets and objectives to assess progress and insure the implementation of measures is a key step of the process. Similarly, it is crucial to progressively review these objectives and the policy instruments to implement according to new knowledge and experience.

19.2.2 Obstacles to regional and local adaptation policies

According to the literature, different categories of obstacles and barriers, often intertwined, can hinder regional and local adaptation action (Ribeiro et al., 2009; Corfee-Morlot et al., 2010; ONERC, 2010; Biesbroeck et al., 2011; Carter, 2011; EEA, 2012). First of all, sub-national authorities are in some cases faced with jurisdictional and institutional obstacles such as a lack of mandate to address climate issues, non-adapted institutional designs for vertical and/or horizontal coordination and the existence of national or regional laws, rules or regulations that lead to maladaptation. Regarding this last point, such policies and actions are those that increase vulnerability, by being short-term sighted and/or inflexible (Carter, 2011). Examples are policies that accept or encourage development in exposed locations or unregulated water consumption in drought-prone areas. Political factors constitute a second category of possible barriers, for instance: short-term electoral cycles that do not favour long-time risk management, pressuring to maintain 'business as usual' development pathways and a lack of leadership and willingness to accept costs and behavioural change. Third, the lack of resources and funding may also constrain adaptation as well as inter-sectoral competition over budgeting and difficulties to mainstream adaptation into budget lines (cf. also Section 19.3: The Cost and Funding of Adaptation at the Subnational Level). Finally, there are technical and cognitive obstacles. These may consist in a lack of understanding, an inadequate perception or even an ignorance of climate change risks. Scientific uncertainty, the deficiency of technical capacity or access to expertise, the lack of information about vulnerability at the regional or local scale and the weakness of information sharing about best practices of adaptation constitute other constraints of the same kind.

Of course, some of these barriers also exist at the national or supranational levels. However, it seems that these are particularly important at the subnational scales of governance because they are closer to adaptive action (Biesbroek et al., 2011). This is especially the case for the lack of financial resources or the lack of coordination between governments.

19.2.3 Multi-level governance and the need for coordination

Although the role of subnational authorities in adaptation policies is critical, we should not underestimate the importance of vertical integration and coordination of climate change adaptation between local, regional, national and supranational levels. Adaptation to climate change is naturally not a purely local governance issue but, as stated in previous sections, it is a multi-governance issue. Therefore, adaptation measures must be, on one hand, implemented at the different appropriate scales of the decision-making process and, on the other hand, specific regional or local measures must be consistent with the strategies at upper levels (Isoard, 2011) and not be constrained by national processes (Adger et al., 2005; Næss et al., 2005; Juhola and Westerhoff, 2011).

Actually, various sectors that may be affected by climate change are organised at different scales of governance, including subnational, national and supranational levels (such as the EU level), hence the importance of formal or/and informal coordination. Furthermore, some issues, such as water, coastal or mountainous management, even if relevant at the local scale, require cross-boundary actions and thus cooperation between concerned neighbouring countries. Without such coordination mechanisms, policies and all efforts may become ineffective or even counterproductive. Table 19.1 illustrates how multi-level governance of adaptation to climate change may operate in practice here in the urban context.

While this multi-level nature of adaptation governance is necessary and acknowledged as an essential principle of good practice, there are some gaps in implementing it. Structural and operational challenges are indeed numerous, notably in terms of decision-making and budgets structures as well as communication and transparency cultures (EEA, 2012). For example, while most (European) national adaptation strategies recognise the importance of participation and of taking measures at the most appropriate scale, involvement of regional and local representatives in the building of these strategies has not yet become widespread (Biesbroek et al., 2010). In addition, roles and responsibilities at regional and local levels are rarely clearly assigned.

Furthermore, actions at national and global scales of governance can really impact the success of adaptation at regional and local scales, depending on existing institutional frameworks and power structures (Næss et al., 2005; Tompkins and Adger, 2005; Urwin and Jordan, 2008; Corfee-Morlot et al., 2010). For example, the absence of a national adaptation strategy implies that authorities have to operate without central funding, data and expertise and thus, to mobilise other resources (Westerhoff et al., 2011) (cf. also Section 19.4.1. Context of the Emergence of Subnational Adaptation Policies).

Table 19.1 Multi-level governance of adaptation to climate change

Local action	*Regional action*	*National action*	*European action*
← *Implementing action*			
• Planning and implementation of local adaptation strategies • Mainstreaming of adaptation concerns into other policy areas • Spatial integration of adaptation needs through urban planning • Local emergency plans • Allocation of municipal resources and raising of other funds • Upgrading local infrastructure to make it resilient to climate change • Engaging civil society and private actors	• Providing incentives, funding and authorisation to enable local action • Addressing inter-municipal and urban-rural relations of climate change impacts and vulnerabilities • Developing and implementing with cities regional approaches, e.g. in river basins • Ensuring regional coherence of local/municipal plans and measures	• Providing a supportive national legal framework, e.g. appropriate building standards • Mainstreaming of urban adaptation into the different national policy areas and the national adaptation strategy • Funding of local adaptation measures • Providing national information related to climate change and regionally downscaled information • Funding of research and knowledge development for urban adaptation • Supporting boundary organisations who link science and policy to local adaptation needs • Adjusting the degree of decentralisation of competencies and authorities	• Providing a supportive European legal framework • Mainstreaming of urban adaptation needs into the different European policy areas, e.g. cohesion policy • Funding of local adaptation measures as well as knowledge development for urban adaptation • Providing European and global information related to climate change • Enabling and coordinating exchange of knowledge and experience across national borders • Addressing and coordinating cross-border adaptation issues
			Supporting action →

Source: EEA 2012, Table 4.1: 96

19.3 Cost and funding of adaptation at the subnational level

Some climate change adaptation measures can be implemented at relatively low cost. Take for example those linked to behavioural and institutional change[2] such as mainstreaming adaptation into all policy areas, or applying cheap 'safety margins'[3] in calibration of infrastructures in the design phase (e.g. drainage systems, dikes or sea walls; Hallegatte, 2009). Other measures require major investments, especially for building and adapting infrastructures. In addition, costs are not confined to material expenditures: ex-ante and ex-post costs of adaptation policies must be taken into account, such as costs of the analysis of vulnerability and the planning process, communication efforts and monitoring of the policies (ONERC, 2010).

Besides disparities in financial capacity across regional and local authorities, a major problem concerning adaptation is that the methods of economic assessment for the needed investments are still complicated and incomplete. Indeed, calculating the benefits of an adaptation that may occur several decades after action is problematic. Furthermore, it is not always possible to assign monetary value to all costs and benefits. As a consequence, this lack of information may constitute a barrier for implementing adaptation measures, even if it is generally admitted that anticipatory actions can prevent huge future costs (Stern, 2006).

However, some public funding is available at different governance levels and may partly alleviate this operational barrier. It includes local budgets and national or supranational funding. Furthermore, different economic instruments are potentially conceivable, such as climate-based taxes and charges, subsidies and budget allocations, and removal or modification of subsidies or taxes with harmful impacts and of instruments that create incentives to maladaptation (such as insurance systems that work against risk management) (Mickwitz et al., 2009). In the European Member States, adaptation activities in regions, cities and municipalities can be covered by already existing EU financial mechanisms (Ribeiro et al., 2009), for instance the Rural Development Fund in the agriculture and forestry sectors, LIFE+ for environmental and nature conservation projects, INTERREG for very various regional development projects (spatial planning, environment, innovation, etc.) or European Commission Research Funding. Furthermore, the European Commission has proposed, for the period 2014–2020 to allocate at least 20% of the overall Multiannual Financial Framework for climate change mitigation and adaptation (the overall total budget is €960 billion). In this respect, a better communication of these funding opportunities might be useful as subnational authorities are not always aware of them.

Furthermore, private funding constitutes another possible and complementary option. As infrastructures, flood defences and natural disasters management are mostly financed by public authorities, public-private partnerships may offer a solution to overcome operational and financial constraints (Agrawala et al., 2008). These partnerships can also be implemented for research and development (see also Box 19.2: Boundary Science-Policy Organisations). Nevertheless, such market mechanisms are for the time being less developed for climate change adaptation than for mitigation (EEA, 2012).

Finally, regional and local authorities can benefit from existing tools and methodologies developed by other actors, which will avoid some of the costs associated with the adaptation process (ONERC, 2010).

19.4 Regional and local adaptation strategies, plans and programmes: Characteristics and concrete examples

As adaptation to climate change has gained interest at the political level, several industrialised countries have developed national adaptation strategies and policies (Biesbroek et al., 2010; Isoard, 2011; Moser, 2011). Climate change adaptation strategies are defined as *"a general plan of action for addressing the impacts of climate change, including climate variability and extremes. It will include a mix of policies and measures with the overarching objective of reducing the country's* (or region or city/municipality's) *vulnerability*" (Niang-Diop et al., 2005:186 cited in Biesbroek et al., 2010). A similar trend is visible at the regional and local levels, including in cities. Indeed, specific adaptation strategies or plans have been lately elaborated and information on potential climate change impacts has been integrated into some planning or development programmes (Juhola and Westerhoff, 2011). Unlike national top-down initiatives, regional and local adaptation efforts are not always officially described as 'climate change strategies', which may complicate their identification (Ribeiro et al., 2009).

In Europe, the European Commission published a Green Paper in 2007 and a White Paper in 2009 about adaptation to climate change (EC, 2007; EC, 2009) as well as a European adaptation strategy in 2013 (EC, 2013). These documents acknowledge the need to implement such comprehensive strategies at both national and lower levels of governance. In 2009, 31 formal regional and local adaptation strategies were identified in the European Union (Ribeiro et al., 2009), i.e. in France, Germany, the Netherlands, United Kingdom, Sweden and Spain. These initiatives were developed by subnational governments with varying levels of autonomy (for example *länders* in Germany, *comunidades autónomas* in Spain, countries in the unitary state of the UK) and large cities and agglomerations. Forerunner regions in this field were Andalucia (Spain), North-Rhine Westphalia (Germany) and Rhône-Alpes (France) (Isoard, 2011). Since then, other strategies and plans have been prepared in regions (such as in Belgian Regions or Catalonia) and cities (e.g. Copenhagen and Helsinki) (European Climate Adaptation Platform website). Furthermore, other adaptation efforts have been elaborated independently of the formal frameworks of strategies and plans, such as health warning systems and heat action plans.

We can find a similar evolution in industrialised countries outside Europe. In the United States, state governments have started considering adaptation in their climate actions plans and/or developing comprehensive assessments and planning, although these efforts vary considerably across U.S. regions (Moser, 2011). Dispersed initiatives also exist in sector-specific policies, for example in coastal management. In Canada, in the absence of federal leadership and of a national strategy or plan, provinces and municipalities have developed their own plans, strategies and programmes, take for instance Yukon, Nunavut or Quebec (Dickinson et al., 2011). In parallel, municipal initiatives are emerging in partnership with the Federation of Canadian Municipalities and the network 'ICLEI-Local Governments for Sustainability Canada'.

19.4.1 Context of the emergence of subnational adaptation policies

As briefly illustrated by the cases in the previous section, regional and local adaptation strategies and policies have emerged in the presence or absence of official national adaptation strategies or legal frameworks. While some subnational actions were stimulated and/or harmonised by national adaptation initiatives, others were developed before or in parallel with the national process (Ribeiro et al., 2009). Moreover, as already stated, the power and capacity of subnational authorities in implementing adaptation measures depends on the institutional dynamics of each country.

In some cases, regional and local institutions can surpass national-level objectives and be forerunners. This is the case, for example, of the province of Ferrara in the Region of

Emilia-Romagna in Italy, where provinces are competent to design and implement several sectoral regional plans and policies and where no national adaptation strategy has been developed (Westerhoff et al., 2011). In the absence of formal institutions to address climate change adaptation, regional and local governments must be able to access funding, information and best practices, in particular through participation in networks (for example NGO networks, local and municipal governments' associations or EU-funded research programmes). Networks can also facilitate local initiatives in countries that have a national adaptation strategy or plan. For instance, in Finland, the national strategy does not assign responsibility to subnational authorities. Consequently, voluntary regional and local initiatives emerged (Juhola and Westerhoff, 2011) and inter-municipal networks allowed smaller municipalities to benefit from the needed resources (Westerhoff et al., 2011). In the United States, resources, leadership and staff for regional adaptation initiatives come from the state level since the federal government does not provide coordination in this field (Moser, 2011). Some U.S. local initiatives emerged independently, sometimes in relation to regional initiatives, such as New York City, which made use of a regional vulnerability assessment, whilst others were initiated or supported by networks, i.e. the 'Center for Clean Air Policy' and 'ICLEI-Local Governments for Sustainability' (Moser, 2011).

In other cases, national institutions facilitated regional and local approaches to adaptation. For instance, the UK has established a legal framework to allocate funding: the United Kingdom Climate Impact Programme (UKCIP) and the UK Government Department for Environment, Food and Rural Affairs (DEFRA) promote 'Regional Climate Change Partnerships', the multi-actor institutions responsible for regional action. In addition, UKCIP supports projects led by local authorities by providing guidance to them. In this respect, UKCIP plays a central role by providing knowledge, tools and best practices relevant to regional and local decision-making. Furthermore, local authorities are encouraged to consider adaptation through the local government performance framework that contains a national indicator on 'planning to adapt to climate change' (Boyd et al., 2011). As another example, the Australian National Climate Change Adaptation Program provides guidelines, planning tools, and information as well as a funding program ('Local Adaptation Pathways Program') to assist local governments in their adaptation initiatives (Smith et al., 2011).

Furthermore, apart from these linkages between different scales of governance, the elaboration of an adaptation strategy or policy emerged in some cases from wider regional or local (mitigation) climate change or sustainable development strategies, for example in the cities of Madrid, Hamburg and Manchester (Carter, 2011) or in regions and cities in France (Bertrand and Rocher, 2011).

Other motivating factors to develop and implement an adaptation strategy, including climate and non-climate triggers. Climate change (either real or perceived) and experience of extreme weather events constitute a frequent motivation (Ribeiro et al., 2009; Tompkins et al., 2010; ONERC, 2010; Moser, 2011). The proximity with a region that is implementing adaptation measures and the existence of a political leadership in environmental and sustainable development issues are other possible drivers for action (ONERC, 2010; Moser, 2011). In addition, non-climate-change-related legislation may also produce adaptation 'by-products', such as regulations about water efficiency (Tompkins et al., 2010).

The different motives for action are not always explicitly mentioned in strategies and policies (Biesbroek et al., 2010; Tompkins et al., 2010), a situation that might hinder comparison between case studies.

Box 19.2 Boundary science-policy organisations: A support to subnational adaptation policies

As explained, network membership can facilitate the emergence of adaptation strategies and policies at the regional and local levels by providing access to funding, information and best practices. In particular, science-policy interface is a critical issue in response to a deficiency of relevant scientific and technical information at the subnational level (cf. Section 19.2.2: Obstacles to Regional and Local Adaptation Policies). In this respect, the establishment of science-policy organisations, or so-called 'boundary' organisations, in collaboration with upper levels of governance, is often promoted as a possible useful support (EEA, 2012). Indeed such institutions are specifically charged with linking research and policy needs by translating scientific knowledge into information and tools adapted to local demands.

Several boundary organisations and networks have been developed or are in preparation in different countries, for example (Corfee-Morlot, 2010):

- United Kingdom Climate Impact Programme (UKCIP): established in 1997 by the UK government and based at the Environmental Change Institute at the University of Oxford, UKCIP coordinates and influences climate change impacts research and shares the outputs with stakeholders (public and private, at the different scales of governance) in a useful way, by spreading information and developing tools.
- Ouranos (Quebec, Canada): initiated in 2001 by the provincial government of Quebec, Hydro-Quebec (a publicly owned power generation and distribution company) and other regional partners (e.g. academic institutions), this network of multidisciplinary scientists and professionals aims to develop and use climate change knowledge to inform and advise local and regional decisions on adaptation.
- New York City Climate Change Program (U.S.): the result of a science-policy collaborative exchange that has co-produced scientific assessments, drawing on prior work conducted at the national level, notably on the knowledge and the network of experts that was created through a U.S. national assessment effort.
- Club ViTeCC (France): launched in 2008 by the Mission Climat of Caisse des Dépôts (CDC, a national institutional investment bank that also manages public infrastructure investment), in cooperation with two French organizations – Météo-France and the French National Observatory on Climate Change Impacts (ONERC) - ClubViTecCC collects, translates and shares results on both mitigation and adaptation with regions and cities and also companies of public services.
- Competence Centre on Climate Impacts and Adaptation – KomPass (Germany): compiles the latest research findings and uses them to develop target group-specific information products on adaptation activities and policies, through databases and best practice examples.
- Factory for Adaptation Measures Operated by Users at Different Scales – FAMOUS (Austria): project to facilitate the adaptation to climate change in Austrian provinces, regions and cities by better understanding the multi-level governance of adaptation in Austria, and by developing, applying, refining and disseminating tailor-made adaptation toolkits, developed based on experiences with similar guidance tools that can be found across Europe, and in close cooperation with local stakeholders.
- European Climate Adaptation Platform CLIMATE-ADAPT: created in 2012, hosted by the European Environment Agency and developed by the European scientific and policy making community, this online platform has been designed to support policy-makers at EU, national, regional and local levels in the development of climate change adaptation measures and policies. It shares information on current and future vulnerability and expected climate change in European regions and sectors, regional, national and transnational adaptation activities and strategies, adaptation case studies and potential adaptation options, tools that support adaptation planning and research projects.

19.4.2 *Characteristics of subnational strategies and plans*

Adaptation strategies and plans focus on different sectors, depending on regional and local vulnerability. Nonetheless, frequent key sectors are landscape, spatial planning and water management (e.g. flooding, urban water disposal, sea level rise, droughts) and health (e.g. heat stress), followed by biodiversity (Ribeiro et al., 2009; Isoard, 2011; Moser, 2011).

Concerning the development process, the preliminary studies and drafts of the strategies or plans are led internally within public administrations or in some cases subcontracted to consultants (for example in the UK, France or in Wallonia, Belgium), notably depending on whether internal resources and expertise are available. Consultative processes have been mentioned in most regional and local strategies, but there is not always detailed information about the organisation of stakeholder participation. Furthermore, involvement of the wider public is still limited (Ribeiro et al., 2009; Moser, 2011). The scientific background of strategies depends on climate change scenarios provided by the national meteorological services or boundary organisations, vulnerability assessments and other available expert knowledge or studies. As a consequence, climate scenarios and methods are not harmonised.

Stages of progress differ between regional and local initiatives. Most of them are at the stage of diagnosis through vulnerability assessments and of formulation of recommendations, but more concrete action plans are emerging. However, little information is available about implementation of specific policy instruments, assignments of specific responsibilities to different actors, costs and needed resources for adaptation measures, as well as monitoring processes.

Box 19.3 Spatial planning: A key instrument for adaptation to climate change at the subnational level

As mentioned, regional and local adaptation strategies often refer to spatial planning in order to anticipate and prevent climate change impacts. Spatial planning, defined as the "comprehensive, cross-sectoral, coordinating spatially oriented planning by the public sector" (Rannow et al., 2010: 160), is indeed one of the main public policy areas for regions, cities and municipalities in which to integrate adaptation measures into future land use and the different involved sectors.

First, land-use decisions may truly affect, positively or negatively, the vulnerability of urban dwellers and infrastructures to climate change (Corfee-Morlot et al., 2010). For instance, spatial planning plays a role in preventing flood risk by restricting building in flood plains, maintaining flood retention areas and minimising impermeable surfaces. Another possible measure to reduce vulnerability is building networks of green areas. For example, according to the planning law in Copenhagen, all flat roofs must be greened in new buildings. The city of Malmö in Sweden uses a green scoring factor for new urban developments that ensures a certain green space proportion thanks to different solutions, scored according to their efficiency (EEA, 2012).

Second, regional and local authorities usually have spatial or urban planning departments, avoiding the creation of new institutions (Sanchez-Rodriguez, 2009).

In addition, spatial planning is also a key policy area to address mitigation to climate change and consequently has a significant potential to combine both approaches and to search for synergies (Biesbroek et al., 2009).

Note that incorporating climate assessment into planning may generate some direct and indirect costs, such as impacts on land and property prices and increased construction development and insurance costs, even if early action should be more cost-effective than delaying it (Wilson, 2006).

19.5 Conclusions

In this chapter we identified the reasons behind why the subnational level is crucial for elaborating and implementing adaptation policies and presented some examples of domains of action. However, adaptation initiatives are quite recent and the realisation of concrete measures remains at an early stage. In order for regional and local authorities to take their role and responsibility in hand, a number of challenges are still ahead.

We mentioned the different kinds of obstacles that currently slow down the process: jurisdictional-institutional, political, technical-cognitive and financial barriers. However, solutions and success factors exist, though they require leadership and political commitment in fostering adaptation in the policy agenda and investment in building adaptive capacities. The framework in favour of adaptation has to be coherent across the different scales of governance through coordination mechanisms in order to promote synergies in a common effort and avoid conflicts and counterproductive initiatives. This vertical integration is necessary at the different stages of the adaptation process, e.g. the preparation and implementation phases of policy process.

Regional and local authorities have the responsibility to raise awareness in all concerned stakeholders (including the private sector and citizens) of climate change in general and adaptation more specifically, in addition to mitigation. Moreover, funding research on vulnerability and adaptation is necessary, but it is also critical to promote knowledge transfer and translation into relevant and useful information. In this respect, relevant authorities at both the subnational level and upper scales of governance have to engage in bridging mechanisms between science and policy-making, in particular with the support of boundary organisations. The establishment and functioning of such institutions constitute an example where multi-level coordination can operate. Possible outputs of the translation of scientific data are the development of guidance and tools to facilitate vulnerability assessment and choice of adaptation measures. On this matter, the exchange of lessons learned from successful adaptation measures is interesting and needed for guidance. However these lessons must be assessed according to the different contexts of local and regional situations, with regard to demography, climate and environmental impacts, economic characteristics, cultures and values (Isoard, 2011). Developing indicators of vulnerability and of adaptation success is also a key aspect to work on. Indeed, the lack of such indicators, especially those of success, hampers an efficient monitoring process of adaptation policies and measures, which is essential for a flexible adaptive approach.

Finally, funding of adaptation measures still raises lots of unanswered questions, notably who (which governance level) should pay for what (research, infrastructures, etc.). To overcome the obstacle of uncertainty and prevent competition for financial resources, the responsible authorities can choose to implement 'no-regret' or 'low-regret' options that bring benefits in all climate change scenarios, or win-win solutions, that provide benefits to sectors and domains other than climate change adaptation (such as improvement in public-health systems or increase of green spaces). Others options are reversible measures that allow a strategic change in the short-term (for instance, restrictive land use planning), and measures with a safety margin, that reduce vulnerability at low cost, for example in designing infrastructures. Furthermore, it is also crucial that regional and local authorities better mobilise the available opportunities of funding in public budgets or possibly develop new opportunities with private actors.

Notes

1 Maladaptation is increasing risks and vulnerability from adaptation, or more precisely "action taken ostensibly to avoid or reduce vulnerability to climate change that impacts adversely on, or increases the vulnerability of other systems, sectors or social groups" (Barnett et al., 2010: 211).
2 These measures are qualified as 'soft' measures in comparison with 'hard' measures linked to technological and infrastructural options.
3 The design of projects with an additional capacity to cope with climate variability and climate extremes events via a safety margin, for example, by using over-pessimistic climatic scenarios.

References

Aaheim, A., Berkhout, F., McEvoy, D., Mechler, R., Neufeldt, H., Patt, A., Watkiss, R., Wreford, A., Kundzewicz, Z., Lavalla, C., Egenhofer, C. (2008), 'Adaptation to climate change: Why is it needed and how can it be implemented?' CEPS Policy brief, n°161.
Adger, N., Arnell, N., Tompkins, E.L. (2005), 'Successful adaptation to climate change across scales', *Global Environmental Change,* 15, pp. 77–86.
Agrawala, S., Fankhauser, S. (Eds.) (2008), 'Economic aspects of adaptation to climate change: Costs, benefits and policy instruments', Paris: Organisation for Economic Co-operation and Development – OECD.
Barnett, J., O'Neill, S. (2010), 'Maladaptation', *Global Environmental Change,* 20-2, pp. 211–213.
Bertrand, F., Rocher, L. (2011), 'La prise en compte des risques associés au changement climatiques dans les politiques locales' in Labranche, S. (ed.) *Le changement climatique. Du méta-risque à la métagouvernance,* Paris: Lavoisier.
Biesbroek, R., Klostermann, J., Termeer, C., Kabat, P. (2011), 'Barriers to climate change adaptation in the Netherlands', *Climate Law,* 2, pp. 181–199.
Biesbroek, G.R., Swart, R.J., van der Knaap, W.G.M. (2009), 'The mitigation-adaptation dichotomy and the role of spatial planning', *Habitat International,* 33, pp. 230–237.
Biesbroek, G.R., Swart, R.J., Carter, T.R., Cowan, C., Henrichs, T., Mela, H., Morecroft, M.D., Rey, D. (2010), 'Europe adapts to climate change: Comparing national adaptation strategies', *Global Environmental Change,* 20, pp. 440–450.
Boyd, E., Street, R., Gawith, M., Lonsdale, K., Newton, L., Johnstone, K., Metcalf, G. (2011), 'Leading the UK adaptation agenda: A landscape of stakeholders and networked organizations for adaptation to climate change' in Ford, J.D. and Berrang-Ford, L. (eds.) *Climate Change Adaptation in Developed Nations, From Theory to Practice,* Dordrecht: Springer, pp. 85–102.
Carter, J.G. (2011), 'Climate change adaptation in European cities', *Current Opinion in Environmental Sustainability,* 3, pp. 1–6.
Corfee-Morlot, J., Cochran, I., Hallegatte, S., Teasdale, P-J. (2010), 'Multilevel risk governance and urban adaptation policy', *Climatic Change,* 104, pp. 169–197.
Dickinson, T., Burton, I. (2011), 'Adaptation to climate change in Canada : A multi-level mosaic' in Ford, J.D. and Berrang-Ford, L. (eds.) *Climate Change Adaptation in Developed Nations, From Theory to Practice,* Dordrecht: Springer, pp. 103–117.
EC, European Commission (2007), Green Paper. 'Adapting to climate change in Europe – options for EU action', COM(2007) 354 final, Brussels.
EC, European Commission (2009), White Paper. 'Adapting to climate change: Towards a European framework for action', COM(2009) 147 final, Brussels.
EC, European Commission (2013), 'An EU Strategy on adaptation to climate change', COM(2013) 216 final, Brussels.
EEA, European Environment Agency (2012), 'Urban adaptation to climate change in Europe. Challenges and opportunities for cities together with supportive national and European policies', Copenhagen.

Füssel, H.M. (2007), 'Adaptation planning for climate change: Concepts, assessment approaches, and key lessons', *Sustainability Science,* 2, pp. 265–275.

Hallegatte, S. (2009), 'Strategies to adapt to an uncertain climate change', *Global Environmental Change,* 19, pp. 240–247.

Hallegatte, S., Lecocq, F., de Perthuis, C. (2010), 'Economie de l'adaptation au changement climatique', Conseil économique pour le développement durable, Paris.

IPCC (2007), *Climate Change 2007 Impacts, Adaptation and Vulnerability, Appendix I: Glossary,* Working Group II Contribution to the Fourth Assessment, Report of the Intergovernmental Panel on Climate Change, pp. 869–883.

Isoard, S. (2011), 'Perspectives on adaptation to climate change in Europe' in Ford, J.D. and Berrang-Ford, L. (eds.) *Climate Change Adaptation in Developed Nations, From Theory to Practice,* Dordrecht: Springer, pp. 51–68.

Juhola, S., Westerhoff, L. (2011), 'Challenges of adaptation to climate change across multiple scales: A case study of network governance in two European countries', *Environmental Science and Policy,* 14 (3), pp. 239–247.

Mickwitz, P. (2009), *Climate Policy Integration, Coherence and Governance,* Partnership for European Environmental Research, Helsinki.

Moser, S.C. (2011), 'Entering the period of consequences : The explosive us awakening to the need for adaptation' in Ford, J. D. and Berrang-Ford, L. (eds.) *Climate Change Adaptation in Developed Nations, From Theory to Practice,* Dordrecht: Springer, pp. 33–49.

Næss, L.O., Bang, G., Eriksen, S., Vevatne, J. (2005), 'Institutional adaptation to climate change: Flood responses at the municipal level in Norway', *Global Environmental Change* 15 (2), pp. 125–138.

Niang-Diop, I., Bosch, H. (2005), 'Formulating an adaptation strategy' in Lim, B., Spanger-Siegfried, E., Huq, S., Malone, E.L., and Burton, I., *Adaptation Policy Frameworks for Climate Change: Developing Strategies, Policies and Measures,* Cambridge: Cambridge University Press, pp.185–204.

Observatoire National sur les Effets du Réchauffement Climatique – ONERC (2010), *Villes et adaptation au changement climatique,* Paris: La Documentation française.

Puppim de Oliveira, J.A. (2009), 'The implementation of climate change related policies at the subnational level: An analysis of three countries', *Habitat International,* 33, pp.253–259.

Rannow, S., Loibl, W., Greiving, S., Gruehn, D., Meyer, B.C. (2010), 'Potential impacts of climate change in Germany – Identifying regional priorities for adaptation activities in spatial planning', *Landscape and Urban Planning,* 98, pp. 160–171.

Ribeiro, M., Losenno, C., Dworak, T., Massey, E., Swart, R., Benzie, M., Laaser, C. (2009), Final report. 'Design of guidelines for the elaboration of Regional Climate Change Adaptations Strategies', Study for European Commission – DG Environment, Ecologic Institute, Vienna.

Sanchez-Rodriguez, R. (2009), 'Learning to adapt climate change in urban areas. A review of recent contribution', *Current Opinion in Environmental Sustainability,* 1, pp. 201–206.

Smith, T.F., Thomsen, D.C., Keys, N. (2011), 'The Australian experience' in Ford, J.D. and Berrang-Ford, L. (eds.) *Climate Change Adaptation in Developed Nations, From Theory to Practice,* Dordrecht: Springer.

Stern, N. (2006), *Stern Review on the Economics of Climate Change. Part V : Policy Responses for Adaptation,* London: HM Treasury.

Tompkins, E.L., Adger, N.W. (2005), 'Defining response capacity to enhance climate change policy', *Environmental Science and Policy,* 8-6, pp. 562–571.

Tompkins, E.L., Adger, W.N., Boyd, E., Nicholson-Cole, S., Weatherhead, K., Arnell, N. (2010), 'Observed adaptation to climate change: UK evidence of transition to a well-adapting society', *Global Environmental Change,* 20-4, pp. 627–635.

Tubiana L., Gemenne, F., Magnan, A., (2010), *Anticiper pour s'adapter: le nouvel enjeu du changement climatique,* Paris: Pearson.

Urwin, K., Jordan, A. (2008), 'Does public policy support or undermine climate change adaptation? Exploring policy interplay across different scales of governance', *Global Environmental Change,* 18, pp. 180–191.

Westerhoff, L., Keskitalo, E.C.H., Juhola, S. (2011), 'Capacities across scales: Local to national adaptation policy in four European countries', *Climate Policy,*.11-4, pp.1071–1085.

Wilson, E. (2006), 'Adapting to climate change at the local level: The spatial planning response', *Local Environment,* 11-6, pp. 609–625.

Wolf, J. (2011), 'Climate change adaptation as a social process', in Ford, J.D. and Berrang-Ford, L. (eds.) *Climate Change Adaptation in Developed Nations, From Theory to Practice,* Dordrecht: Springer, pp. 21–32.

Websites

Club ViTeCC (France): http://www.cdcclimat.com/Le-Club-ViTeCC-Villes-Territoires.html?lang=fr, accessed 14 September 2012.
European Climate Adaptation Platform (CLIMATE-ADAPT): http://climate-adapt.eea.europa.eu/web/guest/home, accessed 14 September 2012.
KomPass (Germany), national website platform: www.anpassung.net, accessed 14 September 2012.
New York City Climate Change Program: http://www.nyc.gov/html/dep/html/news/climate_change_report_05-08.shtml, accessed 14 September 2012.
Ouranos (Canada): http://www.ouranos.ca/, accessed 14 September 2012.
UKCIP: http://www.ukcip.org.uk/, accessed 14 September 2012.

20

THE ROLE OF TECHNOLOGY IN ADAPTATION

John M. Callaway[1]

[1] UNEP-RISØ CENTRE, ROSKILDE, DENMARK

20.1 Introduction

20.1.1 Background

The term "technology" has many definitions, ranging widely from "knowledge" in general to "applied knowledge" to narrow economic definitions that focus on the processes that transform inputs into outputs. Other definitions also focus on the characteristics of inputs that lead to greater technical efficiency, or exclude processes entirely.

When it comes to climate change, there has traditionally been a strong emphasis on technologies that reduce emissions through various chemical, mechanical and biophysical processes. This emphasis has made it possible to develop emissions-reductions supply curves based, in large part, on the technological effectiveness of different mitigation options and the capital maintenance and operating costs of the individual technologies (IPCC, 2001 and Metz et al., 2007). In some sectors, such as sea-level rise (Linham and Nichols, 2012 and Klein et al., 2001) and to a lesser extent in agriculture (Clements et al., 2011) and water resources (Elliott et al., 2011), there continues to be a strong emphasis on the technological aspects of adaptation to climate change. However, it is generally recognized that adaptation to climate change also has important behavioural, organizational and institutional aspects that are as, or more, important factors than is technological efficiency or effectiveness in reducing the damages due to climate change. The different technology focus is due in large part to two factors: first, the fact that adaptation actions are implemented locally and do not have uniform effects at the global level as do mitigation options and second, the fact that mitigation actions generally involve a lot more top-down governmental planning and action, while adaptation actions are not only highly decentralized and geographically dispersed, but also are expected to be dominated by market-driven incentives, with governments playing more of a facilitative role.

20.1.2 Objectives

The main question faced in writing this chapter was to find an economic framework for characterizing the technological content of adaptation to climate change, while recognizing that the traditional "hard technology" approach used to characterize mitigation of climate change was too narrow. In the end, it was decided to approach this problem from the standpoint of the "production function" in conjunction with the producers' objective of cost minimization to explore the ways in which these two concepts relate to the reduction of climate change damages in the short and long run.

20.1.3 Organization

The remainder of this chapter is divided into four sections. The following section defines the term "technology" with reference to the concept of the production function that is used in economics to characterize the relationship between the inputs and outputs of producers in firms and households. However, it also shows that the production function must be combined with a behavioural objective, such as cost minimization, in order to come up with a more accurate characterization of technology – one which is both physical and allocative in nature. This is presented in terms of the long- and short-run total and marginal cost (or supply) functions. The next section then takes these concepts and shows how they are relevant for explaining how producers in firms and households adapt differently to climate change in the short- and the long run; how technological change contributes to adaptation; and how economists measure (monetize) the benefits and costs of adaptation in terms of changes in welfare. Finally, the last section (before the conclusion) explores the problem of climate risk and how robust technological changes by producers can minimize the risk of economic losses (regrets) in the planning and implementation of long-run adaptation investments when future climate outcomes at the local, sectoral and household levels are as uncertain as they are today.

20.2 Defining "technology" in economic and climate-friendly terms

The challenge, here, is to define "technology" in terms that are useful to characterize the role that it plays in adapting to climate change. We start with the production function, $Q = f(x, k, c, \alpha)$.[1] Q is the output of the firm or of household production; f(.) is the mathematical or process-oriented form of the production function that transforms inputs into outputs. The inputs to production include those that can be adjusted by firms and households. These are divided into those that can be varied, generally, x, which are known as variable inputs and those that are, in some cases fixed, and in others variable. These inputs, k, are called quasi-fixed or fixed factors. The input, c, represents climate. It is exogenous, itself beyond the direct control of the firm or household. These agents can adjust (adapt) to changes in climate, but cannot change it. Finally, the input, α, stands for the parameters of the production function that characterize how Q increases as x, k, and c increase and how x and k can be substituted for one another by producers and households, in response to changes in c (and other "exogenous" influences beyond their direct control).

Given this fairly general representation of a production function, we can define the term technology, at least partly, in terms of the form of the production function and its parameters. The fact that the characteristics of x and k can change over time due to technological change cannot be ignored, but in the example shown later, technological change will be captured by changes in the parameters, α. But this is only part of the big picture, because it doesn't characterize fully or neatly the economically efficient input possibilities that are available to firms and households, given the input possibilities. More narrowly, and more importantly for this chapter, the process aspects of this definition do not relate at all to how producers and households can adjust their inputs to changes in climate, which is what adaptation is mostly about.

To do this, we first have to focus on the objectives of producers and households and how they utilize their inputs to production, given their production functions and the external constraints they face in meeting these objectives. This will be explained within an economic framework in the text in a non-mathematical graphical manner.

Producers and households can have many different objectives in production. Farmers in a developed country, who sell their products into fairly competitive markets, will likely minimize their costs to identify their input possibilities and maximize their profits to determine how they respond to market output price signals. Subsistence farmers in developing countries may have very few monetary costs to minimize and their objectives may be more related to nutritional objectives subject not only to their production function, but also resource scarcity constraints, including land, labour and livestock. Cost minimization is just one (not the only) normative framework that can be used to show how producers and households define their input possibilities. Other objectives will yield different outcomes. However, for this chapter, we will assume that both producers and households minimize the costs of their inputs subject to the constraints implied by their production functions in order to define the economically optimal combination of inputs to use to produce a give given level of output in a market (producers) or given level of household services (households).[2]

When producers minimize their input costs subject to their production functions, the results are total cost functions.[3] These cost function s vary according to whether x and k are variable (long-run) or only x is variable and k is fixed (short-run). Distinctions between these functions play an important role in how producers respond to climate change and how much of the damages caused by climate change can be avoided. Just how this happens will be illustrated using a "Cobb-Douglas" production function. This function is convenient to use for illustrative purposes because closed forms for the long- and short-run supply functions are easy to retrieve and graph from the cost minimization problem. Figure 20.1 illustrates the information contained in these cost curves and how different production function parameters (α) influence the long- and short-run total cost and supply curves in an increasing-cost industry. In the upper panel, there are two pairs of LRTC and SRTC curves for each of the two values of α, a1 and a2. For a1 values, the LRTC curve is a black solid line and the SRTC curve is a black dashed line. For a2 values, the LRTC curve is a grey sold line and the SRTC curve is a grey dashed line. For each pair, the LRTC curve shows the minimum total cost at which different values of Q can be produced for a given technology (and input prices and climate) when both x and k are variable. The SRTC curve shows, for each pair of a (α) values, the short-run minimum total cost, when k is fixed and x variable and also dependent on the fixed value of k (and input prices and climate). The fixed value of k is held constant at the same level in both pairs of SRTC curves. By changing just the fixed value of k in both of the SRTC curves, one could draw an infinite number of SRTC functions, each with a unique fixed value of k, that were perfectly tangent to its companion LRTC curve. Indeed, the LRTC is the envelope of the SRTC curves (Samuelson, 1972 and Silberberg 1999). At this single point of tangency LRTC = SRTC for a unique value of k, holding all other factors constant.

At the outset, it is important to note that the LRTC and SRTC curves do not provide guidance to producers about the optimal levels at which to produce Q. Instead, they only show the minimum total cost possibilities for each level of output in the long run and short run. As such, they represent a mixture of the production function, the prices of the inputs and form

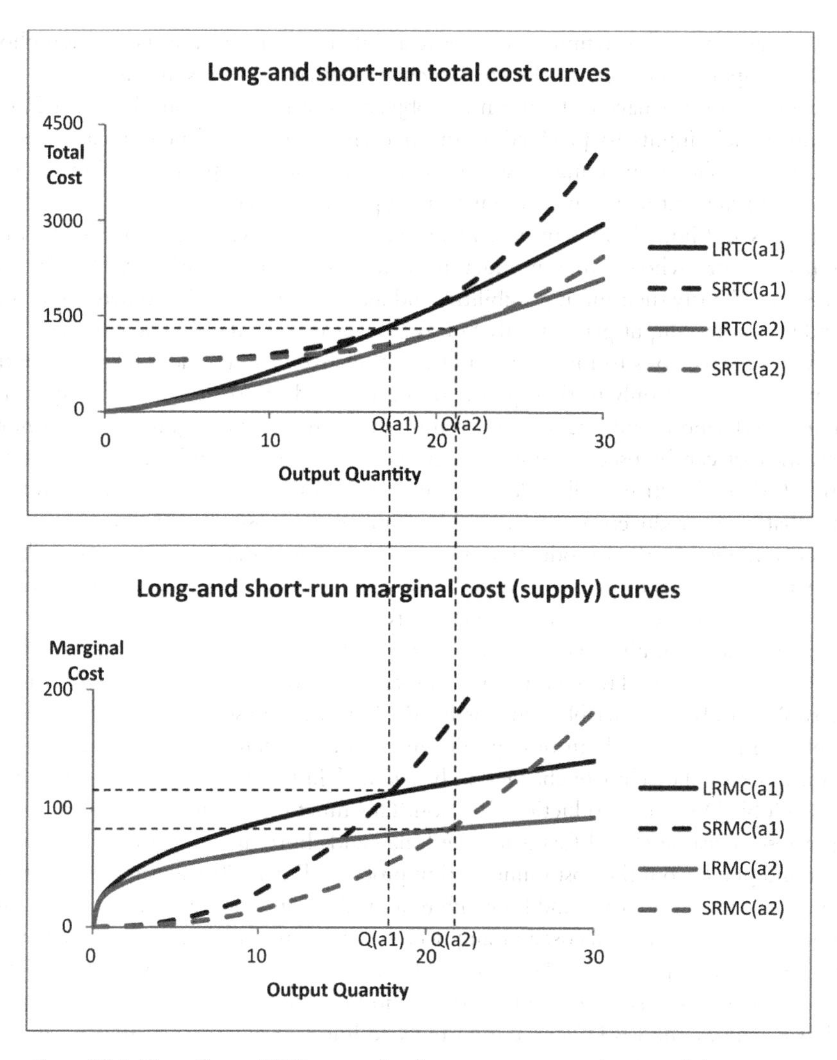

Figure 20.1 The effects of different technology parameters on long- and short-run cost minimization

and parameters of the production function given the objective function of cost-minimization. At any single point along both curves, the optimal input use quantities are an imbedded part of the information they contain. The two pairs of cost curves differ only in the parameters of the production function, a1 or a2, and this results in two unique pairs of cost functions. For a given value of Q except for zero, the LRTC(a1) curve lies below the LRTC(a2) curve. This is also true of the SRTC(a1) curve, although this is hard to see when Q is less than 10. However this is still the case, since a2 represents the more efficient of the two technologies.

The bottom panel of Figure 20.1 shows the short- and long-run supply curves associated with short- and long-run total cost functions in the upper panel. Each point on the two LRMC curves represents the minimum long-run marginal cost at which a given level of Q can be produced, for a different technology parameter when both x and k are variable. Each point on the two short-run supply curves represents the minimum marginal cost at which a given level of Q produced, when k is fixed. For each pair, the two curves intersect at the point of tangency in relation to Q in the upper panel. At this point, LRMC = SRMC. The greater efficiency of the a2 cost curves is exhibited by the fact that these curves both lie below their less efficient (a1) cost-curve counterparts. More generally, the "steepness" of the SRMC curve in relationship to the LRMC curve depends on how much higher the cost of x is than k and how easy it is to for x to replace the lost effect of k in increasing output when k is fixed. The "closer" in space the curvature of the SRMC curve to the LRMC curve, the less dependent the producer will be on their fixed factors in the short run, and this reduces the cost of adapting to external shocks.

However, there is one inconsistency about the relationship between the SRMC and the LRMC curves that is important for measuring total costs, using just these functions and curves. While the area under the LRMC curves between any two values of Q represents the total long-run cost in the upper panel, this same area under a short-run supply (SRMC) curve only yields the short-run variable cost, since the fixed cost part of the SRTC function is not a function of Q. Thus, when transforming short-run marginal costs into total costs via integration, one must also take into account any change in fixed costs, over the range of Q for which costs are being evaluated (Just et al., 1982).

The treatment of total and marginal cost curves shown in Figure 20.1 sheds important light on the economic meaning of technology, without reference to climate change. Given the production function and the various total cost and supply curves, an operating definition of technology that can be adopted for this chapter, at this point, is: *Technology is characterized by the structure of the processes through which inputs are transformed into outputs through production functions, consistent with the objectives of producers, as revealed in cost curves.*

In the remaining sections of this chapter, we will show how climate change affects producers' cost functions and how producers respond (adapt) to these changes both by adjusting their inputs and the structure of the production, also taking into account the effect of what is called "climate risk" in their technology choice decisions. As in the first figure, the illustrations of supply curves are generated using a Cobb-Douglas production function, in concert with the assumptions of cost short- and long-run cost minimization to capture the technology and behaviour of producers of market and household goods as they adapt to climate change.

20.3 The role of technology choice: How producers adapt to climate change

The more or less accepted definition of adaptation was originally put forwarded by Fankhauser (1997). It involved the transition between a climate state in which economic agents adapted to the existing climate, using existing measures to adapt to climate variability and a changed climate state in which economic agents adapted to climate change using measures specifically designed for this purpose. Callaway (2004a and 2004b) modified this definition to suggest that there are two stages of adaptation: short-run and long-run. In doing so, he defined climate change damages as the transition from Fankhauser's initial climate state to one in which economic agents made short-run adjustments, while arguing that a "no adjustment" reference case for measuring adaptation benefits will never be observed, so that estimation of economic welfare changes from this counterfactual reference case would be hard to estimate. He went on to define the measurable net adaptation benefits in terms of long-run adaptation, but failed to bring into play the appropriate long-run cost curves for measuring the costs of producers. This section of the chapter corrects these mistakes.

Consequently, it is possible to think of adaptation to climate change based on the following typology of impacts and adjustments:

- The pure effect of climate change,
- Partial, or short-run, adjustment and
- Complete, or long-run, adjustment.

The pure effect of climate change represents the impact of climate change on the long- and short-run cost, total cost and supply curves of producers, without any adjustment to the climate impacts, except for what is directly caused by the physical effects of climate change. What happens is that adverse forms of climate change will cause the LRTC, LRMC, SRTC and SRMC curves to shift as shown in Figure 20.2 (below). However, producers don't react to these impacts by adjusting their input allocation. As stated above, this effect is ignored by Fankhauser (1997) and Callaway (2004a and 2004b) because it is hard to observe situations in real life when producers do not have economic incentives to adjust to external shocks, except in the case of the very short run in conjunction with some extreme weather events in some sectors. A "no adjustment" scenario makes even less empirical sense in connection with climate change. This is because climate change occurs slowly and is defined in terms of parameter changes in the partial and joint distributions of meteorological variables, and not in terms of weather events.

Nevertheless, even if the pure effect of climate change is an unobservable counterfactual, it still represents a valid concept, not only for valuing the maximum economic damages caused by climate change, but also as a baseline for measuring the benefits and costs of adaptation. This concept can also be of interest to economists because this is how many natural scientists estimate physical climate change impacts – without any adjustments by economic agents.

If producers and households aren't able to untangle the climate signal from the variability in geophysical records of the existing climate, or don't have reliable regional climate model predictions, it is possible that producers will respond to climate change as if it were existing climate variability (Callaway, 2004a and 2004b). That is: they will make short-run adjustments due to the economic risk of making investments in fixed factors whose optimal rate of return will depend on reliable climate information, such that the climate they plan for, ex-ante, has a reliable chance of actually occurring, ex-post. In Fankhauser's original terms, this form of short-run adaptation would involve adapting to climate change using measures to adapt to normal climate variability.

Long-run adaptation consists of making investments in a new production process, infrastructure and other forms of quasi-fixed and fixed inputs: producers move off their short-run supply curves onto their long-run supply curves. From a descriptive standpoint, long-run adaptation can take place for three possible reasons:

1. Producers can detect climate change, or else they have reliable regional climate model predictions, and if such information is not available;

2. These agents focus on making investments that will improve economic efficiency for reasons other than reducing climate change damages, but which will also reduce the damages from climate change, simply by virtue of greater technical efficiency;
3. These agents cope with uncertainty by investing in robust production processes and capital that minimize economic regrets (Matalas and Fiering, 1977) associated with climate risk.

The analysis that follows takes the form of an exercise in which an economist or project planner would attempt, on an ex-ante basis, to value the ex-post economic impacts of climate change and the benefits and costs of adapting to climate change that will occur, ex-post. In doing so, it also contains descriptive elements regarding how economic agents will behave under climate uncertainty and blends that into the ex-ante analysis.

20.3.1 Impact on cost curves: The pure effect of climate change

The pure effect of climate change on the long- and short-run marginal cost curves is illustrated in Figure 20.2. The top panel depicts the base case LRMC and SRMC curves for the two different levels of technical efficiency under the existing climate (as in Figure 20.1), while the bottom panel depicts the effects of climate change on these curves (we have chosen not to show the effects of climate change on the total cost curves, because it is mirrored by the effects on the marginal cost curves). The changes seen in the bottom panel of this figure are purely the result of changing the climate variable, c, in the production function $Q = f(x, k, c, \alpha)$ so that an increase (or a decrease) in c will cause Q to decrease. Input prices of x and k are held constant. The effect of this climate change is to shift all of the long and short-run cost curves up and to the left, such that for any given level of Q on the horizontal axis, both the LRMC and SRMC increase on the vertical axis. The result of these cost curve impacts is to reduce the cost-minimizing levels of Q and increase the long- and short-run marginal costs of producing $Q_1(a1)$ and $Q_1(a2)$ relative to the base case quantities $Q_0(a1)$ and $Q_0(a2)$.[4]

However, the effects of climate change on LRTC and SRTC are somewhat ambiguous, depending on the functional form of the production function and the LRTC and SRTC cost functions derived from it.[5] In the production function used to generate the figures in this chapter, this is not the case, and the result is that the values of x and k are the same at the points e1 (upper panel-before the climate change) and e2 (upper panel-after the climate change). Since these equilibrium points occur where each SRMC curve intersects its companion LRMC curve, there is no change in the LRTC or SRTC at these points in relation to the base case. However, this would not be true if climate change directly damaged fixed inputs. This result is a direct feature of the technology characterized in the production function and may apply in certain cases, such as in empirical crop response functions, where precipitation and temperature are treated as independent variables. However, this same characterization of technology would not be true of firms or households whose livelihoods were exposed to changes in the frequency of extreme events, like typhoons, which destroy infrastructure. This illustrates how the technological characterization of the production function is so important in determining, not only how well existing technologies will respond to climate change, but also how accurately these technologies are modelled for the purposes of estimating the climate change damages and the costs and benefits of adapting to climate change.

In this case, households and producers are assumed to be in long- and short-run equilibrium under the initial climate, at either e_1 or e_2 in Figure 20.2, depending on the efficiency

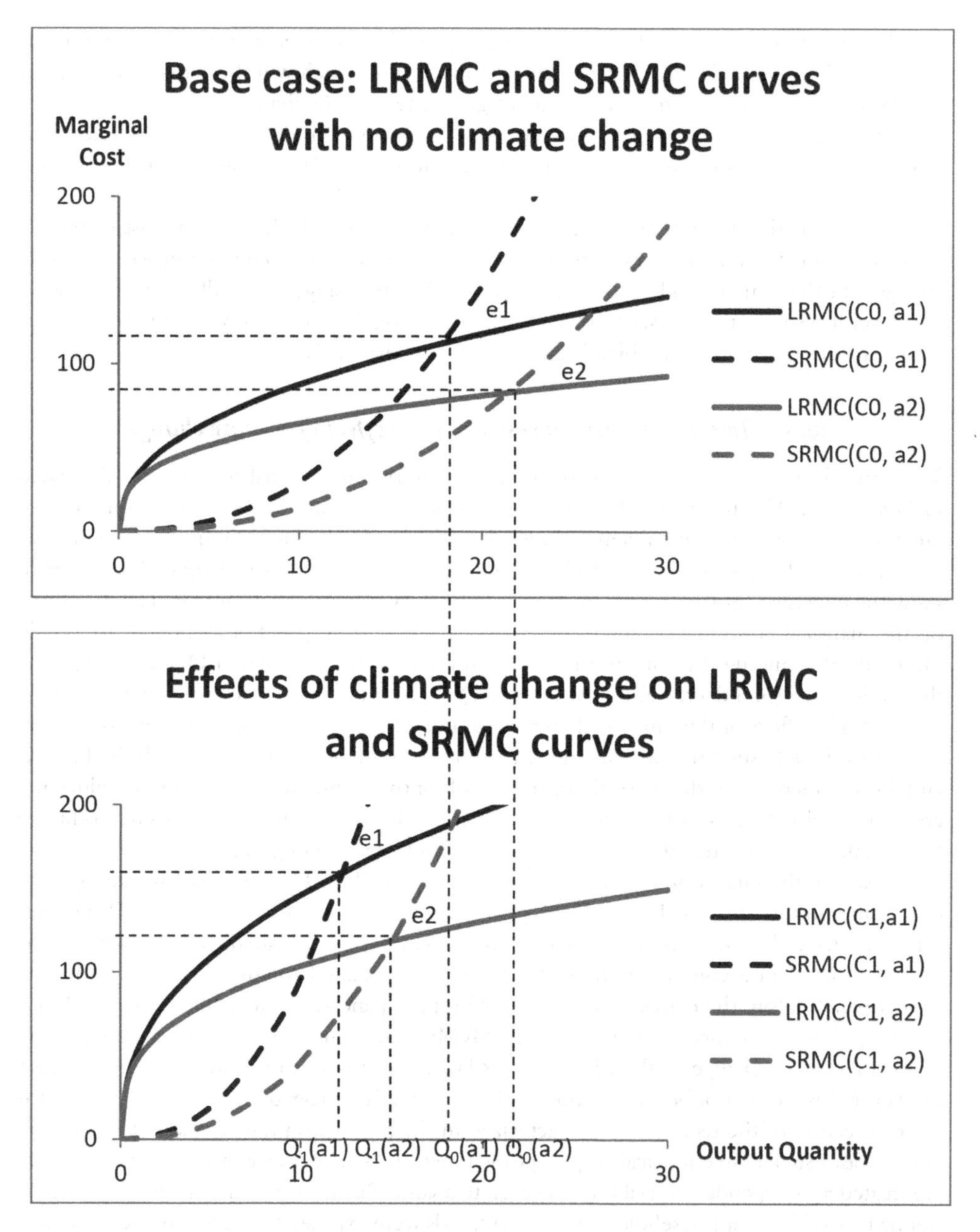

Figure 20.2 Illustration of the pure effect of climate change on cost curves

parameter, α. However, this might not be the case in every real-world situation. They could be producing along their short-run supply curve, on a point above or below the long-run supply curve. This would influence both the calculation of the climate change damages due to the pure effect of climate change, and the resulting short- and long-run benefits of adaptation benefits, which have the effect of reducing these damages. However, the assumption maintained here is the one most generally used in planning studies, and we will stick with it.

Figure 20.3 illustrates the valuation of climate change damages due to the pure effect of climate change. This figure is different than Figure 20.2 in that the market demand for the output Q has been added to the supply side of the picture, and it only focuses on the a1 technology. This is necessary to show how climate change influences equilibrium product prices. The Marshallian market demand function for Q is shown by the downward sloping line, labelled MKT DMND. The intersection of this demand curve and the point e_0, where LRMC(C_0, a1) and SRMC(C1, a1) intersect, represents a stable, long-run price-quantity equilibrium point in a competitive market, at which producers equate their LRMC to the market price, P_0* and, in doing so, aggregate produce Q_0* units of output. The impacts of climate change are illustrated by the two arrows, which shift the LRMC and SRMC curves, respectively, up and to the left to the point where they intersect at e_1. At this point climate change has caused Q to decrease from Q_0* to Q_1" and both the LRMC and SRMC of producing Q from P_0* to MC_1". In a competitive market, this equilibrium would not be stable since producers would have incentives to reduce the gap between the demand price of Q, P_1*, and its associated supply price, MC_1", until either the SRMC or LRMC, depending on the adjustment path, equalled the demand price along the demand curve. However, it is precisely because there is no such adjustment that this equilibrium point serves as the reference for the pure effect of climate change.

Figure 20.3 also contains labels for geometric areas that coincide with economic values. All of these areas have an economic meaning behind them, and these will be explained as we look at how the welfare of economic agents is affected by climate change and, in later figures, by short- and long-run adaptation.

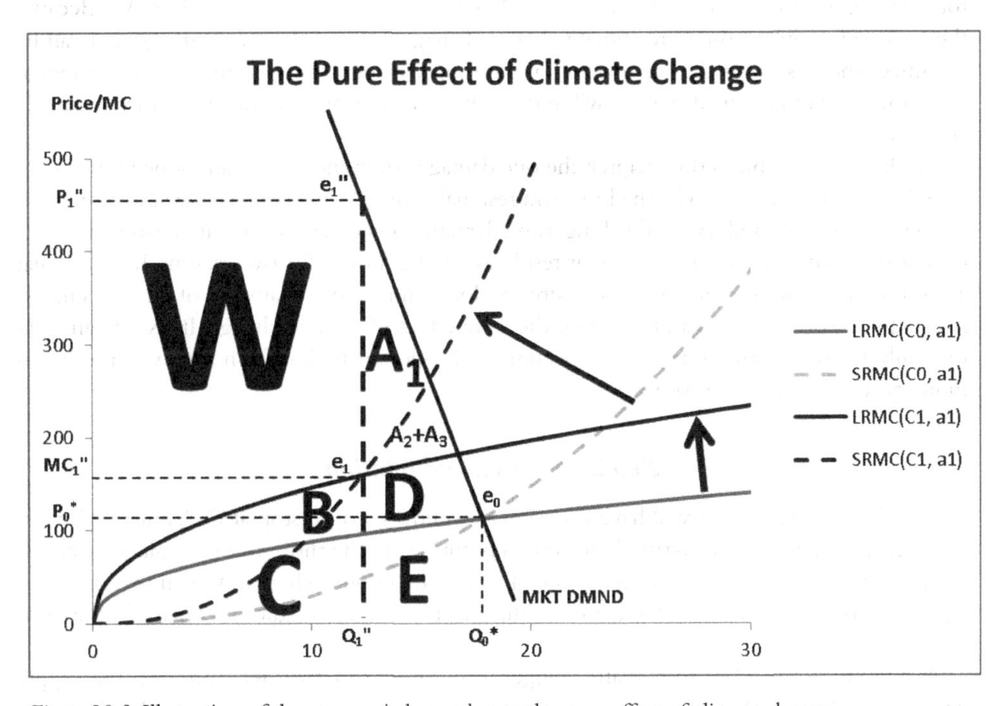

Figure 20.3 Illustration of the economic losses due to the pure effect of climate change

The net economic value of the damages associated with the pure effect of climate change can be measured by the economic losses sustained by consumers and producers as a result of moving from the market equilibrium point, e_0, to the non-stable equilibrium point e_1. This change can be calculated by subtracting the change in the aggregate long-run costs of producers from the change in the total willingness to pay of consumers, or by adding the changes if costs are expressed in negative terms. This welfare measure – the change in the aggregate surpluses of producers and consumers – will be used to compute all of the economic losses and gains due to climate change and adaptation.[6]

The calculation of this change in total surplus is illustrated as follows. First, on the supply side, the initial LRTC is equal to the area under the supply curve, LRMC(C_0, a1), from 0 to Q_0*, which is equal to the sum of the geometric areas – (C + E). Since the equilibrium point, e_0, occurs at the intersection of the LRMC(C0, a1) and SRMC(C_0, a1) curves, LRTC = SRTC at this point. The LRTC associated with the shift in the long- and short-run supply curves, where they cross one another at point e_1, is equal to the area under this supply curve from 0 to Q_1", which is equal to the sum of areas, – (C + B). The change in LRTC is, therefore equal to – (C + E) – (– C – B) = E – B.[7]

In a Marshallian framework, the change in the total welfare of consumers (buyers) of a product can be measured by the change in total willingness to pay for a good in the market for Q. This is illustrated by the change in the area under the Marshallian demand curve between Q_0* and Q_1". The change in total willingness to pay (an economic loss) is equal to the sum of the geometric areas $-A_1 - A_2 - A_3 - D - E$.[8]

The economic value of the pure climate change damages is illustrated (and calculated) by summing the changes in costs (since costs are expressed in negative terms) with the changes in total willingness to pay, or $E - B - A_1 - A_2 - A_3 - D - E = -(B+D) - (A_1 + A_2 + A_3)$. Because the economic value of the pure climate change damages is the maximum damage that can be sustained (there is no further adjustment along supply or demand curves), it stands to reason that short- and long-run adaptation will reduce the value of some, but not all, of these damages components.

Looking ahead a bit in this chapter, the one damage component that cannot be reduced is – (B + D), which turns out to be the long-run residual damages of climate change after long-run adaptation has occurred; even after long-run adaptation occurs, these climate change damages cannot be avoided. This is an important result, as will be shown, for two reasons. It means that all of the information required to measure the long-run residual damages of climate change is contained within the measurement of the pure effect of climate change. It also means that one only has to determine the effect of climate change on the long-run supply curves of an industry to calculate these damages.

20.3.2 Short-run adaptation

As stated earlier, producers will have economic incentives, not to stop at Q_1", but to adapt by moving up either their short-run demand to a point where the short-run marginal cost equals the demand price, or up their long-run supply curve to a point where long-run marginal cost equals the demand price. We look now at the short-run path of adjustment, as illustrated in Figure 20.4.

When producers adapt to climate change in the short-run, they will move up the supply curve SRMC (C1, a1) and consumers will move down the demand curve until the short-run

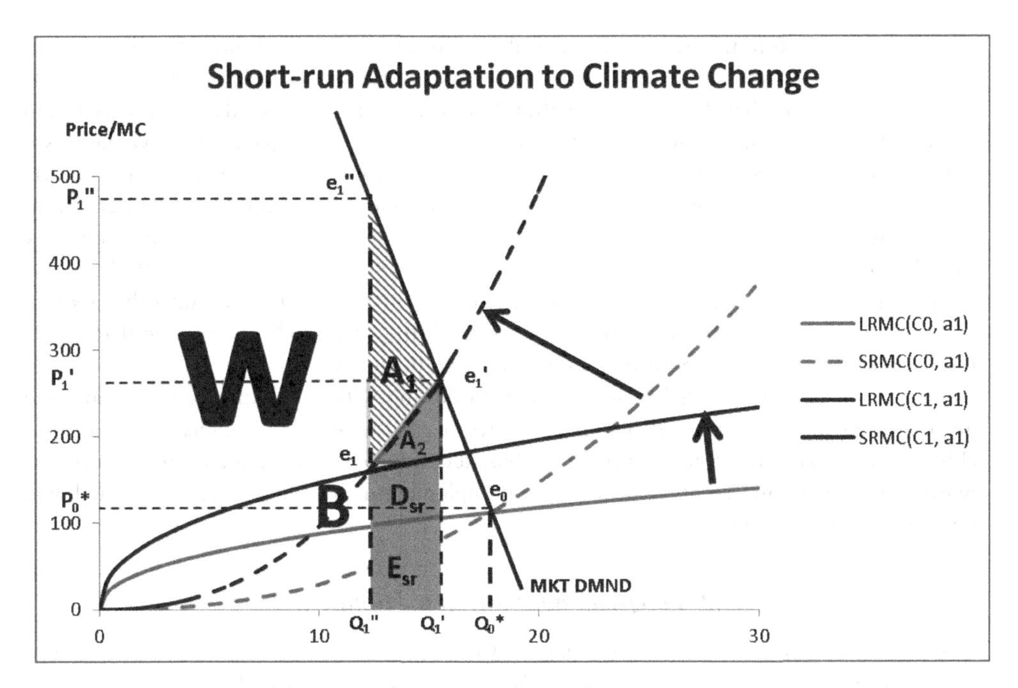

Figure 20.4 Illustration of short-run adaptation to climate change

marginal cost equals the demand price. This equilibrium occurs at point e_1', where output is Q_1' and the market price is P_1'.

This short-run adaptation behaviour on the part of consumers and producers is decomposed in Figure 20.4 into short-run adaptation benefits, costs and the difference between the two, the net benefits of short-run adaptation. The short-run adaptation cost is illustrated by the grey area under the short-run supply curve SRMC(C1, a1), between Q_1'' and Q_1', which consists of a short-run cost increase of $A_2 + D_{sr} + E_{sr}$. D_{sr} and E_{sr} are part of the area B and D, respectively, in Figure 20.3.

The total short-run adaptation benefits, as measured by the change in total willingness to pay, are equal to the sum of the cross-hatched area, A_1, plus the grey area $A_2 + D_{sr} + E_{sr}$. Therefore the short-run net benefits of adaptation are equal to the difference between short-run total benefits and costs, as measured by the area, A_1. This area represents the pure climate change damages that are avoided by short-run adaptation, a reduction in the pure damages of climate change from $-(B + D + A_1 + A_2 + A_3)$ to $-(B + D + A_2 + A_3)$. This loss is equal to $-(B + D) - (A_2 + A_3)$. It represents the economic value of the pure climate change damages that cannot be avoided by short-run adaptation over this interval, otherwise known as the residual damages of short-run adaptation.

One thing that is important to remember is that, while short-run adaptation represents a behavioural and not a technological adjustment, producers in the market will adjust along their short-run supply curves. These supply curves are based on the existing technology in the production function, as embodied in the short-run supply curves with k fixed. Thus, the short-run responses of these agents will be limited by their technology and, if the performance of the technology with which they make the short-run response is climate sensitive, then the range of adjustments to the new climate will also be limited. The more limited the shift in their short-run supply curves caused by climate change, the less they will need to adapt in the short-run. Does that mean that short-run adaptation has to always involve exactly the same type and quality of inputs, the same production function form and the same parameters? While this is what is assumed in the analysis thus far, the answer is no, not necessarily. This will depend on how one defines variable, quasi-fixed and fixed factors (and how congruent these definitions are with reality); and what kind of flexibility is built into the production function technology to make short-run adjustments when some factors are fixed. It is also associated with the size of the cost involved, since ex-ante uncertainty about the ex-post climate introduces the risk that any change in technology may not be very effective, if the ex-post climate regime that occurs is different than the one used for investment planning purposes

These issues will be tackled to some extent in dealing with long-run adaptation and to a greater extent in the sections that follow it, covering no regrets/low regrets adjustments to climate change and the issue of choosing robust technologies that reduce ex-post economic regrets when economic agents are planning and implementing investments, ex-ante, which will operate long into the future.

20.3.3 Long-run adaptation to climate change

The process of long-run adaptation is depicted in Figure 20.5 as an alternative to short-run adaptation, not as an incremental form of adaptation, although this is what may occur if producers respond to climate change as if it were climate variability and only later make a long-run adjustment when sufficiently reliable information is available about climate change.

The short-run supply curve, SRMC(C1, a1, adapt) represents a new short-run supply curve reflecting a new fixed level of k on LRMC(C1, a1) at the long-run equilibrium point, e_1*, where LRMC = SRMC and LRTC = SRTC. It represents the short-run supply curve along which producers can adjust to changes in product demand, but is actually not needed in the context of the analysis presented here.[9]

However, producers do not adjust just their variable inputs along this short-run curve. Instead, they adjust both variable and fixed inputs in response to climate change along LRMC(C1, a1), starting at the point e_1 and ending at the point e_1*, where the long-run cost equals the demand price. At this point, the aggregate output level of Q is Q_1* and the market price is P_1*.

The total long-run adaptation benefits of this adjustment of both variable and fixed factors are measured at the market level by the change in total willingness to pay of producers and consumers from Q_1" to Q_1*. This is equal to the sum of the areas $A_1 + A_2 + A_3 + D_{lr} + E_{lr}$. The long-run adaptation cost to producers of making this adjustment is equal to area under the long-run supply curve LRMC(C1, a1) over the same interval, or the sum of the areas $D_{lr}+E_{lr}$. The difference between the changes in total benefits and total long-run costs over the adjustment interval is, thus, equal to $A_1 + A_2 + A_3\ D_{lr} + E_{lr} - (D_{lr} + E_{lr}) = A_1 + A_2 + A_3$, the net benefits of long-run adaptation. These net benefits are larger than the net benefits of short-run adaptation by an amount equal to the sum of the areas $(A_2 + A_3)$.

Thus, long-run adaptation reduces the economic value of the pure climate change damages from $-(B + A_1 + A_2 + A_3 + D)$ to $-(B + D)$. This remainder, $-(B + D)$, is the value of the climate change damages that cannot be avoided by short- or long-run adaptation, otherwise known as residual damages. However, these unavoidable, or the residual damages after long-run

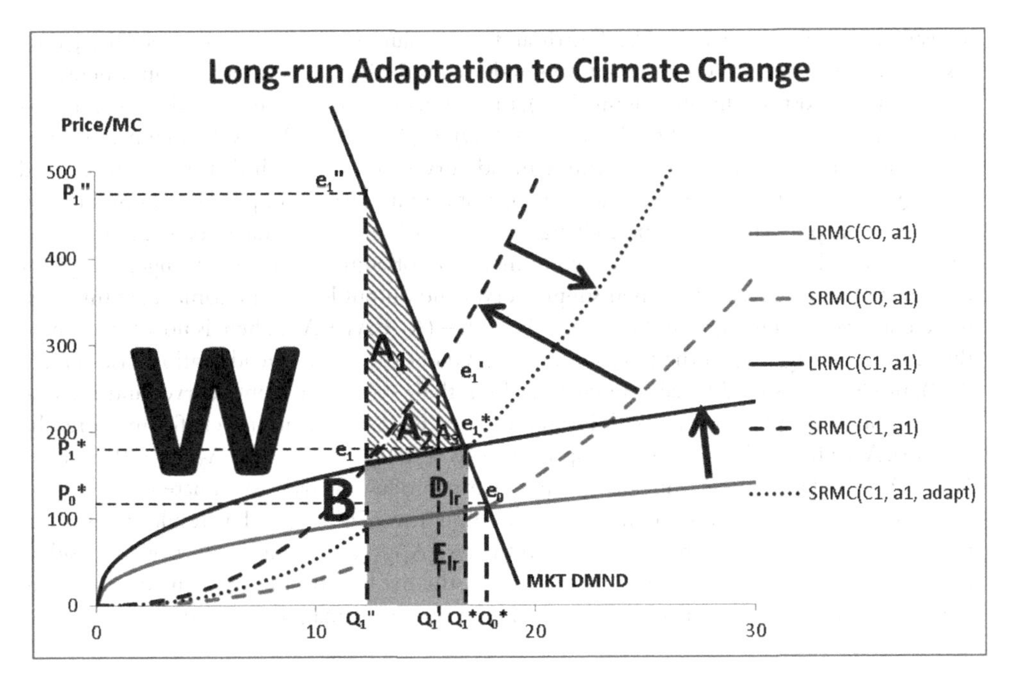

Figure 20.5 Illustration of long-run adaptation to climate change

adaptation has taken place can be more simply expressed, as the area between the two long-run supply curves LRMC(C1, a1) and LRMC(C0, a1) and to the left of the market demand curve.

The changes in the net benefits of adaptation and residual damages over the intervals associated with the pure effect of climate change, long-run and short-run adaptation are presented in Table 20.1.

The first column shows the output level for the base case, the pure climate change damage case, and the long- and short-run adaptation cases. The second column shows the net welfare for each case, while the third column shows the net welfare loss/gain for the three types of

Table 20.1 Welfare accounting for climate change adaptation

Output	*Net welfare*	*Net loss/gain in welfare*	*Type loss/gain*	*Residual damages*
Q*(C0)	$B + D + A_1 + A_2 + A_3$	(Base Case)	(Base Case)	(Base Case)
Q"(C1)	W	$-(B + D + A_1 + A_2 + A_3)$	Pure climate change damages	$-(B + D + A_1 + A_2 + A_3)$
Q'(C1)	$W + A_1$	$+ A_1$	Net short-run adaptation benefits	$-(B + D + A_2 + A_3)$
Q*(C1)	$W + A_1 + A_2 + A_3$	$+ A_1 + A_2 + A_3$	Net long-run adaptation benefits	$-(B + D)$

Reference: Figures 20.3–20.5

adjustment to climate change. The fourth and fifth columns identify, respectively, the gains/losses in column three in conceptual terms and the residual damages after the adjustment.

Under the existing climate (Figure 20.3), the net economic welfare of producers and consumers in the market is equal to the area, $(B + D) + (A_1 + A_2 + A_3)$. When climate change occurs, it shifts the short- and long-run demand curves in a way, such that output is reduced from $Q_0{}^*$ to $Q_1{}''$, but input use remains constant (at least in the cases implied by our production function). The pure effect of climate change (Figure 20.3) and the resultant damages are simulated by assuming that producers and consumers do not adjust to climate change, except for the effect of climate change on their supply curves and output levels. The damages at this stage represent a welfare loss equal to the area, $-(B + D) - (A_1 + A_2 + A_3)$. There is no adaptation, so the residual damages equal the pure welfare loss. Under the short-run adaptation case (Figure 20.4), producers respond to climate change along their short-run supply curves and increase their output from $Q_1{}''$ to $Q_1{}'$. By doing so, they receive net adaptation benefits measured by the area, A1. The residual damages drop from $-(B + D) - (A1 + A_2 + A_3)$ to $-(B + D) - (A_2 + A_3)$. Under the long-run adaptation scenario, producers adapt to climate change along their long-run supply curves and increase their output from $Q_1{}'$ to $Q_1{}^*$. The result of this is that they receive net adaptation benefits equal to the area $A_1 + A_2 + A_3$ and a decrease in residual damages from $-(B + D + A_2)$ to $-(B + D)$. Thus, the net benefits of long-run adaptation are higher than those for short-run adaptation and the residual damages, correspondingly less.

20.3.4 Long-run adaptation with efficiency changes

In the previous analysis of adaptation to climate change, the form and parameters of the production function were held constant. Even so, the technology embedded in the long- and short-run supply curves played a primary role in determining the economic value of the damages due to the pure climate change effect and the benefits and costs of adaptation.

Now we look at the impact of adopting more efficient production technology (High Technology) on the economic impacts of climate change. There are many good sectoral examples of this, both in developed and developing countries. In the energy sector, for example, reducing electricity demand through improvements in the efficiency of energy using durables such as household appliances, commercial lighting and air-conditioning and buildings' technology can result in lower energy use. The same is true for the demand for water by industry, households and irrigated agriculture. On the production side, which is what we will focus on here, power plants in developing countries tend to have lower technical efficiencies than in developed countries and poor maintenance also reduces their operating performance and reliability. The same is true in developing country agriculture and forestry where average household and commercial production yields per hectare are well below those in developed countries, both in gardens farmed by households and commercial farms and plantations. In these, and many other cases, there are good reasons to improve output efficiency based on objectives that are unrelated to avoiding the damages caused by adverse climate changes.

Long-run changes that producers make in technological efficiency for reasons other than avoiding climate change damages will generally result in long-run supply curves that exhibit cost reductions under both an existing climate and climate change. Other things being equal, this will result in increased aggregate output at lower market prices, producing higher aggregate net welfare. In climate change terms, this would at least mean that lower damages associated with the pure effect of climate change, higher long-run net benefits of adaptation and a reduc-

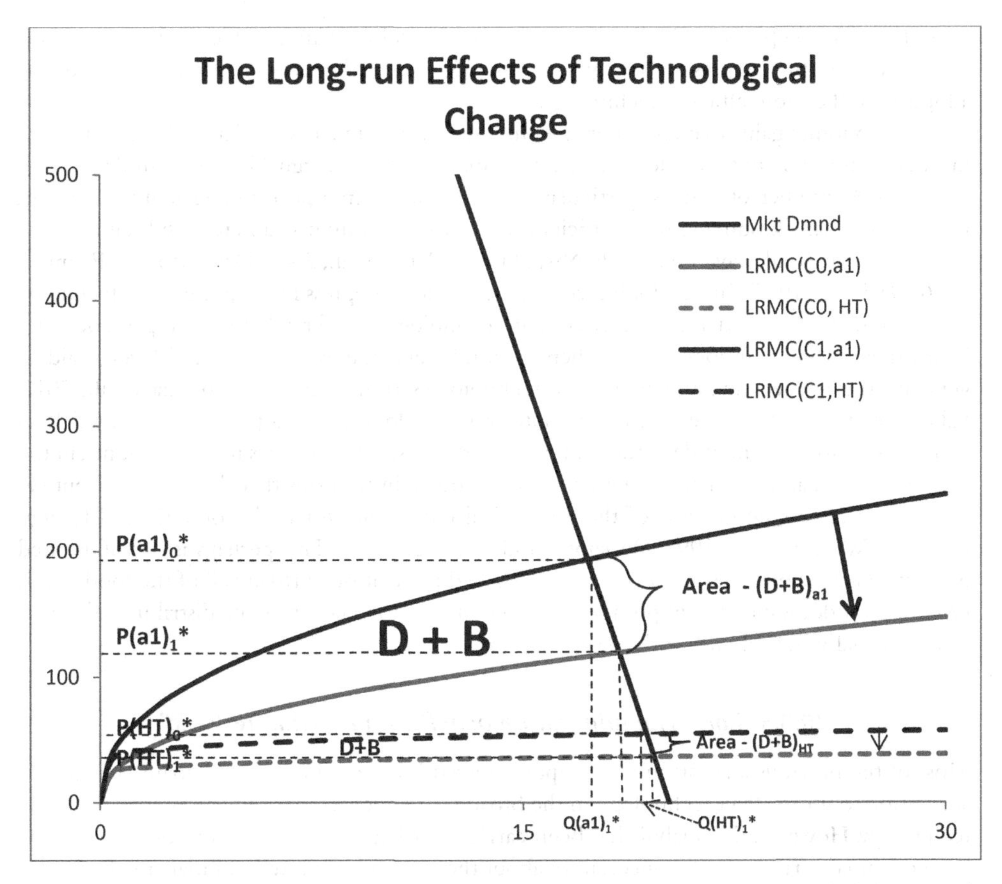

Figure 20.6 Illustration of the long-run effect of more efficient technology on the residual damages of climate change

tion in long-run residual damages. This is illustrated in Figure 20.6, based on the results shown in Figures 20.3 to 20.5. To simulate the technology change, the output elasticity of both variable and fixed inputs and the price of the fixed factor were increased by a factor of 1.125. The LRMC and SRTC curves for the new, High Technology (HT) are illustrated with dashed lines and the a1 LRMC and SRTC curves (a1) are shown by solid lines.

The adoption of the new technology results in LRMC curves that are much flatter and approach constant-cost conditions, as opposed to the increasing-cost character of the a1 LRMC curves. They also lie well below the a1 LRMC curves, indicating that for any given level of Q along the HT LRMC curves, the long-run marginal cost of producing Q is lower than along the a1 LRMC curves. This is reflected by the fact that $P(a1)_0$* and $P(a1)_1$* lie well above $P(HT)_0$* and $P(HT)_1$*. The observably small difference between the output levels, holding either the technology or climate constant between the LRMC curves, is due to the fact that the market demand curve is very inelastic at the lower end of the demand curve. But most importantly, the effect of changing to a more efficient technology reduces the long-run residual damages of climate change, as measured by the areas between each of the two sets of LRMC curves – $(B+D)_{a1}$ and – $(B+D)_{HT}$. The difference between the two, or $- (B + D)_{HT} - [- (B + D)_{a1}]$ is the reduction in residual damages due to the technology change. This, in turn, represents the increase in long-run net benefits of adaptation due to the adoption of the more efficient technology.

The economic gains that are observable in Figure 20.6 are based on large changes in technical efficiency. The potential for such gains in many sectors is, arguably, largest in developing countries. A number of studies, particularly in the energy, transport and agricultural sectors, have shown that introducing more efficient technology and inputs can create substantial economic benefits (Callaway et al., 2006, Nwaobi 2004, Vlek et al., 2004, DeCastro and Rahman 1996, Davidson, 1993). In developing countries, the slow progress in adopting these technologies is sometimes viewed as an equity issue in technology transfer, where developed countries hoard their new technology and/or there is insufficient domestic savings and foreign aid to generate the needed investment in new technologies (Karekezi 1994, Sokona et al., 2012, Laborte et al., 2012). However, the slow rate of technological development in some of these countries is probably more dependent upon underlying structural issues in the economy, in the legal, financing, and governance frameworks and other institutions that distort the incentives to reap the economic benefits of these technical efficiencies for both domestic and foreign investors (Winters et al., 1998, De Janvry and Sadoulet 2002). This perhaps is best illustrated in South Africa where, despite a large impoverished population, virtually all of the food that is produced for domestic consumption is produced on commercial farms and distributed through a commercial wholesale network to supermarkets

20.3.5 Long-run adaptation benefits under climate "risk"

Most of the previous analysis in this chapter showed how the impacts of climate change and adaptation to it can affect technology in the broad terms adopted here and elsewhere to define technology. However, the analysis has been carried out largely in a deterministic fashion. The issue of climate "risk" – really uncertainty about the rate and magnitude of man-made climate change at local levels – and how it may affect the way economic agents adapt to climate change in the long-run and short-run has been addressed in one way in the previous section. Increasing technical efficiency for reasons other than reducing climate change damages is one approach of reducing the regrets associated with climate risk/uncertainty as long as the technical efficiency of the new technology is less, or no more, sensitive to climate change than the one which it replaced.

This section builds on that analysis and looks more carefully at how economic agents can adapt to climate risk in long-run situations where planners try to minimize the economic losses associated with incorrect predictions of future climates. This analysis is based on a more elaborate definition of "regrets" that can be used in the analysis of adaptation to climate change that goes back to the early literature on this subject, primarily by water resource engineers.

The problem, as stated by Lettenmaier and Burges (1978), is that climate change is hard to detect in stream flow records by observation given their relatively short length in relation to the variability in these. Different types of models can fit these data equally well (or poorly) and yet have dramatically different consequences for the design capacity of storage and flood control reservoirs. Matalas and Fiering (1977) added a second dimension to this problem by noting that, due to uncertainty about the rate and magnitude of climate change, there is also likely to be considerable uncertainty about what future stream flows water resource planners need to plan for. As a result, water resource planners run the risk that a reservoir design that is optimal for the "planned" climate will not be optimal for the climate that actually occurs. The inability of the reservoir to meet the original design parameters, in its future operation, represents a form of regrets. The way to avoid regrets, they argued, was to plan for a reservoir that minimized the expected value of these regrets, using a regrets-based criterion (without specifying which one). A robust design would not be optimal for any single climate, but would minimize regrets over a large number of expected climates.[10]

Both of these papers were written when Global Climate Models (GCMs) were in a stage of infancy and Regional Climate Models (RCMs) were still an afterthought. Despite the large amount of progress in state-of-the art in GCMs and RCMs, the same two problems still exist: RCM climate predictions vary widely for runoff in single watersheds in the same future period (Feyen and Watkiss 2011 and Feyen 2009), and the regrets from basing a flood control or water supply dam from a design based on any single prediction can be large. This same situation exists in any sector where a development project must be planned and built in one period and operate for a much longer: in the planning of electrical generating capacity, agricultural and forestry developments, transport infrastructure and almost any form of land-use development that on which future uses are climate-sensitive.

Following up on Matalas and Fiering's (1977) work, Hashimoto et al., (1982) indicated three climate-related risk parameters for which estimates of economic damage costs could be developed: reliability – how frequently a "system" fails to meet its design goals; resiliency – how quickly the system recovers from a failure; and vulnerability – how sever the consequences of failures. They suggested that planners could translate these risk parameters into economic costs, aggregate them based on explicit risk trade-off preferences, and then compute the aggregate the regrets costs by comparing the costs of ex-ante, ex-post planning "mistakes" for a number of system designs over a number of expected climates and select the design that minimized some over-all regrets-based decision criteria. The approach suggested by Hashimoto et al. (1982) can be extended, more generally, to economic losses measured in welfare terms in the previous discussion.

To these terms, we have to add the concept of "flexibility" (Fuss and McFadden 1978) to characterize the production technologies that will lead to robust system designs. This entails developing production process and inputs, both variable and fixed, that will reduce the economic impact of the pure effect of climate change and/or make it easier to adapt to climate change in the short- and long run by compressing the distance both supply curves shift in response to climate change and/or by making them more price-sensitive to price changes caused by climate change in markets.

An example of how this could be done, for a development project in the framework of ex-ante minimization of ex-post regret minimization, is shown in Tables. 20.2 and 20.4. Table 20.2 presents the ex-ante expected net present value (NPV) for each of the three projects, taking the dimensions of risk into account as a part of project costs. In the ex-ante planning stage, D*(1) through D*(3) are optimal for the ex-post Climates (C1) through (C3), respectively, as shown by the diagonal cell-entries. D(Robust) is a design that is not optimal for any climate state (compared to the other three designs), but performs relatively well under all three ex-post climate scenarios. For all the designs, the occurrence of other ex-post climates to the right or left of the diagonal entry will result in reduced NPV. The Robust design exhibits the same general characteristics, except that the spread of NPV values from the maximum value, and resulting losses, are smaller. It also ranks second-highest on the basis of a maximum ordering of mean NPV values for each project, without any probability weighting.

Table 20.2 Ex-post net present value for an ex-ante generic project with four designs and three climate states (10^6 €)

Ex-ante project designs	*Expected (ex-post) climates*			*Mean and max. (rank) for all ex-post climates*
	C1	*C2*	*C3*	
D*(1)	*120*	40	5	55.00 (1)
D*(2)	50	*70*	20	46.67 (4)
D*(3)	10	50	*70*	43.33 (3)
D(Robust)	50	55	50	51.67 (2)

Table 20.3 shows the ex-post economic regrets calculated for the four ex-ante project designs, computed from the information in Table 20.2. The values of the ex-post economic regrets for the first three designs are the off-diagonal column differences between the optimal NPV value and NPV the values above and/or below it. The robust project, on the other hand, is not optimal for any single climate and so there are economic regrets under all three ex-post climate states. The mean value of the regrets for each design is shown in the last column. As shown, the mean economic regrets values for the designs are grouped fairly tightly, but this is just a hypothetical example. The robust design ranks second in terms of minimum regrets.

The final Table, 20.4, shows the results of a simulation analysis for a small number (n = 30) of climate outcomes (C1,..., C3) with different associated probabilities, p1,..., p3 = 1 for each climate state for each of the 30 trials. In actual practice, the number of simulations could be much larger, consistent with the climate model information available. Surprisingly, or not, this set of trials resulted in a reordering of the mean minimum regrets, such that the robust project minimized the expected value of the economic regrets over all simulations and climates . Thus, it represents the most robust of the designs according to that risk criteria.

Different risk criteria and more sophisticated (Bayesian) approaches for calculating regrets could have been used in the analysis. The point is just to show the general approach to an ex-ante, ex-post analysis of robustness and how it can result in a reordering of the designs based on regrets-related criteria. Unfortunately, at the current time there is no convincing way to generate the probabilities of climate outcomes predicted by GCMs and RCMs. While the capability to simulate "ensembles" of climate outcomes, over time, for a given emissions scenario with a given model exists (van der Linden and Mitchell 2009), computation of the probabilities of these outcomes is problematic from the standpoint statistical theory for several reasons:

Table 20.3 Ex-post regrets matrix for an ex-ante generic project with four designs and three climate states (10^6 €)

Ex-ante project designs	*Expected (ex-post) climates*			*Mean and min. (rank) for all ex-post climates*
	C1	*C2*	*C3*	
D*(1)	*0*	-30	-65	-31.67 (1)
D*(2)	-70	*0*	-50	-40.00 (3)
D*(3)	-110	-20	*0*	-43.33 (4)
D(Robust)	-70	-15	-20	-35.00 (2)

Table 20.4 Simulated mean ex-post regrets matrix for a generic project with four designs and three climate states (10^6 €)

Ex-ante project designs	*Expected (ex-post) climates*			*Mean and min. (rank) for all ex-post climates*
	C1	*C2*	*C3*	
D*(1)	0	-9.20	-25.13	-11.44 (2)
D*(2)	-20.06	0	-19.33	-13.13 (4)
D*(3)	-31.53	-6.13	0	-12.55 (3)
D(Robust)	-20.06	-4.60	-7.73	-10.8 (1)
Mean	-17.91	-4.98	-13.05	-11.98

- The emissions scenarios that drive climate scenarios have no known distribution and are based on "story-lines", not observation-based trends,
- The physical processes represented in the GCMs and parts of the RCMs are deterministic,
- Many of the GCM parameters that are varied to produce ensembles are not related to criteria that can be observed empirically, or are stochastic, and
- Using uniform random distributions as a basis for weighting the climate outcomes is likely (as we found out) to reduce the amount of variability in the regrets outcomes, such that the ranking of the unweighted regrets results (Table 20.3) does not change very much.

Thus, while robustness analysis in an economic framework has the potential to address the two problems identified by Lettenmaier and Burges (1978) and Matalas and Fiering (1977), it can only yield sensible results under a stable climate where parameters in the partial and joint distributions of observed geophysical variables can be fit to known distributions as Matalas and Fiering (1977) indicated. However, in the case of climate change, the want of statistical underpinnings to generate these same types of distributions make the approach promising, but problematic.

20.4 Conclusions

While there has been an increasing emphasis for multi- and bi-lateral donors to focus on technology issues related to climate change, this has not prompted economists to focus very much on the technological aspects of climate change, nor on the importance of long-run versus short-run adaptation to climate change (Callaway 2004b and Callaway et al., 2006). One of the important conclusions of this chapter is that the adjustments made by producers from the short run to the long run is, in and of itself, an economically efficient adaptation to climate change, even if there is no change in technological efficiency. In the same vein, increases in technological efficiency can reduce climate change damages even further. One fruitful avenue of further investigation involves looking at how technology can reduce the potential costs of shifting from the short run to the long run when the future climate is characterized by great uncertainty and adaptation investments are lumpy and long-lived.

One finding that came as a bit of a surprise is that there is hardly any theoretical literature about how changes in exogenous, non-economic inputs in production functions (such as climatic variables) influence long- and short-run total, variable and marginal costs. In fact, it turns out that for some common production functions, a change in a climatic variable that shifts long-run and short-run total and marginal costs to the left will, in some well-defined situations, have no influence on optimal variable input levels or long-run and short-run total costs, but will result in decreased output and welfare. This probably means that economists need to take a closer look at the implications of how they model climate change impacts in structural terms, particularly in econometric studies and in market-driven integrated assessment models.

This chapter also has identified a theoretical construct, "the pure (physical) effect of climate change". This represents the welfare loss by producers and consumers when the only adjustment to climate change that occurs is through the initial movement of the short- and long-run total, and marginal cost curves, reducing physical output but holding the variable and fixed factors constant. The ability to measure the welfare loss that occurs as a result of this effect makes it possible to decompose effects of climate change into three categories: no adjustment, short-run and long-run adaptation, where the latter two adjustments take place as a result of movements along the short- and long-run supply curves and the demand curve.

Finally, the chapter looked at the problem that producers face when they must plan and implement long-run adaptation before the climate changes and therefore face the risk that actual climate change may not be the one they used to plan the long-run adaptation. When this occurs, producers will face economic regrets because the adaptation is no longer optimal. A "flexible" approach will involve undertaking a long-run adjustment, which is not optimal for any single climate change, but which minimizes expected economic regrets over a large number of future climate scenarios. Here, too, economists will benefit by combining the concept of design flexibility, as presented by Matalas and Fiering (1977), with economic concept of flexibility vs. efficiency as presented by Fuss and McFadden (1978).

Notes

1 Two classic works on the production function are Shephard (1970) and Frisch (1965). Blackorby et al. (1979) show how the structure of production and cost functions influence how inputs to the production function can be substituted for one another, while Diewert (1982) and McFadden (1978) explore the duality between production, cost and profit functions.

2 Producers will be used from this point on to refer to firms engaged in the production of market services and households engaged in the production of household services.

3 That is: Total Cost = $\Psi(Q, w, c, \alpha)$ = Min $\{w_x{*}x + w_k{*}k$ s.t. $f(x, k, c, \alpha) \geq Q\}$, where w_x and w_k are the prices of the variable and fixed factors, respectively.

4 Figure 20.2 shows the various cost curves for the same pair of production function parameters (α = a1 and a2) as in Figure 20.1.

5 Given a change in climate from c_0 to c_1, it is sufficient that the production function $Q=f(x, k, c)$, be separable in the form $\varphi(Q, c) = \psi(x, k)$ and LRTC = C[Min] = $g[\varphi(Q, c)]{*}h(w)$ – where w is a vector of input prices – for $LRTC(c_1) = SRTC(c_1) = LRTC(c_0) = SRTC(c_0)$. This condition is satisfied by many production function used in economic analysis, including the Cobb-Douglas, Leontief, CES, exponential and Mitscherlich functions, and many others (Griffen et al., 1987).

6 Changes in the distribution of this total surplus will also take place, and while they can and should be calculated, doing so illustratively with a limited number of figures is difficult.

7 As previously noted, there are cases when C+E = C+B and for these cases E=B, in which case there will be no change in costs. However, this condition does not have to be applied to get more general results.

8 $TWTP_1'' - TWTP_0 = (W+B+C) - (W + B + C + A_1 + A_2 + A_3 + D + E) = -A_1 - A_2 - A_3 - D - E$, where W denotes the area under the market demand curve above the LRMC(C1, a1) curve from Q_1'' to Q_0^* that is common to both levels of Q.

9 The short-run benefits and costs shown in Figure 20.4 can be estimated using information from this curve, as shown by Callaway (2004a), with the result as in Figure 20.4.
10 Not all decision criteria will lead to robust systems in the same way. For example, minimizing the maximum regret (avoiding the worst case) to avoid catastrophic outcomes, in particular, will be robust only over catastrophic outcomes, while minimizing the sum of expected regrets will be robust over a wider range of climates.

References

Blackorby, C., Primont, D. and Russell, R. (1979). *Duality, Separability and Functional Structure: Theory and Economic Applications*. North-Holland, Amsterdam.
Callaway, J.M. (2004a). "Adaptation Benefits and Costs: Are They Important in the Global Policy Picture and How Can We Estimate Them", *Global Environmental Change*, 14 (2) 273–282.
Callaway, J.M. (2004b). "The Benefits and Costs of Adapting to Climate Variability and Change", in *The Benefits and Costs of Climate Change Policies: Analytical Framework and Issues*. Organisation of Economic Cooperation and Development, Paris.
Callaway, J.M., Louw, D.B., Nkomo, J.C., Hellmuth, M.E. and Sparks, D.A. (2006). "Chapter 3. Benefits and Costs of Adapting Water Planning and Management to Climate Change and Water Demand Growth in the Western Cape of South Africa", in Leary, N., Conde, C., Kulkarni, J., Nyong, A. and Pulhin, J. (eds.). *Climate Change and Vulnerability and Adaptation*. Earthscan Books, London, 53–70.
Clements, R., Haggar, J., Quezada, A. and Torres, J. (2011). *Technologies for Climate Change Adaptation: Agriculture Sector*. X. Zhu (ed.). UNEP Risø Centre, Roskilde.
Davidson, O.R. (1993). "Opportunities for Energy Efficiency in the Transport Sector", *Energy Options for Africa*, 1 (1) 106–127.
DeCastro, A. and Rahman, S. (1996). "Efficient Electrical End-Use Technologies for Mitigating Greenhouse Gas Emissions", *Africa Energy Sources*, 18 (4) 407–418.
De Janvry, A. and Sadoulet, E. (2002). "World Poverty and the Role of Agricultural Technology: Direct and Indirect Effects", *Journal of Development Studies*, 38 (4) 1–26.
Diewert, W.E. (1982). "Duality Approaches to Microeconomic Theory," in *Handbook of Mathematical Economics*, vol. 2, K.J. Arrow and M.D. Intriligator (eds.) North Holland, Amsterdam.
Elliott, M., Armstrong, A., Lobuglio, J. and Bartram, J. (2011). *Technologies for Climate Change Adaptation: The Water Sector*. T. De Lopez (ed.). UNEP Risø Centre, Roskilde.
Fankhauser, S. (1997). "The Costs of Adapting to Climate Change", *Working Paper, No. 13*, Global Environmental Facility, Washington.
Feyen, L. (2009). "River Floods Assessment", in Ciscar, J-C. (ed.) *Climate Change Impacts in Europe*. Institute for Prospective Technological Studies, JRC, European Commission, Seville, 45–47.
Feyen, L. and Watkiss, P. (2011). "Technical Policy Briefing Note 3. The Impacts and Economic Costs of River Floods in Europe, and the Costs and Benefits of Adaptation. Results from the EC RTD ClimateCost Project" in P. Watkiss (ed.), The *ClimateCost Project*. Final Report. Published by the Stockholm Environment Institute, Sweden,
Frisch, R. (1965). *Theory of Production*. Rand McNally, Chicago.
Fuss, M. and McFadden, D. (1978). "Chapter II.4, Flexibility versus Efficiency in Ex Ante Plan Design", in Fuss, M. and McFadden, D, (eds.), *Production Economics: A Dual Approach to Theory and Applications, Volume I: The Theory of Production*. North Holland, Amsterdam.
Griffen, R., Montgomery J. and Rister, M.E. (1987). "Selecting Functional Form in Production Function Analysis", *Western Journal of Agricultural Economics*, 12 (2) 216–227.
Hashimoto, T., Stedinger, J.R. and Loucks, D.P. (1982). "Reliability, Resiliency, and Vulnerability Criteria for Water Resource System Performance Evaluation", *Water Resources Research*, 18 (1) 14–20.
Intergovernmental Panel on Climate Change (IPCC) (2001). *Climate Change 2001: Mitigation*. Cambridge University Press, Cambridge and New York.
Just, R.E., Hueth, D.L. and Schmitz, A. (1982). *Applied Welfare Economics and Public Policy*. Prentice-Hall, Englewood Cliffs.
Karekezi, S. (1994). "Energy policy issues in Africa", *Resources, Conservation and Recycling*, 12 (1–2) 23–29.

Klein, R., Nicholls, R.J., Ragoonaden, S., Capobianco, M., Aston, J. and Buckley, E.N. (2001). "Technological Options for Adapting to Climate Change in Coastal Zones", *Journal of Coastal Research*, 17 (3) 531–543.
Laborte, A.G., de Bie, K.C.A.J.M., Smaling, E.M.A., Moya, P.F., Boling, A.A. and Van Ittersum, M.K. (2012). "Rice yields and Yield Gaps in Southeast Asia: Past Trends and Future Outlook", *European Journal of Agronomy*, 36 (1) 9–20.
Lettenmaier, D. and Burges, S. (1978). "Climate Change: Detection and Its Impact on Hydrologic Design", *Water Resources Research*, 14 (4) 679–687.
Linham, M.W. and Nicholls, R.J. (2012). "Adaptation Technologies for Coastal Erosion and Flooding: a Review", *Proceedings of the ICE - Maritime Engineering*, 165 (3) 95–112.
Matalas, N.C. and Fiering, M.B. (1977). *"Water Resource Systems Planning" in National Academy of Sciences, Climate Change and Water Supply.* Government Printing Office, Washington.
McFadden, D. (1978). "Cost, Revenue, and Profit Functions" in M. Fuss and D. McFadden (eds.), *Production Economics: A Dual Approach to Theory and Applications, Volume: The Theory of Production.* North-Holland, Amsterdam.
Metz, B., Davidson, O.R., Bosch, P.R., Dave, R. and Meyer, L.A. (eds.) (2007). *Contribution of Working Group III to the Fourth Assessment Report of the Intergovernmental Panel on Climate Change*, 2007. Cambridge University Press, Cambridge and New York.
Nwaobi, G.C. (2004). "Emission Policies and the Nigerian Economy: Simulations from a Dynamic Applied General Equilibrium Model", *Energy Economics*, 26 (5) 921–936.
Samuelson, P.A. (1972). "Jacob Viner, 1892–1970", *Journal of Political Economy*, 80 (1) 5–11.
Shephard, R.W. (1970). Theory of Cost and Production Functions. Princeton: Princeton University Press.
Silberberg, E. (1999). "The Viner-Wong Envelope Theorem", *Journal of Economic Education*, 30 (1) 75–79.
Sokona, Y., Mulugetta, Y. and Gujba H. (2012). "Widening Energy Access in Africa: Towards Energy Transition", *Energy Policy*, 47 (2012) 3–10.
van der Linden P., and Mitchell, J.F.B. (eds.) (2009). *ENSEMBLES: Climate Change and its Impacts: Summary of Research and Results from the ENSEMBLES Project.* Met Office Hadley Centre, Exeter, UK.
Vlek, P.L.G., Rodríguez-Kuhl, G. and Sommer, R. (2004). "Energy Use and CO2 Production in Tropical Agriculture and Means and Strategies for Reduction or Mitigation", *Environment, Development and Sustainability*, 6 (1–2) 213–233.
Winters, P., Murgai, R., Sadoulet, E., de Janvry, A. and Frisvold, G. (1998). "Economic and Welfare Impacts of Climate Change on Developing Countries", *Environmental and Resource Economics*, 12 (1) 1–24.

21

DISASTER RISK MANAGEMENT AND ADAPTATION TO EXTREME EVENTS

Placing disaster risk management at the heart of national economic and fiscal policy

Tom Mitchell[1,2], *Reinhard Mechler*[3,4] *and Katie Peters*[1]

[1] OVERSEAS DEVELOPMENT INSTITUTE
[2] CLIMATE AND DEVELOPMENT KNOWLEDGE NETWORK
[3] INTERNATIONAL INSTITUTE FOR APPLIED SYSTEMS ANALYSIS (IIASA)
[4] VIENNA UNIVERSITY OF ECONOMICS AND BUSINESS

21.1 Introduction

The year 2011 was the costliest year on record for disasters, with estimated global losses of US$ 380bn dominated by disasters associated with the Japanese earthquake, tsunami and nuclear disasters, Christchurch earthquake, Thailand floods and extreme weather in the U.S. Not only were aggregate losses extremely high, but the distributional and knock-on effects were also large. Recent data suggests that the 2011 floods in Thailand reduced Japan's industrial output by 2.6% in November 2011 compared to the previous month, a result of disruption to electronics and automotive just-in-time supply chains. Japanese car makers, whose production facilities are spread across the region, lost US$ 500m.[1] Lloyds of London reported that the disaster caused the third biggest loss in its market's 324-year history (US$ 2.2bn).[2]

The 2011 losses are an extension of a trend that has seen global average economic disaster losses rise by 200% over the last 25 years in inflation-adjusted terms (see Figure 21.2). The trend is predominantly the result of more people and assets being located in areas exposed to natural hazards. The population exposed to flood risk in Asia in 2030 will be more than two and a half times the figure in 1970 (Handmer et al., 2012) and 50% higher than in 2000 (Vafeidis et al., 2011). While recent data suggest disasters are already hampering economic growth in low and middle-income countries, a continuation of the current upward trend in disaster losses poses a severe threat to national and regional macroeconomic outlook in such countries. Tackling this problem involves placing measures to address disaster risk at the heart of national economic and fiscal policy as well as embedding it within sector-based economic and land-use planning. At a country level, reducing losses or even stemming their increase will require investing in vulnerability reduction of people and infrastructure located in exposed areas or enacting policies that, over time, result in a safer spatial distribution of assets.

While a number of countries such as those highlighted later in this chapter have already started this process, progress is highly variable, with the balance of efforts focused on building budgetary reserves and establishing sovereign insurance mechanisms to help bolster and speed up the flow of money to support ex-post emergency assistance and relief processes (Surminski and Oramas-Dorta, 2011).[3] Yet, insurance reduces the variability around outcomes ("the rainy day") and any follow-on impacts, if losses exceed people's, business' and governments' ability to cope. Insurance, however, does not directly reduce risk in particular that related to frequently occurring events.

Direct and indirect losses are also difficult to measure accurately as they can extend across borders, through regional and global supply chains, as well as to intangibles, such as the cultural value of historic buildings or artifacts or the detrimental health impacts associated with stress and anxiety, which are difficult to insure. The effort, therefore, needs to be focused on holistically managing risk, losses and impacts through economic and fiscal policy. But how can this be achieved?

By drawing on examples, this chapter starts by examining whether disasters genuinely do inhibit economic development and whether rising disaster losses will have a greater impact on developing economies in the future. It then considers whether economic and fiscal planning at the national level can reduce exposure to disasters, before considering the necessary steps countries need to take if they are to achieve economic development in a way that is more resilient.

21.2 Do disasters affect economic development?

Not only do disasters directly lead to immense human suffering and loss, they may also directly and indirectly affect economic outcomes in the medium to longer term. Disasters can lead to microeconomic (household, business level) effects, as well as macroeconomic consequences. The macroeconomic, aggregate impacts, which are the focus of this chapter, comprise effects on gross domestic product, consumption, savings, investment and inflation due to the effects of disasters, as well as due to the reallocation of public financial resources post-event to relief and reconstruction efforts (Handmer et al., 2012).

The impact of disasters on aggregate economic performance and development has been examined by several studies over the last four decades based on empirical and statistical analysis as well as modelling exercises. While the earlier studies addressed predominantly developed economies and focussed on sectoral and distributional impacts of disasters, in recent years more emphasis has been given to developing countries. These studies generally find very limited aggregate macroeconomic impacts in developed countries, but important regional economic and distributional effects (Okuyama, 2009). In terms of developing countries, disasters have been found to lead to important adverse macroeconomic and developmental impacts and affect the pace and nature of socioeconomic development (Otero and Marti, 1995; Benson and Clay, 2004; ECLAC, 2003; Charveriat, 2000; Mechler, 2004; Raddatz, 2007; Kellenberg and Mobarak, 2008; Hochrainer, 2009; Noy, 2009; Cavallo and Noy, 2009; World Bank and UN, 2010; Handmer et al., 2012). As one example, Honduras following Hurricane Mitch in 1998 suffered setbacks to macroeconomic performance, and rural poverty increased by more than 5 percentage points to a staggering 75% (Morris et al., 2002) (see Box 21.1). However, as a recent review by Handmer et al., (2012) suggests, there is only medium confidence in these findings as a few studies have also found positive effects such as increases to GDP post-disaster (Albala-Bertrand, 1993; Skidmore and Toya, 2002). Yet, it is generally argued that these 'positive' findings can be attributed to the lack of systematic and robust GDP counterfactuals (what would have happened to GDP if the disaster had not occurred?) in these studies, a lack of accounting for informal sector effects, a lack of accounting for financial inflows (insurance and aid), and importantly the problem that national accounting generally measures flows rather than stocks. The latter point means that the relief and reconstruction effort (measured as a flow in terms of increased consumption and investment) shows up positively in national statistics, whereas the destruction (a loss to capital stock) does not enter the accounting at all.

Despite these reservations, there is consensus that macro effects are much more pronounced in lower income countries (Mechler, 2004, Lal et al., 2012). Handmer et al. (2012) suggest that developing countries exhibit higher economic vulnerability due to their (a) reduced resilience and dependence on natural capital and disaster-sensitive activities (such as tourism and agriculture); (b) lack of developed assessment processes and techniques of responding to disasters including preparedness, financing, information, and risk management; and (c) governance shortcomings. Countries with one or more of the following characteristics are considered particularly at risk of significant macroeconomic consequences post-disaster (see Mechler, 2004): (i) high natural hazard exposure; (ii) economic activity clustered in a limited number of areas with key public infrastructure exposed to natural hazards; and (iii) tight constraints on tax revenue and domestic savings, shallow financial markets, and high indebtedness with little access to external finance.

Box 21.1 The macroeconomic impacts of Hurricane Mitch on Honduras

Honduras may be considered a good illustration of a country subject to high disaster risk (severe exposure to hurricanes, flooding, drought and earthquakes), limited economic diversification with a reliance on cash crops such as bananas, and tight financial and fiscal constraints due to high indebtedness and high prevalence of poverty. Honduras was heavily hit by Hurricane Mitch at the end of 1998, killing 6,000 people, leaving an estimated 20% of the population homeless and causing assets losses of about two billion USD, or 18% of capital stock. Important macroeconomic effects ensued, and the figure shows actual GDP in absolute terms (solid line) as well as two pre-disaster projections (dashed lines). GDP growth in Honduras became negative in the year after the event (shown as the downward spike of GDP in absolute terms in the chart), but then rebounded later on with substantial inflow of foreign assistance, which increased by about 500 million USD or from about 6% pre-disaster to close to 16% of GDP post-disaster. For determining the impact on longer-term growth, the gap to the counterfactual projections without a disaster event can be taken as one indicator.

Using this approach for Honduras, a "GDP gap" can be identified. For example, in 2004, about six years after the event, this gap can be considered to have, ceteris paribus, amounted to about 6% of potential GDP given linear extrapolations of pre-disaster GDP with a 4-year average growth rate, and to about 9% based on another projection. Of course, the total cost as the sum of gaps over time would be much larger.

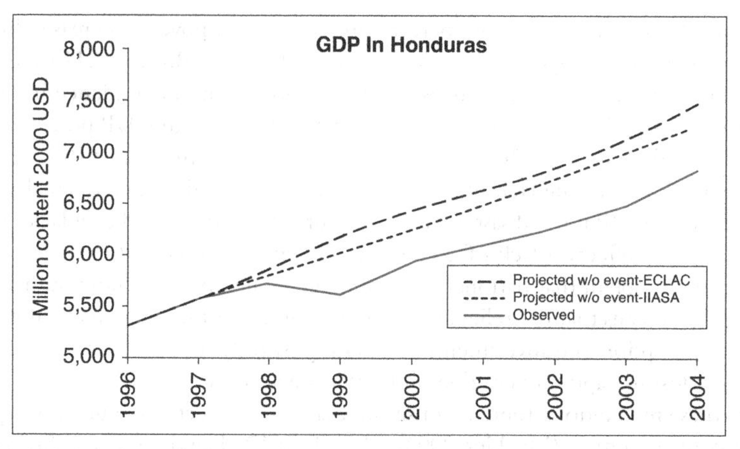

Figure 21.1 Observed GDP in Honduras with events vs. projected growth without events
Source: Mechler et al., 2006

21.3 Impact of disasters on future economic development

While there is high inter-annual variability in global insured and uninsured losses due to disasters, when adjusted for increases in wealth, disaster losses have in fact increased over the last few decades. As indicated in Figure 21.2, recent data show that 2011 was by some margin the costliest year ever recorded for disaster losses, amplified by losses associated with the earthquake, tsunami and nuclear disaster in Japan, flooding in Thailand and the earthquake in Christchurch, New Zealand.

However, economic exposure to disasters is not evenly distributed, with loss risk as a proportion of annual GDP concentrated in middle-income countries with rapidly developing economies and particularly exposed low-income countries, such as small island developing states (O'Brien et al., 2006; Kellenberg and Mobarak, 2008; Pelham et al., 2011). Figure 21.3 shows the largest ever events in terms of loss to assets ratio, and many affected countries are small and/or low-income countries. The burden of losses in middle-income countries has been increasing, with average losses of 1% of GDP from 2001–2006 (compared to 0.1% for

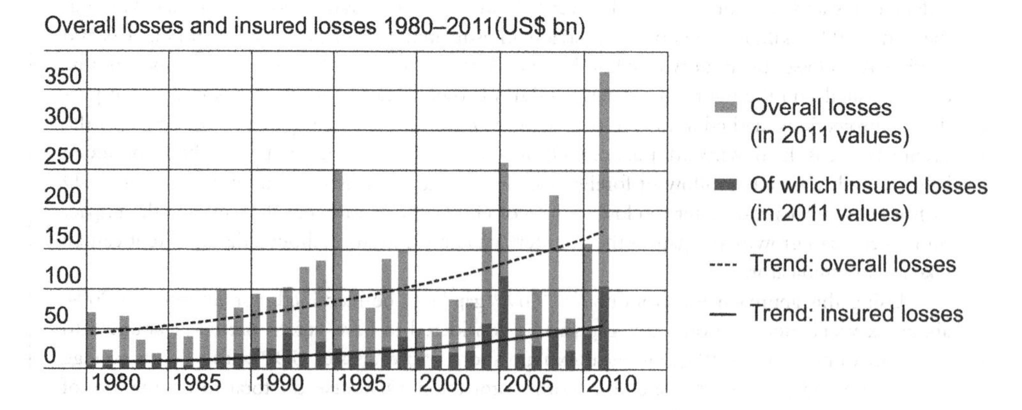

Figure 21.2 Increasing disaster losses. (Munich Re, 2012)

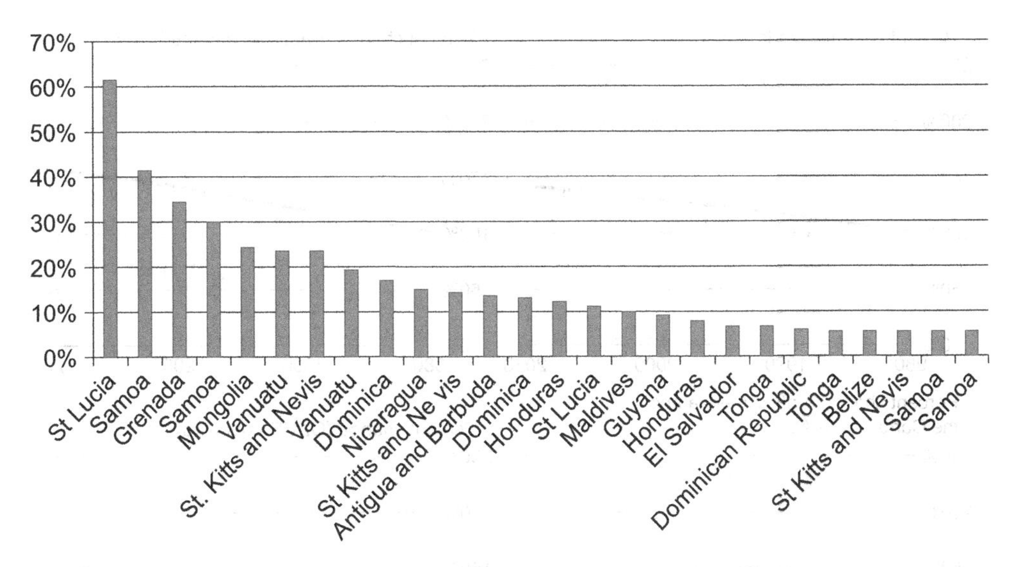

Figure 21.3 Largest monetary disaster losses since 1960 as measured in terms of assets destroyed
Source: Mechler, 2009

high income countries) (Cummins and Mahul, 2009). However, economic loss of GDP is even more concentrated, with China and countries of South Asia accounting for more than 49% of global annual loss since the 1970s (UNISDR, 2009). Given the influence of the economies of India, China and other Asian countries on global growth, future trends in disaster losses in such countries are of particular concern for other countries in the region and globally. Asia suffered disproportionally in 2008, a year regarded as one of the worst for natural and man-made disasters. A total of US$ 269bn were lost in Asia in that year, with emerging economies being particularly exposed because of high urban migration, intensification of natural resource use without adequate management and population growth (Swiss Re, 2009).

Without a significant recalibration of economic policy to take account of rising disaster risk, disaster losses will likely rise more rapidly than economic growth in the future in many low- and middle income-countries (Bouwer et al., 2007). This is because the rate at which economic exposure to disasters is increasing is faster than upward trends in wealth creation and of efforts to adequately protect people and assets (UNISDR GAR, 2011). The trend is caused by rapid urbanisation, which concentrates exposure, a global move of people and assets to coastal locations and degradation or loss of natural ecosystem buffers. These are coupled with a lack of appropriate legislation and land use planning. Figure 21.4 shows that this trend of increasing exposure and with it disaster risk, is particularly pronounced in Latin America and in South Asia, where risks are estimated to have risen. Whereas in East Asia and the Pacific and in OECD countries risks are declining due to progress on reducing vulnerability, even though exposure is still growing. The models used do not capture endemic or extensive risks well; particularly those associated with food security and slow onset disasters, meaning that the accuracy of assessments for Sub-Saharan Africa is limited. More research is needed to understand at what point disaster losses and economic growth decouple and the key drivers of this, but deployment of the type of tools introduced in section 21.4 are likely to be important, along with a greater progress in reducing poverty and upgrading infrastructure.

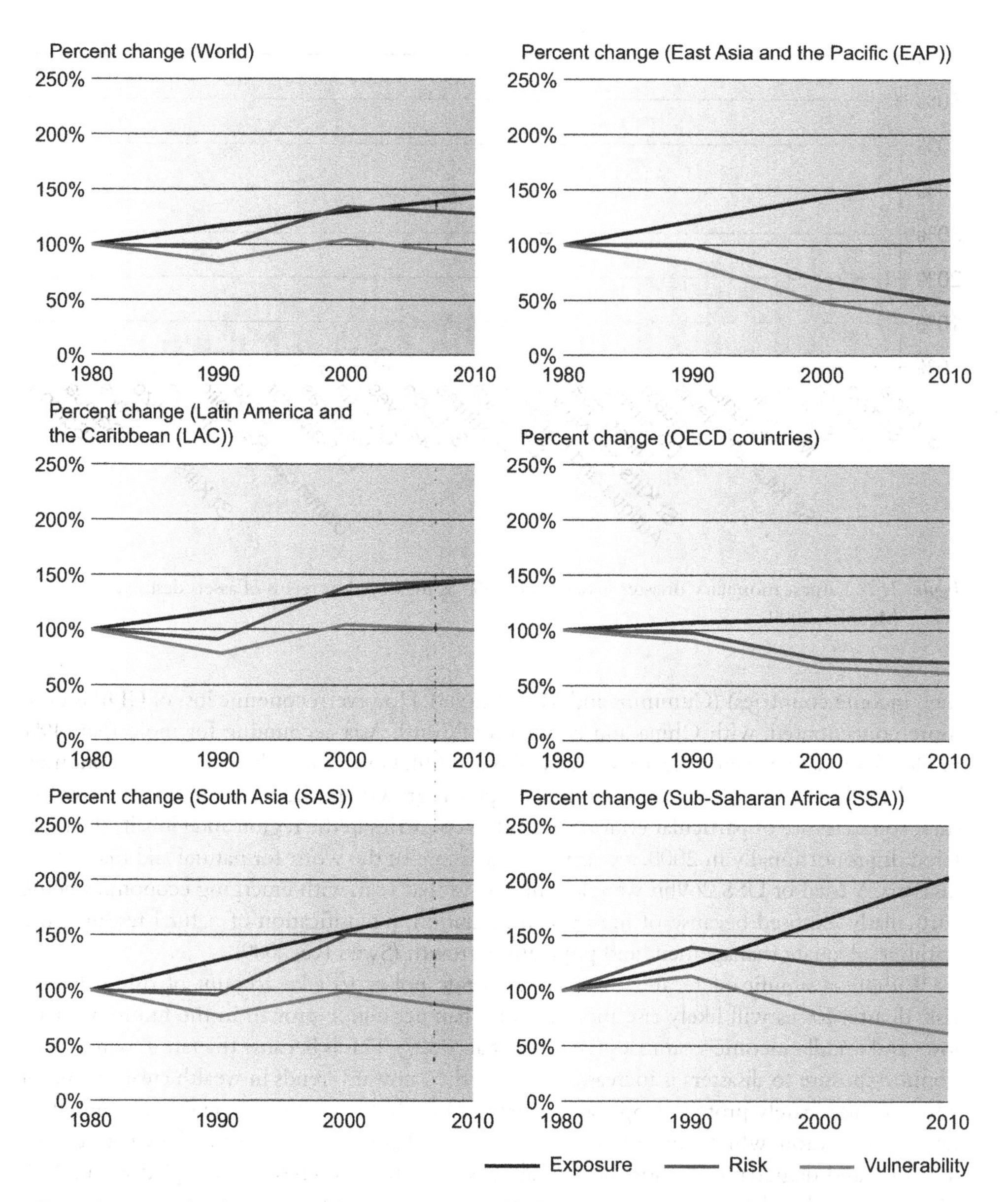

Figure 21.4 Graphs showing trends in disaster risk, exposure and vulnerability between 1980 and 2010 (UNISDR GAR, 2011)

21.3.1 Impact of climate change-related disasters on future losses

Although we cannot yet know the full effects of climate change on disaster risk, it is estimated that losses from disasters associated with weather events are doubling globally every 12 years.[4] While the recent rise in economic losses is mainly attributed to changes in exposure and asset values, natural climate variability and anthropogenic climate change could impact future catastrophe losses. Therefore, future predictions cannot be based solely on historic patterns, but need to take into account future trends.

Many global projections of future disaster loss estimates are based on analysis from insurance and reinsurance companies, who are concerned that the changing frequency and intensity of climate extremes, alongside increases in exposure, will dramatically affect the industry in years to come. In 2007, the Association of British Insurers estimated worldwide annual losses from events associated with hurricanes and windstorms will increase by two-thirds by 2080 in inflation-adjusted terms (ABI, 2004). However, the industry is starting to adjust to these trends, warning about future uninsurability if risks remain unmanaged. As Chief Economist of Swiss Re Asia, Clarence Wong (2009) states, 'ex-post' risk financing is unsustainable. Investing in disaster risk management is the only way to reduce the burden on public budgets and build the foundation for more sustainable risk financing arrangements. Concerns have already been raised by the European Environment Agency (2008) that uninsured disaster losses are likely to rise as obtaining insurance gets more difficult in places experiencing increased disaster risk. And that those likely to experience uninsured disaster losses are those from socially deprived groups and increasingly those in countries where insurance markets will not operate or where premiums are too high to afford. In such cases, public-private partnership insurance and risk financing is one option which is attracting increased attention as a means to develop the insurance industry and support both public and private investments (European Environment Agency, 2008; Swiss Re, 2009). The Caribbean sovereign insurance scheme discussed in Boxes 21.2 and 21.5 is one example. Another interesting case in point is Mexico, where the government insured its disaster liabilities in the international markets using traditional and alternative insurance products. This arrangement built heavily on Mexico's' far-reaching expertise in modelling and managing risk, technical assistance provided by the World Bank and expertise provided by a large reinsurer in terms of acting as a joint bookrunner (see also Box 21.4).

Such ideas are also being considered by the United Nations Framework Convention on Climate Change under the Work Programme on Loss and Damage (2010–2012), which recognises that climate change mitigation and adaptation is sometimes not enough to avoid losses associated with climate change. Among other issues, the Work Programme is considering whether some form of insurance mechanism may be developed to ensure developing countries can be 'compensated' for such losses, where the public investments necessary to support such insurance markets could come from countries with higher historical greenhouse gas emissions.

21.4 Tools and approaches for integrating disaster risk management into economic and fiscal policy

Given the concepts discussed above and the experience presented, what are the tools and approaches available that countries, NGOs and the international community can draw upon? This section considers a set of tools and approaches that can be used to integrate disaster risk management into national economic and fiscal policy, with the goal of proactively reducing economic exposure to disasters and ultimately reversing the trend of rising disaster losses, especially in middle- and low-income countries. A recent assessment by the OECD (2012) on economic growth in the context of climate change finds that building in resilience to climate change impacts by integrating adaptation to climate change in development planning and infrastructure design is critical to the growth prospects of low-income countries.

21.4.1 Conducting comprehensive risk assessments as part of economic policy decisions

The first key step in managing risk is to assess and characterize it. In terms of risk factors, disaster risk is commonly defined by three elements: the hazard, exposure of elements, and vulnerability (Swiss Re, 2000; Kuzak, 2004; Grossi and Kunreuther, 2005). Thus, understanding risk involves observing and recording hazards and hazard analysis, studying exposure and drivers of vulnerability, vulnerability assessment and finally arriving at an estimate in terms of probabilistic risk. In a next step, disaster risk assessments can be factored into national development strategies, sector plans, the budget, regulations, programs and projects.

Comprehensive disaster risk assessments are dependent on having a clear baseline and strong time series information covering hazard, exposure and vulnerability data so it is easier to see what is changing (Lal et al., 2012). Given the dynamic nature of disaster risk, underlined by the possible effects of natural climate variability and anthropogenic climate change and attendant uncertainties, regular updates of information are necessary (Lal et al., 2012). For national level assessments, it is particularly important to have data on the distribution of national assets and their values (derived from inventories and census data) and also on institutional and organizational capacity at different scales (see Table 21.1).

While considerable advances have been made in terms of data availability, including in developing countries, many countries lack relevant datasets, and assessments are not regular practice. Nonetheless, a variety of disaster risk assessment tools and methods have been developed, including the Comprehensive Approach for Probabilistic Risk Assessment (CAPRA). CAPRA is a GIS platform where probabilistic assessments are made of hazards and then combined with exposure and vulnerability data, enabling users to examine multiple hazards and different risks simultaneously. Since its inception in 2008, it has faced problems of building the necessary capacity to use the tool and to collect enough data on national assets to ensure the exposure module is accurate (GFDRR, 2009). It has been applied in Central America, and has been designed to facilitate decision-making, including of risk transfer instruments and the evaluation of cost-benefit ratios for risk reduction strategies. A roll-out of CAPRA is currently underway in South Asia (Cardona et al., 2010).

Box 21.2 includes examples from the Caribbean and Central Asia on how risk assessments are being used to shape economic policy.

21.4.2 Integrating disaster risk management into fiscal policy and budget planning

Based on such comprehensive risk assessments, a key entry point for governments to better deal with disasters is to budget for disaster risk and to include strategies for managing disaster risk in wider fiscal policy. Historically, losses in developing countries have been financed by diversions of funds from the national budget, loans and donations by the international community. Yet these sources are often insufficient, and ex-post gaps in necessary financing of disaster losses are frequently encountered. When stimulus is most needed, such lack of timely funding can lead to important follow-on effects. As one example, in Honduras after Hurricane Mitch, aid measured in terms of GDP almost tripled from 6% to 15% or from US$ 0.3 to US$ 1 billion, yet it remained far below what would have been necessary to support effective relief and recovery. As

Table 21.1 List of information requirements for selected disaster risk management and climate change adaptation activities

Activities		*Examples of information needs*
Cross-cutting	Climate change modelling	Time series information on climate variables – air and sea surface temperatures, rainfall and precipitation measures, wind, air circulation patterns, and greenhouse gas levels
	Hazard zoning and 'hot spot' mapping	Georeferenced inventories of landslide, flood, drought, and cyclone occurrence and impacts at local, sub-national and national levels
	Human development indicators	Geospatial distribution of poverty, livelihood sources, access to water and sanitation
	Disbursement of relief payments	Household surveys of resource access, social well-being, and income levels
	Seasonal outlooks for preparedness planning	Seasonal climate forecasts; sea surface temperatures; remotely sensed and in situ measurements of snow cover/depth, soil moisture, and vegetation growth; rainfall-runoff; crop yields; epidemiology
	A system of risk indicators reflecting macro and financial health of national, social and environmental risks, human vulnerability conditions, and strength of governance (Cardona et al., 2010)	Macroeconomic and financial indicators (Disaster Deficit Index) Measures of social and environmental risks Measures of vulnerability conditions reflected by exposure in disaster-prone areas, socioeconomic fragility, and lack of social resilience in general Measures of organizational, development, and institutional strengths
Flood risk management	Early warning systems for fluvial, glacial, and tidal hazards	Real-time meteorology and water-level telemetry; rainfall, stream flow, and storm surge; remotely sensed snow, ice, and lake areas; rainfall-runoff model and time series; probabilistic information on extreme wind velocities and storm surges
	Flooding hot spots, and structural and non-structural flood controls	Rainfall data, rainfall-runoff, stream flow, floods, and flood inundation maps Inventories of pumps, stream gauges, drainage and defense works; land use maps for hazard zoning; post-disaster plan; climate change allowances for structures; floodplain elevations
	Artificial draining of proglacial lakes	Satellite surveys of lake areas and glacier velocities; inventories of lake properties and infrastructure at risk; local hydro-meteorology
Drought management	Traditional rain and groundwater harvesting, and storage systems	Inventories of system properties including condition, reliable yield, economics, ownership; soil and geological maps of areas suitable for enhanced groundwater recharge; water quality monitoring; evidence of deep-well impacts
	Long-range reservoir inflow forecasts	Seasonal climate forecast model; sea surface temperatures; remotely sensed snow cover; in situ snow depths; multi-decadal rainfall-runoff series
	Water demand management and efficiency measures	Integrated climate and river basin water monitoring; data on existing systems' water use efficiency; data on current and future demand metering and survey effectiveness of demand management

From Lal et al., 2012, adapted from Wilby, 2009

another example, Mexico has been one of the foremost countries to recognise these liabilities and integrated disaster risks with fiscal planning. Box 21.3 describes Mexico's approach to fiscal planning for extreme events.

Box 21.2 Experience of risk assessment in economic policy: Central Asia and the Caribbean

Central Asia has been identified as one of the most vulnerable regions to the impacts of climate change. While prone to extreme temperatures and rainfall-related landslides, recurrent drought in the first decade of the 21st century has affected hydropower generation, water supply for irrigation, rain-fed cropland and pasture productivity. Electric power generation shortages in Kyrgyzstan and Tajikistan stalled industrial growth in both countries as well as deprived millions of people of access to heat and electricity in severe winter conditions, resulting in a humanitarian crisis (see http://www.npr.org/templates/story/story.php?storyId=18784716, accessed 25/04/12).

A new partnership between UNDP's Central Asia Climate Risk Management Programme and CDKN is focused on improving capacity and methodologies for comprehensive and integrated climate risk assessment tools. The partnership has found that systematic risk assessments require a broad range of expertise given the range of data needs. A successful process depends on adequate capacity to conduct assessments, interpret results and identify appropriate actions.

In the **Caribbean**, in 2009, CARICOM Leaders established the *Liliendaal Declaration*, which recognises that countries in the region need to take decisive and potentially transformative action to build disaster resilient, low carbon economies. A risk assessment approach to disaster risk management in macro-economic planning is a central component to its implementation plan.

As well, the Caribbean Catastrophe Risk Insurance Facility (CCRIF), discussed further below, is supporting the development of country risk profiles and their integration with economic and fiscal planning. The country risk profiles offered via the Multi-Peril Risk Evaluation System (MPRES) catastrophe risk modelling platform provide a systematic basis and entry point for more detailed information. These data have been generated and used to underpin CCRIF policies since the 2010/11 policy year and represent a valuable regional public-good resource informing holistic disaster risk management. As a next step, the CCRIF will be emphasising the integration of country risk profiles with economic and fiscal planning (CCRIF, 2011).

Box 21.3 Fiscal planning for extreme events in Mexico

Mexico has been a pioneer in planning for disaster risk and using sovereign risk financing instruments to reduce the public costs of bearing risk. More than 9,000 people lost their lives in the 1985 Mexico City earthquake, and estimates put the direct economic cost of the disaster at about $8 billion (in 2010 prices). Lying as it does within one of the world's most active seismic regions and in the path of hurricanes and tropical storms, Mexico's population and economy are highly exposed to natural hazards. Severe natural disasters (the type likely to occur infrequently but at great cost) imply large fiscal liabilities for the Mexican government. In the case of a disastrous event, the Mexican government is responsible for providing emergency aid and economic support for its low-income population. According to Mexican law, public assets are to be insured and thus reconstruction will be financed by insurance claims. In the past, severe disasters have created large fiscal liabilities and imbalances. Given its financial vulnerability, over the last few years, the Mexican government has been working to improve its fiscal and debt management to reduce the costs imposed by natural disasters and other shocks. Alerted by the Mexico City catastrophe, in 1996 the national government authorities created a budgetary program called FONDEN (Fund for Natural Disasters) to enhance their country's financial preparedness for natural disasters. As a budgetary item, FONDEN is established at the start of each fiscal year by the Mexican parliament as part of the federal government budget plan and provides last-resort funding for uninsurable losses, such as emergency response and disaster relief. In addition to the budgetary program, in 1999 a reserve trust fund was created, which is filled by the surplus of the previous year budget item. FONDEN's objective is to prevent imbalances in the federal government finances derived from outlays caused by natural catastrophes.

But the series of natural disasters that occurred in recent years forced them to look at alternative risk management strategies. From 2000, the Mexican government began collecting data to assess the exposure of its assets to losses from earthquakes and to analyse different financial instruments that could be used to transfer the risk. In 2006, it became the first emerging economy to transfer part of its public-sector natural catastrophe risk to the reinsurance and capital markets – and thus out of the country. This decision came just over 20 years after the 1985 Mexico City earthquake had highlighted the shortcomings of after-the-event approaches for coping with disasters and associated losses. The public sector risk management strategy in Mexico is strongly informed by risk analysis, including modelling as well as economic assessment. Lessons learnt are that data collection over a longer time horizon is crucial to inform sound planning. Building research and analytical capacity domestically has paid off with universities leading the data collection and modelling efforts.

Source: see Linnerooth-Bayer et al., 2011.

While governments are right to consider financial mechanisms, including reserve funds and insurance products, that offer financial protection against different parts of their residual disaster risk portfolios, it is crucial that with rising disaster risk, further ex-ante measures, such as physical infrastructure, building codes and improved preparedness, are taken to protect lives and livelihoods and to increase the affordability of insurance. Taking disaster risk assessments as a starting point, it is crucial that decision makers can make informed decisions about the relative costs and benefits of investing in measures to reduce risk compared to preparing for and financing residual risk. Further, it is desirable that the disaster risk financing strategy incentivises such investment through explicitly costing risk and providing for cost savings on insurance premiums after the further implementation of risk reduction strategies. Figure 21.5 highlights the range of policy options open to decision makers in the way they manage risks.

However, the process of evaluating policy choices to manage the disaster risk portfolio can be complicated, with considerable attendant uncertainties associated with the frequency and severity of hazards, challenges in projecting the distribution of vulnerability into the future and a competing range of budgetary and investment priorities. Box 21.4 describes a model to support decision makers navigating these challenges, particularly with relevance to fiscal policy.

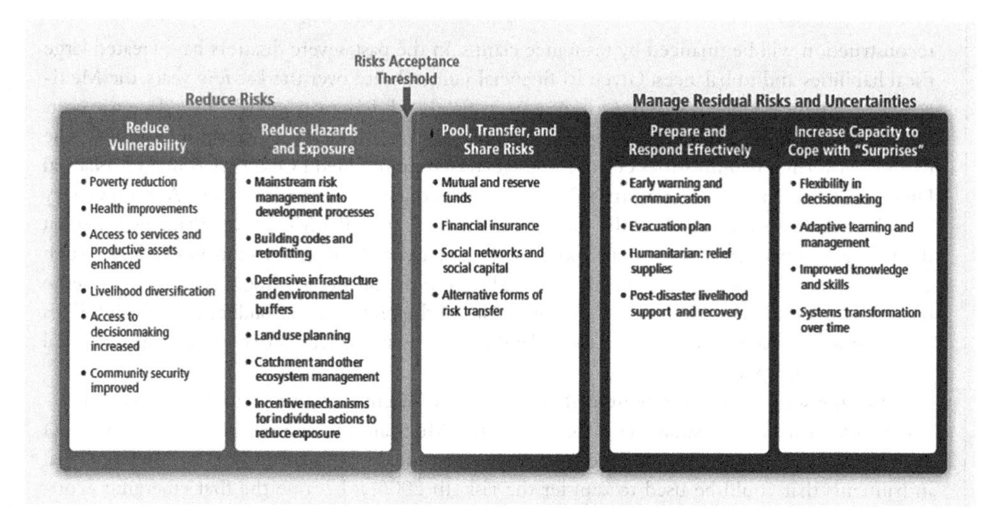

Figure 21.5 Complementary policy options for managing disaster risk portfolios at the national level (Lal et al., 2012)

Box 21.4 Modelling fiscal vulnerability, the liquidity gap and risk

The CATastropheSIMulation (CATSIM) model, developed by IIASA, is a risk-based economic framework for evaluating economic disaster impacts and the costs and benefits of measures for reducing those impacts. CATSIM uses stochastic simulation of disaster risks by randomly and repeatedly generating disaster events in a specified region and examines the ability of the government and private sector to finance relief and recovery. The model is interactive in the sense that the user can change parameters and test different assumptions about hazards, exposure, vulnerability, general economic conditions and the government's ability to respond. As a capacity-building tool, it can illustrate the trade-offs and choices government authorities are confronted with for increasing their economic resilience to the impacts of catastrophic events. The model can be used for supporting policy planning processes for the allocation of resources between ex-ante spending on disaster risk management (such as prevention, national reserve funds, sovereign insurance) and ex-post spending on relief and reconstruction.

For example, CATSIM and other tools are now being used to support a multi-sector steering committee in Madagascar to simulate the impacts of hazards and disasters on the budget and assessing costs and consequences of financial solutions adopted in terms of important indicators such as economic growth or debt. The tool will thus allow the development of comprehensive funding strategies for disaster risk, stimulated by the 2008 cyclone season which saw economic losses of 4% of GDP through impacts on housing, agriculture, trade, tourism and transport. The project called 'Mainstreaming Disaster Risk Management and Climate Change in Economic Development', is being financed by the Global Facility for Disaster Reduction and Recovery (GFDRR)

The model compares asset loss distribution with fiscal resilience, defined as the total of ex-post and ex-ante risk financing (see Figure 21.6).

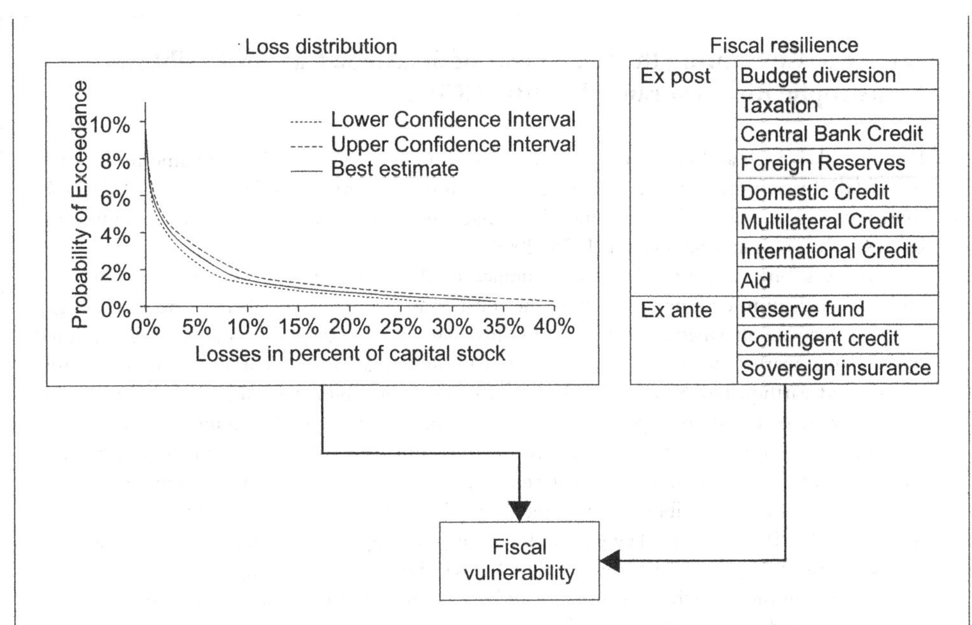

Figure 21.6 Modelling fiscal vulnerability and resilience to natural hazards
Source: Mechler et al., 2010

Using this approach, countries can be identified that exhibit high fiscal and economic vulnerability, and can help quantify the 'fiscal gap' (Hochrainer, 2006; Mechler et al., 2010).

The model, applied globally, highlights the following countries to be particularly fiscally vulnerable: (i) some small island developing states in the Caribbean and Pacific, (ii) countries in Latin America (Honduras, Nicaragua, El Salvador and Bolivia), Africa (Madagascar, Mozambique, Zimbabwe, Sudan, Nigeria and Mauritania) and Nepal in Asia. These countries are prime candidates for stepping up activities to plan, reduce and manage risks in order to reduce serious human and financial loss burden to exposure populations, business and wider macroeconomic health.

However, while risk financing products are becoming increasingly popular, the IPCC (2011, p. 8) concluded that there is medium confidence in the findings that "risk sharing and transfer mechanisms at local, national, regional and global scales can increase resilience to climate extremes". Such mechanisms provide a means to finance relief and recovery of livelihoods and reconstruction, reduce vulnerability and provide knowledge and incentives for reducing risk (see Box 21.5). However, the IPCC also concedes that such mechanisms – if not well integrated with risk reduction and economic planning and policy – can actually provide disincentives for reducing disaster risk by focussing money and effort on the more infrequent risks covered under the policies and thus providing a feeling of safety, while drawing attention from frequent and dynamically increasing risk.

Box 21.5 Containing the fiscal costs of disasters: Case of Caribbean Catastrophe Risk Insurance Facility (CCRIF)

Disaster risk is high and prevalent in the Caribbean. On an annual basis it can amount to up to 6% of GDP, when these direct asset losses are measured in terms of GDP (see Figure 21.7). In most instances, climate and socioeconomic changes are projected to increase this risk despite and because of economic development (ECA, 2009).

The Caribbean Catastrophe Risk Insurance Facility (CCRIF) was established in 2007 as a regional mechanism designed to contain the fiscal costs of disasters and bridge the liquidity gap in the immediate aftermath. It is the world's first regional catastrophe insurance pool, reinsured in the capital markets, to provide governments with immediate liquidity in the aftermath of hurricanes and earthquakes. Sixteen Caribbean countries contribute resources ranging from US$ 200,000 to US$ 4 million, depending on the exposure of their specific country to earthquakes and hurricanes (Young, 2010). The CCRIF requires comprehensive and sound risk analysis, and a number of Caribbean countries have started budgeting for disaster risk, which represents a shift in mindset in terms of Caribbean governments treating risk pre- rather than post-event. The next step with CCRIF is to tackle key gaps in terms of providing stronger linkages to risk reduction and economic policy. As one course of action, CCRIF is currently investing further in developing country risk profiles, which by helping to study the reduction of risk (and fiscal costs) over time effectively provides incentives for building down risk.

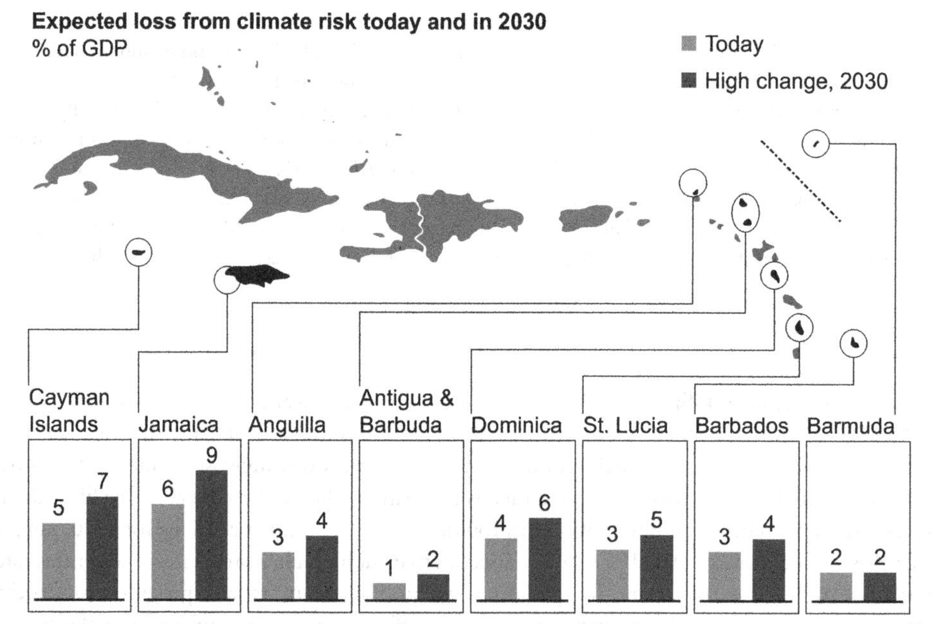

Figure 21.7 Current and future disaster risk measured as a share of GDP in the Caribbean
Source: ECA, 2009

21.4.3 Mainstreaming disaster risk management in sector planning

In order to reduce disaster losses, the balance of effort needs to shift towards reducing exposure through national and sector-based economic planning that takes detailed account of risk assessments. Such strategies benefit from being part of a national strategic risk management processes. In Kiribati, the Kiribati Adaptation Program (KAP) is designed to 'develop and demonstrate the systematic diagnosis of climate-related problems and the design of cost-effective measures, while continuing the integration of climate risk awareness and responsiveness into economic and operational planning' (World Bank, 2011, p. iv). On the basis of extensive consultations and detailed risk assessments, risk management has been integrated across national development strategies and ministry operational plans for all relevant sectors. All of the Programme's investments are tied directly to priorities and activities identified in government planning documents, guided by a National Strategic Risk Management Unit in the Office of the President. However, there are few documented, detailed examples in developing countries of where sector-based investments and planning have been systematically shaped by comprehensive disaster risk assessments and national-level strategic risk management plans. In this regard, early stage work on energy sector risk management tools and methods is being supported by CDKN in Central Asia and West Africa.

One of the most critical implementation issues surrounds the extent to which disaster risk assessments influence the location of critical infrastructure and other important economic assets. As UNISDR (2010) state, reducing climate-related risk will involve protecting critical infrastructure, such as schools and health facilities, retrofitting buildings, relocating settlements and restoring ecosystems, or better yet, avoiding risky development in the first place. The majority of countries now have some form of disaster risk management legislation (Llosa and Zodrow, 2011), but few developing countries have the capacity (or political will) to develop disaster risk zonation and then implement the necessary regulatory and compliance processes. This is because of competing pressures, lack of incentives, a common practice of unplanned development and difficulties involved in blocking access to areas that often offer rich livelihood resources, such as flood plains, volcanic slopes or the coastal strip. Based on Llosa and Zodrow's (2011) assessment of disaster risk management legislation, the most effective in reducing risk are those (i) that are coherent with other legislation and policies across scales and sectors, (ii) that allocate sufficient finance across levels of government, (iii) that clarify institutional arrangements, (iv) that are based on up-to-date risk assessments and mandate periodic reassessments and (v) that establish regulatory and accountability mechanisms and associated penalties (Llosa and Zodrow, 2011; Lal et al., 2012). Reducing exposure will require legislation to be in place that either blocks building on exposed areas or ensures building standards are commensurate with current and future risk profiles. This will require clear communication, mechanisms to integrate this into planning and comprehensive enforcement and penalties. Box 21.6 describes the current situation in Nepal.

Box 21.6 Disaster risk management legislation to reduce exposure: Case of Nepal

Recent work has demonstrated the high vulnerability of Nepal and the potentially large economic and development losses should a major disaster affect the country (ADPC, 2010). One factor behind the high vulnerability is insufficient uptake of building codes. While national building codes do exist in Nepal, there is currently no clear mechanism for implementing these. This means that high-risk buildings continue to be constructed, including in heavily populated seismically active zones. Even where implementation has started, no city administration has managed to implement building regulations through prior approval, inspection and enforcement. Land use planning is not clearly regulated, and its responsibility is split between the Ministry of Physical Planning and Public Works and municipal authorities. New developments occur without approval and there is no clear mechanism for ensuring these meet safety standards or do not occur on land at high risk from natural hazards. There is no legal mechanism to relocate individuals or communities from highly exposed land, though this has been implemented ad hoc, primarily following a disaster.

It is hoped that a new national Disaster Management Act, currently under development, will address some of these issues. It will supersede the 1982 Natural Calamity (Relief) Act, which is focused on response and relief. The Act will establish new and more broadly representative disaster management institutions at national, regional, district and local levels, including (i) a National Commission for Disaster Risk Management chaired by the Prime Minister; (ii) a National Authority for Disaster Risk Management as the implementation authority; (iii) specialist committees on rescue and relief, preparedness and mitigation, resourced by the Ministries of Home Affairs (MoHA), Local Development (MoLD) and Physical Planning and Works (MoPPW), respectively; and (iv) regional, district and local disaster management committees involved in both planning and implementation.

Source: IFRC (2011)

21.4.4 Challenges of managing risk through economic and fiscal policy

While entry points exist for reducing exposure through risk-sensitive economic and fiscal planning, experience suggests that reducing risk via this mechanism remains challenging. Few countries do in fact budget for risk and contingent liabilities, and the capacity is often limited, requiring major efforts in terms of training and building the knowledge and skills. The recent and ongoing financial crisis has shown that it is unaffordable to continue leaving risks unattended in countries exposed to high risk (disaster, fiscal and financial). Turning these hidden and unrecognised liabilities into explicit budget items (and costs) is a first step to being better prepared and more effectively managing potential shocks to the system. As most governments do not deal with risk directly, significant effort would be necessary to generate the additional capacity necessary for assessing and effectively managing risks.

In order to do so, a key related challenge is integrating economic and fiscal risk planning with comprehensive disaster risk management. Although both efforts lead in a similar direction as they identify and manage risks, the framing of risk management by the different experts involved differs importantly. Economic planning is by definition a top-down effort involving macroeconomists and public finance experts, whereas disaster risk management experts are predominantly working on local scales (villages and communities) and employ more bottom-up processes to identify vulnerable households and communities. Bridging the gaps between these different discourses and integrating risk management expertise across layers of government decision-making requires considerable efforts.

Beyond these technical and procedural challenges, a systemic challenge remains in terms of the political economy of disaster response. While this is not the focus of the chapter, effective disaster risk reduction requires engagement, empowerment and leadership across layers of government and in way that supports marginalized people. This is difficult to achieve given the political capital attached to 'heroic action' in disaster response, the media focus on this element of a disaster and the considerable increase in funding flows, each provide a disincentive to significant investment in ex-ante action.

As we demonstrate, making the economic case for disaster risk management is possible, but difficult. Although many entry points and improved data and technologies are being made available, these difficulties will not easily go away. Innovation in the field of disaster risk financing is currently outpacing demand. Responding to risk, even when quantified, remains a hard sell for politicians, particularly in resource-constrained environments. It remains tempting for policy makers to rely on a retroactive and myopic "wait and see" approach and provide relief and reconstruction assistance after an event, which can be easily and effectively promoted through the mass media, and thus creates high visibility with potential voters.

21.5 Conclusions: Promoting disaster resilience for climate-compatible development

As this chapter has demonstrated, disaster losses are rising and threaten future economic development, especially given the associated threat of climate change. A policy solution is to integrate risk management into economic and fiscal policy with the goal of reducing exposure and vulnerability over time. To be successful, it would be well worth considering the following:

- Ensure disaster risk assessments are included in economic projections and economic planning across key sectors. Tools and indicators are available to support such exercises, including ways to include assessments and scenarios in growth diagnostic and economic competitiveness tools. However, examples of detailed sector-based approaches are limited.
- Possibly create a frequently updated and accessible national risk atlas, which includes probabilistic assessments of natural hazards, current and projected distributions of assets and people and their associated vulnerability and capacity. This needs to inform economic decisions at all levels through its inclusion in impact assessments for new investments and in country-wide and provincial and local development planning. Such assessments need to consider how the investment will influence the distribution of people and other assets, for example a new road will likely attract people and services, magnifying potential losses if the road passes through highly exposed areas.
- Enact suitable legislation and adequate enforcement measures that seek to carefully manage exposure, for example by establishing suitable building codes and in some cases prohibiting development in flood plains or in low-lying coastal areas for example.
- Integrate government risk financing schemes with risk reduction and economic planning. Bridging the gap in government insurance with risk reduction and economic planning would provide incentives for monitoring and reducing risk, as well as adequately put a cost on risk in economic planning as a way to incentivise investment in risk management.

More broadly this will require a shift in focus from ex-post relief and reconstruction financing to ex-ante investment at national and sub-national levels, recognising that the former is unsustainable.

How can uncertainties associated with dynamic hazards, catalytic events and complex system-wide responses, such as impacts through international supply chains, be factored into economic models and cost-benefit analysis? How can remote disasters that disrupt supply chains and impact prices be factored in? Can models of society accurately predict future patterns of exposure and vulnerability? Is a 'green' economy inherently more resilient to disasters (and other shocks and stresses) than a more traditional model of economic development? These questions need to be taken up by researchers, decision makers and the private sector in particular, who would benefit from improved data collection nationally. Further linking disaster risk reduction, climate change adaptation, development planning and climate change mitigation institutionally and analytically through an integrated climate-compatible development approach offers a promising avenue to better calculate the trade-offs and benefits of action and may help with political buy-in and improving longer-term fiscal and economic development planning.

Acknowledgement

Support from the Climate Development Knowledge Network (CDKN) is gratefully acknowledged.

Notes

1 http://www.bloomberg.com/news/2011-12-27/japan-factory-output-falls-on-global-slump.html
2 http://www.guardian.co.uk/business/feedarticle/10092592
3 From an insurance perspective, Surminski and Oramas-Dorta (2011) found 123 disaster risk transfer schemes in developing countries. Of these, 12 could be classed as sovereign schemes, with 7 of those operation and 5 under development. See: http://www2.lse.ac.uk/GranthamInstitute/publications/Policy/briefingNotes/2011/sustainable-risk-transfer-economies.aspx
4 http://www.unep.org/Documents.Multilingual/Default.asp?DocumentID=485&ArticleID=5422&l=en

References

ABI (2004) A changing climate for insurance: a summary report for chief executives and policymakers. London: Association of British Insurers.

ADPC (2010). *Nepal Hazard Risk Assessment*. ADPC, Bangkok.

Albala-Bertrand, J.M. (1993). *Political Economy of Large Natural Disasters with Special Reference to Developing Countries*. Clarendon Press, Oxford.

Benson, C. and Clay, E. (2004). *Understanding the Economic and Financial Impacts of Natural Disasters*. Disaster Risk Management Series. No. 4. The World Bank, Washington, D.C.

Bouwer, L.M., Crompton, R.P., Faust, E., Höppe, P. and Pielke Jr., R.A. (2007). 'Confronting disaster losses', *Science* 318, 753.

Cardona, O., Ordaz, M., Reinoso, E., Yamin, L. and Barbat, A.H. (2010). *Comprehensive Approach for Probabilistic Risk Assessment (CAPRA): International Initiative for Disaster Risk Management Effectiveness*. 14th European Conference on Earthquake Engineering, Ohrid, Macedonia, August 30–September 3, 2010.

Caribbean Catastrope Risk Insurance Facility (2011). Quarterly Report. Sepember–November. Caribbean Catastrope Risk Insurance Facility.

Cavallo E. and Noy, I. (2009). *The Economics of Natural Disasters: A Survey*. RES Working Papers 4649. Inter-American Development Bank, Research Department, Washington, D.C.

Charveriat, C. (2000). *Natural disasters in Latin America and the Caribbean: An Overview of Risk*. RES Working Papers 434. Inter-American Development Bank, Research Department, Washington, D.C.

Cummins, J. and Mahul, O. (2009). *Catastrophe Risk Financing in Developing Countries: Principles for Public Intervention*. World Bank, Washington, D.C.

ECA (2009). *Shaping Climate-Resilient Development: A Framework for Decision Making Study*. Report of the Economics of Climate Adaptation Working Group, New York, NY, 159 pp.

ECLAC (2003). *Handbook for Estimating the Socio-economic and Environmental Effects of Disaster.* ECLAC, Mexico City.

European Environment Agency (EEA) (2008). *Direct Losses from Weather Disasters* (CLIM 039) 2008. Copenhagen: European Environment Agency.

GFDRR (2009) *GFDRR Case Study: Central American Probabilistic Risk Assessment (CAPRA).* Global Facility for Disaster Reduction and Recovery, Washington, D.C.

Grossi, P. and Kunreuther, H. (eds.) (2005). *Catastrophe Modeling: A New Approach to Managing Risk.* Springer, New York.

Handmer, J., Honda, Y., Kundzewicz, Z.W., Arnell, N., Benito, G., Hatfield, J., Mohamed, I.F., Peduzzi, P., Wu, S., Sherstyukov, B., Takahashi, K. and Yan, Z. (2012). 'Changes in impacts of climate extremes: human systems and ecosystems'. In: *Managing the Risks of Extreme Events and Disasters to Advance Climate Change Adaptation.* Field, C.B., Barros, V., Stocker, T.F., Qin, D., Dokken ,D.J., Ebi, K.L., Mastrandrea, M.D., Mach, K.J., Plattner, G.-K., Allen, S.K., Tignor, M., and Midgley, P.M. (eds.). A Special Report of Working Groups I and II of the Intergovernmental Panel on Climate Change (IPCC). Cambridge University Press, Cambridge, UK, and New York, NY, USA, pp. 231–290.

Hochrainer, S. (2006). *Macreoeconomic Risk Management Against Natural Disasters.* German University Press, Wiesbaden.

Hochrainer, S. (2009). *Assessing Macroeconomic Impacts of Natural Disasters: Are There Any?* Policy Research Working Paper, 4968. World Bank, Washington D.C.

IFRC (2011). *Analysis of Legislation Related to Disaster Risk Reduction in Nepal.* Prepared by Picard, M. International Federation of Red Cross and Red Crescent Societies, Geneva, Switzerland.

IPCC (2011). 'Summary for policymakers'. In: *Intergovernmental Panel on Climate Change Special Report on Managing the Risks of Extreme Events and Disasters to Advance Climate Change Adaptation.* Field, C.B., Barros, V., Stocker, T.F., Qin, D., Dokken ,D.J., Ebi, K.L., Mastrandrea, M.D., Mach, K.J., Plattner, G.-K., Allen, S.K., Tignor, M., and Midgley, P.M. (eds.). Cambridge University Press, Cambridge, United Kingdom and New York, NY, USA.

Kellenberg, D. and Mobarak, A. (2008). 'Does rising income increase or decrease damage risk from natural disasters?' *Journal of Urban Economics,* 63(3), 788–802.

Kuzak, D. (2004). 'The application of probabilistic earthquake risk models in managing earthquake insurance risks in Turkey'. In: E. Gurenko (ed.) *Catastrophe Risk and Reinsurance: A Country Risk Management Perspective.* Risk Books, London.

Lal, P.N., Mitchell, T., Aldunce, P., Auld, H., Mechler, R., Miyan, A., Romano, L.E. and Zakaria, S. (2012). 'National systems for managing the risks from climate extremes and disasters'. In: *Managing the Risks of Extreme Events and Disasters to Advance Climate Change Adaptation* Field, C.B., Barros, V., Stocker, T.F., Qin, D., Dokken ,D.J., Ebi, K.L., Mastrandrea, M.D., Mach, K.J., Plattner, G.-K., Allen, S.K., Tignor, M., and Midgley, P.M. (eds.). A Special Report of Working Groups I and II of the Intergovernmental Panel on Climate Change (IPCC). Cambridge University Press, Cambridge, UK, and New York, NY, USA, pp. 339–392.

Linnerooth-Bayer, J., Hochrainer, S. and Mechler, R. (2011). 'Insurance against Losses from natural disasters in developing countries. evidence, gaps and the way forward'. *Journal of Integrated Disaster Risk Management,* doi:10.5595/idrim.2011.0013

Llosa, S. and Zodrow, I. (2011). 'Disaster risk reduction legislation as a basis for effective adaptation'. *Global Assessment Report on Disaster Risk Reduction.* United Nations International Strategy for Disaster Reduction.

Mechler, R. (2004). *Natural Disaster Risk Management and Financing Disaster Losses in Developing Countries.* Verlag fuer Versicherungswissenschaft, Karlsruhe.

Mechler, R., Linnerooth-Bayer, J. Hochrainer, S. and Pflug, G. (2006). Assessing financial vulnerability and coping capacity: The IIASA CATSIM Model'. In: J. Birkmann (ed.). *Measuring Vulnerability and Coping Capacity to Hazards of Natural Origin: Concepts and Methods.* United Nations University Press, Tokyo, 380–398.

Mechler, R. (2009). 'Disasters and economic welfare: can national savings help explain post-disaster changes in consumption?' World Bank Policy Research Working Paper Series No. 4988. World Bank, Washington D.C.

Mechler, R., Hochrainer, S., Pflug, G., Lotsch, A. and Williges, K. (2010). 'Assessing the financial vulnerability to climate-related natural hazards'. Background Paper for the World Development Report 2010 "Development and Climate Change." Policy Research Working Paper 5232. World Bank, Washington, D.C.

Morris, S., Neidecker-Gonzales, O., Carletto, C. Munguía M. and Medina, J. (2002). 'Hurricane Mitch and the livelihoods of the rural poor in Honduras', *World Development*, 30(1), 49–60.

Munich Re (2012). *Münchener Rückversicherungs-Gesellschaft, Geo Risks Research, NatCatSERVICE.*

Noy, I. (2009). 'The macroeconomic consequences of disasters'. *Journal of Development Economics*, 88(2), 221–231.

O'Brien, G., O'Keefe, P., Rose, J. and Wisner, B. (2006). 'Climate change and disaster management'. *Disasters*, 30, 64–80.

OECD (2012). *Enabling Local Green Growth: Addressing Climate Change Effects on Employment and Local Development.* Green Growth Strategy. OECD.

Okuyama, Y. (2009). 'Critical review of methodologies on disaster impacts estimation'. Background paper for World Bank/UN Report Unnatural Disasters. World Bank, Washington D.C.

Otero, R.C. and Marti, R.Z. (1995). 'The impacts of natural disasters on developing economies: Implications for the international development and disaster community'. In: M. Munasinghe and C. Clarke (eds.). *Disaster Prevention for Sustainable Development: Economic and Policy Is*sues, World Bank, Washington D.C., 11–40.

Pelham, L., Clay, E., and Braunholz, T. (2011). *Natural Disasters: What is the Role for Social Safety Nets?* The World Bank, Washington D.C.

Raddatz, C. (2007). 'Are external shocks responsible for the instability of output in low-income countries?' *Journal of Development Economics,* 84, 155–187.

Skidmore, M. and Toya, H. (2002). 'Do natural disasters promote long-term growth?' *Economic Inquiry* 40(4), 664–687.

Surminski, S. and Oramas-Dorta, D. (2011). 'Building effective and sustainable risk transfer initiatives in low- and middle-income economies: What can we learn from existing insurance schemes?' Briefing note of Policy Paper: December 2011. (www2.lse.ac.uk/GranthamInstitute/publications/Policy/briefingNotes/2011/sustainable-risk-transfer-economies.aspx).

Swiss Re (2000). *Storm over Europe. An Underestimated Risk.* Swiss Reinsurance Company, Zurich.

Swiss Re (2009). 'Disaster risk financing: a paradigm shift?' Clarence Wong, Swiss Re's Chief Economist Asia online article. Posted October 2009. Available at: http://www.swissre.com/clients/insurers/property_casualty/disaster_risk_financing_a_paradigm_shift.html.

UNISDR (2009). *Applying Disaster Risk Reduction for Climate Change Adaptation: Country Practices and Lessons.* ISDR, Geneva.

UNISDR (2010). *Policy Brief 3: Strengthening Climate Change Adaptation through Disaster Risk Reduction.* United Nations, Geneva.

UNISDR (2011) *Revealing risk, redefining development. Global Assessment Report on Disaster Risk Reduction (GAR).* United Nations International Strategy for Disaster Reduction, Geneva, Switzerland.

Vafeidis, A., Neumann, B., Zimmerman, J. and Nicholls, R.J. (2011). *Analysis of Land Area and Population in the Low-Elevation Coastal Zone.* Paper for the Foresight Study on Migration and Global Environmental Change.

Young, S. (2010). *CCRIF: A Natural Catastrophe Risk Insurance Mechanism for Caribbean Countries Insurance, Reinsurance and Risk Transfer.* Presentation at IDB Capacity Building Workshop on Climate Change Adaptation and Water Resources in the Caribbean. March 22–23, 2010, Trinidad and Tobago.

Wilby, R.L. (2009). *Climate for Development in South Asia (Climdev-Sasia): An Inventory of Cooperative Programmes and Sources of Climate Risk Information to Support Robust Adaptation.* Report prepared on behalf of DFID. Department for International Development, UK.

World Bank (2011). *Kiribati – Second Adaptation Program (Pilot Implementation Phase II) Project.* The Worldbank, Washington D.C.

World Bank and United Nations (2010). *Natural Hazards, Unnatural Disasters: The Economics of Effective Prevention.* The World Bank, Washington, D.C.

INDEX

Page numbers for figures, tables and boxes are in italics.

9780415633116